Railway Transportation (Engineering Essentials)

Railway Transportation (Engineering Essentials)

Editor: Callen Lucas

MURPHY & MOORE
www.murphy-moorepublishing.com

Published by Murphy & Moore Publishing
1 Rockefeller Plaza,
New York City, NY 10020, USA
www.murphy-moorepublishing.com

Railway Transportation (Engineering Essentials)
Edited by Callen Lucas

International Standard Book Number: 978-1-63987-471-2 (Hardback)

Cataloging-in-Publication Data

Railway transportation (engineering essentials) / edited by Callen Lucas.
p. cm.
Includes bibliographical references and index.
ISBN 978-1-63987-471-2
1. Railroads. 2. Transportation engineering. I. Lucas, Callen.
TF153 .R35 2022
625.1--dc23

Contents

Preface

Railway transportation refers to the conveyance of passengers or the transfer of goods by means of trains that are designed to run along railroads or railways. Tracks, or railroads, guide the directional motion of rail vehicles. Since these offer lower frictional resistance, they can facilitate passage to longer trains. Most tracks are built with a signalling system which offers safe transit of vehicles. Rail gauge can be standard gauge, broad gauge or narrow gauge. In addition, the tracks are laid in conformation with a loading gauge that determines the maximum height, width and load of a railway vehicle for safe passage through tunnels, bridges and other structures. All vehicles operating on a rail network have gears that are compatible with the track gauge. The different forms of railway locomotives include steam locomotives, electric locomotives, diesel locomotives and high-speed rails. Regardless of whether a train is steam-propelled or electrically-propelled, the propulsion may be provided by individual motors in self-propelled multiple units or by a separate locomotive. This book presents the complex subject of railway transportation in the most comprehensible and easy to understand language. Different approaches, evaluations, methodologies and advanced studies on railway transportation have been included herein. This book will help new researchers by foregrounding their knowledge in this domain.

This book unites the global concepts and researches in an organized manner for a comprehensive understanding of the subject. It is a ripe text for all researchers, students, scientists or anyone else who is interested in acquiring a better knowledge of this dynamic field.

I extend my sincere thanks to the contributors for such eloquent research chapters. Finally, I thank my family for being a source of support and help.

Editor

1

A Sparse Optimization Approach for Energy-Efficient Timetabling in Metro Railway Systems

Ziyan Luo,[1] Xiaoyu Li ,[2] and Naihua Xiu[2]

[1]State Key Laboratory of Rail Traffic Control and Safety, Beijing Jiaotong University, Beijing 100044, China
[2]Department of Mathematics, Beijing Jiaotong University, Beijing 100044, China

Correspondence should be addressed to Xiaoyu Li; 16121607@bjtu.edu.cn

Academic Editor: Luca Pugi

In this paper, we propose a sparse optimization approach to maximize the utilization of regenerative energy produced by braking trains for energy-efficient timetabling in metro railway systems. By introducing the cardinality function and the square of the Euclidean norm function as the objective function, the resulting sparse optimization model can characterize the utilization of the regenerative energy appropriately. A two-stage alternating direction method of multipliers is designed to efficiently solve the convex relaxation counterpart of the original NP-hard problem and then to produce an energy-efficient timetable of trains. The resulting approach is applied to Beijing Metro Yizhuang Line with different instances of service for case study. Comparison with the existing two-step linear program approach is also conducted which illustrates the effectiveness of our proposed sparse optimization model in terms of the energy saving rate and the efficiency of our numerical optimization algorithm in terms of computational time.

1. Introduction

With the rapid increase in industry and population in cities, the metro railway system has been playing a more and more important role in urban economic and social development for metropolis due to its high loading capacity and low pollution. Despite the inherent efficiency, the energy consumption of metro railway systems is of a huge magnitude and the efficient energy management for such systems is then of global importance. It is known that the energy consumption of trains is largely influenced by the driving strategy and the arrivals and departures in the timetable, which inspires the extensive study on energy-efficient train control and energy-efficient train timetabling. A careful treatment on such topics can be found in the nice survey paper by Scheepmaker et al. in [1], with focuses on reviewing the mathematical problem formulations and solution approaches in the literature. Related work on energy optimization for rail traffic systems such as tramway systems has also been well studied; see, e.g., [2, 3].

Among all the involved energy saving strategies in metro railway systems, an energy-efficient timetable with regenerative braking which schedules the arrival time and the departure time of each train to and from the platforms it visits for the minimization of energy consumption and the maximization of produced regenerative energy utilization, has gained considerable academical attention in recent years. In 2012, Peña-Alcaraz et al. [4] proposed a large-scale mixed integer programming model to maximize the utilization of regenerative braking energy by means of maximizing the total duration of all possible synchronization processes of acceleration and regenerative braking between all possible train pairs. The simulation model is applied to line three of the Madrid metro system and the energy savings of the optimized timetable are reported to be about 7% on average. By merely choosing those train pairs suitable for regenerating energy transfer other than all possible train pairs in [4], Das Gupta et al. developed a more tractable mixed integer programming model in [5]. In 2013, Yang et al. [6] presented a cooperative integer programming model to maximize the time overlaps of nearby accelerating and braking trains and applied a genetic algorithm to approximately solve the resulting optimization model. For the case study on Beijing Metro Yizhuang Line, they reported 22.1% and 15.2% increase of effective time overlaps during peak hours and off-peak hours, respectively. A two-objective timetable optimization model for maximizing the regenerative braking energy with consideration of the passenger waiting time was developed by Yang et al. in [7] where the genetic algorithm was again adopted to solve the optimization model. An integrated energy-efficient operation model was proposed by Li & Lo in [8] with the speed profile part to minimize the net energy consumption and the timetabling part with headway constraints to synchronize the accelerating and regenerative braking for the reuse of regenerative energy. About 25% energy savings were reported for Beijing Metro Yizhuang Line with the headway between trains being 90 seconds. Li & Lo further developed an integrated energy-efficient timetable and speed profile optimization model to determine the cycle time, the headway time, and the speed profiles for a metro line to minimize the energy consumption [9]. The resulting mathematical model is a strictly convex quadratic programming problem with simple bound constraints and an analytical approach based on the first-order optimality condition (i.e., the KKT condition). An integrated timetable and speed profile optimization model with Taylor approximation were also considered by Yang et al. in [10] where the active set method was employed to handle the resulting strictly convex quadratic programming with linear constraints. Under the assumption that trains are operating in the optimal speed profiles, Yang et al. [11] proposed a mixed integer programming model aiming at synchronizing the arrivals and departures of the trains such that regenerative braking energy can be used as much as possible during the whole day service. A two-step linear programming model with the first step to minimize the total energy consumed by all trains and the second step to maximize the utilization of regenerative energy produced by braking trains for a whole day service in metro railway networks was proposed by Das Gupta et al. in [12]. To ensure the transfer of maximum possible regenerative energy from braking trains to accelerating trains, a time overlap vector was introduced which requires to be as close as possible to the zero vector. To effectively quantify such a requirement, the ℓ_0-norm and the ℓ_1-norm were employed in [12]. By adopting the ℓ_1-norm heuristic for the original ℓ_0-norm (i.e., the number of nonzero components in a vector), together with epigraph approach that equivalently transforms the convex ℓ_1-norm minimization to a linear programming, the resulting relaxation problem in the second step was efficiently solved by Gurobi Optimizer. Numerical experiments on different instances of service PES2-SFM2 of line 8 of Shanghai Metro network spanning a full service period of one day were conducted and a significant reduction in effective energy consumption with the worst case being 19.27% was reported by applying the proposed approach.

The original ℓ_0-norm minimization problem proposed in [12] is well-known to be NP-hard due to the discontinuity and nonconvexity of the ℓ_0-norm. Such a type of optimization problems is coined by sparse optimization in the optimization community and has a wide range of applications in compressed sensing, signal processing, statistical regression, financial portfolio, and so on. Various tractable relaxation strategies are proposed to replace the ℓ_0-norm, such as the popular convex surrogate ℓ_1-norm [13], and some weighted ℓ_1-norm variants [14]. Numerical algorithms are also developed for the resulting relaxation models, such as the greedy-type methods [15], the first-order optimization methods [16], and second-order methods for some special structured models [17].

Inspired by the development of sparse optimization methods, in this paper, a sparse optimization model with the ℓ_0-norm and the squared Euclidean norm as the objective function is built to maximize the utilization of the regenerative braking energy. Motivated by the adaptive LASSO model in the high-dimensional statistical regression [14], we propose a weighted ℓ_1-norm relaxation scheme. A two-stage alternating direction method of multipliers (TSADMM) is then proposed with the first stage for calculating an appropriate weight vector and an initial point for the iterative algorithm in the second stage. To illustrate the effectiveness of our new mathematical model, the linear programming model in [12], a quadratic model merely considering the squared Euclidean norm, and our weighted-ℓ_1 companied with the squared Euclidean norm are applied to Beijing Metro Yizhuang Line for comparison. The numerical results on energy savings show that our model is the best among them. To demonstrate the efficiency of our proposed TSADMM, the academical version of CVX (http://cvxr.com/cvx/) is called to solve our model as well. The numerical results in computation time show that our algorithm outperforms that used in CVX.

The rest of this paper is organized as follows. In Section 2, the sparse optimization model for energy-efficient timetabling in metro railway systems is built. In Section 3, the weighted ℓ_1 norm relaxation model is proposed and a two-stage alternating direction method of multipliers (TSADMM) is designed to get an approximate optimal solution. Theoretical global convergence is established as well. By applying our proposed approach to Beijing Metro Yizhuang Line, we demonstrate the effectiveness of our proposed model and the efficiency of our proposed TSADMM by the numerical results in Section 4. Conclusions and future research are given in Section 5. For reference, the appendix provides the proof of the global convergence theorem for our algorithm.

2. Model Formulation

2.1. Notation. The notation that will be used throughout the paper is listed in Table 1 for the convenience of the subsequent model formulation.

2.2. Constraints. The set of constraints in the metro train networks including constraints of the trip time, the dwell time, the headway time, the total travel time, and the domain of event times will be described in this subsection, following from [12].

(i) Trip time constraint: it includes the trip time constraints with tracks and with crossing-overs. For the former one, when a train $t \in \mathcal{T}$ has a trip from platform $i \in \mathcal{N}^t$ to platform $j \in \mathcal{N}^t$ along the track $(i, j) \in \mathcal{K}^t$, the departure time d_i^t and the arrival

Table 1: A list of notations.

a_i^t	The arrival time of train t at platform i
d_i^t	The departure time of train t from platform i
$\mathcal{T}$	The set of all trains
$\lvert\cdot\rvert$	The cardinality of a set, i.e., the number of elements in a set
$[\underline{tr}_{ij}^t, \overline{tr}_{ij}^t]$	The trip time window for train t from platform i to j
ϕ	The set of all the crossing-overs
$\mathcal{O}_{ij}$	The set of all train pairs associated with the crossing-over (i, j)
$[\underline{\kappa}_{ij}^{tt'}, \overline{\kappa}_{ij}^{tt'}]$	The trip window for train t on the crossing-over (i, j)
$[\underline{dw}_i^t, \overline{dw}_i^t]$	The dwell time window for train t at platform i
$\mathcal{S}$	The set of all platform pairs at the same interchange station
$\mathcal{C}_{ij}$	The set of all connection train pairs for a platform pair (i, j)
$[\underline{c}_{ij}^{tt'}, \overline{c}_{ij}^{tt'}]$	The connection window between the train t at the platform i and the train t' at platform j
$\mathcal{H}_{i,j}$	The set of train-pairs that move along the track (i, j)
$[\underline{h}_i^{tt'}, \overline{h}_i^{tt'}]$	The headway window between train t and t' arrival at or departure from platform i
$[\underline{tt}^t, \overline{tt}^t]$	The window of total travel time constraint for train t
f_{ij}	Energy consumption associated with the trip $(i, j) \in \mathcal{A}_{tr}$
$\mathcal{P}$	The set of all platform pairs powered by a same electrical substation
$\overrightarrow{t}$	Temporally closest train to the right of train t
$\overleftarrow{t}$	Temporally closest train to the left of train t
∇_i^t	The relative distance from a_i^t to regenerative alignment point
Δ_i^t	The relative distance from d_i^t to the consumption alignment point
$\overrightarrow{\varepsilon}$	A set of ε containing elements of the form $(i, j, t, \overrightarrow{t})$
$\overleftarrow{\varepsilon}$	A set of ε containing elements of the form $(i, j, t, \overleftarrow{t})$
$\mathcal{N}$	The set of all platforms
$\mathcal{K}$	The set of all tracks
$\mathcal{K}^t$	The set of tracks visited by a train t
$\mathcal{N}^t$	The set of platforms visited by a train t
$\mathcal{K}_{tr}$	The set of all arcs associated with trip time constraints
$\delta_S(\cdot)$	The indicator function with respect to the set S
$\Pi_S(\cdot)$	The projection operator onto the set S
$\mathrm{sign}(\cdot)$	The sign function
$\circ$	The componentwise product, also known as the Hadamard product
A^T	The transpose of the matrix A
$vec(X)$	The column vector generated by stacking all columns of the matrix X

time a_j^t are bounded by $\underline{tr}_{ij}^t$ and $\overline{tr}_{ij}^t$, respectively. The corresponding constraints are

$$\underline{tr}_{ij}^t \le a_j^t - d_i^t \le \overline{tr}_{ij}^t, \quad \forall t \in \mathcal{T}, \ \forall (i,j) \in \mathcal{K}^t. \tag{1}$$

For the latter one, one knows that a crossing-over connecting two train-lines is a special track, where a train-line is a path with some nonopposite platforms and tracks. The crossing-over is to connect the terminal platform of a train-line with the first platform of another. A train t turns around by a crossing-over and departs from the first platform to travel through the train-line after it arrives at the final platform at another train-line, and then the same train is labelled as t' in order to treat it as two different trains [20]. Let ϕ represent the set of all crossing-overs. Consider a crossing-over $(i, j) \in \phi$, we denote the set of all train pairs associated with a same physical train through the crossing-over (i, j) by $\mathcal{O}_{ij}$. The trip time to travelling through the crossing-over $(i, j) \in \phi$ is the time from the departure from the platform i (train is labelled as t) to the arrival at the platform j (train is labelled as t'), which has to stay within a window $[\underline{\kappa}_{ij}^{tt'}, \overline{\kappa}_{ij}^{tt'}]$. And the corresponding constraints can be described as below:

$$\underline{\kappa}_{ij}^{tt'} \le a_j^{t'} - d_i^t \le \overline{\kappa}_{ij}^{tt'}, \quad \forall (i,j) \in \phi, \ \forall (t,t') \in \mathcal{O}_{i,j}. \tag{2}$$

(ii) Dwell time constraint: when a train $t \in \mathcal{T}$ arrives at a platform $i \in \mathcal{N}^t$, it should dwell there in a time interval denoted by $[\underline{dw}_i^t, \overline{dw}_i^t]$ to make sure that passengers can get off and get on the train before it departures from the platform i. The dwell time constraint can be written as follows:

$$\underline{dw}_i^t \le d_i^t - a_i^t \le \overline{dw}_i^t, \quad \forall t \in \mathcal{T}, \ \forall i \in \mathcal{N}^t. \tag{3}$$

(iii) Connection constraint: an interchange station used by connecting trains is to make passengers transfer between trains on different lines. Any platform pair $(i, j) \in \mathcal{S}$ consists of two platforms at a same interchange station. For a connection train pair $(t, t') \in \mathcal{C}_{ij}$, we assume that the train t arrives at platform i and another train t' departs from platform j. The difference of both times can be in a window $[\underline{c}_{ij}^{tt'}, \overline{c}_{ij}^{tt'}]$ to maintain that passengers can get off from the former train and get on the latter. The constraint of connection can be described as follows:

$$\underline{c}_{ij}^{tt'} \le d_j^{t'} - a_i^t \le \overline{c}_{ij}^{tt'}, \quad \forall (i, j) \in \mathcal{S}, \ \forall (t, t') \in \mathcal{C}_{ij}. \tag{4}$$

(iv) Headway time constraint: similar to the trip time constraints, the headway time constraints includes those with tracks and with crossing-overs. In any subway system, a minimum interval between the departures and arrivals of sequenced trains on the same track is maintained, called headway time. To meet the needs of the passengers, railway networks in many cities designate an upper bound between the departures and arrivals of consecutive trains, in order to ensure passengers do not wait for a long time before the arrival of next train. Define $\mathcal{H}_{ij}$ as the set of pairs of successive trains that pass by track $(i, j) \in \mathcal{K}$. Consider $(t, t') \in \mathcal{H}_{ij}$, and let $[\underline{h}_i^{tt'}, \overline{h}_i^{tt'}]$ and $[\underline{h}_j^{tt'}, \overline{h}_j^{tt'}]$ be the time intervals that maintained the departures and arrivals of consecutive trains t and t' from and to platform i and platform j, respectively. For any $(i, j) \in \mathcal{K}$ between platform i and platform j, headway time constraint with tracks can be described as follows:

$$\begin{aligned} &\underline{h}_i^{tt'} \le d_i^t - d_i^{t'} \le \overline{h}_i^{tt'}, \\ &\underline{h}_j^{tt'} \le a_j^t - a_j^{t'} \le \overline{h}_j^{tt'}, \\ &\qquad \forall (i, j) \in \mathcal{K}, \ (t, t') \in \mathcal{H}_{ij}. \end{aligned} \tag{5}$$

In the other case, consider two consecutive trains t_1 and t_2 that go through a crossing-over $(i, j) \in \phi$ after departing from the final platform i of a train-line. They will be labelled as t_1' and t_2' when they arrive at the first platform j of some other train-line. Let $\widetilde{\mathcal{O}}_{ij}$ be the set of all such two-train pairs $((t_1, t_1'), (t_2, t_2'))$. There is an interval time window $[\underline{h}_i^{t_1 t_2}, \overline{h}_i^{t_1 t_2}]$ between the departures of trains t_1 and t_2 from platform i and a window $[\underline{h}_j^{t_1' t_2'}, \overline{h}_j^{t_1' t_2'}]$ between the arrivals of trains t_1' and t_2' at platform j. We can then write the headway constraints with crossing-overs as follows:

$$\begin{aligned} &\underline{h}_i^{t_1 t_2} \le d_i^{t_1} - d_i^{t_2} \le \overline{h}_i^{t_1 t_2}, \\ &\underline{h}_j^{t_1' t_2'} \le a_j^{t_1'} - a_j^{t_2'} \le \overline{h}_j^{t_1' t_2'}, \\ &\qquad \forall (i, j) \in \phi, \ \forall ((t_1, t_1'), (t_2, t_2')) \in \widetilde{\mathcal{O}}_{ij}. \end{aligned} \tag{6}$$

Besides, the headway time is related to the passenger demand and the number of trains. Let $\overline{m}'$ be the number of trains that are in service per hour, c_t be the train capacity, u be the train utility rate, and D be the passenger demand. Then, $D = c_t \times u \times \overline{m}'$ [8]. Let h be the headway time. Since $h = 3600/\overline{m}'$, we have

$$h = \frac{3600 \times c_t \times u}{D}. \tag{7}$$

Here we assume that u and c are constant while the passenger demand varies. In other words, the headway time of trains changes at different periods in a day.

(v) Total travel time constraint: when a train $t \in \mathcal{T}$ has a trip, traversing all platforms and tracks in chronological order, it can have the total travel time between $\underline{tt}^t$ and $\overline{tt}^t$, to meet the needs of passengers in subway systems. Total travel time constraints can be written as follows:

$$\underline{tt}^t \le d_{\mathcal{N}^t(|\mathcal{N}^t|)}^t - a_{\mathcal{N}^t(1)}^t \le \overline{tt}^t, \quad \forall t \in \mathcal{T}, \tag{8}$$

(vi) Domain of event times: we set zero second as the time of the departure of the first train from a platform in a day when the railway is in service. By adding the maximum possible values of all trip times and dwell times, we can set an upper bound for arrival of the last train at the final platform. And we denote the upper bound by $m_T \in Z_{++}$. So, we can write the domain of the event time constraints as follows:

$$\begin{aligned} &0 \le a_i^t \le m_T, \\ &0 \le d_i^t \le m_T, \\ &\qquad \forall t \in \mathcal{T}, \ \forall i \in \mathcal{N}^t. \end{aligned} \tag{9}$$

2.3. Minimizing the Energy Consumption. The major energy consumption of a train is from the acceleration phase. To efficiently minimize the total energy consumption of trains in a metro railway network, Das Gupta [12] proposed an optimization model with linear inequality constraints to get a feasible timetable. The involved optimization model takes the form of

$$\begin{aligned} \min \quad & \sum_{(i,j)\in \mathcal{K}_{tr}, t\in \mathcal{T}} f_{ij}\left(a_j^t - d_i^t\right) \\ \text{s.t.} \quad & (1)\text{–}(9) \end{aligned} \tag{10}$$

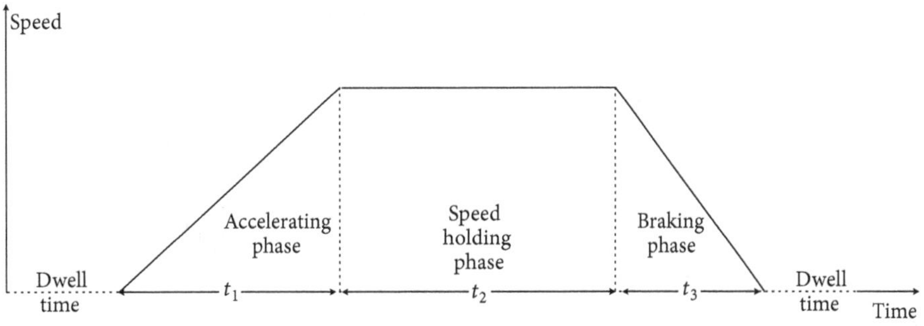

FIGURE 1: Three working phases (accelerating-cruising-braking) of speed profile.

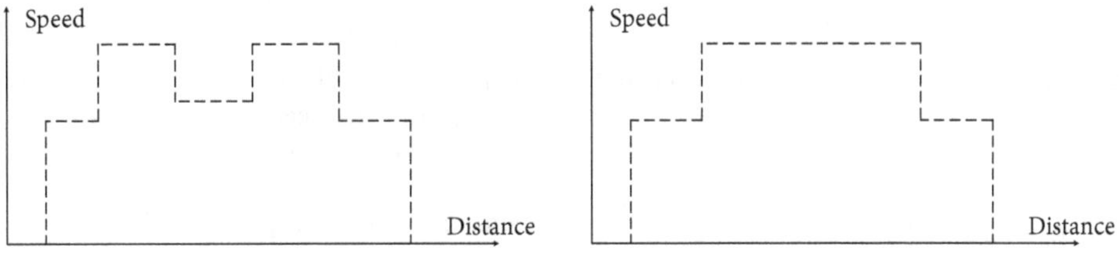

FIGURE 2: The speed limit of a train with a multiphase section.

where $f_{ij} : \mathbf{R}_{++} \longrightarrow \mathbf{R}_{++}$ with argument $(a_j^t - d_i^t)$ being the energy consumption function that is unknown in advance and some best possible affine approximate function in the sense of least-squares by using the measured energy in practice is then utilized. By denoting $x := \text{vec}(X^T)$ with

$$X = \begin{pmatrix} a_1^1 & a_1^2 & \cdots & a_1^{|\mathcal{T}|-1} & a_1^{|\mathcal{T}|} \\ d_1^1 & d_1^2 & \cdots & d_1^{|\mathcal{T}|-1} & d_1^{|\mathcal{T}|} \\ a_2^1 & a_2^2 & \cdots & a_2^{|\mathcal{T}|-1} & a_2^{|\mathcal{T}|} \\ d_2^1 & d_2^2 & \cdots & d_2^{|\mathcal{T}|-1} & d_2^{|\mathcal{T}|} \\ \vdots & \vdots & \ddots & \vdots & \vdots \\ a_{|\mathcal{N}|}^1 & a_{|\mathcal{N}|}^2 & \cdots & a_{|\mathcal{N}|}^{|\mathcal{T}|-1} & a_{|\mathcal{N}|}^{|\mathcal{T}|} \\ d_{|\mathcal{N}|}^1 & d_{|\mathcal{N}|}^2 & \cdots & d_{|\mathcal{N}|}^{|\mathcal{T}|-1} & d_{|\mathcal{N}|}^{|\mathcal{T}|} \end{pmatrix} \tag{11}$$

the approximate counterpart of (10) can be formulated by the following linear programming:

$$\begin{aligned} \min \quad & \sum_{(i,j)\in\overline{\mathcal{K}}_{tr}} c_{ij}\left(x_j - x_i\right) \\ s.t. \quad & l_{ij} \le x_j - x_i \le u_{ij}, \quad (i,j) \in \overline{\mathcal{K}} \\ & 0 \le x_i \le m_T, \quad i = 1,2,\ldots,n, \end{aligned} \tag{12}$$

where $n = 2|\mathcal{T}||\mathcal{N}|$, $\overline{\mathcal{K}}$ is a subset of $\{(i,j) \mid i,j = 1,2,\ldots,n\}$ that collects all those indices according to constraints (1)-(8), and $\overline{\mathcal{K}}_{tr}$ is a subset of $\overline{\mathcal{K}}$ that collects all those indices according to constraints (1)-(2). By utilizing its optimal solution $\overline{x}$, we can get $\overline{a}_j^t$'s and $\overline{d}_j^t$'s such that

$$a_j^t - d_i^t = \overline{a}_j^t - \overline{d}_i^t, \quad \forall t \in \mathcal{T},\ \forall (i,j) \in \mathcal{K}^t. \tag{13}$$

$$a_j^{t'} - d_i^t = \overline{a}_j^{t'} - \overline{d}_i^t, \quad \forall (i,j) \in \phi,\ \forall \left(t,t'\right) \in \mathcal{O}_{i,j}. \tag{14}$$

For all other constraints, lower and upper bounds are allowed to vary as described in (3)-(9).

2.4. Maximizing the Energy Regeneration: A Sparse Optimization Model. We adopt the speed profile by maximum accelerating, speed holding, maximum braking strategy from [18] with a multi-phase-speed-limit section, as described in Figures 1 and 2.

Let t_1, t_2, and t_3 be the time of maximum accelerating, speed hold, and maximum braking phases, and a_{acc} and a_{bra} are the accelerating rate of the accelerating and braking phase. It is known that, with regenerative braking, kinetic energy can be converted into electricity which can be fed back to the power supply system to be used by other nearby trains in the same electric substation. Denote the conversion factor from electricity to kinetic energy as θ_{acc} and the conversion factor from kinetic energy to regenerative electricity as θ_{bra} and the resistance rate at the speed holding phase as γ. For any two adjacent stations, we adopt the formula from [18] to calculate the energy consumption for accelerating and the regenerative

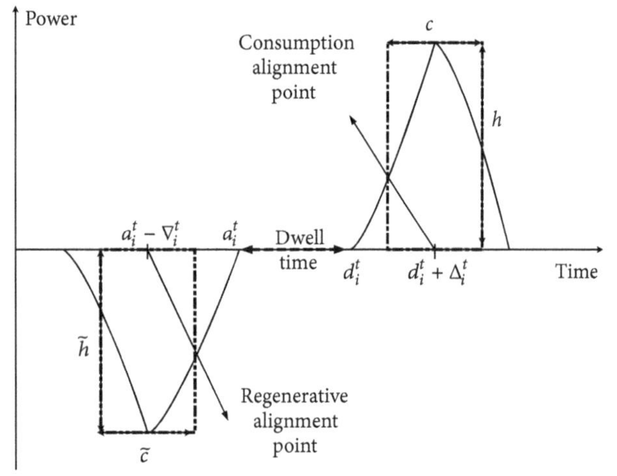

FIGURE 3: Use rectangles instead of energy consumption and regenerative energy.

energy produced during braking, denoted by E^{con} and E^{reg}, respectively, in the way as below:

$$\begin{aligned} E^{con} &= \frac{a_{acc}^2 t_1^2}{2\theta_{acc}} + \int_{t_1}^{t_1+t_2} \frac{\gamma^2 (t-t_1)}{\theta_{acc}} dt \\ &= \frac{a_{acc}^2 t_1^2}{2\theta_{acc}} + \frac{\gamma^2 t_2^2}{2\theta_{acc}} \\ \text{and } E^{reg} &= \frac{\theta_{bra} a_{bra}^2 t_3^2}{2}. \end{aligned} \tag{15}$$

Here E^{con} and E^{reg} are both for the unit mass. A simple heuristic method is applied to approximate the power graph as shown in Figure 3, where c and $\tilde{c}$ are the width of the rectangles for accelerating and braking and h and $\tilde{h}$ are heights of the rectangles. Similar to those in [12], the midpoint of the width in the rectangle is called the regenerative or consumptive alignment point, where Δ_i^t represents the distance between d_i^t and the regenerative alignment point and ∇_i^t represents the distance between a_i^t and the consumption alignment point.

Here, we align the distance between regenerative and consumption alignment points to maximum the utilization. Before introducing such a distance, we recall the suitable train pairs as introduced in [12]. Assume that platform pairs are opposite to each other at the same electrical substation. Let $\mathcal{P}$ be the set of all platform pairs. For any pair $(i, j) \in \mathcal{P}$ and the set $\mathcal{T}_i \subseteq \mathcal{T}$ contains all trains that arrives at, dwells, departs from platform i. When the train $t \in \mathcal{T}_i$ departs from the platform i, it is a best choice to find a train $\overrightarrow{t}$ at the platform j which are producing the regenerative braking energy using the following definition.

Definition 1 (see [12]). Consider any $(i, j) \in \mathcal{P}$. For every train $t \in \mathcal{T}_i$, the train $\overrightarrow{t} \in \mathcal{T}_j$ is called the temporally closest train to the right of t if

$$\overrightarrow{t} = \underset{t' \in \{x \in \mathcal{T}_j : 0 \le (\overline{a}_j^x + \overline{d}_j^x)/2 - (\overline{a}_i^t + \overline{d}_i^t)/2 \le \beta\}}{\arg\min} \left\{ \left| \frac{\overline{a}_i^t + \overline{d}_i^t}{2} - \frac{\overline{a}_j^{t'} + \overline{d}_j^{t'}}{2} \right| \right\} \tag{16}$$

where β is an empirical parameter determined by the timetable designer and is much smaller than the time horizon of the entire timetable [12]. Similarly, when a train $t \in \mathcal{T}_i$ arrives at platform i, it produces the regenerative energy. To find a suitable train $\overleftarrow{t}$ departing from platform j for transferring the regenerative braking energy, we recall the following definition from [12].

Definition 2 (see [12]). Consider any $(i, j) \in \mathcal{P}$. For every train $t \in \mathcal{T}_i$, the train $\overleftarrow{t} \in \mathcal{T}_j$ is called the temporally closest train to the left of t if

$$\overleftarrow{t} = \underset{t' \in \{x \in \mathcal{T}_j : 0 \le (\overline{a}_i^t + \overline{d}_i^t)/2 - (\overline{a}_j^x + \overline{d}_j^x)/2 \le \beta\}}{\arg\min} \left\{ \left| \frac{\overline{a}_i^t + \overline{d}_i^t}{2} - \frac{\overline{a}_j^{t'} + \overline{d}_j^{t'}}{2} \right| \right\}. \tag{17}$$

Denote the sets $\overleftarrow{\varepsilon}$ and $\overrightarrow{\varepsilon}$ whose components are $(i, j, t, \overleftarrow{t})$ and $(i, j, t, \overrightarrow{t})$, respectively. Consider any $(i, j, t, \overleftarrow{t}) \in \overleftarrow{\varepsilon}$. To maximize the transfer of the regenerative energy from the braking train $\overleftarrow{t}$ by the accelerating train t, we attempt to make the term $(d_i^t + \Delta_i^t - a_j^{\overrightarrow{t}} + \nabla_j^{\overrightarrow{t}})$ to be zero or as close to zero as possible. Similarly, for $(i, j, t, \overrightarrow{t}) \in \overrightarrow{\varepsilon}$, our goal is to make the term $(d_j^{\overleftarrow{t}} + \Delta_j^{\overleftarrow{t}} - a_i^t + \nabla_i^t)$ to be zero or as close to zero as possible. Following from [12], we define an auxiliary variable $y \in \mathbf{R}^n$ with $n = 2\overline{n}\,\overline{m}$, $\overline{n} = |\mathcal{N}|$ and $\overline{m} = |\mathcal{T}|$ by

$$y = \left(\left(d_i^t + \Delta_i^t - a_j^{\overrightarrow{t}} + \nabla_j^{\overrightarrow{t}} \right)_{(i,j,t,\overrightarrow{t}) \in \overrightarrow{\varepsilon}}, \left(d_j^{\overleftarrow{t}} + \Delta_j^{\overleftarrow{t}} - a_i^t + \nabla_i^t \right)_{(i,j,t,\overleftarrow{t}) \in \overleftarrow{\varepsilon}} \right). \tag{18}$$

The components of y represent difference of the time between the alignment points of accelerating and braking. Once the running time is fixed, Δ_i^t and ∇_i^t are two constants.

The ultimate energy consumption can be approximately calculated as follows (see Figure 4):

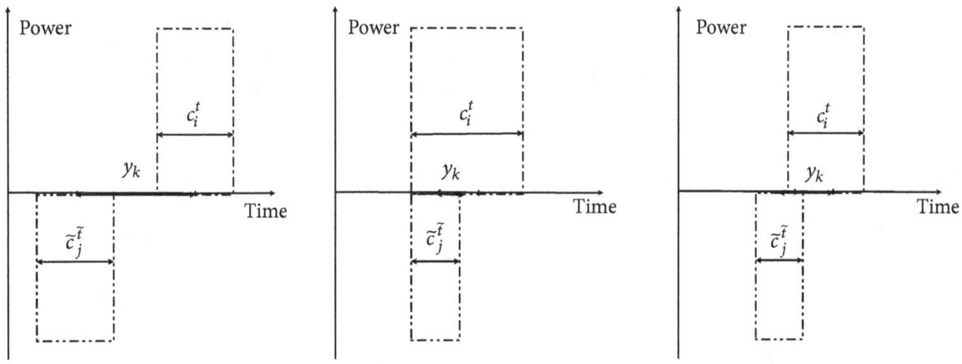

FIGURE 4: Three cases for net energy consumption function.

$$E_{(i,j,t,\tilde{t})} = \begin{cases} E_k^{t,con}, & \text{if } |y_k| \geq \dfrac{\left(\tilde{c}_j^{\tilde{t}} + c_i^t\right)}{2}, \\ E_k^{t,con} - \min\left\{\tilde{c}_j^{\tilde{t}}, c_i^t\right\} * \min\left\{h_i^t, \tilde{h}_j^{\tilde{t}}\right\}, & \text{if } |y_k| \leq \left|\dfrac{\left(c_i^t - \tilde{c}_j^{\tilde{t}}\right)}{2}\right|, \\ E_k^{t,con} - \left(\dfrac{\left(\tilde{c}_j^{\tilde{t}} + c_i^t\right)}{2} - y_k\right) * \min\left\{h_i^t, h_j^{\tilde{t}}\right\}, & \text{otherwise.} \end{cases} \tag{19}$$

where train t is accelerating and train $\tilde{t}$ is braking.

By sorting all possible consecutive platforms for braking energy absorbing, a 0-1 vector $\alpha \in \mathbf{R}^n$ is introduced to indicate all those locations in y with regeneration energy usage when the components of α take the value 1. In order to make the utilization of the regenerative energy as much as possible, we then turn to minimizing the number of nonzero elements of $\alpha \circ y$ which corresponds to minimizing $\|\alpha \circ y\|_0$ and making nonzero elements in $\alpha \circ y$ as close to zero as possible which corresponds to minimizing $\|\alpha \circ y\|_2^2$. The resulting sparse optimization model turns out to be

$$\begin{aligned} \min_{x,y} \quad & \lambda_1 \|\alpha \circ y\|_0 + \frac{\lambda_2}{2} \|\alpha \circ y\|_2^2 \\ s.t. \quad & (3)\text{–}(9),\ (13),\ (14),\ (18). \end{aligned} \tag{20}$$

where $x, y \in \mathbf{R}^n$ are decision variables. Similar to (12), (20) can be rewritten as

$$\min_{x,y} \left\{ \lambda_1 \|\alpha \circ y\|_0 + \frac{\lambda_2}{2} \|\alpha \circ y\|_2^2 : y - Ax - b = 0,\ Fx - g = 0,\ f - Ex \geq 0. \right\} \tag{21}$$

where $A \in \mathbf{R}^{n\times n}$, $E \in \mathbf{R}^{m\times n}$, $F \in \mathbf{R}^{l\times n}$ are

$$E = \begin{pmatrix} E_1 \\ -E_1 \\ E_2 \\ -E_2 \\ E_3 \\ -E_3 \\ E_4 \\ -E_4 \end{pmatrix},$$

$$F = \begin{pmatrix} O_{\overline{m}\times\overline{m}} & \widehat{E}_1 & O_{\overline{m}\times 2\overline{m}} & \cdots & O_{\overline{m}\times 2\overline{m}} & O_{\overline{m}\times\overline{m}} \\ O_{\overline{m}\times\overline{m}} & O_{\overline{m}\times 2\overline{m}} & \widehat{E}_2 & \cdots & O_{\overline{m}\times 2\overline{m}} & O_{\overline{m}\times\overline{m}} \\ \vdots & \vdots & \vdots & \ddots & \vdots & \vdots \\ O_{\overline{m}\times\overline{m}} & O_{\overline{m}\times 2\overline{m}} & \cdots & O_{\overline{m}\times 2\overline{m}} & \widehat{E}_{\overline{n}-1} & O_{\overline{m}\times\overline{m}} \end{pmatrix} \tag{22}$$

with

$$E_1 = \begin{pmatrix} \widehat{E}_1 & O_{\overline{m}\times 2\overline{m}} & \cdots & O_{\overline{m}\times 2\overline{m}} \\ O_{\overline{m}\times 2\overline{m}} & \widehat{E}_2 & \cdots & O_{\overline{m}\times 2\overline{m}} \\ \vdots & \vdots & \ddots & \vdots \\ O_{\overline{m}\times 2\overline{m}} & O_{\overline{m}\times 2\overline{m}} & \cdots & \widehat{E}_{\overline{n}} \end{pmatrix},$$

$$\widehat{E}_i = \left(-I_{\overline{m}\times\overline{m}} \quad I_{\overline{m}\times\overline{m}}\right),$$

$$E_2 = \left(-I_{\overline{m}\times\overline{m}} \quad O_{\overline{m}\times(2\overline{n}\,\overline{m}-2\overline{m})} \quad I_{\overline{m}\times\overline{m}}\right),$$

$$E_3 = I_{2\overline{n}\,\overline{m}\times 2\overline{n}\,\overline{m}},$$

$$E_4 = \begin{pmatrix} \widetilde{E}_1 & O_{(\overline{m}-1)\times\overline{m}} & \cdots & O_{(\overline{m}-1)\times\overline{m}} \\ O_{(\overline{m}-1)\times\overline{m}} & \widetilde{E}_2 & \cdots & O_{(\overline{m}-1)\times\overline{m}} \\ \vdots & \vdots & \ddots & \vdots \\ O_{(\overline{m}-1)\times\overline{m}} & O_{(\overline{m}-1)\times\overline{m}} & \cdots & \widetilde{E}_{2\overline{n}} \end{pmatrix},$$

$$\widetilde{E}_i = \begin{pmatrix} -1 & 1 & 0 & \cdots & 0 \\ 0 & -1 & 1 & \cdots & 0 \\ \vdots & \vdots & \ddots & \ddots & \vdots \\ 0 & 0 & \cdots & -1 & 1 \end{pmatrix}, \tag{23}$$

$m = 2(2\overline{n}(\overline{m}-1) + \overline{n}\overline{m} + \overline{m} + n)$ and $l = (\overline{n}-1)\overline{m}$, b, g, and f are the corresponding column vectors from constraints (18), (13), (14), and (3)-(9). Apparently, $y - Ax - b = 0$ is the reformulation of the constraint (18), $Fx - g = 0$ is the condensed form of constraints (13) and (14), and $f - Ex \geq 0$ is the representation of all the inequality constraints distributed in (3)-(9).

3. Model Relaxation and Solution Algorithm

3.1. Relaxation Model. Problem (21) is generally NP-hard due to the combinatorial property of the involved ℓ_0-norm [21]. By adopting the well-known convex surrogate ℓ_1-norm initiated in compressed sensing [13, 22], we get the following $\ell_1 + \ell_2^2$ relaxation model:

$$\min_{x,y} \left\{ \lambda_1 \|\alpha \circ y\|_1 + \frac{\lambda_2}{2} \|\alpha \circ y\|_2^2 : y - Ax - b = 0,\ Fx - g = 0,\ f - Ex \geq 0. \right\} \tag{24}$$

Inspired by Oracle property beneficial from the adaptive lasso strategy proposed by Zou et al. [14], we will further utilize a reweighted ℓ_1-norm term instead of the ℓ_1-norm in (24) and get the following relaxation counterpart:

$$\min_{x,y} \left\{ \lambda_1 \|\alpha \circ w \circ y\|_1 + \frac{\lambda_2}{2} \|\alpha \circ y\|_2^2 : y - Ax - b = 0,\ Fx - g = 0,\ f - Ex \geq 0. \right\} \tag{25}$$

where w is some appropriate positive column vector to encourage a low sparsity of y.

3.2. A Two-Stage Alternating Direction Method of Multipliers. By introducing an auxiliary vector $z \in \mathbf{R}^m$, (25) can be rewritten as the following equality constrained optimization problem:

$$\min_{x,y,z} \left\{ \lambda_1 \|\alpha \circ w \circ y\|_1 + \frac{\lambda_2}{2} \|\alpha \circ y\|_2^2 + \delta_{\mathbf{R}_+^m}(z) : y - Ax - b = 0,\ Fx - g = 0,\ z - f + Ex = 0, \right\} \tag{26}$$

where $\delta_{\mathbf{R}_+^m}(h)$ is the indicator function with respect to $\mathbf{R}_+^m$ which takes the value 0 if $h \in \mathbf{R}_+^m$, and $+\infty$ otherwise. The augmented Lagrangian function of (26) takes the form of

$$\begin{aligned} L_\sigma^A(x, y, z; u, v, s) &= \lambda_1 \sum_{i=1}^n w_i |\alpha_i y_i| + \frac{\lambda_2}{2} \|\alpha \circ y\|_2^2 \\ &+ \delta_{\mathbf{R}_+^m}(z) - \langle u, y - Ax - b \rangle - \langle v, Fx - g \rangle - \langle s, z - f + Ex \rangle + \frac{\sigma}{2} \left(\|y - Ax - b\|_2^2 + \|Fx - g\|_2^2 + \|z - f + Ex\|_2^2 \right), \end{aligned} \tag{27}$$

where $\sigma > 0$ is the penalty parameter, and $u \in \mathbf{R}^n$, $v \in \mathbf{R}^l$, $s \in \mathbf{R}^m$ are Lagrangian multipliers corresponding to those three equality constraint systems in (26). The well-known alternating direction method of multipliers (ADMM) proposed by Glowinski and Marroco [23] and Gabay and Mercier [24] is then adopted for handling (26) with the iterative scheme for $k = 0, 1, \ldots$,

$$\begin{aligned} \left(y^{k+1}, z^{k+1}\right) &= \arg\min \left\{ L_\sigma^A\left(x^k, y, z; u^k, v^k, s^k\right) \mid y \in \mathbf{R}^n,\ z \in \mathbf{R}^m \right\}; \\ x^{k+1} &= \arg\min \left\{ L_\sigma^A\left(x, y^{k+1}, z^{k+1}; u^k, v^k, s^k\right) \mid x \in \mathbf{R}^n \right\}; \\ \left(u^{k+1}, v^{k+1}, s^{k+1}\right) &= \left(u^k, v^k, s^k\right) - \tau\sigma \left(y^{k+1} - Ax^{k+1} - b, Fx^{k+1} - g, z^{k+1} - f + Ex^{k+1}\right); \end{aligned} \tag{28}$$

where $\tau \in (0, (1+\sqrt{5})/2)$ is the dual stepsize. Note that $L_\sigma^A(x, y, z; u, v, s)$ is separable of y and z. Thus the first subproblem in (28) can be easily solved by the corresponding first-order optimality condition which leads to the following closed-form solution:

$$y_i^{k+1} = \begin{cases} \left(Ax^k + b + \dfrac{u^k}{\sigma}\right)_i & \text{if } \alpha_i = 0, \\ \operatorname{sign}(q_i) \max\left\{ |q_i| - \dfrac{\lambda_1 w_i}{\lambda_2 + \sigma}, 0 \right\} & \text{otherwise,} \end{cases} \tag{29}$$

with $q_i = (1/(\lambda_2 + \sigma))(u^k + \sigma Ax^k + \sigma b)_i$ for all $i = 1, \ldots, n$, and

$$z^{k+1} = \Pi_{\mathbf{R}_+^m}\left(f - Ex^k + \frac{1}{\sigma}s^k\right). \tag{30}$$

The update for x^{k+1} from the second subproblem can be easily obtained by solving the following linear system:

$$Hx = rhs, \tag{31}$$

where $rhs = A^T(y^{k+1} - b - (1/\sigma)u^k) + F^T(g + (1/\sigma)v^k) + E^T(f - z^{k+1} + (1/\sigma)s^k)$ and $H = A^TA + F^TF + E^TE$. Note that E is full

Require: Choose an initial point $(x_0^0, y_0^0, z_0^0, u_0^0, v_0^0, s_0^0)$, accuracy parameters $\varepsilon_1, \varepsilon_2 > 0$, the parameters $\lambda_1^0, \lambda_2^0, \sigma, r, \epsilon$.
Ensure: (x^*, y^*, z^*);
Stage I Compute $(\widehat{x}, \widehat{y}, \widehat{z}, \widehat{u}, \widehat{v}, \widehat{s})$.
Step 1.1 Set $\lambda_1 = \lambda_1^0, \lambda_2 = 0, (x^0, y^0, z^0, u^0, v^0, s^0) = (x_0^0, y_0^0, z_0^0, u_0^0, v_0^0, s_0^0)$, $w = e$ and $k = 0$;
Step 1.2 Compute $(x^{k+1}, y^{k+1}, z^{k+1}, u^{k+1}, v^{k+1}, s^{k+1})$ by (31), (29), (30) and (28);
Step 1.3 Set $k = k + 1$. If some criterion is satisfied to the accuracy ε_1, then stop. Otherwise, go to Step 1.2.
Set $(\widehat{x}, \widehat{y}, \widehat{z}, \widehat{u}, \widehat{v}, \widehat{s}) = (x^k, y^k, z^k, u^k, v^k, s^k)$ and update $w = (1/(|\widehat{y}|^r + \epsilon))$;
Stage II Solve Problem (26) and output (x^*, y^*, z^*).
Step 2.1 Set $\lambda_1 = \lambda_1^0, \lambda_2 = \lambda_2^0, (x^0, y^0, z^0) = (\widehat{x}, \widehat{y}, \widehat{z}), (u^0, v^0, s^0) = 0$ and $k = 0$;
Step 2.2 Compute $(x^{k+1}, y^{k+1}, z^{k+1}, u^{k+1}, v^{k+1}, s^{k+1})$ by (31), (29), (30) and (28);
Step 2.3 Set $k = k + 1$. If some criterion is satisfied to the accuracy ε_2, then set $(x^*, y^*, z^*) = (x^k, y^k, z^k)$ and stop. Otherwise, go to Step 2.2.

ALGORITHM 1: A two-stage alternating direction methods of multipliers (TSADMM).

column rank. Thus, $E^T E$ is positive definition and also is H. This indicates that (31) admits a unique solution:

$$x^{k+1} = H^{-1} rhs. \quad (32)$$

To effectively reduce the computation cost resulting from the large size of the involved matrices in H, the preconditioned conjugate gradient (PCG) method is employed to numerically solve the linear system (31) instead of the direct way as in (32).

It is worth mentioning that we can utilize the above ADMM method to solve an ℓ_1 problem which is exactly (26) by taking $w = e$ and $\lambda_2 = 0$ to a relatively low accuracy. The resulting numerical solution $(\widehat{x}, \widehat{y}, \widehat{z})$ can not only be served as the initial point, but also produce the weight vector w in the way that $w = (1/(|\widehat{y}_i|^r + \epsilon))$ with $r \in (0, +\infty)$ and $\epsilon > 0$ sufficiently small for a low sparsity promotion when solving (26). The two-stage ADMM algorithmic framework then can be summarized in Algorithm 1.

3.3. Global Convergence. Algorithm 1 admits a global convergence property as stated in the following theorem whose proof will be established in the appendix.

Theorem 3. *Let $\{(x^k, y^k, z^k, w)\}$ be generated from Algorithm 1. If $\tau \in (0, (1+\sqrt{5})/2)$, then the sequence $\{(x^k, y^k, z^k)\}$ converges to an optimal solution to (26).*

4. Numerical Experiments

Numerical experiments on Beijing Metro Yizhuang Line will be conducted in this section to evaluate our proposed model and to illustrate the performance our designed TSADMM algorithm. The procedure of the numerical study is shown in Figure 6. All the computational results are obtained by running Matlab (version 2016b) on a windows desktop (Intel(R) Core(TM) i7-6700 CPU @ 3.40GHz 3.40GHz RAM 16.0 G).

Beijing Metro Yizhuang Line, opened on December in 2010, is one of the 20 lines in Beijing Metro Networks. It has two train lines with 14 stations as shown in Figure 5. The total length of Yizhuang Line is 23.23 km, with the average distance of two station being 1.8 km, the minimum distance being 1 km, and the maximum being 2.6 km. In the section, we calculate the energy conversion when a train has a trip under unit mass.

4.1. Inputs and Algorithm Initialization. For (26), $\alpha \in \mathbf{R}^n$ in the objective function is a 0-1 vector defined in Section 2.4. And the weight vector is denoted by $w = (1/(|\widehat{y}|^r + \epsilon))$, where $\widehat{y}$ is an approximate solution generated by Stage I in Algorithm 1. Here we set $r = 0.5$ and $\epsilon = 10^{-20}$. In the equality constraint $y - Ax - b = 0$, $A \in \mathbf{R}^{n \times n}$ with $n = 2\overline{n}\,\overline{m}$, $\overline{n} = |\mathcal{N}| = 28$, $\overline{m}$ the number of trains in one day service, $b \in \mathbf{R}^n$ with components $(\Delta_i^t + \nabla_j^{\overline{t}})_{(i,j,t,\overline{t})}$ sorting in the appropriate way, where train t is accelerating and $\overline{t}$ is braking, and platforms i and j are opposite to each other, satisfying the equation $i + j = 29$. $Fx - g = 0$ in (26) is an equality constraint about trip times. $F \in \mathbf{R}^{l \times m}$ is the coefficient matrix in (26), $l = (\overline{n} - 1)\overline{m}$. The components of $g \in \mathbf{R}^l$ are the trip times. $E \in \mathbf{R}^{m \times n}$ in $z - f + Ex = 0$ is the coefficient matrix of inequality constraints, and $m = 2(2\overline{n}(\overline{m}-1)+\overline{n}\,\overline{m}+\overline{m}+n)$. The elements in the vector $f \in \mathbf{R}^m$ are upper and lower bounds of dwell time constraints, headway time constraints, total travel time constraints, and domain of event times constraints. For testing purpose, elements of f are given in the following way. The bounds of dwell time constraints are given in Table 2. Let D be the passenger demand for one-day service (16 hours) and $u^D \in \mathbf{R}^{16}$ be the distribution of D for each hour which is presented in Table 3. Let c_t be the capacity of a train, which takes the value 1440, and $u \in \mathbf{R}^{16}$ be the train utility rate for different hours as stated in Table 3 as well. By utilizing formula (7), we can get the headway time for each different period of time (per hour). Specifically, denote the headway time vector by $h \in \mathbf{R}^{16}$. Then $h_k = (3600 \times 1440 \times u_k)/(D \times u_k^D)$ for all $k = 1, 2, \ldots, 16$. Set the windows for the headway time constraints to be $h_i \pm 8$ for all possible train pairs in the ith hour. The total travel time constraints are related

TABLE 2: The windows of the dwell time constraints.

Platforms	SJZ1	XC1	XHM1	JG1	YZQ1	YZWHY1	WYJ1	RJDJ1	RCDJ1	TJNL1	JHL1	CQN1	CQ1	YZHCZ1
Lower bounds	25	25	25	25	25	25	25	25	25	25	25	25	25	25
Upper bounds	45	45	45	45	45	45	45	45	45	45	45	45	45	45
Platforms	SJZ2	XC2	XHM2	JG2	YZQ2	YZWHY2	WYJ2	RJDJ2	RCDJ2	TJNL2	JHL2	CQN2	CQ2	YZHCZ2
Lower bounds	25	25	25	25	25	25	25	25	25	25	25	25	25	25
Upper bounds	45	45	45	45	45	45	45	45	45	45	45	45	45	45

TABLE 3: The train utility rate and distribution of D for one-day (16 hours) services.

	Hours															
Periods	1	2	3	4	5	6	7	8	9	10	11	12	13	14	15	16
u	0.3	0.8	0.8	0.6	0.4	0.4	0.5	0.5	0.3	0.6	0.6	0.8	0.8	0.4	0.3	0.3
u^D	0.03	0.11	0.13	0.07	0.05	0.05	0.04	0.05	0.04	0.05	0.07	0.13	0.12	0.03	0.02	0.02

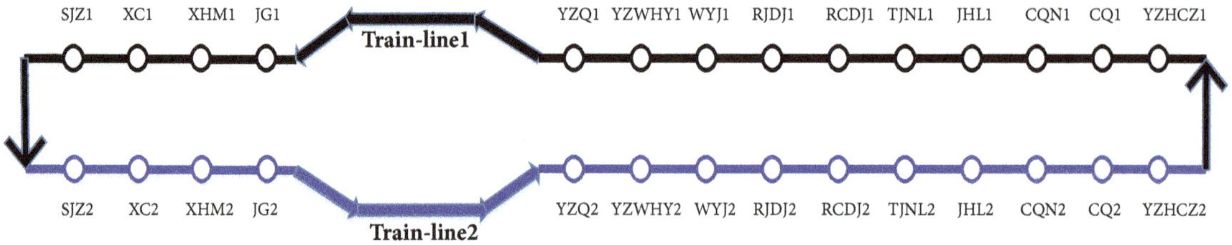

FIGURE 5: Railway network considered for numerical experiment.

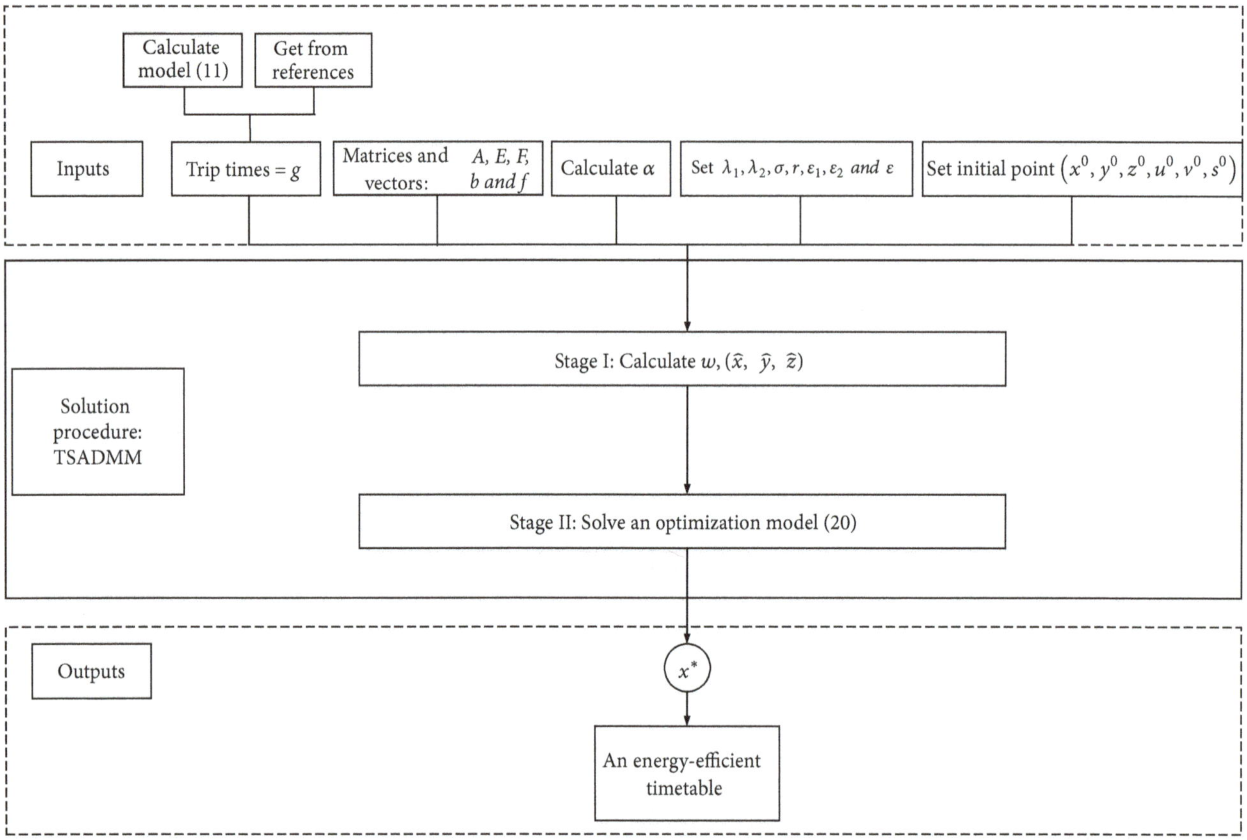

FIGURE 6: The procedure of the numerical experiments.

to the dwell time and trip time. The bounds of trip times are calculated by Table 4 as reference. For each train t, the bounds of trip time are more than $\sum_{q=1}^{5}(s_q/v_q)$ for all sections (ij), $i = 1,2,\ldots,27$, $j = i+1$. The bounds of trip time constraints are presented in Table 5. The following formulae are adopted for calculating the windows of total travel time constraints,

$$\underline{tt} = r_1 \times \left(\sum_{i=1}^{28} \underline{dw}_i + \sum_{\substack{i=1 \\ i\neq 14}}^{27} \underline{tr}_{ij} + \sum_{i=14} \underline{\kappa}_{ij} \right)$$
$$\overline{tt} = r_2 \times \left(\sum_{i=1}^{28} \overline{dw}_i + \sum_{\substack{i=1 \\ i\neq 14}}^{27} \overline{tr}_{ij} + \sum_{i=14} \overline{\kappa}_{ij} \right) \quad (33)$$

where $r_1 = 1.1$ and $r_2 = 0.85$. For the domain of event times constraints, we set $m_T = 16 \times 3600$.

The initial point (u^0, v^0, s^0) in Algorithm 1 is chosen to be 0. To calculate the energy consumption and regenerative energy, we set the maximum acceleration $a_{acc} = 0.5m/s^2$ in accelerating, $a_{bra} = -0.8m/s^2$ for braking, the resistance acceleration $\gamma = -0.05m/s^2$ during the speed holding phase, the conversion factor from electricity to kinetic energy $\theta_{acc} = 0.7$, and the ratio of kinetic energy converting to regenerative energy $\theta_{bra} = 0.5$ as used in [18]. For testing purpose, we approximately calculate the energy consumption for accelerating and the regenerative energy produced during braking in the following way. As presented in Section 2.4, the maximum accelerating, speed holding, maximum braking strategy from [18] is adopted and the parameters t_1, t_2, and t_3 are set in Table 6. The parameters c and $\tilde{c}$ in Figure 3 which represent

TABLE 4: Speed limits of every section on Beijing Metro Yizhuang Line following from [18].

Sections	Segment v_1(m/s)	Segment s_1(m)	Segment v_2(m/s)	Segment s_2(m)	Segment v_3(m/s)	Segment s_3(m)	Segment v_4(m/s)	Segment s_4(m)	Segment v_5(m/s)	Segment s_5(m)
SJZ-XC	13.89	0-150	23.61	150-480	18.06	480-1161	23.61	1161-2501	16.67	2501-2631
XC-XHM	16.67	2631-2643	23.61	2643-2797	20.83	2797-3534	23.61	3534-3780	16.67	3780-3905
XHM-JG	16.67	3905-3918	23.61	3918-5808	20.83	5808-6141	23.61	6141-6271	-	-
JG-YZQ	16.67	6271-6281	23.61	6281-8122	16.67	8122-8254	-	-	-	-
YZQ-YZWHY	16.67	8254-8265	23.61	8265-9116	16.67	9116-9246	-	-	-	-
YZWHY-WYJ	16.67	9246-9256	23.61	9259-10655	16.67	10655-10785	-	-	-	-
WYJ-RJDJ	16.67	10785-10797	23.61	10797-11933	16.67	11933-12065	-	-	-	-
RJDJ-RCDJ	16.67	12065-12077	23.61	12077-13289	16.67	13289-13419	-	-	-	-
RCDJ-TJNL	16.67	13419-13431	23.61	13419-14649	20.83	14649-15426	23.61	15426-15624	16.67	15624-15756
TJNL-JHL	16.67	15756-15768	23.61	15768-17891	16.67	17891-18021	-	-	-	-
JHL-CQN	16.67	18021-18033	23.61	18033-19982	16.67	19982-20107	-	-	-	-
CQN-CQ	16.67	20107-20120	23.61	20120-21264	16.67	21264-21394	-	-	-	-
CQ-YZHCZ	16.67	21394-21406	23.61	21406-22569	16.67	22569-22728	-	-	-	-

Table 5: The windows of the trip time constraints.

Sections	1	2	3	4	5	6	7	8	9	10	11	12	13	14
Lower bounds	160	92	150	130	80	100	100	100	140	130	130	100	100	170
Upper bounds	190	110	170	150	110	125	120	120	170	160	145	120	120	200
Sections	15	16	17	18	19	20	21	22	23	24	25	26	27	-
Lower bounds	100	100	130	130	140	100	100	100	80	130	150	92	160	-
Upper bounds	120	120	145	160	170	120	120	125	110	150	170	110	190	-

Table 6: Parameters for calculating E^{con} and E^{bra}.

t_1	t_2	t_3	c	$\tilde{c}$
20%	45%	35%	14%	21%

Table 7: Results of numerical experiments on Beijing Metro Yizhuang Line: A1 stands for the linear programming model (36) solved by Gurobi; A2 stands for the quadratic programming (37) solved by CVX; A3 stands for our proposed model (26) solved by Algorithm 1.

$\overline{m}$	$\eta(\%)$ A1	A2	A3
357	25.54	24.52	**29.92**
368	23.67	23.44	**28.36**
380	25.56	23.01	**28.54**
391	24.50	23.80	**28.49**
403	24.78	24.17	**27.47**
414	27.02	26.57	**30.05**
426	27.73	27.07	**30.52**
437	27.59	26.94	**30.46**
449	28.26	27.54	**30.84**

the width of the rectangles for E^{con} and E^{bra} (in percentage of the trip time) are also given in Table 6 for approximately calculating the energy consumption for accelerating E^{con} and the regenerative energy produced during braking E^{bra}.

4.2. Stopping Criteria. To measure the accuracy of an approximate optimal solution (x, y, z) of Problem (26) from (x, y, z, u, v, s) generated by Stage II in Algorithm 1, the relative primal and dual infeasibilities will be adopted. Specifically, denote

$$\eta_p = \max\left\{\frac{\|y - Ax - b\|}{1 + \|y\|}, \frac{\|Fx - g\|}{1 + \|g\|}, \frac{\|z - f + Ex\|}{1 + \|z\|}\right\}, \quad (34)$$

$$\eta_d = \frac{\|A^T u - F^T w - E^T s\|}{1 + \|A^T u\| + \|F^T w\| + \|E^T s\|}.$$

For a given tolerance $\varepsilon_2 > 0$, we will stop the tested algorithms when both η_p and η_d are less than ε_2. The algorithms will also be terminated when they reach a given maximum number of iterations $MaxIter_2$. Here we set $\varepsilon_2 = 10^{-3}$ and $MaxIter_2 = 10^4$. For Stage I in Algorithm 1, it will be stopped when both η_p and η_d are less than ε_1 or the maximum iteration reaches $MaxIter_1$. Here we set $\varepsilon_1 = 10^{-1}$ and $MaxIter_1 = 10^3$.

4.3. Experiment Results. To evaluate our model with ℓ_0-norm plus ℓ_2^2-norm as the object, the other two possible models are computed for comparison. The first model is the ℓ_0-norm plus ℓ_1-norm proposed by Das Gupta et al. in [12] which takes the form of

$$\min_x \{\|y\|_0 + \lambda \|y\|_1 : y - Ax - b = 0,\ Fx - g = 0,\ Ex \le f\} \quad (35)$$

which is handled by approximately solving its linear programming relaxation problem

$$\min_x \{e^T s : y \le s,\ y \ge -s,\ y - Ax - b = 0,\ Fx - g = 0,\ Ex \le f\}. \quad (36)$$

The second possible model is the squared ℓ_2-norm related quadratic programming,

$$\min_x \left\{\frac{1}{2}\|y\|_2^2 : y - Ax - b = 0,\ Fx - g = 0,\ Ex \le f\right\}, \quad (37)$$

where only the squared Euclidean norm is employed to measure the vanishing of y. Since we focus on the reduction of the effective energy consumption, we define the energy saving rate η as follows:

$$\eta = \frac{\sum_{t=1}^{|\mathcal{T}||\mathcal{N}|} E_{(i,j,t,\bar{t})}}{\sum_{t=1}^{|\mathcal{T}||\mathcal{N}|} E_t^{con}}. \quad (38)$$

Energy savings by using these three different models are listed in Table 7 with different choices of $\overline{m}$.

As shown in Table 7, the energy savings by using our model are better than those by the other two in all the testing instances with at least 2.58 percentage higher in the energy saving rate η.

Other choices for the initial trip times from [6, 9–11, 19] with $\overline{m} = 368$ are also considered as inputs. The comparison results in terms of the energy saving rate η are stated in Table 8.

Table 8: Energy saving rates for $\overline{m} = 368$: A1, A2, and A3 are the same as in Table 7.

Ref.	$\eta(\%)$ A1	A2	A3
[19]	19.82	21.34	**24.94**
[9]	18.85	18.83	**23.11**
[10]	18.99	18.88	**23.11**
[11]	19.02	18.85	**23.12**
[6]	18.97	18.78	**23.29**

As one can see in Table 8, the utilization of regenerative energy of our approach is the best among these three models

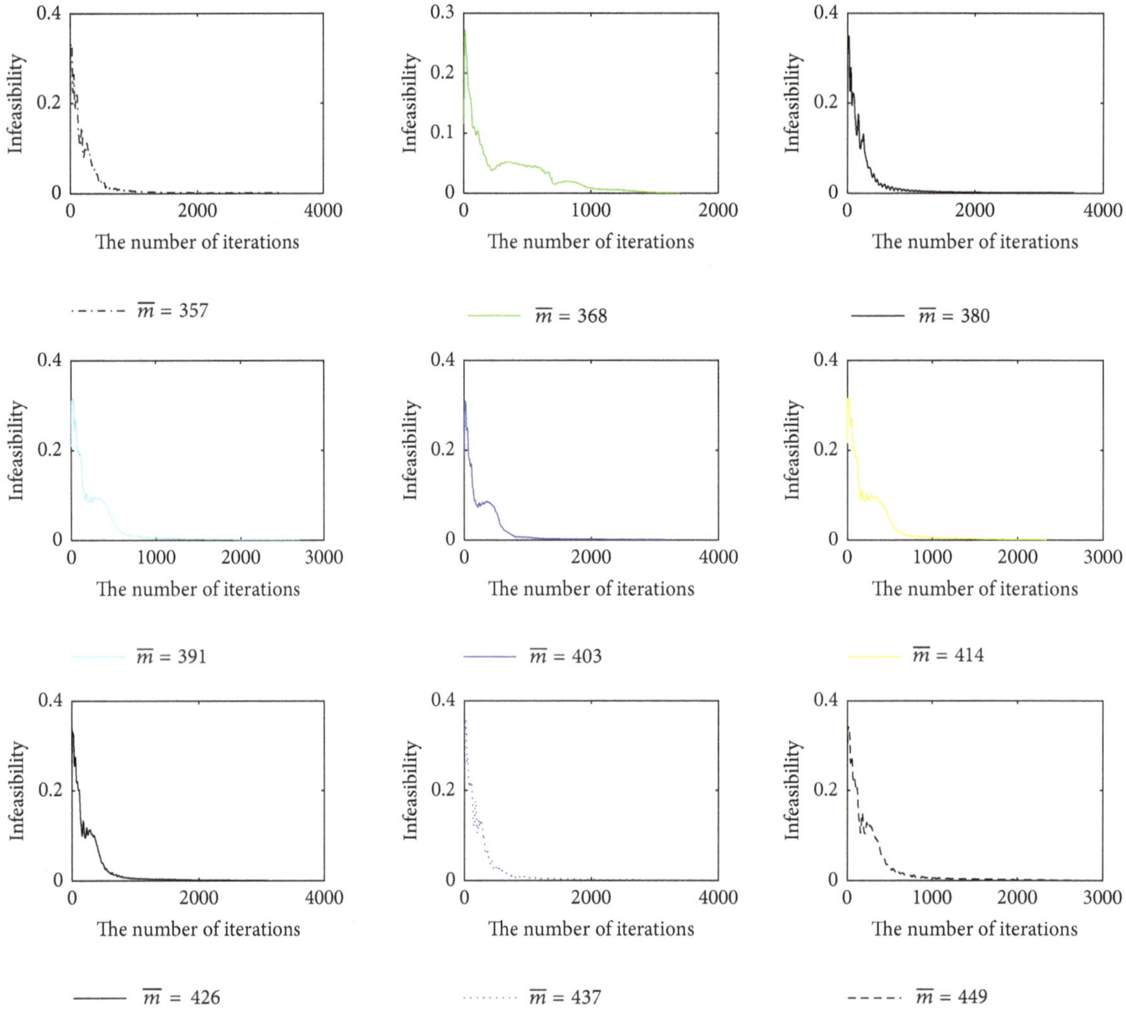

FIGURE 7: The numerical convergence behavior of TSADMM with different choices of $\overline{m}$.

as well with the energy saving rate at least 3.60 percent higher than the other two. These evidently show the effectiveness of our proposed model.

In order to demonstrate the efficiency of our proposed TSADMM algorithm, we call the academical CVX to solve ours model for comparison. The time comparisons for all the aforementioned instances are presented in Tables 9 and 10, respectively.

As can be seen in Tables 9 and 10, our proposed TSADMM outperforms the CVX solver for the model (26) in terms of computation time. Particularly, for the cases of $\overline{m}$ equaling 368 and 449 in Table 9, the computation time for TSADMM is less than one-third of those for CVX. Besides the theoretical global convergence of our TSADMM, the numerical convergence behaviors in terms of the infeasibility measure for all the above 14 instances are presented in Figures 7 and 8. One can see that for each instance, the infeasibility decreases rapidly as the number of iterations increases and it meets the required accuracy 10^{-3} within less than 4000 iterations.

5. Conclusion

In this paper we have proposed a sparse optimization model using ℓ_0-norm and squared ℓ_2-norm as objective function so as to maximize the utilization of regenerative energy in metro railway system. To overcome the NP-hardness resulting from the ℓ_0-norm, the weighted ℓ_1-norm by invoking the adaptive lasso strategy in statistical regression has been introduced to make the relaxation counterpart computationally tractable and numerically effective. A two-stage alternating direction method of multipliers (TSADMM) has been designed to efficiently solve the proposed mathematical model and the global convergence of the algorithm has been established as well. Numerical experiments have been conducted on Beijing Metro Yizhuang Line for case study. The comparison results with the linear programming proposed by Das Gupta et al. and a quadratic programming with least-squares loss function have illustrated the effectiveness of our proposed model in terms of the energy saving rate, and the timing comparison results with the CVX toolbox have demonstrated the efficiency of our proposed TSADMM algorithm.

Table 9: Time comparison between CVX and TSADMM with different choices of $\overline{m}$.

Alg.	$\overline{m}$							
	357	368	380	391	403	414	426	437	449
					time(s)				
CVX	47.17	47.06	54.33	56.90	31.09	50.63	56.18	70.72	71.14
TSADMM	24.35	12.80	31.06	21.26	25.39	19.36	27.61	30.80	23.66

Table 10: Time comparison between CVX and TSADMM with different trip times from [6, 9–11, 19].

Alg.	Ref. [19]	[9]	[10]	[11]	[6]
			time(s)		
CVX	55.61	57.86	55.45	51.25	47.81
TSADMM	20.36	29.22	28.21	27.64	31.84

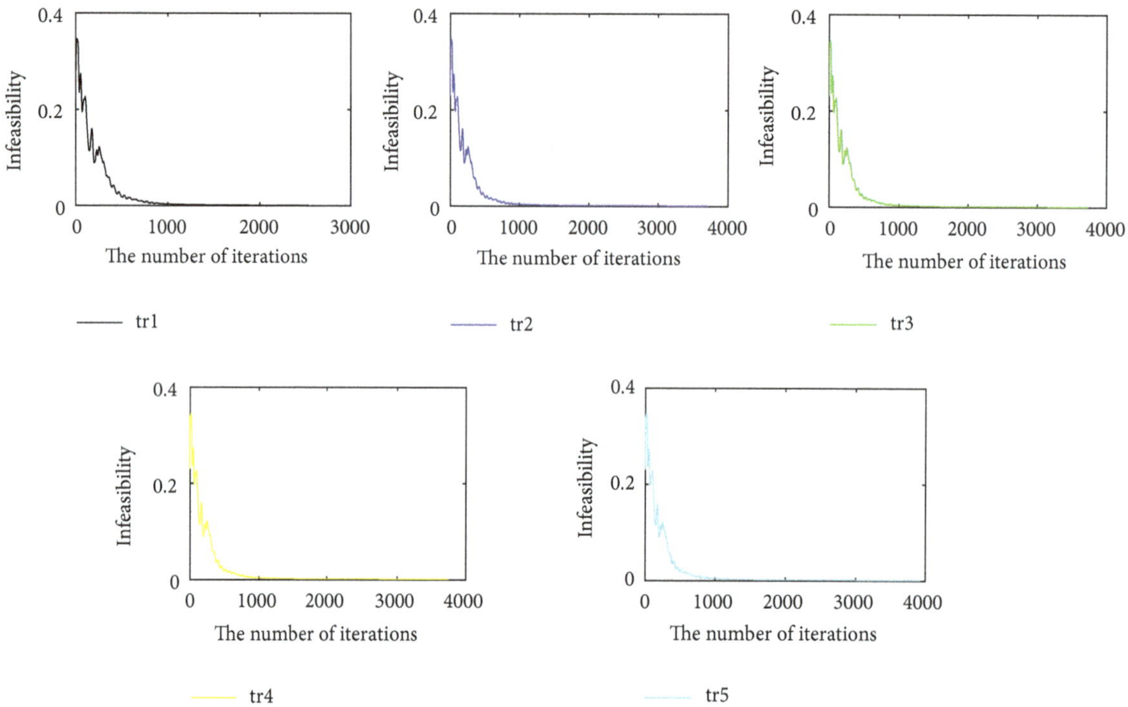

Figure 8: The numerical convergence behavior of TSADMM for different trip times: tr1,...,tr5 stand for the cases for different trip times proposed in [6, 9–11, 19].

Since the weighted ℓ_1-norm is a convex approximation of the ℓ_0-norm, the optimal solution to the (26) is a proximal optimal solution to the original sparse optimization model. How to design some more effective nonconvex relaxation scheme to get a better proximal optimal solution so as to get a better energy saving rate will be one of our future work. Meanwhile, the delay perturbations of the metro trains and the randomness of passenger demands have not been considered in our static timetabling. How to use advanced optimization techniques for handling such cases will also be our future research issue.

Appendix

The Proof of Theorem 3

Proof. Let $\overline{x} = ([x^{(1)}]^T, \ldots, [x^{(2\overline{n})}]^T)^T \in \mathbf{R}^n$ be an optimal solution of Problem (12). Then, the optimality and the feasibility of $\overline{x}$ imply that $F\overline{x} - g = 0$ and $f - E\overline{x} \geq 0$. By virtue of the expressions of E and F in Section 2.4, together with the time window constraints as stated in f, we can find some $\widetilde{x}$ in a neighborhood $N(\overline{x}, \delta)$ with $\delta < \min_{i: f_i-(E\overline{x})_i>0} f_i-(E\overline{x})$, such that $F\widetilde{x} - g = 0$ and $f - E\widetilde{x} > 0$. Set $\widetilde{y} = A\widetilde{x} + b$ and $\widetilde{z} = f - E\widetilde{x}$. It is easy to verify that $(\widetilde{x}, \widetilde{y}, \widetilde{z}) \in (\mathbf{R}^n \times \mathbf{R}^n \times \mathbf{R}^m_{++}) \cap P$ with P being the feasible region of Problem (26). By employing [25, Theorem B.1], we can get the desired global convergence result in Theorem 3. □

Conflicts of Interest

The authors declare that they have no conflicts of interest regarding the publication of this paper.

Acknowledgments

This work was partly supported by the State Key Laboratory of Rail Traffic Control and Safety, Beijing Jiaotong University [Grants nos. RCS2017ZJ001 and RCS2016ZZ01], and National

Natural Science Foundation of China [Grants nos. 11771038 and 11431002].

References

[1] G. M. Scheepmaker, R. M. Goverde, and L. Kroon, "Review of energy-efficient train control and timetabling," *European Journal of Operational Research*, vol. 257, no. 2, pp. 355–376, 2017.

[2] S. Barsali, P. Bolognesi, M. Ceraolo, M. Funaioli, and G. Lutzemberger, "Cyber-physical modelling of railroad vehicle systems using modelica simulation language," in *Railways*, Corsica, Ajaccio, France, 2014.

[3] A. Capasso, R. Lamedica, A. Ruvio, M. Ceraolo, and G. Lutzemberger, "Modelling and simulation of electric urban transportation systems with energy storage," in *Proceedings of the 16th International Conference on Environment and Electrical Engineering (EEEIC '16)*, Florence, Italy, June 2016.

[4] M. Peña-Alcaraz, A. Fernández, A. P. Cucala, A. Ramos, and R. R. Pecharromán, "Optimal underground timetable design based on power flow for maximizing the use of regenerative-braking energy," *Proceedings of the Institution of Mechanical Engineers, Part F: Journal of Rail and Rapid Transit*, vol. 226, no. 4, pp. 397–408, 2011.

[5] S. D. Gupta, L. Pavel, and J. K. Tobin, "An optimization model to utilize regenerative braking energy in a railway network," in *Proceedings of the American Control Conference (ACC '15)*, pp. 5919–5924, July 2015.

[6] X. Yang, X. Li, Z. Gao, H. Wang, and T. Tang, "A cooperative scheduling model for timetable optimization in subway systems," *IEEE Transactions on Intelligent Transportation Systems*, vol. 14, no. 1, pp. 438–447, 2013.

[7] X. Yang, B. Ning, X. Li, and T. Tang, "A two-objective timetable optimization model in subway systems," *IEEE Transactions on Intelligent Transportation Systems*, vol. 15, no. 5, pp. 1913–1921, 2014.

[8] X. Li and K. Lo Hong, "An energy-efficient scheduling and speed control approach for metro rail operations," *Transportation Research Part B: Methodological*, vol. 64, pp. 73–89, 2014.

[9] X. Li and H. K. Lo, "Energy minimization in dynamic train scheduling and control for metro rail operations," *Transportation Research Part B: Methodological*, vol. 70, pp. 269–284, 2014.

[10] S. Yang, J. Wu, X. Yang, H. Sun, and Z. Gao, "Energy-efficient timetable and speed profile optimization with multi-phase speed limits: theoretical analysis and application," *Applied Mathematical Modelling: Simulation and Computation for Engineering and Environmental Systems*, vol. 56, pp. 32–50, 2018.

[11] X. Yang, A. Chen, X. Li, B. Ning, and T. Tang, "An energy-efficient scheduling approach to improve the utilization of regenerative energy for metro systems," *Transportation Research Part C: Emerging Technologies*, vol. 57, pp. 13–29, 2015.

[12] S. D. Gupta, J. K. Tobin, and L. Pavel, "A two-step linear programming model for energy-efficient timetables in metro railway networks," *Transportation Research Part B: Methodological*, vol. 93, pp. 57–74, 2016.

[13] D. L. Donoho, "Compressed sensing," *IEEE Transactions on Information Theory*, vol. 52, no. 4, pp. 1289–1306, 2006.

[14] H. Zou, "The adaptive lasso and its oracle properties," *Journal of the American Statistical Association*, vol. 101, no. 476, pp. 1418–1429, 2006.

[15] S. S. Chen, D. L. Donoho, and M. A. Saunders, "Atomic decomposition by basis pursuit," *SIAM Review*, vol. 43, no. 1, pp. 129–159, 2001.

[16] A. Beck and M. Teboulle, "A fast iterative shrinkage-thresholding algorithm for linear inverse problems," *SIAM Journal on Imaging Sciences*, vol. 2, no. 1, pp. 183–202, 2009.

[17] X. Li, D. Sun, and K.-C. Toh, "A highly efficient semismooth Newton augmented Lagrangian method for solving lasso problems," *SIAM Journal on Optimization*, vol. 28, no. 1, pp. 433–458, 2018.

[18] S. Yang, J. Wu, H. Sun, X. Yang, Z. Gao, and A. Chen, "Bi-objective nonlinear programming with minimum energy consumption and passenger waiting time for metro systems, based on the real-world smart-card data," *Transportmetrica B*, vol. 4, pp. 1–18, 2017.

[19] Z. Tian, P. Weston, N. Zhao, S. Hillmansen, C. Roberts, and L. Chen, "System energy optimisation strategies for metros with regeneration," *Transportation Research Part C: Emerging Technologies*, vol. 75, pp. 120–135, 2017.

[20] L. W. P. Peeters, *Cyclic Railway Timetable Optimization*, Erasmus University Rotterdam, 2003.

[21] B. K. Natarajan, "Sparse approximate solutions to linear systems," *SIAM Journal on Computing*, vol. 24, no. 2, pp. 227–234, 1995.

[22] E. J. Candes and T. Tao, "Decoding by linear programming," *IEEE Transactions on Information Theory*, vol. 51, no. 12, pp. 4203–4215, 2005.

[23] R. Glowinski and A. Marroco, "Sur lapproximation, par elements finis dordre un, et la resolution, par penalisation-dualite', dune classe de problems de dirichlet non lineares," *Revue Francaise d'Automatique, Informatique, et Recherche Ope'rationelle*, vol. 9, pp. 41–76, 1975.

[24] D. Gabay and B. Mercier, "A dual algorithm for the solution of nonlinear variational problems via finite element approximation," *Computers & Mathematics with Applications*, vol. 2, no. 1, pp. 17–40, 1976.

[25] M. Fazel, T. K. Pong, D. Sun, and P. Tseng, "Hankel matrix rank minimization with applications to system identification and realization," *SIAM Journal on Matrix Analysis and Applications*, vol. 34, no. 3, pp. 946–977, 2013.

2

Optimal Operation of High-Speed Trains Using Hybrid Model Predictive Control

Yingze Yang, Zheng Xu, Weirong Liu, Heng Li, Rui Zhang, and Zhiwu Huang

School of Information Science and Engineering, Central South University, Changsha 410083, China

Correspondence should be addressed to Zhiwu Huang; hzw@csu.edu.cn

Academic Editor: Paola Pellegrini

The high-speed train operation process is highly nonlinear and has multiple constraints and objectives, which lead to a requirement for the automatic train operation (ATO) system. In this paper, a hybrid model predictive control (MPC) framework is proposed for the controller design of the ATO system. Firstly, a piecewise linear system with state and input constraints is constructed through piecewise linearization of the high-speed train's nonlinear dynamics. Secondly, the piecewise linear system is transformed into a mixed logical dynamical (MLD) system by introducing the auxiliary binary variables. For the transformed MLD system, a hybrid MPC controller is designed to realize the precise control under hard constraints. To reduce the online computation complexity, the explicit control law is computed offline by employing the mixed-integer linear programming (MILP) technique. Simulation results validate the effectiveness of the proposed method.

1. Introduction

With rapid development of high-speed railway, the operation safety, punctuality, and energy consumption of train have received more and more attentions. Nowadays, in China, the high-speed railway generally adopts distributed traction power to improve speed and traction efficiency, where locomotives and wagons are marshalled together to form a high-speed train [1]. For distributed traction power trains, there are some critical issues that need to be solved, such as vehicle traction and brake force allocation, operation safety, punctuality, and energy consumption [2].

To achieve the safety and multiobjective optimization, automatic train control system is developed for high-speed railway, and it is one of the key on-board equipment with high safety assurance [3]. The automatic train operation (ATO) system is the essential component that plays a key role in the train operation process [4]. The ATO system controls the train traction/brake force to follow the reference speed. Moreover, ATO system can reduce the energy consumption and improve riding comfort. Now, it is still a challenge to develop an efficient modeling and control method for ATO system [5].

Currently, the researches of automatic train control are mainly based on single-point model and multipoint model [6]. For single-point model, the whole high-speed train is simplified as a single rigid mass point, which ignores the coupling characteristics among the vehicles and the allocation of the distributed traction/brake force. The dynamic properties of single-point model are simple and easy to design. However, high-speed train adopts distributed traction power mode; thus the single-point model is obviously not able to describe the coupling characteristics and hydrodynamic dispersion characteristics of the train. Thus multipoint model is favoured to study the automatic train control strategy.

Yang and Sun [7, 8] analyzed the coupling characteristics among vehicles of high-speed train and established train multipoint dynamical model. For both push-pull driving and distributed driving train types, the hybrid H_2/H_∞ automatic train controller was designed and was compared by H_2 controller and H_∞ controller, respectively. Lin et al. [9] proposed a control strategy based on improved sliding mode control to overcome the effect of the disturbance in the process of train operation, which had good robustness and effectively restrained the high frequency chattering phenomenon. Thus the work reduced the impact of the frequent switching of control input and increased riding comfort. To simplify the controller design, Song et al. [10] utilized the geometric topology to reduce multiple positions to one position for multipoint model. In their work, to address the nonlinear saturation constraints of traction and brake forces, the robust adaptive controller was designed with low computing load, which had robustness for train disturbance and uncertainty of model parameters.

Based on the multipoint model, Zhuan and Xia [11] have achieved constructive results on the automatic train control strategy of the heavy-duty train in South Africa. Specifically, Zhuan and Xia [12] put forward three kinds of offline open-loop optimal operation strategies, whose goals are to minimize the workshop bonding force, ensure the safety of train driving, and reduce the maintenance costs of the coupler buffer. Then, Chou and Xia [13] proposed a closed-loop cruising linear quadratic regulator controller to minimize the running cost of electric air brake system in heavy-haul train, in which the control objective includes the speed tracking, coupling force, and energy consumption. In order to overcome the communication limitation, the vehicle barrier was introduced and the controller was reconstructed based on the current orbital slope. In addition, Zhuan and Xia [14] adopted the output feedback adjustment method to regulate the speed of the heavy train and verified the application conditions of their proposed method.

As a summary, Scheepmaker et al. [15] presented an extensive review on energy-efficient train control and the related topic of energy-efficient train timetabling, from the first simplest model of a train running on a level track to the advanced models and algorithms of the last decade dealing with varying gradients and speed limits and including regenerative braking. And there have appeared various theories and technologies to realize the efficient and safety train operation, such as the passivity-based cruise control [16], the robust adaptive control [17], and the iterative learning and fault detection approaches [18, 19].

Li et al. [20] considered the optimal guaranteed cost of cruise control for high-speed train movement. The sufficient condition for the existence of guaranteed cost cruise control law is given in terms of linear matrix inequalities. And a convex optimization problem is formulated to determine the optimal guaranteed cost that minimizes the performance upper bound of the cruise control law. Ye and Liu [21] propose a novel approach to solve the complex optimal train control problems in closed loop by introducing some simplifications. The operation sequence consists of maximum traction, speed holding, coasting, and maximum braking on each subsection of the track or a constant force is applied on each subsection. Zhao et al. [22] proposed a new cluster consensus technique to design the distributed control law, by which the trains can track the desired speeds asymptotically and the distance between the neighboring cars can be kept in the ideal range.

However, with the improvement of the requirement for ATO control performance, the above methods lack the ability to address the nonlinear, multiconstraint, and multiobjective problem based on the multipoint model, which leads to their limitations in practical applications. Especially for online operation process of high-speed trains, it raises the nonlinear resistance force which makes many classical control methods based on the linear resistance force model or the single equilibrium point linearizable model difficult to be implemented. And it is necessary to consider the safe speed, the saturation characteristics of traction and braking units, and the work conditions such as maximization of bonding forces, operating punctuality, energy-efficiency, and reduction of coupling wear forces. The model predictive control has the advantages of fully considering the input and state constraints of the system and the multiobjective optimization problems in the design of the controller, which is more suitable for the design of ATO controller [23, 24].

In this paper, an optimal automatic train control strategy is proposed for high-speed trains based on model predictive control. Due to the high computation complexity of nonlinear train model, the common method is to linearize the nonlinear system model at the equilibrium points [25]. However, if the nonlinear system has a wide range of operating condition, there are many deviations on entire running process for singular linearized model with one equilibrium point. Thus, the piecewise linear systems are utilized to approximate the original system, which can describe complex nonlinear systems accurately [26]. In this paper, the nonlinear running resistance is fitted by multiple line segments to construct the piecewise linear system model of train.

In the traditional controller design based on piecewise linear model, the linear submodel will be determined according to the initial state of receding horizon; then the predicted control is solved based on the determined sublinear model, which is fixed in the entire receding horizon [27]. However, for each actual control step, the system state changes and the corresponding submodel switch may also occur. The fixed submodel is difficult to approximate the original nonlinear model, especially near the model switching point. Therefore, the traditional model prediction control just according to current linear submodel can result in a large prediction error.

In order to solve the above problems, the logic variables are introduced to describe the switch of different linear submodels in this paper. By using this method, the piecewise linear system can be transformed into a hybrid logic dynamical system [28, 29]. For the hybrid logic dynamical system, the model switch can be described by changing the logical variables. Then the hybrid model predictive control (MPC) can be utilized to generate the control inputs based on the constructed mixed logic dynamical system.

The hybrid MPC controller uses the mixed logic dynamical model to predict the future evolution of the system at each time step, and a certain performance index is optimized under operating constraints with respect to a sequence of future input moves. The first of such optimal solution is the control action applied to the plant. For the hybrid MPC controller, the optimal problem that needs to be solved is a mixed integer linear program (MILP) due to the hybrid system description. Because the mixed logic dynamical model ensures that the submodel can switch on the entire receding horizon. Thus, more accurate approximation can be obtained by hybrid MPC and the prediction error is reduced.

Compared to the existing work, the proposed hybrid MPC makes a superior trade-off between the model complexity and control performance. The proposed controller can obtain the optimal input sequence without excessive

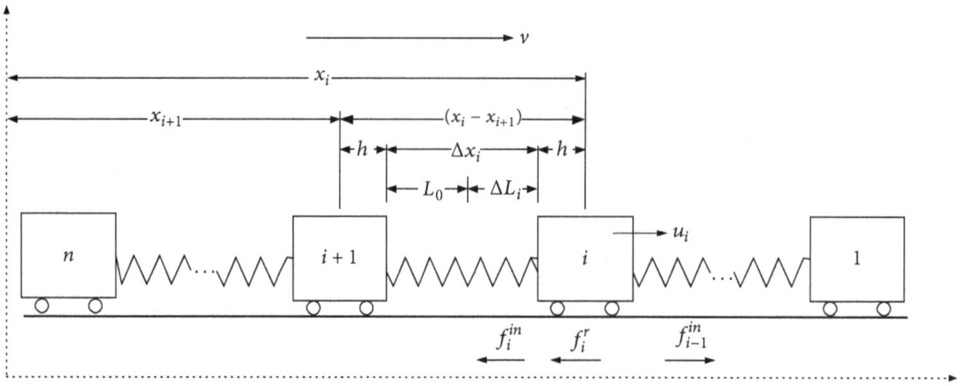

FIGURE 1: The multiple mass-point model structure of the electric multiple units.

simplifying assumption, while the piecewise linear model reduces the computation complexity under the sufficient precision compared to the classical MPC method. The main contributions in this paper can be summarized as follows.

(1) The nonlinear multiple mass point dynamical model of high-speed trains is constructed. In the model, the complex nonlinear characteristics of running resistance are piecewise linearized and the piecewise linear description is transformed to mixed logic dynamical system.

(2) Based on the MLD system model, a hybrid MPC framework is introduced and the MLD model is used as the prediction model. Based on the MLD prediction model, the control problem can be formulated as a unified mixed integer linear programming (MILP) problem, even the piecewise linear characteristics of train model.

(3) For the formulated MILP problem containing logical variables and continuous variables, the branch and bound method (*B&B*) is utilized to transform the original MILP problem into linear programming (LP) by relaxing the logic variable constraints.

The remainder of the paper is organized as follows. In Section 2, we establish the piecewise linear model of the high-speed train and transform the piecewise linear model to MLD model. In Section 3, a mixed logical dynamical system is developed. In Section 4, a hybrid MPC controller is designed by analyzing the performance indices and constraints of high-speed trains. Then, the simulation results are provided to validate the effectiveness of the controller in Section 5. Finally, the major conclusion of this paper is given in Section 6.

2. The Model of High-Speed Train

In this section, we establish the nonlinear multiple mass-point dynamic model of high-speed trains by analyzing their dynamical characteristics. Because the nonlinear model is complex for designing the system controller, the nonlinear curve of the running resistance is piecewise linearized and the piecewise linear model of the train is constructed.

2.1. The Multiple Mass-Point Model of Trains. Figure 1 presents the multiple mass-point model structure for the simplified electric multiple power units. The total number of locomotives and wagons is denoted by n. These vehicles are connected by the vehicle hook couplers. The mechanical properties of the couplers are usually described as the "elastic-damping" system. Thus, the hook coupler force between vehicle i and vehicle $i + 1$ is as follows:

$$f_i^{in} = k_i \Delta L_i + d_i \Delta \dot{L}_i, \quad (1)$$

where $\Delta L_i = x_i - x_{i+1} - h_{i+1} - h_i - L_0$ is the offset value of the relative equilibrium position of the hook coupler between vehicle i and vehicle $i+1$, L_0 is the natural length of the coupler and h_i is the half length of vehicle i, and x_i is the displacement of vehicle i. Since the length of each vehicle is essentially the same and constant (denoted by h) and the hook coupler has the same type, the balance position between two vehicles can be denoted by $L_e = 2h + L_0$. Thus $\Delta L_i = x_i - x_{i+1} - L_e$. And the hook coupler force (1) can be transferred to

$$f_i^{in} = k\left(x_i - x_{i+1} - L_e\right) + d\left(\dot{x}_i - \dot{x}_{i+1}\right). \quad (2)$$

The multiple mass-point dynamic equation of electric multiple power units can be described as

$$\begin{aligned} m_1 \dot{v}_1 &= u_1 - f_1^{in} - f_1^r, \\ m_i \dot{v}_i &= u_i + f_{i-1}^{in} - f_i^{in} - f_i^r, \quad (i = 2, 3, \ldots, n-1) \\ m_n \dot{v}_n &= u_n + f_{n-1}^{in} - f_n^r, \\ \dot{x}_i &= v_i, \quad (i = 1, 2, \ldots, n), \end{aligned} \quad (3)$$

where u_i is the control input of vehicle i. When the train is in traction state or cruise state, $u_i > 0$ is traction force. When the train is in brake state, $u_i < 0$ is the brake force. When the train is in coasting state, $u_i = 0$, implying that there is no traction force and brake force. For locomotive vehicles, they can provide both traction force and brake force, while for wagon vehicles, they can just provide brake force. Assume that f_i^r denotes the running resistance of vehicle i; that is,

$$\begin{aligned} f_i^r &= f_i^b + f_i^a \\ &= m_i c_0 + m_i c_v v_i + m_i c_a v_i^2 + m_i g \cdot \sin\theta_i + 0.004 \\ &\quad \cdot m_i D_i, \end{aligned} \quad (4)$$

where f_i^b and f_i^a are the basic resistance and appended resistance, respectively, and m_i and v_i are the mass and

TABLE 1: The basic resistance parameters of Chinese electric multiple units high-speed train CRH3.

Parameter	Value	Unit
m_i	47.5	t
c_0	7.75×10^{-3}	Nkg^{-1}
c_v	2.28×10^{-4}	$\mathrm{Ns(mkg)}^{-1}$
c_a	1.66×10^{-5}	$\mathrm{Ns^2(m^2kg)}^{-1}$

TABLE 2: The piecewise linear parameters of basic resistance.

j	α_j (s^{-1})	β_j (Nkg^{-1})	Speed ($\mathrm{m{\cdot}s}^{-1}$)
1	0.0008	0.0044	[0, 35]
2	0.0020	−0.0363	[35, 70]
3	0.0031	−0.1100	[70, 100]

velocity of vehicle i, θ_i is the ramp angle, and D_i is the curvature of rail.

2.2. The Multiple Mass-Point Piecewise Linearized Model of Trains. According to the dynamic analysis of high-speed trains, the basic resistance of the train is composed of mechanical resistance and air resistance. When the train is running at low speed, for instance, below 44.44 m/s (160 km/h), the basic resistance is mainly mechanical resistance, and air resistance can be ignored. However, when the train runs at high speed, air resistance will be dominant, and it is related to the square item of the train speed with nonlinear characteristics. As the speed of high-speed trains becomes higher, this nonlinear relationship is becoming stronger. In order to facilitate the controller design of nonlinear model, we consider linearization of the nonlinear model of the train.

For nonlinear systems, the linear model is usually obtained by using the equilibrium linearization method. However, for high-speed trains with varying speeds, it is not appropriate to choose one equilibrium point. Piecewise linear systems approximate the original system by using multiple linear subsystems, which can approximate most complex nonlinear systems accurately. Thus, the least square method is adopted to fit the train running resistance linearly. By using the method, the piecewise linear system model of the train is established.

The train running parameters are selected from Chinese electric multiple units high-speed train CRH3, which are shown in Table 1. Basic resistance after the piecewise linearization is presented in Figure 2, where the train speed range is from 0 m/s to 100 m/s (0–360 km/h). Through lots of experiments, we set $v_a = 35$ m/s and $v_b = 70$ m/s as the boundaries of piecewise linearization. The original nonlinear equations are divided into three linear equations, and the piecewise linear parameters of basic resistance are shown in Table 2.

It can be seen from Figure 2 that the running resistance of the train unit after linearization is already a good approximation to the original nonlinear curve. In fact, using more sublinear model is better for the approximation effect of the original nonlinear model, but it will increase the calculation of the controller. In order to further verify the approximate effect of the linearized train running resistance, the error of the approximate linear curve and the original curve is drawn, which is shown in Figure 3. From Figure 3, the error between the piecewise linearization of running basic resistance and the

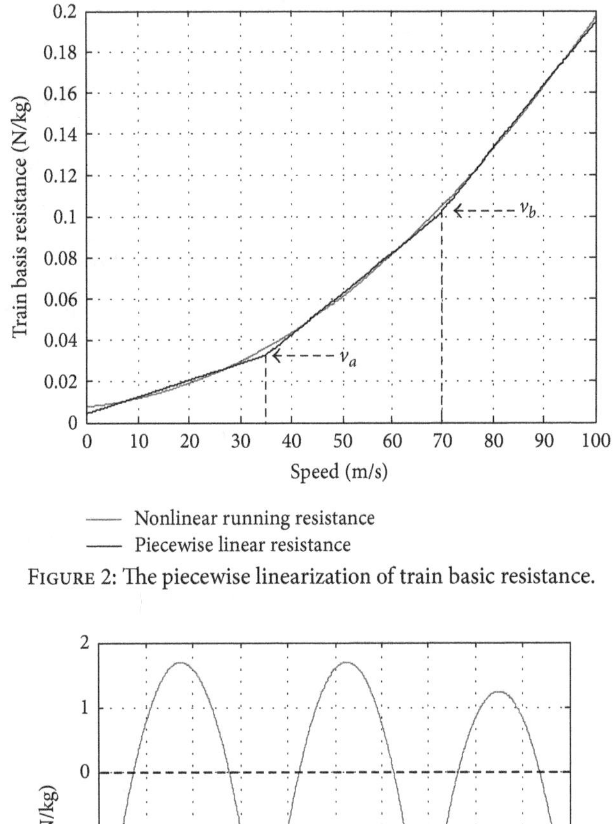

FIGURE 2: The piecewise linearization of train basic resistance.

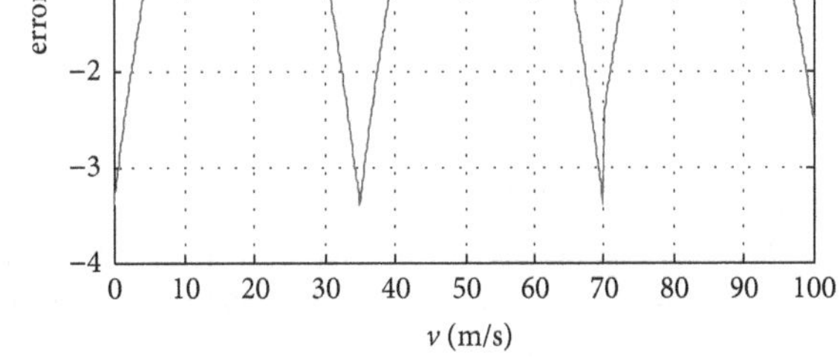

FIGURE 3: The error between the piecewise linearization of train basic resistance and the original true basic resistance.

practical basic resistance is less than 3.3 N/kg, which is just 0.55% compared to the biggest traction force of single vehicle, 600 N/kg.

Thus, the linearization of train basic running resistance can be approximated as

$$\omega_0 = \alpha_j v + \beta_j, \quad j = 1, 2, 3. \tag{5}$$

In fact, due to the force constraints of hook couplers, there is small difference among the operative velocities of vehicles. Thus, the velocity of vehicle 1 can be taken as the piecewise point of piecewise linear model. Taking it into the train multiple mass-points model, the piecewise linear dynamic model of train can be described as

$$\begin{aligned} \dot{x}(t) &= A_1^c x(t) + B^c u(t) + f_1^c \quad v_i \le v_a \\ \dot{x}(t) &= A_2^c x(t) + B^c u(t) + f_2^c \quad v_a < v_i \le v_b \\ \dot{x}(t) &= A_3^c x(t) + B^c u(t) + f_3^c \quad v_i > v_b, \end{aligned} \tag{6}$$

where the state variable $x = [v_1, v_2, \ldots, v_n, x_1, x_2, \ldots, x_n]^T$, control input $u = [u_1, u_2, \ldots, u_n]^T$, system matrix $A_j^c \in \mathfrak{R}^{2n\times 2n}$, $B^c \in \mathfrak{R}^{2n\times n}$, and $f_j^c \in \mathfrak{R}^{2n\times 1}$ are as follows:

$$A_j^c = \begin{bmatrix} A_{11}^j & A_{12}^j \\ I_{n\times n} & 0_{n\times n} \end{bmatrix},$$

$$B^c = \begin{bmatrix} B_{11} \\ 0_{n\times n} \end{bmatrix},$$

$$f_j^c = \left[f_j^1, f_j^2, \ldots, f_j^n, 0, 0, \ldots, 0\right]^T,$$

$$A_{11}^j = \begin{bmatrix} -\frac{d}{m_1} - \alpha_j & \frac{d}{m_1} & 0 & \cdots & 0 & 0 & 0 \\ \frac{d}{m_2} & -\frac{2d}{m_2} - \alpha_j & \frac{d}{m_2} & \cdots & 0 & 0 & 0 \\ \vdots & \vdots & \vdots & \ddots & \vdots & \vdots & \\ 0 & 0 & 0 & \cdots & \frac{d}{m_{n-1}} & -\frac{2d}{m_{n-1}} - \alpha_j & \frac{d}{m_{n-1}} \\ 0 & 0 & 0 & \cdots & 0 & \frac{d}{m_n} & -\frac{d}{m_n} - \alpha_j \end{bmatrix}, \tag{7}$$

$$A_{12}^j = \begin{bmatrix} -\frac{k}{m_1} & \frac{k}{m_1} & 0 & \cdots & 0 & 0 & 0 \\ \frac{k}{m_2} & -\frac{2k}{m_2} & \frac{k}{m_2} & \cdots & 0 & 0 & 0 \\ \vdots & \vdots & & \ddots & \vdots & \vdots & \vdots \\ 0 & 0 & 0 & \cdots & \frac{k}{m_{n-1}} & -\frac{2k}{m_{n-1}} & \frac{k}{m_{n-1}} \\ 0 & 0 & 0 & \cdots & 0 & \frac{k}{m_n} & -\frac{k}{m_n} \end{bmatrix},$$

and $B_{11} = \mathrm{diag}\{1/m_1, 1/m_2, \ldots, 1/m_n\}$, and $f_j^i = k_1 L_e/m_i - \beta_j - 9.98\sin\theta_i - 0.004D_i$. In practical controller design, L_e can be eliminated by state transformation, so the equilibrium position is set to zero for each subsystem.

Then, the zero-order holder is used to discretize the continuous time model. By setting the sampling period as T_s, the discrete-time piecewise linear model can be obtained as

$$x(k+1) = A_j^d x(k) + B_j^d u(k) + f_j^d, \quad j = 1, 2, 3, \tag{8}$$

where $A_j^d = e^{A_j^c T_s}$, $B_j^d = \int_0^{T_s} e^{A_j^c \tau} d\tau B^c$, and $f_j^d = \int_0^{T_s} e^{A_j^c \tau} d\tau f_j^c$.

3. Hybrid MPC Based ATO Design

A discrete-time piecewise linear model of high-speed trains has been established, which consists of multiple piecewise linear submodels. For the traditional predictive controller design based on the piecewise linear model, the linear submodel based on the current state is determined at the beginning of the prediction and then the predictive control is operated according to the determined linear submodel. However, the predictive submodel has been fixed and does not change throughout the prediction horizon, which cannot reflect the system nonlinear characteristics by only using a fixed submodel. Actually, the predictive system state may change, and the corresponding submodel may also switch during the solution process. Using a fixed submodel is not enough to express the nonlinearity of the original model, especially near the switching point between submodels. Therefore, the traditional predictive control based on piecewise linear model may produce large prediction errors.

In order to address the above issues, a mixed logic dynamical system is introduced in this paper, which is proposed by Bemporad and Morari in the context of MPC technology [28]. By introducing logical variables to represent different linear submodels, the piecewise linear system can be transformed into a mixed logic dynamical system, which is a global dynamical system covering the whole nonlinear dynamics. Based on the model predictive control of the mixed logic dynamical system, the switching of the model can be expressed by the change of the logic quantity in the model. The model can be switchable throughout the predicted horizon, which ensures more accurate approximation of the original nonlinear model and reduces the prediction error. According to the propositional transformation of the mixed logic dynamic system and mixed integer linear inequality conversion rules, we transform the train piecewise linear model into a mixed integer dynamic model by introducing the binary variables [28].

Two binary variables $\delta_1(k)$ and $\delta_2(k)$ are given by

$$\begin{aligned} [v_i \le v_a] &\Longleftrightarrow [\delta_1(k) = 1] \\ [v_i \le v_b] &\Longleftrightarrow [\delta_2(k) = 1]. \end{aligned} \tag{9}$$

Based on $v_i \in [v_{\min}, v_{\max}]$, (9) can be written as follows:

$$\begin{aligned} \delta_1(k)(v_a - v_{\min}) &\ge v_i(k) - v_{\min} \\ \delta_1(k)(v_a - v_{\max}) &\le v_a - v_i(k) - \varepsilon \\ \delta_2(k)(v_b - v_{\min}) &\ge v_i(k) - v_{\min} \\ \delta_2(k)(v_b - v_{\max}) &\le v_b - v_i(k) - \varepsilon, \end{aligned} \tag{10}$$

where ε is a small positive number and generally takes 10^{-6}. It is used to transform a strict equality into a nonstrict inequality.

By defining $\Phi_i(k) = A_i^d x(k) + B_i^d u(k) + f_i^d$, the relationship between $\delta_1(k)$, $\delta_2(k)$, and $x(k+1)$ can be shown in Table 3.

Then, the piecewise linear system (8) can be described as

$$\begin{aligned} x(k+1) = {} & \delta_1(k)\delta_2(k)\Phi_1(k) \\ & + (1-\delta_1(k))\delta_2(k)\Phi_2(k) \\ & + (1-\delta_1(k))(1-\delta_2(k))\Phi_3(k). \end{aligned} \tag{11}$$

We introduce a new variable $\delta_3(k) = \delta_1(k)\delta_2(k)$, which satisfies

$$\begin{aligned} -\delta_1(k) + \delta_3(k) &\le 0 \\ -\delta_2(k) + \delta_3(k) &\le 0 \\ \delta_1(k) + \delta_2(k) - \delta_3(k) &\le 1. \end{aligned} \tag{12}$$

Table 3: The relationship between $\delta_1(k)$, $\delta_2(k)$, and $x(k+1)$.

$\delta_1(k)$	$\delta_2(k)$	$x(k+1)$
1	1	$\Phi_1(k)$
0	1	$\Phi_2(k)$
0	0	$\Phi_3(k)$

In addition, the auxiliary mixed logical vectors and variables are introduced, that is, $z_1^x(k) = \delta_1(k)x(k)$, $z_2^x(k) = \delta_2(k)x(k)$, $z_3^x(k) = \delta_3(k)x(k)$, $z_1^u(k) = \delta_1(k)u(k)$, $z_2^u(k) = \delta_2(k)u(k)$, and $z_3^u(k) = \delta_3(k)u(k)$, which can be expressed as

$$\begin{aligned} x_{\min}\delta_i(k) &\le z_i^x(k) \le x_{\max}\delta_i(k) \\ x(k) - x_{\max}(1-\delta_i(k)) &\le z_i^x(k) \\ &\le x(k) - x_{\min}(1-\delta_i(k)) \\ u_{\min}\delta_i(k) &\le z_i^u(k) \le u_{\max}\delta_i(k) \\ u(k) - u_{\max}(1-\delta_i(k)) &\le z_i^u(k) \\ &\le u(k) - u_{\min}(1-\delta_i(k)), \end{aligned} \tag{13}$$

for $i = 1, 2, 3$, where $x \in [x_{\min}, x_{\max}]$ and $u \in [u_{\min}, u_{\max}]$. The maximum tract input $u_{\max}$ is a nonincreasing function of speed, which is approximated by a piecewise linear, quadratic, and/or hyperbolic function of speed [15]. Thus it can be described as a group of hyperbolic or parabolic formulas and each formula approximates the actual traction force for a certain speed interval [30], such as

$$u_{\max}(v) = c_{1,j} + c_{2,j}v + c_{3,j}v^2, \quad v \in [v_j, v_{j+1}] \tag{14}$$

or

$$u_{\max}(v) = \frac{c_{h,j}}{v}, \quad v \in [v_j, v_{j+1}], \text{ for } j = 1, 2, \ldots, M-1, \tag{15}$$

where the piecewise point v_j and v_{j+1} and the parameters c_{1j}, c_{2j}, c_{3j}, and c_{hj} can be determined by train operation experiments.

Then (8) can be transformed into the following mixed logical dynamical (MLD) system:

$$\begin{aligned} x(k+1) = {} & Ax(k) + B_1u(k) + B_2\delta(k) \\ & + B_3z(k) + B_4 \\ E_2\delta(k) + E_3z(k) \le {} & E_1u(k) + E_4x(k) + E_5, \end{aligned} \tag{16}$$

where $z(k) = [z_1^x(k)^T\ z_2^x(k)^T\ z_3^x(k)^T\ z_1^u(k)\ z_2^u(k)\ z_3^u(k)]^T$, $\delta(k) = [\delta_1(k)\ \delta_2(k)\ \delta_3(k)]^T$, $A = A_3^d$, $B_1 = B_3^d$, $B_4 = f_3^d$, $B_2 = [-f_3^d, f_2^d - f_3^d, f_3^d - f_2^d + f_1^d]$, and $B_3 = [-A_3^d, A_2^d - A_3^d, A_1^d - A_2^d + A_3^d, -B_3^d, B_2^d - B_3^d, B_1^d - B_2^d + B_3^d]$.

Specifically, the resulting MLD model (14) is obtained by using the HYSDEL compiler [31]. Matrices $E_1, \ldots, E_5$ include constraints (16). They can be obtained and analyzed in Matlab using, for example, the Hybrid Toolbox [32].

4. Controller Design

Based on the MLD system model, a hybrid MPC framework is introduced in this section. Within this, the MLD model is used as the prediction model. Then the control problem is formulated as a constrained finite time optimal control problem and converted into mixed integer linear programming (MILP) problem, for which efficient solvers are available.

4.1. Formulation of Optimal Control Problem. Automatic train control needs to track the desired velocity profile quickly such that the timeliness of the interstation travel can be ensured. With the development of high-speed railway, energy-saving driving has become an important goal of automatic train control. As the running distance of the train between stations is fixed, the energy consumption directly related to the traction and braking force. Thus, the control input is used to express energy consumption through this paper. In addition, the in-train force should be kept in the desired range to ensure the safety of the train.

Therefore, the optimal problem can be established as follows, which considers the speed tracking error, control input, and the in-train force.

$$J(u_N(k)) = \sum_{i=0}^{N-1} \|\varepsilon((k+i)\mid k)\|_Q^1 + \|u((k+i)\mid k)\|_R^1 + \|x_{in}((k+i)\mid k)\|_H^1, \tag{17}$$

where k is the current sampling time; $u_N(k) = [u(k), \ldots, u(k+N-1)]$ is the sequence of N control inputs. $\varepsilon((k+i)\mid k) = v((k+i)\mid k) - v^r(k+i) \in \mathbb{R}^{n\times 1}$ represents the difference between actual speed v and reference speed v^r, $u((k+i)\mid k) = [u_1(k+i), u_2(k+i), \ldots, u_n(k+i)]^T$ represents the control input, and $x_{in}((k+i)\mid k) = (x_{in}^1(((k+i)\mid k), \ldots, x_{in}^{n-1}(((k+i)\mid k))$ represents the relative displacements between neighboring cars. 1 means 1-norm and Q, R, and H are the weight matrixes, respectively.

In order to ensure the safe and efficient operation of the train, some constraints must be satisfied as follows.

(1) Constraint on Control Input. The tractive and braking forces are bounded due to the nature physical characteristics of the traction motor. Thus, the control input is set as $-F_{\max}^B \le u_i(k) \le F_{\max}^T$.

(2) Constraint on Maximum Operational Velocity. The maximum speed of the train depends first on the maximum speed allowed by the train itself. Second, when the line conditions and operating conditions change, the maximum allowable speed of the train may change.

(3) Constraint on In-Train Force. The in-train force should be in the range $[-f_{\max}^{in}, f_{\max}^{in}]$ to keep the operational safety of the trains.

(4) Constraint on Coupler Deformation. The coupler deformation should be in the range $[-\Delta L_{\max}, \Delta L_{\max}]$ to keep the operational safety of the trains.

Then, the constraints considered can be formulated by

$$\begin{gathered} -F_{\max}^B \le u_i(k) \le F_{\max}^T \\ 0 \le v_i(k) \le v_{\max} \\ -f_{\max}^{in} \le f_{i,i+1}^{in} \le f_{\max}^{in} \\ -\Delta L_{\max} \le \Delta L_i \le \Delta L_{\max}. \end{gathered} \tag{18}$$

4.2. Hybrid MPC Controller Design. Based on (17) and (18), a constrained optimal control problem can be constructed as follows:

$$\begin{aligned} \min \quad & J(u_N(k)) \\ & = \sum_{i=0}^{N-1} \|\varepsilon((k+i)\mid k)\|_Q^1 + \|u((k+i)\mid k)\|_R^1 \\ & \quad + \|x_{in}((k+i)\mid k)\|_H^1 \\ \text{s.t.} \quad & x(k+1) \\ & = Ax(k) + B_1u(k) + B_2\delta(k) + B_3z(k) + B_4 \\ & E_2\delta(k) + E_3z(k) \le E_1u(k) + E_4x(k) + E_5 \\ & -F_{\max}^B \le u_i(k) \le F_{\max}^T \\ & 0 \le v_i(k) \le v_{\max} \\ & -f_{\max}^{in} \le f_{i,i+1}^{in} \le f_{\max}^{in} \\ & -\Delta L_{\max} \le \Delta L_i \le \Delta L_{\max}. \end{aligned} \tag{19}$$

In order to facilitate the solution of the optimal problem, (19) can be transformed to the linear objective function (20) by introducing the auxiliary variables $w^\varepsilon(k+i\mid k) = \|\varepsilon((k+i)\mid k)\|_Q^1$, $w^u(k+i\mid k) = \|u((k+i)\mid k)\|_R^1$, and $w^{x_{in}}(k+i\mid k) = \|x_{in}((k+i)\mid k)\|_H^1$.

$$J(u_N(k)) = w^\varepsilon(k+N\mid k) + \sum_{i=0}^{N-1} w^\varepsilon(k+i\mid k) + w^u(k+i\mid k) + w^{x_{in}}(k+i\mid k) \tag{20}$$

with some linear inequalities:

$$\begin{gathered} w^\varepsilon(k+i\mid k) \ge \pm\varepsilon((k+i)\mid k) \\ w^u(k+i\mid k) \ge \pm u((k+i)\mid k) \\ w^{x_{in}}(k+i\mid k) \ge \pm x_{in}((k+i)\mid k). \end{gathered} \tag{21}$$

In addition, the state update equation (22) can be obtained by the MLD model.

$$\begin{gathered} x(k+i) = A^i x(k) + \sum_{j=0}^{i-1} A^j \big(B_1u(k+i-j-1) + B_2\delta(k+i-j-1) + B_3z(k+i-j-1)\big) \\ E_2\delta(k) + E_3z(k) \le E_1u(k) + E_4x(k) + E_5. \end{gathered} \tag{22}$$

TABLE 4: Definition of some notations.

Notation	Definition
P	The MILP problem
J_0^*	The minimum upper bound of optimal performance of train operation
V^*	The optimal solution vector of optimal performance index for train operation
Π	The object table of recording all LP subproblems that need to be solved
$J(\overline{V})$	The optimal performance of the current relaxed LP problem
$\overline{V}$	The optimal solution vector of the relaxed LP subproblem
Q_{num}	Number of LP subproblems that have been solved
Maxnum $LP(j)[P_0], LP(j)[P_1]$	The maximum number of LP problems

Then, based on (20), (21), and (22), the optimal control problem can be converted to a mixed integer linear programming (MILP) problem as follows:

$$J(V,\theta) = \min_{V} \quad C^T V \tag{23a}$$

$$\text{st.} \quad GV \le W + S\theta(k), \tag{23b}$$

where $\theta(k) = [x(k)^T, v^r(k), \dots, v^r(k+N_p-1)]^T$ is commonly represented as the parameter vector and $V = [u'_{N_p}(k), \tilde{\delta}(k)^T, \tilde{z}(k)^T, \tilde{w}(k)^T]^T$ is the optimization vector. Let

$$\tilde{\delta}(k) = \begin{bmatrix} \delta(k) \\ \delta(k+1) \\ \vdots \\ \delta(k+N_p-1) \end{bmatrix},$$

$$\tilde{z}(k) = \begin{bmatrix} z(k) \\ z(k+1) \\ \vdots \\ z(k+N_p-1) \end{bmatrix}, \tag{24}$$

$$\tilde{w}(k) = \left[w^{\varepsilon}(k), w^{u'}(k), \dots, w^{\varepsilon}(k+N_p-1), w^{u'}(k+N_p-1)\right]^T;$$

$C^T = [0, \dots, 0, 1, \dots, 1]$. The matrices G, W, and S are obtained according to (16), (18), and (21).

At each sampling step k, the MILP problem is solved based on the current parameter vector $\theta(k)$, and an optimal sequence of future control input $u'_{N_p}(k)$ can be obtained. At time $k+1$, the parameter vector is updated to $\theta(k+1)$, and a new MILP problem is solved to get the new optimal control input. This procedure is repeated at each sampling step, and the problems can be solved in real time. In this paper, only $N_c = N_p$ is considered.

4.3. Solution to MILP Problem. The problem of automatic train control is established as a constrained finite time optimal control problem and further transformed into mixed integer linear programming (MILP) problem by analyzing the operation target and constraints. Since the constructed MILP problem contains logical variables and continuous variables, this section uses the branch and bound method (B&B) to solve this problem so that, in each control cycle, the optimal traction of each vehicle or braking force output can be obtained.

The basic idea of branch and bound method is to relax the logic variable constraints in the optimization problem. $\delta_i(k) \in \{0,1\}$ in the MLD model of formula (16) can be relaxed as $\delta_i(k) \in [0,1]$, which transforms the original MILP problem into linear programming (LP). Then, the traditional linear programming method is used to solve the optimal subproblem. Algorithm 1 is presented to solve the MILP problem and some notations are defined and presented in Table 4.

5. Performance Evaluation

In this section, we will evaluate the performance of the proposed hybrid MPC ATO algorithm using numerical results. We will first introduce the parameter settings of the ATO system. Specifically, the parameters of the train model are from the CRH-3 high-speed train in China, which are given in Table 5. Therein, F_{max}^T and F_{max}^B are the maximum tractive force and braking force, respectively.

5.1. Simulation Scenario. In order to evaluate the performance of the controller, the desired velocity curve including accelerating, decelerating, velocity step increase, velocity step decrease, and constant-velocity stages is characterized by the following equation:

$$v_r(t) = \begin{cases} 0.4t & 0 \le t \le 100 \\ 40 & 100 \le t \le 400 \\ 75 & 400 \le t \le 900 \\ 50 & 900 \le t \le 1150 \\ -t+1200 & 1150 \le t \le 1200. \end{cases} \tag{25}$$

In (25), the acceleration is 0.4 m/s, and the deceleration is 0.5 m/s, both of which are within the threshold of acceleration and velocity. In the operation of trains, the basic resistance is the main resistance for the trains. Thus, we only consider the basic resistance in this paper.

```
initialize $J_0^* = +\infty$, $Q_{\text{num}} = 0$, $\Pi = \text{Null}$;
for each $i \in [1, n]$ do
    $\delta_i(k) \in [0, 1] \leftarrow \delta_i(k) \in \{0, 1\}$;
    obtain LP-subproblem LP($i$);
    $\Pi \leftarrow \Pi \cup \text{LP}(i)$
end for
if there are no $\text{LP}(i) \in \Pi$ that need to be solved then
    obtain optimal performance index $J_0^*$;
    obtain optimal solution vector $V^*$;
    obtain optimal control input $u(k) \leftarrow V^*$;
else
    while $Q_{\text{num}} <$ Maxnum do
        for each $j \in [n, 1]$ do
            extracting LP($j$) from $\Pi$;
            if LP($j$) has feasible solution then
                obtain $J(\overline{V})$ and $\overline{V}$;
            else
                if LP($j$) satisfies $\delta_i(k) \in \{0, 1\}$ and $J(\overline{V}) <$
                $J_0^*$ then
                    $J_0^* \leftarrow J(\overline{V})$ and $V^* \leftarrow \overline{V}$;
                    $Q_{\text{num}} = Q_{\text{num}} + 1$;
                end if
                if $J(\overline{V}) > J_0^*$ then
                    $Q_{\text{num}} = Q_{\text{num}} + 1$;
                end if
                if LP($j$) dose not satisfy $\delta_i(k) \in \{0, 1\}$ and
                $J(\overline{V}) < J_0^*$ then
                    choose $\delta_i(k)$;
                    $\text{LP}(j)[P_0] \leftarrow \delta_i(k) = 1$;
                    $\text{LP}(j)[P_1] \leftarrow \delta_i(k) = 0$;
                    $\Pi \leftarrow \Pi \quad \text{LP}(j) \cup \text{LP}(j)[P_0] \cup \text{LP}(j)[P_1]$;
                else
                    $Q_{\text{num}} = Q_{\text{num}} + 1$;
                end if
                $Q_{\text{num}} = Q_{\text{num}} + 1$;
            end if
        end for
    end while
end if
```

ALGORITHM 1: Solution of MILP problem.

The designed MPC based ATO controller considers the train velocity tracking performance, energy consumption, and intertrain force. With the train parameters, we provide two simulation scenarios to evaluate the performance of the proposed algorithm under different operating conditions. Specifically, we evaluate the performance of the controller with different train types: all-locomotive train and partial-locomotive train.

5.1.1. Scenario A. In this scenario, we evaluate the performance of the ATO controller with the all-locomotive train. To simplify the simulation, we choose the number of trains as $n = 4$, and the train is comprised of all locomotives. This implies that each locomotive has the tracking motor and braking unit and thus can provide both the tracking force and the braking force. The parameters of the train are shown in Table 5, where m_i is the mass of locomotive i. Since the numbers of passengers in each locomotive may be different, we choose $m_1 = 48.5t$, $m_2 = 52.5t$, $m_3 = 51.0t$, $m_4 = 47.5t$. For the MPC controller, the prediction horizon N_p and the control horizon N_c also affect the controller performance. Without loss of generality, we choose $N_p = N_c = 3$. The weighting matrix gain is chosen as

$$Q = \operatorname{diag}\Big(\underbrace{4, \dots, 4}_{n}\Big),$$
$$R = \operatorname{diag}\Big(\underbrace{1, \dots, 1}_{n}\Big), \tag{26}$$
$$H = \operatorname{diag}\Big(\underbrace{1, \dots, 1}_{n-1}\Big).$$

5.1.2. Scenario B. In this scenario, we evaluate the effectiveness of the ATO controller in partial-locomotive trains. To

TABLE 5: Parameters of the CRH-3 high-speed train.

Parameter	Value	Unit
m_i	47.5	t
c_0	7.75×10^{-3}	Nkg^{-1}
c_v	2.28×10^{-4}	Ns(mkg)^{-1}
c_a	1.66×10^{-5}	$\text{Ns}^2(\text{m}^2\text{kg})^{-1}$
$F^T_{\max}$	300	kN
$F^B_{\max}$	300	kN
k	1×10^4	kNm^{-1}
d	1000	kNsm^{-1}
θ	0	–
D	0	–
$\Delta L_{\max}$	0.2	m
$f^{in}_{\max}$	2000	kN

simplify the simulation, we choose the number of trains as $n = 4$, and the train is comprised of 2 locomotives and 2 wagons. The locomotive can provide both the tracking force and the braking force, while the wagons can only supply the brake forces. In order to guarantee the total force the same as Scenario A, the maximum tracking force of each locomotive is set as $F^T_{\max} = 600$ kN, and the other train parameters are the same as in Table 5.

5.2. Simulation Results. We will provide simulation results of the ATO controller in different operating conditions. The simulations are conducted in Matlab/SIMULINK. We fist consider the simulations in Scenario A, which is based on all-locomotive train. Then we will further consider the simulation results in Scenario B, which is based on partial-locomotive trains. And in Scenario C, we focus on more realistic condition where the track force limit is dynamic.

5.2.1. Scenario A. The velocity curve of each locomotive is shown in Figure 4, where x-axis represents the time and y-axis represents the velocity. From the figure, we can see that the velocities of four locomotives almost remain the same during the operation time. Specifically, in the accelerating and decelerating stages, each locomotive can operate correspondingly according to the given acceleration and deceleration. At $t = 400$ s, we notice that the reference velocity has the step increase. From the zoomed-in figure, we can find that the locomotives move with the almost same velocity, where the maximum velocity error is 0.003 m/s. From the zoomed-in figure from $t = 470$ s to $t = 490$ s, the velocities of locomotives reach the steady value at about $t = 490$ s, where the settling time is 90 s and the steady error is 0.5 m/s. In the steady state, the velocities of locomotives remain fixed. Based on the simulation results, we can find that the train can track the time-varying reference with small settling time and steady error, which verifies the effectiveness of the algorithm.

The tracking and braking forces of locomotives are shown in Figure 5. In the accelerating stage, the resistance of the train increases with the velocity. In order to guarantee the constant acceleration, the tracking forces of locomotives increase gradually. In the deceleration stage, the resistance of

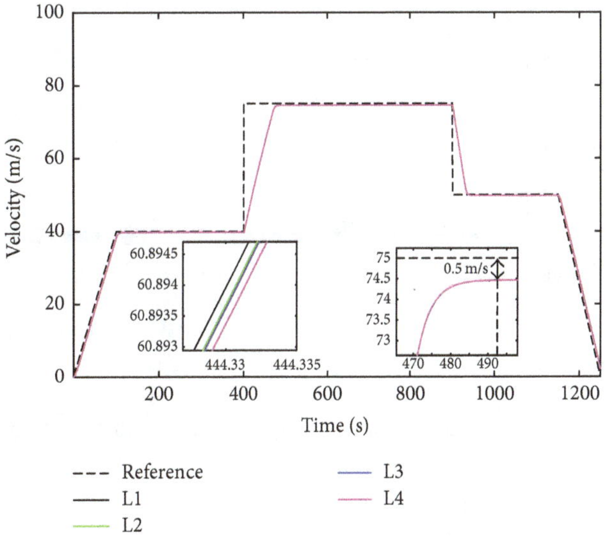

FIGURE 4: The velocity curve of each locomotive in Scenario A.

the train decelerates with the velocity. In order to guarantee the constant deceleration, the braking forces of locomotives increase gradually. Compared with the acceleration or deceleration stages, the tracking forces of locomotives are small and fixed during the constant velocity stage. This is because the resistance of train keeps having a relatively small value during the constant velocity stage. At $t = 400$ s, the reference has a step increase. In order to accelerate to the desired velocity, the tracking forces of locomotives increase to the maximum rapidly and maintain the maximum tracking force for about 72 s. At $t = 900$ s, the braking forces increase to the maximum to decrease the velocity and maintain about 32 s to decrease to the desired velocity. From the figure, we can find that the control outputs of locomotives are almost the same and thus guarantee the reasonable allocation of braking force and tracking force. When the reference velocity curve changes, each locomotive can regulate the tracking force and braking force based on the difference between the actual velocity and desired velocity. The forces for locomotives are

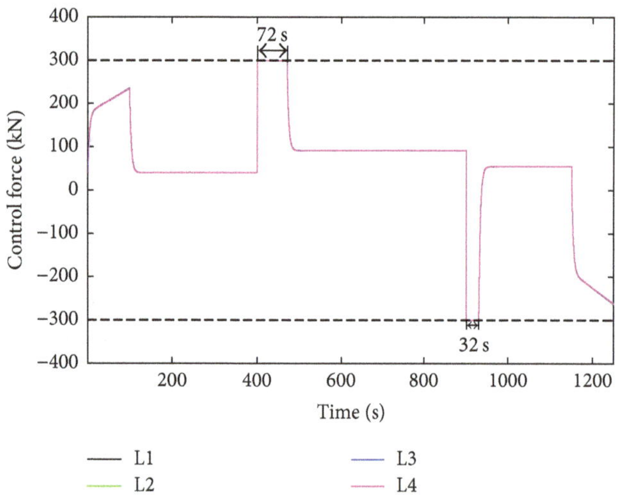

FIGURE 5: The force output of each locomotive in Scenario A.

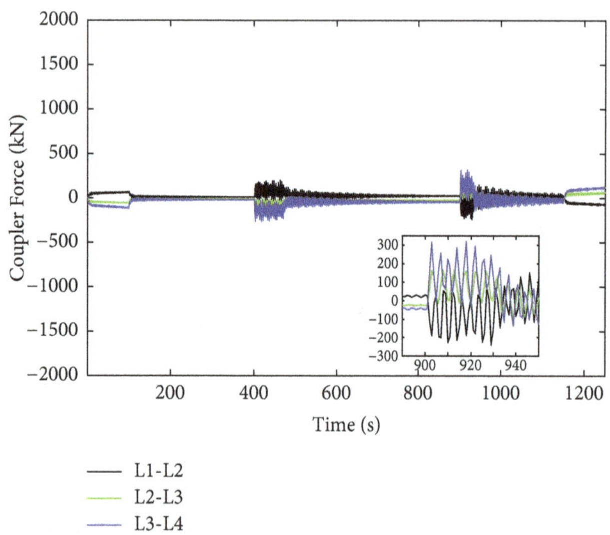

FIGURE 6: The coupler force among locomotives in Scenario A.

almost the same and also satisfy the constraints, which verify the effectiveness of the proposed strategy.

The coupler force among each locomotive is shown in Figure 6. When the coupler force is larger than zero, it means that coupler is pulled. When the coupler force is smaller than zero, the coupler is pushed. From the figure, we can see that the coupler forces among locomotives are always within the safety range. Moreover, the coupler among L1 and L2 is mainly pulled, while the coupler among L3 and L4 is mainly pushed. This is because when we choose the masses of locomotives, the masses of L3 and L4 are relatively smaller than L1 and L2. In order to guarantee the velocity consensus of locomotives, the total forces should be the same and the coupler force can balance the total force of each locomotive. It can be found that, during the constant-velocity stage, constant-accelerating stage, and constant-decelerating stage, the coupler force among locomotives is small. At $t = 200$ s and $t = 400$ s, the coupler force changes dramatically,

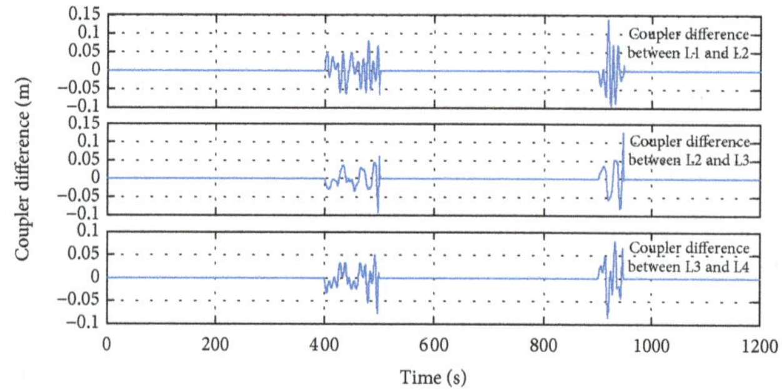

FIGURE 7: The coupler deformation between two neighbouring locomotives in Scenario A.

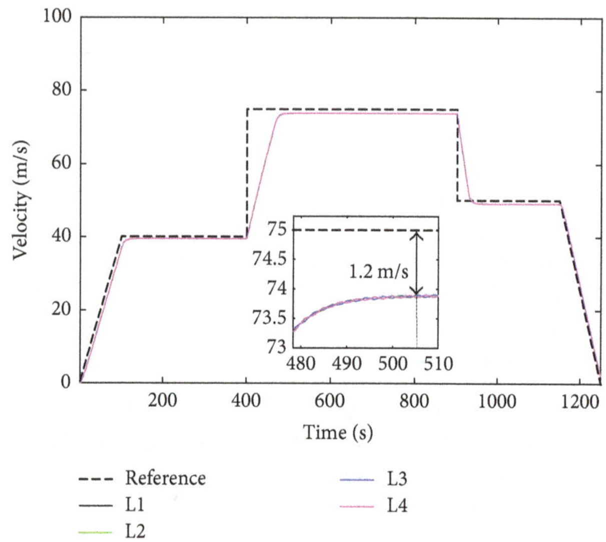

FIGURE 8: The velocity curve of each vehicle in Scenario B.

with the largest coupler pulling force 317 kN and the largest coupler pushing force 241 kN. From the figure, we can see that the ATO controller can guarantee that the coupler force is within the safety range, which verifies the effectiveness.

The operation safety of high-speed trains can be evaluated by the position difference (coupler deformation) between neighboring locomotives. If the deformation is larger than the maximum 0.2 m or less than the minimum −0.2 m, the coupler may have been broken so the safety cannot be guaranteed. The operation safety of high-speed trains can be further illustrated from Figure 7, which shows that the coupler deformation is always with the threshold [−0.2 m, 0.2 m]. This is because the in-train force guarantees that the position difference between neighboring locomotives cannot be too large or too small.

5.2.2. Scenario B. The simulation in Scenario A verifies the effectiveness of the hybrid model predictive controller for all-locomotive case. The metrics include velocity tracking, energy consumption, and coupler force.

In Scenario B, we verified the effectiveness of the ATO controller with 2 locomotives and 2 wagons. The velocity curves are shown in Figure 8. With the same controller

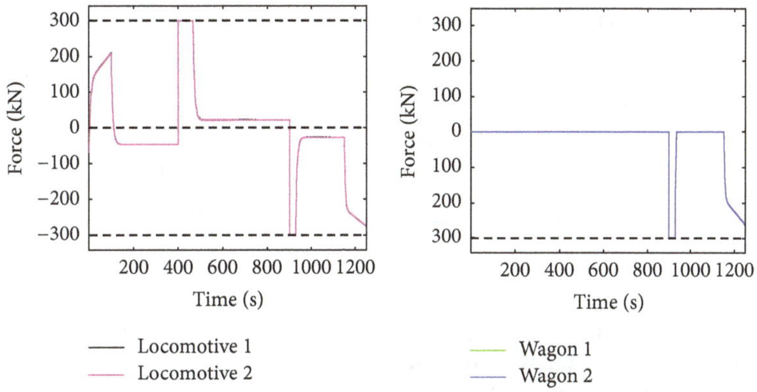

FIGURE 9: The forces of locomotives and wagons in Scenario B.

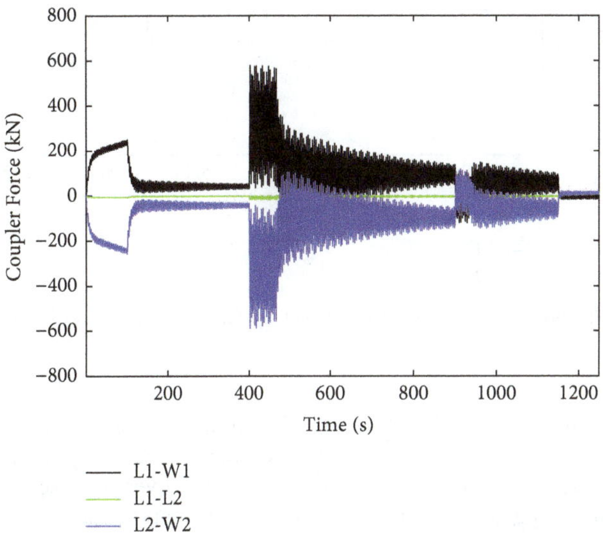

FIGURE 10: The coupler force among vehicles in Scenario B.

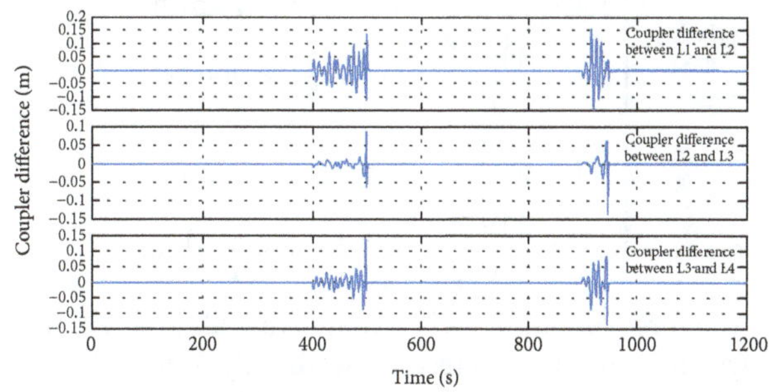

FIGURE 11: The coupler deformation between neighboring locomotives/wagons in Scenario B.

parameters, at $t = 400$ s, where the reference velocity has a step change, the convergence time of the controller from the change at $t = 400$ s to the steady state at $t = 508$ s implies that the convergence time increases by 18 s compared with the all-locomotive case. The steady error is 1.2 m/s, which increases by 0.7 m/s compared with the all-locomotive case. Moreover, the velocities at the steady stage have some oscillations around the steady value.

The force output of each vehicle is shown in Figure 9, where the left subfigure shows the force output of the locomotive, and the right subfigure shows the force output of the wagon. From the figure, we can see that, during the acceleration and constant-velocity stages, only the locomotive has the tracking force output, and the tracking force of wagons is zero. In the deceleration stage, both the locomotives and wagons supply the braking forces.

Figure 10 shows the coupler force among vehicles. Compared with the all-locomotive trains, the coupler forces between L1 and W1 and between L2 and W2 are kept at large values. This is because locomotive L1 supplies the pulling force and locomotive L2 supplies the pushing force, while wagons W1 and W2 cannot supply the tracking force. The maximum coupler force is 578 kN, which is within the safety threshold. From the above simulations, we can find that the proposed hybrid MPC ATO controller has a good control effect.

The operation safety of high-speed trains can be illustrated from Figure 11, which depicts the spring deformation between neighboring locomotives/wagons. We can see that the largest deformation is 0.14 m, and the smallest deformation is −0.15 m, which is with the safety threshold [−0.2 m, 0.2 m]. As illustrated in Figure 7, this is because the in-train force guarantees that the position difference between neighboring locomotives cannot be too large or too small. Simulation results show that the proposed ATO controller can guarantee the safe operation of high-speed trains.

5.3. Further Discussions. In this subsection, we further discuss the performance of the proposed ATO controller in terms of practicality, computation efficiency, superiority, and robustness. For simplicity, we only consider Scenario A, where the train is comprised of four locomotives. Scenario B can also be analyzed similarly. The train parameters are set the same as in Table 5, unless otherwise specified.

5.3.1. Practicality. In the previous simulation setup, we assume that the maximum traction force $F^T_{\max}$ and the maximum braking force $F^B_{\max}$ are constants. Despite simplifying the implementation, the constant maximum traction/braking

TABLE 6: Computational time (s) of each MPC step with and without linearization.

	MPC with linearization			MPC without linearization		
	Min.	Max.	Average	Min.	Max.	Average
$N_c = 3$	$1.6e-3$	$3.8e-3$	$2.5e-3$	$2.1e-1$	$3.6e-1$	$2.4e-1$
$N_c = 5$	$1.8e-3$	$6.4e-3$	$3.5e-3$	$2.5e-1$	$8.3e-1$	$4.7e-1$
$N_c = 7$	$1.9e-3$	$7.3e-3$	$4.6e-3$	$3.9e-1$	$9.8e-1$	$6.5e-1$

force are not realistic as the operation is subject to the power limit of the train. This implies that the maximum traction and braking forces evolve with the operation speed; that is, the maximum traction and braking forces for the CRH-3 train are represented as in [33].

$$F_{\max}^{T} = \begin{cases} 300 - 0.284v, & 0 \le v \le 119.7\,\text{km/h} \\ 266 \times \dfrac{119.7}{v}, & 119.7 < v \le 300\,\text{km/h} \end{cases}$$
$$F_{\max}^{B} = \begin{cases} 300 - 0.281v, & 0 \le v \le 106.7\,\text{km/h} \\ 266 \times \dfrac{119.7}{v}, & 106.7 < v \le 300\,\text{km/h}. \end{cases} \tag{27}$$

The force output of each locomotive with power limit is shown in Figure 12, where we find that the output force of each locomotive evolves with time. It is interesting to note that at $t = 400$ s the maximum traction force decreases at the transition stage, where the speed of the train increases from 40 m/s to 75 m/s. Due to the constant-power operation, the maximum traction force will decrease with the increase of velocity. At $t = 900$ s, it is shown that the maximum braking force increases when the velocity decreases from 75 m/s to 50 m/s.

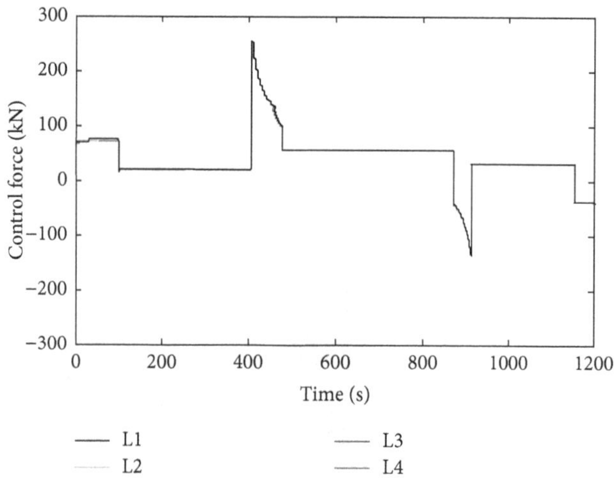

FIGURE 12: The force output of each locomotive with power limit.

5.3.2. Computational Time. In this subsection, we consider the computation time of the proposed method with different prediction horizons.

Within different control horizons, the nonlinear MPC and the proposed piecewise MPC are both implemented on the original nonlinear system. The simulations are performed under Windows 7 operating system with Intel Core i5-2400 CPU, 4 GB RAM, on a desktop computer. Both the nonlinear MPC and the proposed piecewise MPC provide comparable dynamic performance. Their computational times are summarized in Table 6. From Table 6, it can be seen that the computational time of each proposed MPC step is less than the nonlinear MPC on all control horizon lengths. It is more obvious when the horizon length is extended. The comparative results present the computation efficiency of the proposed method. To obtain the average computational time of each MPC step, the repeated numbers of MPC steps are 1200.

5.3.3. Superiority. In this subsection, we verify the superiority of the proposed method through comparison with the classical PID method.

The velocity difference between each locomotive and the reference velocity value is plotted in the following figures. Figures 13(a) and 13(b) show the simulation results under PID and the proposed MPC methods, respectively. From Figure 13(a), the velocities of four locomotives have severe chattering at time 400 s and 900 s when the reference velocity varies, which can further cause drastic changes of the coupler and endanger safe operation of high-speed trains. The proposed method in this paper presents good control performance as shown in Figure 13(b). From this figure, we can see that the velocities of four locomotives do not change frequently and can converge to the desired velocity quickly. This control performance is conducive to the safe operation of the train.

5.3.4. Stochastic Perturbations. In this subsection, we consider the performance of the proposed method with external stochastic perturbations. The stochastic perturbations are added in the system dynamics (6). The maximum amplitude of the perturbation is assumed to be 1. Considering the constant term $\beta_j = 0.0044$, the perturbation is large enough to evaluate the robustness of the proposed method. The external stochastic perturbations are produced by using the Matlab function randn(n), which can create an n-by-n matrix of normally distributed random numbers with mean equal to zero and standard deviation equal to 1. The simulation result of the system with the stochastic perturbations is shown in Figure 14. From this figure, we can see that the practical velocities of locomotives can track the reference values well, though the stochastic perturbations exist. The simulation results show the good robustness of the proposed method.

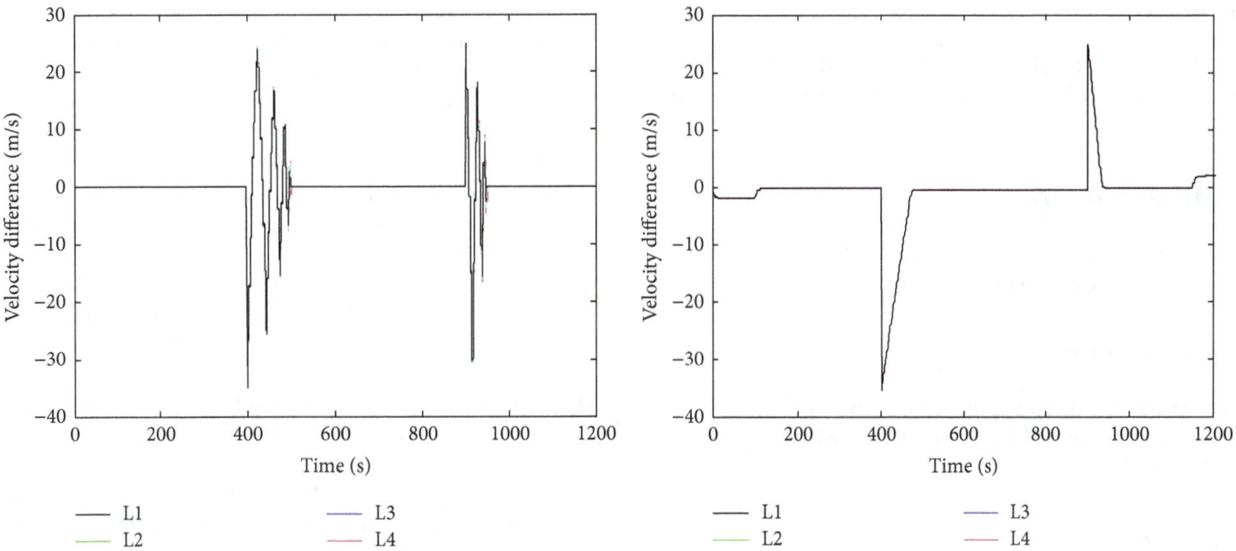

(a) PID method

(b) MPC method

FIGURE 13: Comparison of MPC method with PID method in velocity error.

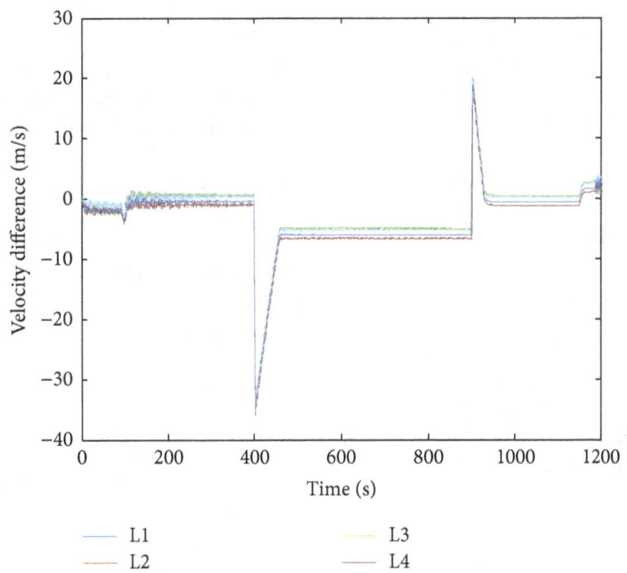

FIGURE 14: The velocity of each vehicle under MPC method with external stochastic perturbations.

6. Conclusion

In this paper, we investigated the automatic cruising control problem of high-speed trains with the aim of tracking the reference speed and minimizing the energy consumption and guaranteeing the safe operation of high-speed trains. First, we built the piecewise linearization of train resistance and developed the piecewise multiple-point model of trains. Due to the multiple constraints and performance metrics, we proposed a model predictive control method to design the ATO controller. In order to improve the accuracy of the method, we introduced the binary variable and transformed the piecewise model into a mixed logical dynamical model and guarantee that the submodel in the prediction process is still switchable. Then we analyze the operation objective and constraints and transformed the tracking control problem into a constrained finite-time optimal control problem. The optimization problem is solved by the integer linear planning method, and the optimal control solution is obtained. Simulations in the Matlab verified the effectiveness of the proposed algorithm.

In this work, we just consider the nonlinear characteristics of resistance when constructing the train model and neglect the control delay. In the future work, more realistic train models will be investigated to incorporate more nonlinear characteristics and the controller robustness will be improved under time delay.

Disclosure

Part of this paper [2] was presented at the 29th Chinese Control and Decision Conference, Chongqing, China, May 2017.

Conflicts of Interest

The authors declare that there are no conflicts of interest regarding the publication of this paper.

Acknowledgments

This work is partially supported by the National Natural Science Foundation of China (Grants nos. 61379111, 61672537, and 61672539).

References

[1] W. Song, J. Ma, L. Zhou, and X. Feng, "Deadbeat Predictive Power Control of Single-Phase Three-Level Neutral-Point-Clamped Converters Using Space-Vector Modulation for Electric Railway Traction," *IEEE Transactions on Power Electronics*, vol. 31, no. 1, pp. 721–732, 2016.

[2] Z. Xu, Z. Huang, K. Gao, S. Li, R. Zhang, and W. Liu, "Optimal operation of high-speed train using hybrid model predictive control," in *Proceedings of the 2017 29th Chinese Control And Decision Conference (CCDC)*, pp. 3642–3647, Chongqing, China, May 2017.

[3] Y. Luo and H. Xu, "Direct robust adaptive control of high speed train based on nonlinear and time-varying models," *International Journal of Control and Automation*, vol. 6, no. 4, pp. 397–410, 2013.

[4] W. Zhao and C. Gao, "A New Control Method of Automatic Train Operation in Urban Rail Transit Based on Improved Generalized Predictive Control Theory," in *Proceedings of the 2013 International Conference on Electrical and Information Technologies for Rail Transportation (EITRT2013)-Volume I*, vol. 287 of *Lecture Notes in Electrical Engineering*, pp. 567–573, Springer Berlin Heidelberg, Berlin, Heidelberg, 2014.

[5] Y. Wang, R. Luo, Z. Yu, and B. Ning, "Study on ato control algorithm with consideration of atp speed limits," *Journal of the China Railway Society*, vol. 5, p. 017, 2012.

[6] A. Albrecht, P. Howlett, P. Pudney, X. Vu, and P. Zhou, "The key principles of optimal train control—Part 1: Formulation of the model, strategies of optimal type, evolutionary lines, location of optimal switching points," *Transportation Research Part B: Methodological*, vol. 94, pp. 482–508, 2016.

[7] C.-D. Yang and Y.-P. Sun, "Robust cruise control of high speed train with hardening/softening nonlinear couple," in *Proceedings of the American Control Conference*, vol. 1999, pp. 2200–2204.

[8] C. Yang and Y. Sun, "Mixed H_2/H_∞ cruise controller design for high speed train," *International Journal of Control*, vol. 74, no. 9, pp. 905–920, 2001.

[9] C-J. Lin, S-H. Tsai, C-L. Chen, and T.-H. S. Li, "Extended sliding-mode controller for high speed train," in *Proceedings of the 2010 International Conference on System Science and Engineering (ICSSE)*, pp. 475–480, Taipei, Taiwan, July 2010.

[10] Q. Song, Y.-D. Song, T. Tang, and B. Ning, "Computationally inexpensive tracking control of high-speed trains with traction/braking saturation," *IEEE Transactions on Intelligent Transportation Systems*, vol. 12, no. 4, pp. 1116–1125, 2011.

[11] X. Zhuan and X. Xia, "Optimal scheduling and control of heavy haul trains equipped with electronically controlled pneumatic braking systems," *IEEE Transactions on Control Systems Technology*, vol. 15, no. 6, pp. 1159–1166, 2007.

[12] X. Zhuan and X. Xia, "Cruise control scheduling of heavy haul trains," *IEEE Transactions on Control Systems Technology*, vol. 14, no. 4, pp. 757–766, 2006.

[13] M. Chou and X. Xia, "Optimal cruise control of heavy-haul trains equipped with electronically controlled pneumatic brake systems," *Control Engineering Practice*, vol. 15, no. 5, pp. 511–519, 2007.

[14] X. Zhuan and X. Xia, "Speed regulation with measured output feedback in the control of heavy haul trains," *Automatica*, vol. 44, no. 1, pp. 242–247, 2008.

[15] G. M. Scheepmaker, R. M. Goverde, and L. . Kroon, "Review of energy-efficient train control and timetabling," *European Journal of Operational Research*, vol. 257, no. 2, pp. 355–376, 2017.

[16] M. Faieghi, A. Jalali, S. K. Mousavi Mashhadi, and D. Baleanu, "Passivity-based cruise control of high speed trains," *Journal of Vibration and Control*, vol. 24, no. 3, pp. 492–504, 2018.

[17] M. Faieghi, A. Jalali, and S. K. E. D. M. Mashhadi, "Robust adaptive cruise control of high speed trains," *ISA Transactions*, vol. 53, no. 2, pp. 533–541, 2014.

[18] T. Tao and H. Xu, "Adaptive fault-tolerant cruise control for a class of high-speed trains with unknown actuator failure and control input saturation," *Mathematical Problems in Engineering*, vol. 2014, Article ID 481315, 2014.

[19] L. Fan, "Iterative learning and adaptive fault-tolerant control with application to high-speed trains under unknown speed delays and control input saturations," *IET Control Theory & Applications*, vol. 8, no. 9, pp. 675–687, 2014.

[20] S. Li, L. Yang, Z. Gao, and K. Li, "Optimal Guaranteed Cost Cruise Control for High-Speed Train Movement," *IEEE Transactions on Intelligent Transportation Systems*, vol. 17, no. 10, pp. 2879–2887, 2016.

[21] H. Ye and R. Liu, "Nonlinear programming methods based on closed-form expressions for optimal train control," *Transportation Research Part C: Emerging Technologies*, vol. 82, pp. 102–123, 2017.

[22] Y. Zhao, T. Wang, and H. R. Karimi, "Distributed cruise control of high-speed trains," *Journal of The Franklin Institute*, vol. 354, no. 14, pp. 6044–6061, 2017.

[23] L. J. Zhang and X. T. Zhuan, "Optimal operation of heavy-haul trains equipped with electronically controlled pneumatic brake systems using model predictive control methodology," *IEEE Transactions on Control Systems Technology*, vol. 22, no. 1, pp. 13–22, 2014.

[24] S. Li, B. De Schutter, L. Yang, and Z. Gao, "Robust Model Predictive Control for Train Regulation in Underground Railway Transportation," *IEEE Transactions on Control Systems Technology*, vol. 24, no. 3, pp. 1075–1083, 2016.

[25] A. Bemporad, F. Borrelli, and M. Morari, "Model predictive control based on linear programming the explicit solution," *Institute of Electrical and Electronics Engineers Transactions on Automatic Control*, vol. 47, no. 12, pp. 1974–1985, 2002.

[26] E. D. Sontag, "Nonlinear regulation: the piecewise linear approach," *Institute of Electrical and Electronics Engineers Transactions on Automatic Control*, vol. 26, no. 2, pp. 346–358, 1981.

[27] C. E. García, D. M. Prett, and M. Morari, "Model predictive control: theory and practice—a survey," *Automatica*, vol. 25, no. 3, pp. 335–348, 1989.

[28] A. Bemporad and M. Morari, "Control of systems integrating logic, dynamics, and constraints," *Automatica*, vol. 35, no. 3, pp. 407–427, 1999.

[29] X.-g. Li, D.-j. Gao, and Y.-h. Wang, "Predictive control for hybrid system based on mixed logic dynamic model," *Control and Decision*, vol. 17, no. 3, pp. 315–319, 2002.

[30] Y. Wang, B. De Schutter, T. J. J. van den Boom, and B. Ning, "Optimal trajectory planning for trains - A pseudospectral method and a mixed integer linear programming approach," *Transportation Research Part C: Emerging Technologies*, vol. 29, pp. 97–114, 2013.

[31] F. D. Torrisi and A. Bemporad, "HYSDEL—a tool for generating computational hybrid models for analysis and synthesis problems," *IEEE Transactions on Control Systems Technology*, vol. 12, no. 2, pp. 235–249, 2004.

[32] A. Bemporad, *Hybrid Toolbox-Usern's Guide*, 2003, Hybrid toolbox-userns guide.

[33] L. Li, W. Dong, Y. Ji, Z. Zhang, and L. Tong, "Minimal-energy driving strategy for high-speed electric train with hybrid system model," *IEEE Transactions on Intelligent Transportation Systems*, vol. 14, no. 4, pp. 1642–1653, 2013.

3

Stop Plan of Express and Local Train for Regional Rail Transit Line

Qin Luo,[1,2] Yufei Hou,[2] Wei Li,[1,3] and Xiongfei Zhang[1]

[1]*College of Urban Traffic and Logistics, Shenzhen Technology University, Lantian Road 3002, Shenzhen, China*
[2]*Shenzhen Key Laboratory of Urban Rail Transit, Shenzhen University, Nanshan Ave 3688, Shenzhen 518000, China*
[3]*Key Laboratory of Optoelectronic Devices and Systems of Ministry of Education and Guangdong Province, College of Optoelectronic Engineering, Shenzhen University, Nanshan Ave 3688, Shenzhen, China*

Correspondence should be addressed to Wei Li; aliweibl@126.com

Academic Editor: Taku Fujiyama

The urban rail transit line operating in the express and local train mode can solve the problem of disequilibrium passenger flow and space and meet the rapid arrival demand of long-distance passengers. In this paper, the Logit model is used to analyze the behavior of passengers choosing trains by considering the sensitivity of travel time and travel distance. Then, based on the composition of passenger travel time, an integer programming model for train stop scheme, aimed at minimizing the total passenger travel time, is proposed. Finally, combined with a certain regional rail line in Shenzhen, the plan is solved by genetic algorithm and evaluated through the time benefit, carrying capacity, and energy consumption efficiency. The simulation result shows that although the capacity is reduced by 6 trains, the optimized travel time per person is 10.34 min, and the energy consumption is saved by about 16%, which proves that the proposed model is efficient and feasible.

1. Introduction

With the continuous expansion of urban scale and the development of urban rail transit network, the round-trip passenger flow between city centers and suburbs grows gradually. As the central city is saturated basically, more and more companies and residential buildings tend to transfer to the suburbs, which further promotes the swelling of the traffic flow. Besides, due to the extension of urban rail transit line, it forms long-distance regional rail transit lines [1]. If the train stops frequently, it will inevitably cause the increase of travel time, thereby reducing the operational efficiency and being not conductive to the efficient development of rail transit. And long-distance commuters want to arrive their destinations as soon as possible and passengers at stations along the route wish to reach the destinations without transfer. Therefore, cities need to use express and local train operating mode to meet the demands for the commuters [2]. On the one hand, from the passenger's point of view, this model can shorten the travel time of commuting passenger flow and improve the passenger service level, but at the same time, it will increase the travel time of some local passengers to a certain extent and affect their travel experience. On the other hand, for operating enterprises, the mode will save the train energy consumption, thereby enhancing operational efficiency, but the system carrying capacity will be reduced. For this reason, how to ensure the stop scheme of express trains reasonably and save the passenger travel time to an extreme is a significant project at present.

In view of the demand for commuters, scholars mainly study the mode of the express and local train and the plan of the full-length and short-run route. Based on the express and local train operation mode, the literature mainly considers the travel demand of passengers. For instance, L. Wang [3] considered the shortest total travel time of the passengers to build up an optimization model. R. Tang [4] constructed a linear programming model with the goal of maximizing passenger satisfaction, which is reducing average passenger waiting time. Hassannayebi E et al. [5] studied the optimal time table of express/local train and built up a model in order to reduce the passenger waiting time by using variance and penalty function. Chun J H et al. [6] come up with the express/local train mode, which can solve the problem of passenger crowding and train delay with building the model of minimizing the number of serviceable cars. Y Lai et al. [7] took transfer passengers into account and then established an innovative network structure to study the train stop plan. Y. Gao et al. [8] proposed a biobjective programming model to achieve the simultaneous minimization of the energy consumption and the travel time. However, the above literature sources do not consider the travel process comprehensively. In the combination mode, there is a situation that express trains cross through local trains. And this part of time should not be ignored. Therefore, Canca D et al. [9, 10] also mainly considered meeting the travel demand of the passenger flow from the passenger angle by adjusting the train schedule including the overtaking factor. However, they ignored the commuting passengers' sensitivity to travel time, whose travel purpose is to arrive the destination quickly. Therefore, D. Di et al. [11, 12] established a passenger path-choice model to determine the passenger principle in path choice by considering perception difference of time without considering the effect of travel distance.

In the condition of the full-length and short-run route, the recent studies mainly focused on the selection of short-run route, service level, traffic capacity, and rolling stock [13]. For example, J. Q. Liao [14], respectively, devised the operation scheme in condition of the single route and the full-length and short-run route aiming at the maximum travel saving time of all passengers. Q. Hu et al. [15] analyzed the section of short-run route from the use of the rolling stock and capacity. However, the operation scheme plan of mixed-routing cannot save the travel time for long-distance passengers and solve the problem of commuting effectively.

To sum up, based on the analysis of the composition of passenger travel time and passenger selection behavior, this paper makes a reasonable stop scheme of express trains aiming to reduce the total travel time to maximize the operational efficiency of express/local train for regional rail transit line.

The remainder of the paper is organized as follows. Section 2 analyzes the passenger flow characteristics and the effect of travel time and carrying capacity in the condition of express and local train operating mode. In Section 3, a stopping plan optimization model of trains based on minimizing the total passenger travel time is established, and then the model is solved by genetic algorithm. In Section 4, an example of a regional line in Shenzhen is provided to optimize the train stop plan by using the model and algorithm above and evaluate it by the travel time, the capacity, and the energy consumption compared with the general plan. The final section draws the conclusions and proposes some issues for future.

2. Analysis of Passenger Flow Characteristics and Effects

2.1. Passenger Flow Characteristics. The long-distance regional rail transit line is connected in series with urban areas, suburbs, and satellite towns, and the intercity passenger flow is not only large, but also imbalanced in time direction. During the morning rush, the traffic flow moves from satellite towns or suburbs to the downtown, and at the evening peak, the situation is opposite. Meanwhile, with different levels of development along the rail transit, the passenger volume of boarding, and alighting at stations are also inconsistent, and overall, they show the disproportion of the section flow. Consequently, two types of the fluctuating passenger flow have demands for regional express rails: commuters prefer to take the through train from downtown area to the suburban, while short-distance passengers want to reach any two stations directly.

2.2. Travel Time Cost. Although operating express/local train prolongs the total trip time of passengers who take local train because of overtaking, it saves the time of those passengers who take express train.

2.3. Carrying Capacity. Express and local train scheme will reduce the capacity of lines [16]. In the traditional pattern, the line capacity n_r is only related to the minimum headway I, and its calculating method is as formula (1). However, the carrying capacity of express/local train takes the deduction coefficient method for calculation and analysis. The deduction coefficient of the express train refers to the number of the local cars deducted from the parallel operation chart because of drawing an express train, and the formula is as (2) [17].

$$n_r = \frac{3600}{I} \tag{1}$$

$$\varepsilon_e = 1 + \left(\frac{t_d}{I}\right) - k \tag{2}$$

where ε_e is the deduction coefficient of the express train, t_d is the difference of travel time of express/local train, and k means the times which are crossed of slow trains.

2.4. Train Energy Consumption. In daily operation, in addition to the maintenance of infrastructure, the energy consumption during the train operation also accounts for a large proportion [18]. And compared to the normal train mode, where trains stop at all stations, the express train only stops at some stations, so the operating scheme of the express and local train reduces the energy consumption of traction and braking at start and stop, and it can achieve the energy saving and emission reduction. In terms of the operating enterprise, it has great benefit [19]. And the equations of energy consumption calculation and parameters (Table 1) are as follows:

$$E = E_t + E_b \tag{3}$$

$$E_t = \sum_{m=1}^{f_1+f_2} \sum_{i=1}^{n} \left(\int_0^{l_t} F_t dl \right) \tag{4}$$

$$E_b = \sum_{m=1}^{f_1+f_2} \sum_{i=1}^{n} \left(\int_0^{l_b} F_b dl - E_r \right) \tag{5}$$

$$E_r = \sum_{m=1}^{f_1+f_2} \sum_{i=1}^{n} \sum_{k=m+1}^{f_1+f_2} x_{m,k} \cdot \int_0^{l_t} F_b dl \cdot \xi \cdot \frac{t_c}{t_b} \tag{6}$$

where (3) means the total energy consumption, (4) is the train traction energy consumption, (5) is the energy consumption of train braking, and (6) shows the energy saved in regenerative braking.

3. Optimization Model and Algorithm

3.1. Problem Description. First, operating express/local train combination plan is mainly for commuting problem, so this paper chooses the morning peak as the optimization period to study. Second, two kinds of trains own the same technical parameters. Third, although the line is divided into the upstream and downstream direction, the model only discusses the one-way situation where the passenger flow is larger. And then, the regional rail transit line which this paper studies can meet the demand of passenger flow without the situation of passenger retention. The last assumption is that passengers' arriving can all be considered to follow the normal distribution.

3.2. Modeling. Passenger travel time t includes waiting time t_w and running time t_r like formula (7), in which the total passenger waiting time consists of two parts: the first part is the normal passenger waiting time when the foregoing local train is not overtook, and the second part is the increased passenger waiting time at stations caused by overtaking. And t_r includes the interval running time and station dwell time.

$$t = t_w + t_r. \tag{7}$$

3.2.1. Analysis of Passenger Selecting Behavior for Train. The research object is commuting passenger flow with early morning rush hour. Considering that passengers are generally unwilling to transfer, this paper assumes that passengers at most experience one transfer in the same line and same direction. Besides, when OD are both express train stations, passengers will not choose the transfer way from express to local train [20]. Therefore, according to the different nature of origin and destination, all kinds of the travel selection behavior are shown in Table 2.

The behavior of passengers selecting trains can be analyzed by Logit model like formula (8) [21]. And (9) shows the passenger flow who selects some kind of travel mode [22].

$$p_{ij}^{(k)} = \frac{\exp\left(-\theta \cdot c_{ij}^{(k)}\right)}{\sum_{k\in S} \exp\left(-\theta \cdot c_{ij}^{(k)}\right)} \tag{8}$$

where $p_{ij}^{(k)}$ is the passenger ratio choosing the mode k to travel from station i to j, θ is the passenger expectation variable based on travel impedance which is inversely proportional to the passenger expectation, $c_{ij}^{(k)}$ is the impedance choosing the mode k to travel from station i to j, and S is collection of travel modes.

$$q^{(k)} = q_{ij} \cdot P_{ij}^{(k)} \tag{9}$$

where $q^{(k)}$ is the passengers choosing mode k to travel and q_{ij} is the passengers flowing from i to j.

Passenger travel impedance is mainly affected by both the train running time t_y and travel distance l, and the combined impedance is determined by the weight of both.

(1) Train Running Time. When passengers travel from local to express station or just the reverse, they need to think about transferring. The principles are as follows.

(1) $t_{(trans)} > t_{(l)}$, where $t_{(trans)}$ is the total time of changing trains and $t_{(l)}$ only refers to the time of local train. In addition, the transfer time is as

$$t_{(trans)} = t_w^1 + t_r^1 + t_w^2 + t_r^2 \tag{10}$$

where t_w^1 and t_r^1, respectively, represent the waiting time and running time including dwell time of the first train for passenger; t_w^2 and t_r^2, respectively, represent the same meaning for the second train.

(2) When passengers consider taking local train to express, they not only need to take the total time into account, but also need to think of whether the middle has express station to supply to transfer.

(2) Travel Distance. The operation ratio of express and local train can be estimated by the number of passengers whose travel distance is greater than five sections and less than five sections, like the following formula [23]:

$$p = \frac{\sum_{|i-j|>5} q_{ij}}{\sum_{|i-j|\le 5} q_{ij}}. \tag{11}$$

From the above two parts, the combined impedance is as (12) and the flow chart of passengers train-choice behavior is as shown in Figure 1.

$$c_{ij}^{(k)} = \mu_1 \cdot t_y + \mu_2 \cdot l \tag{12}$$

where μ_1 and μ_2 are the weight of train running time and travel distance, and they depend on the specific situation.

3.2.2. Objective Function. Total travel time t is calculated as formula (13) and the model parameters are listed in Table 3.

$$t = \left(\sum_{m=1}^{f_1} q_{ij-m}^{(l)} + x_q \cdot \sum_{m=1}^{f_2} q_{ij-m}^{(e)}\right) \cdot \left(\frac{\mathrm{T}}{2 \cdot (c_1 f_1 + c_2 f_2)} + \sum_{k=1}^{i-1} c_1 h_k x_k + t_{ij} + \sum_{w=i+1}^{j-1} (t_s + c_1 x_w h_w)\right). \tag{13}$$

In conclusion, when the number of stations is n, the paper establishes the following objective function to achieve the minimization of the total passenger travel time as follows:

$$\min Z = \min\left(\sum_{i=1}^{n-1} \sum_{j=i+1}^{n} t\right)$$
$$\text{s.t.} \quad q_{ij-m} \le \eta_{\max} \cdot a$$
$$f_d \ge 1 \quad (d = 1, 2, 3). \tag{14}$$

TABLE 1: Energy consumption model parameters.

Parameters (unit)	Explanation
E (KJ)	The total energy consumption
E_t (KJ)	The traction energy consumption
E_b (KJ)	The braking energy consumption
E_r (KJ)	The regenerating energy
l_t (m)	The distance of traction
F_t (KN)	The net force of traction process
$x_{m,k}$	Boolean variable. When the mth and k^{th} train are in the same power supply section, the value is 1; otherwise, the value is 0.
l_b (m)	The distance of braking
F_b (KN)	The net force of braking process
ξ	The energy transformation ratio
t_c (s)	The overlapping time of the acceleration time of the mth and the braking time of the kth.
t_b (s)	The braking time of the kth train.

TABLE 2: Passenger selection behavior.

OD	Local train station	Express train station
Local train station	Local train	Local train Local train→Express train
Express train station	Local train Express train→Local train	Local train Local train→Express train Express train

TABLE 3: Model parameters.

Parameters (unit)	Explanation
f_1 (time/h)	The operating frequency of the local train
f_2 (time/h)	The operating frequency of the express train
x_q	Boolean variable. When the i station and j station are both local stations, the value is zero.
$q_{ij-m}^{(l)}$ (person)	The number of passengers waiting for the mth train which is the local train
$q_{ij-m}^{(e)}$ (person)	The number of passengers waiting to the mth train which is the local train
T (min)	The length of the optimized period
c_1/c_2	Boolean variable. When passengers take the local train, the value of c_1 is one, c_2 is zero, and when passengers take the express train, the value of c_1 is zero, c_2 is one.
h_k (min)	The increased delay time because the slow train is crossed by the express train at preceding stations
x_k	Boolean variable. When the slow train is crossed by the express train at the kth station, the value is one; otherwise, the value is zero.
t_{ij} (min)	The interval running time of trains
t_s (min)	The dwelling time at stations
x_w	Boolean variable. When the local train is crossed by the express train at the wth station, the value is one; otherwise, the value is zero.
h_w (min)	The increased dwell time in the trains because the local train is crossed by the express train at this station
q_{ij-m} (min)	The number of passengers waiting for the mth train
$\eta_{\max}$	Maximum section load factor
a	Passenger capacity of the train

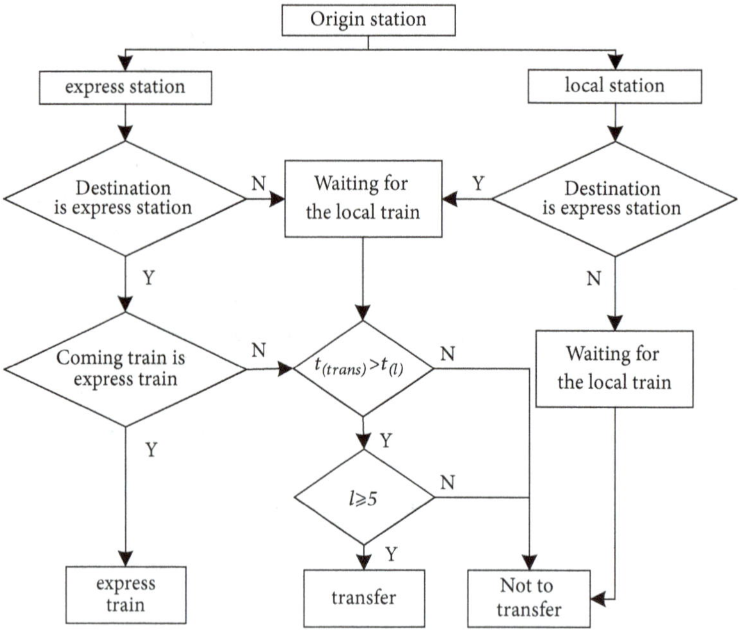

FIGURE 1: The flow chart of passenger train-choice behavior.

3.3. *Solution Algorithm.* The evolutionary algorithm, which is represented by genetic algorithm, uses crossover and mutation operator to realize the information exchange and local search between individuals in the group and provides each individual with the opportunity of optimization [24]. Through the competitive selection mechanism of survival of the fittest, the evolutionary algorithm guides the population to the better. Exactly, the unique binary coding rule of genetic algorithm is suitable for the value of 0-1 of the train stop strategy. Besides, instead of starting from a point and optimizing according to a fixed route, it searches the whole feasible solution space at the same time, so it can avoid falling into a local optimum and find the global optimal results. The steps are as follows.

(1) Population initialization. It generates N binary coded individuals randomly, and the length is the number of stations, where in code, 1 means that express train will stop at the corresponding station and 0 means no stopping. What is more, because the express train must stop at the original and terminal stations, the encoding is 1.

(2) Ensuring the fitness function. It constructs the fitness function as formula (15). The greater the fitness value, the better the individual.

$$u(x) = M - f(x) \tag{15}$$

where M means the total time of general mode and $f(x)$ is the time of combination mode with every plan.

(3) Taking the operations of selecting, crossover, and mutation. The selection operation adopts roulette method, the crossover operator is uniform crossover, and the mutation operator is a single point mutation.

(4) After an iteration, the worst individual is replaced by the best individual in the current generation, that is also called "survival of the fittest".

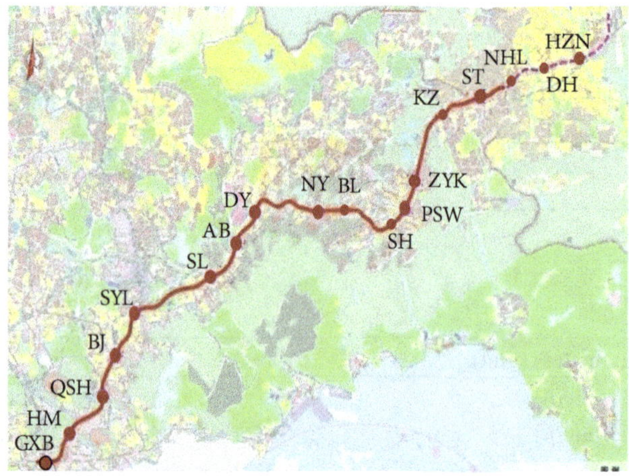

FIGURE 2: The regional rail transit line in Shenzhen.

(5) With the requirement of iterations, steps 4 and 5 are repeated.

4. Case Analysis

4.1. *Line Profile.* Taking a certain line in Shenzhen as an example (Figure 2), this paper analyzes and verifies the optimization of the express and local train stop problem. The data required in the example analysis is shown in Table 4, the OD passenger flow during the morning peak (8:00~9:00) is shown in Table 5, and the length of interval and pure runtime of trains are followed in Table 6. Besides, according to the OD data, the passenger flow of boarding and alighting and the distribution of section flow of up direction (GXB→HZN) and down direction (HZN→GXB) are, respectively, shown in Figures 3, 4, and 5.

TABLE 4: Values of case parameters.

Parameters (unit)	Value	Parameters (unit)	Value
Length (km)	53.07	Train composition	A8
The number of stations	18	Dwell time (s)	30
Minimum headway (s)	120	Maximum running time (km/h)	120
Passenger capacity (person)	2480	Maximum section load factor	100%
Passenger expectation variable θ	0.9	Weight μ_1	0.6
Weight μ_2	0.4		

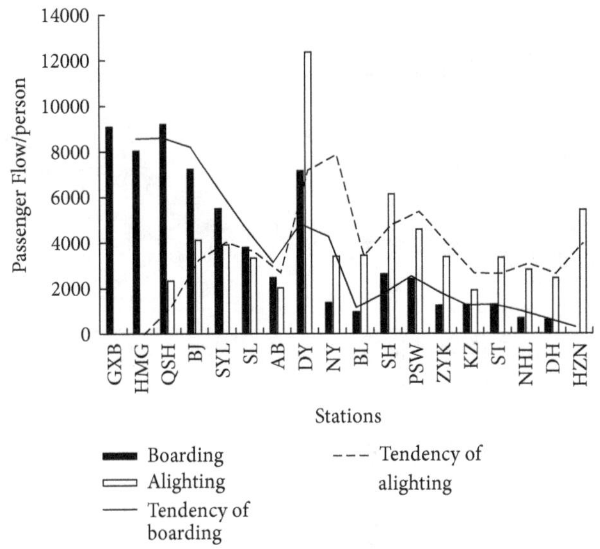

FIGURE 3: The passenger flow of boarding and alighting of up direction.

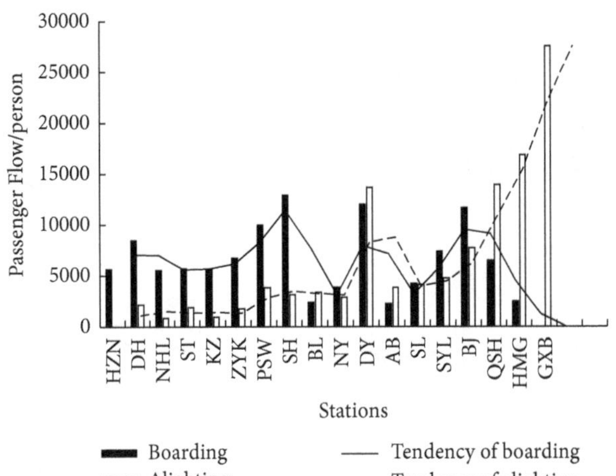

FIGURE 4: The passenger flow of boarding and alighting of down direction.

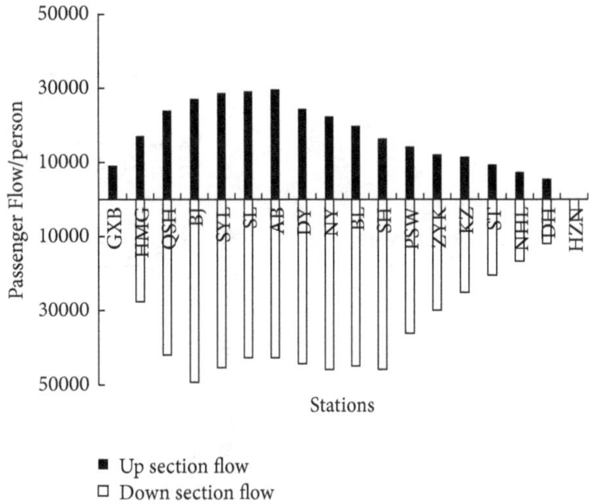

FIGURE 5: The distribution of section passenger flow.

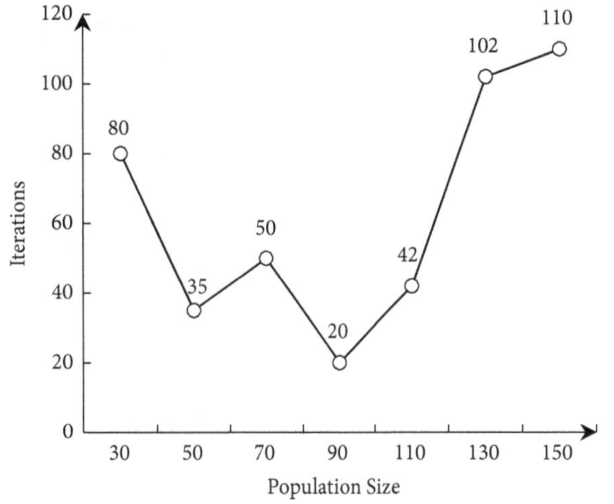

FIGURE 6: The relationship between the population size and number of iterations.

4.2. Solution. This paper uses the genetic algorithm to solve the model. Different parameters will affect the speed and results of the algorithm. The paper uses two of the population sizes, crossover probability and mutation probability, as the independent variables and the other two as the dependent variables to analyze the influence of the values on algorithm results. When the values of the population size were from 30 to 150, the results showed that the final fitness value did not change with the population size, while the rate of convergence will be affected as shown in Figure 6. And according to the current literature that the range of the crossover rate is 0.4 to 0.99 and 0.001 to 0.1, respectively [25], the paper supposes that the crossover rate steps and the mutation rate steps are 0.5 and 0.005. The results show that although the number of

TABLE 5: The value of OD passenger flow during the morning rush hour.

Station	GXB	HMG	QSH	BJ	SYL	SL	AB	DY	NY	BL	SH	PSW	ZYK	KZ	ST	NHL	DH	HZN
GXB	0	1133	1708	1464	597	420	179	2661	261	161	525	391	224	111	118	66	63	131
HMG	2480	0	624	1179	751	462	211	2509	282	185	580	419	238	120	122	74	147	132
QSH	5399	1085	0	1473	1941	543	325	1883	361	322	569	435	276	130	158	129	461	173
BJ	5914	3702	2044	0	621	961	514	1740	440	358	858	629	342	142	172	85	172	183
SYL	1763	1686	2959	945	0	954	488	1190	453	422	658	463	284	115	148	97	66	156
SL	935	807	777	843	829	0	307	1334	338	370	390	263	206	93	126	128	71	164
AB	444	391	395	417	337	246	0	1024	215	242	271	195	139	63	85	87	52	111
DY	3774	2956	1565	1030	528	939	1225	0	1054	1204	1717	918	739	204	456	230	146	474
NY	526	468	446	380	322	267	242	1193	0	180	318	284	162	68	98	93	57	126
BL	293	252	250	217	182	155	140	782	96	0	231	254	122	56	74	71	48	97
SH	1844	1725	1405	1229	766	561	524	3484	628	757	0	337	338	302	622	365	195	475
PSW	1116	1074	999	721	628	396	427	2680	669	853	391	0	302	317	489	433	247	643
ZYK	932	833	679	575	362	288	248	1549	316	389	270	280	0	206	353	240	129	324
KZ	428	375	419	259	186	284	247	868	279	349	568	913	430	0	324	317	175	471
ST	444	382	399	267	183	227	204	892	234	295	767	714	469	211	0	387	245	674
NHL	431	367	421	226	150	272	208	790	243	270	443	707	321	222	443	0	183	497
DH	562	576	971	426	196	352	259	980	307	328	511	852	385	333	834	572	0	610
HZN	226	186	192	156	83	117	90	443	108	116	188	354	167	174	603	274	2138	0

Table 6: The length of interval and pure runtime of trains.

Interval	Length (km)	Runtime (min)	Interval	Length (km)	Runtime (min)
GXB-HMG	2.905	2.563	BL-SH	4.213	2.922
HMG-QSH	4.279	3.363	SH-PSW	1.566	1.521
GSH-BJ	2.494	2.113	PSW-ZYK	2.007	1.714
BJ-SYL	3.535	2.84	ZYK-KZ	4.926	3.339
SYL-SL	6.364	4.173	KZ-ST	3.229	2.367
SL-AB	3.428	2.503	ST-NHL	1.623	1.634
AB-DY	2.413	1.915	NHL-DH	1.745	1.795
DY-NY	6.037	4.35	DH-HZN	1.238	1.241
NY-BL	2.072	1.756			

Table 7: Values of algorithm parameters

Parameters	Value	Parameters	Value
Population	50	Crossover probability	0.8
Iterations	150	Mutant probability	0.05

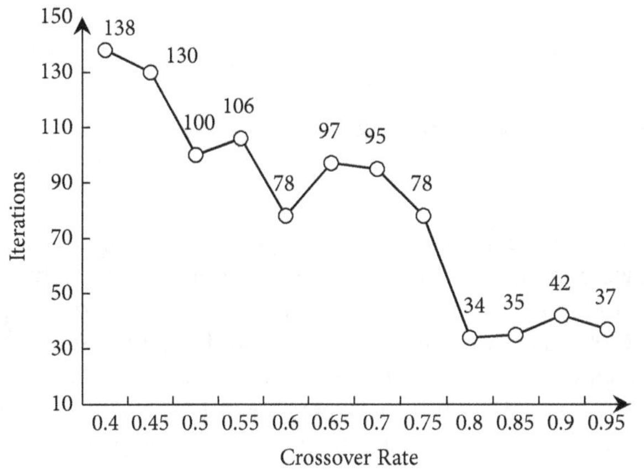

Figure 7: The relationship between the crossover rate and number of iterations.

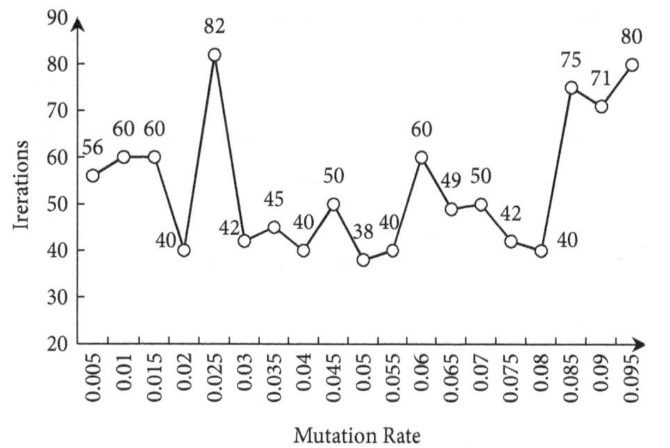

Figure 8: The relationship between the mutation rate and number of iterations.

iterations changes, iterations have little difference from each other and the final fitness value is invariable as Figures 7 and 8 show. Therefore, the final algorithm parameter values are shown in Table 7.

According to the case data, the maximum cross-section passenger flow appears at the BJ station in the downward direction, with 49333 people, so that the number of departure times during the optimization period is 20 pairs/h. As the downward direction of the cross-section of the passenger flow is much larger than the upward direction, it is designed to operate the express and local train mode in the down direction. And in this situation, the optimal result is different with the different proportion of the express/local train. Based on the OD in Table 5 and formula (11), the ratio is approximately 1:2. However, in order not to leave out all the plans, the paper takes 3:1, 2:1, 1:1, 1:2, 1:3 as alternatives.

The case is calculated and analyzed according to the algorithm parameter in Table 7 and fitness function set as

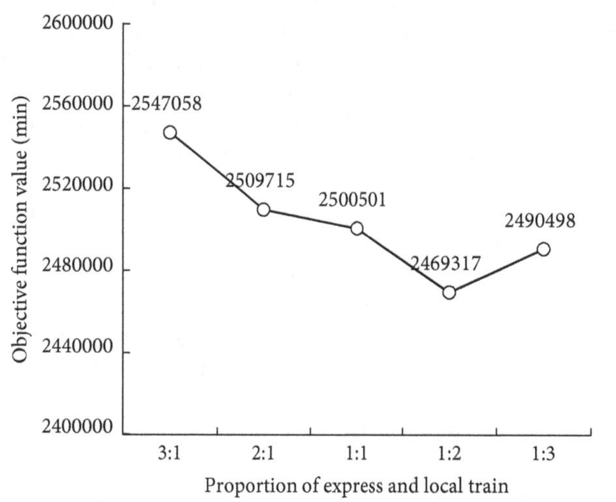

Figure 9: Variation of total travel time.

(15). Figure 9 shows the changing of travel time value. When the proportion is 1:2, that is to say, operating 6 express trains and 14 local trains, the passenger travel time is optimal. Corresponding express station stop scheme is shown in Table 8, in which 1 stands for stopping at station and 0 for no stopping, and express train overtakes the local at station ST and PSW without stopping.

TABLE 8: The best express train stop scheme.

Station	HZN	DH	NHL	ST	KZ	ZYK	PSW	SH	BL	NY	DY	AB	SL	SYL	BJ	QSH	HMG	GXB
Plan	1	0	1	0	0	0	0	1	1	1	1	1	1	1	1	0	0	1

TABLE 9: Total travel time comparison.

Compared content (unit)	General mode	Combination mode	Time benefit	Per capita benefit
Total travel time (min)	3 638 656	2 469 317	1 169 339	10.34 min

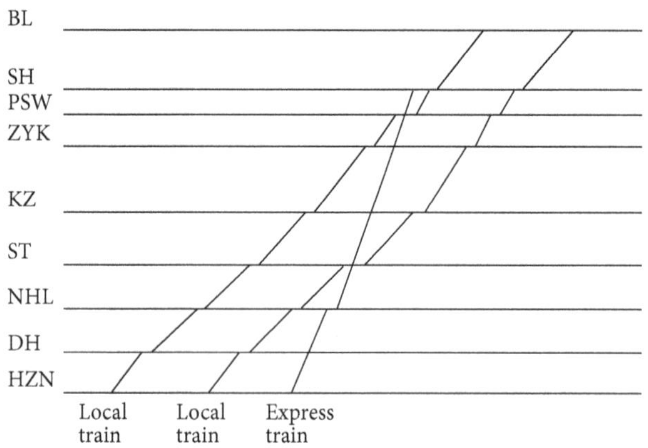

FIGURE 10: The sketch of train operation.

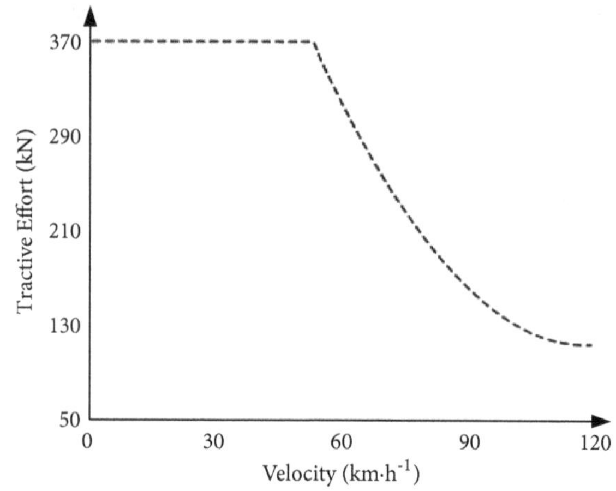

FIGURE 11: The sketch of traction curve.

4.3. Results Analysis

4.3.1. Time Benefit Analysis. Compared with the general mode which only operates the local trains, the combination saves the passenger travel time with taking the express trains, whereas part of the passengers' time is increased because of slow trains. In the condition of the same train frequency, comparing the total travel time before and after the plan, like Table 9, we can see that in the specific passenger flow conditions, the time benefit is quite obvious.

4.3.2. Capacity Analysis. It has an effect on carrying capacity due to operating the express/local train. According to the parameters in the foregoing case, when the minimum headway is 120s and the ratio of express and local train is 6:14 (roughly 1:2), where the capacity of general mode is 30 pairs/h, while in condition of the combination mode, the capacity depresses to 24 pairs/h.

From the stop scheme, we draw a train graph like Figure 10. With not considering the train turnaround temporarily, here we only draw the sketch of two sets of express/local train combination which run from the station HZN to the BL, and the start order is the local train, local train, and the express train, respectively. In a triple of the combination, the first local train is crossed at station PSW, and the second is overtaken at ST without others.

4.3.3. Energy Consumption Comparison. The difference between the energy consumption of express trains and slow trains is mainly caused by the express train running directly at some stations without stopping and experiencing a kind of uniform speed at a longer distance in the interval. The train traction curve, brake curve, and speed curve diagrammatic drawing are as shown in Figures 11, 12, and 13, respectively. Traction energy consumption and braking energy consumption are calculated by (4) and (5), and (6) represents the regenerative braking energy. Above all, the total consumption is as formula (3). In this case, there are three substations along the rail line. Under the above optimal stopping plan and the series calculation of train energy consumption, the results are compared in Table 10. It shows that the express/slow train mode has great advantages in energy saving and ensures the efficiency of the operating enterprises.

5. Conclusion

This paper studied the problem of the train stopping schedule in the condition of express and local train operation mode. With the analysis of the passengers' requirement to minimize the travel time, travel distance, and mainly the stations, the paper established a passenger train-choice model and used the entropy method to calculate the weight coefficient scientifically. On this basis, a 0-1 integer programming mathematical model aimed at minimizing the total travel time was set up, and the genetic algorithm was designed to solve the model. Finally, the algorithm result was insensitivity with the parameters. Besides, the travel time and energy consumption were both saved effectively through the case analysis, resulting in a considerable social benefits. However, the combination of express and local train operation reduced the line capacity, and the actual operation can change the ratio of express/local train to get relatively large capacity.

TABLE 10: Train energy consumption comparison.

Compared content (unit)	General mode	Combination mode	benefit	Benefit rate
Energy consumption (KJ)	1.41×10^8	1.19×10^8	2.26×10^7	16%

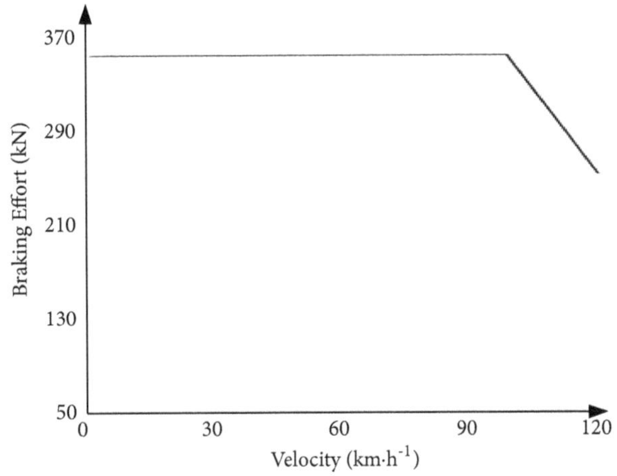

FIGURE 12: The sketch of brake curve.

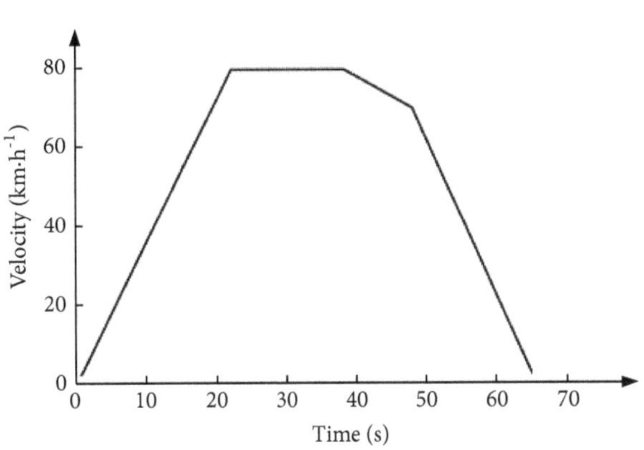

FIGURE 13: The sketch of speed curve.

Conflicts of Interest

The authors declare that they have no conflicts of interest regarding the publication of this paper.

Acknowledgments

This paper is supported by the Research Projects of the Social Science and Humanity on Young Fund of the Ministry of Education under Grant no. 15YJCZH108, the Research Projects of Natural Science Foundation of Guangdong Province under Grant no. 2015A030310341, and the Research Project of Shenzhen Technology University.

References

[1] Z. Rashedi, M. Mahmoud, S. Hasnine, and K. N. Habib, "On the factors affecting the choice of regional transit for commuting in Greater Toronto and Hamilton Area: Application of an advanced RP-SP choice model," *Transportation Research Part A: Policy and Practice*, vol. 105, pp. 1–13, 2017.

[2] C. Hirai, T. Kunimatsu, S. Kondou, N. Tomii, and M. Takaba, "Modelling and solving the train stop deployment planning problem," in *Proceedings of the 12th IASTED International Conference on Artificial Intelligence and Soft Computing, ASC 2008*, pp. 67–72, esp, September 2008.

[3] L. Wang and G. J. Wei, "Stopping Schedules of the Express /Local Train for Urban rail Transit Based on GASA Algorithm," *Urban Mass Transit*, vol. 18, no. 3, pp. 96–100, 2015.

[4] R. Tang and Y. Gao, "A Model of Customer Satisfaction-based for Express/Local Train Operation," *Modern Urban Transit*, vol. no. 5, pp. 59–64, 2015.

[5] E. Hassannayebi, S. H. Zegordi, M. R. Amin-Naseri, and M. Yaghini, "Train timetabling at rapid rail transit lines: a robust multi-objective stochastic programming approach," *Operational Research*, vol. 17, no. 2, pp. 435–477, 2017.

[6] J. H. Chun, R. M. Anderson, and D. Paik, "The S-train system: Synchronized express & local trains for urban commuter rail systems," in *Proceedings of the 14th IEEE International Intelligent Transportation Systems Conference, ITSC 2011*, pp. 1586–1591, usa, October 2011.

[7] Y.-C. Lai, M.-C. Shih, and G.-H. Chen, "Development of efficient stop planning optimization process for high-speed rail systems," *Journal of Advanced Transportation*, vol. 50, no. 8, pp. 1802–1819, 2016.

[8] Y. Gao, L. Yang, and Z. Gao, "Energy consumption and travel time analysis for metro lines with express/local mode," *Transportation Research Part D Transport & Environment*, 2016.

[9] D. Canca, E. Barrena, E. Algaba, and A. Zarzo, "Design and analysis of demand-adapted railway timetables," *Journal of Advanced Transportation*, vol. 48, no. 2, pp. 119–137, 2014.

[10] Q. Zhen and S. Jing, "Train rescheduling model with train delay and passenger impatience time in urban subway network," *Journal of Advanced Transportation*, vol. 50, no. 8, pp. 1990–2014, 2016.

[11] D. Di and D. Yang, "Passenger flow analysis model about express/slow train in urban rail transportation corridor," *Tongji Daxue Xuebao/Journal of Tongji University*, vol. 42, no. 1, pp. 78–83, 2014.

[12] X. M. Zhao and Q. X. Sun, "Passenger Choice Behavior for Regional Rail Transit under Express/Local Operation with Overtaking," *Journal of Transportation Systems Engineering and Information Technology*, vol. 16, no. 5, pp. 104–109, 2016.

[13] R. Xu and X. Li, "Optimization of Routing Mode on Regional Express Rail," *Urban Mass Transit*, vol. 9, no. 5, pp. 36–39, 2006.

[14] J. Q. Liao, *Research of Regional Rail Transit Express Train Plan*, Beijing Jiaotong university, 2016.

[15] Q. Hu, "Study on Mixed-routing in Railway Transit for Disequilibrium Passenger Flow," *Value Engineering*, vol. 31, no. 32, pp. 81–83, 2012.

[16] Y. Zhu, B. Mao, Y. Bai, and S. Chen, "A bi-level model for single-line rail timetable design with consideration of demand and capacity," *Transportation Research Part C: Emerging Technologies*, vol. 85, pp. 211–233, 2017.

[17] M. Y. Qu and S. M. Huang, "Study on express/local train plan of urban rail transit," *Railway Transport and Economy*, vol. 34, pp. 77–82, 2012.

[18] Q. Peng, W. Li, Y. Wang, Q. Zhong, and J. Sun, "Study on Operation Strategies for Metro Trains Under Regenerative Braking," *Tiedao Xuebao/Journal of the China Railway Society*, vol. 39, no. 3, pp. 7–13, 2017.

[19] K. Nomura and M. Miyatake, "Scheduling method for minimum energy consumption considering constraints of time intervals between local and express trains," in *Proceedings of the 2016 IEEE International Conference on Intelligent Rail Transportation, ICIRT 2016*, pp. 469–475, gbr, August 2016.

[20] X. M. Zhao, Q. X. Sun, Y. Ding, R. J. Shi, and Y. Chen, "Passenger choice behavior for regional rail transit under express/local operation with overtaking," *Journal of Transportation Systems Engineering and Information Technology*, vol. 16, no. 5, pp. 104–109, 2016.

[21] Y. Wang, T. Tang, B. Ning, T. J. J. van den Boom, and B. De Schutter, "Passenger-demands-oriented train scheduling for an urban rail transit network," *Transportation Research Part C: Emerging Technologies*, vol. 60, pp. 1–23, 2015.

[22] X. Wang, H. Wang, and X. Zhang, "Stochastic seat allocation models for passenger rail transportation under customer choice," *Transportation Research Part E: Logistics and Transportation Review*, vol. 96, pp. 95–112, 2016.

[23] L. R. Huang, *Research on operation organization of urban rail transit of express/local train*, Southwest Jiaotong University, 2017.

[24] H. Niu and X. Zhou, "Optimizing urban rail timetable under time-dependent demand and oversaturated conditions," *Transportation Research Part C: Emerging Technologies*, vol. 36, pp. 212–230, 2013.

[25] W. Li, R. Xu, and W. Zhu, "Multi-line cooperation method for passenger flow disposal in metro transfer station under train delay," *Tongji Daxue Xuebao/Journal of Tongji University*, vol. 43, no. 2, pp. 239–244, 2015.

4

Optimizing High-Speed Railroad Timetable with Passenger and Station Service Demands

Jin Wang, Leishan Zhou, Yixiang Yue, Jinjin Tang, and Zixi Bai

School of Traffic and Transportation, Beijing Jiaotong University, Beijing 100044, China

Correspondence should be addressed to Leishan Zhou; lshzhou@bjtu.edu.cn

Academic Editor: Samiul Hasan

This paper aims to optimize high-speed railroad timetables for a corridor. We propose an integer programming model using a time-space network-based approach to consider passenger service demands, train scheduling, and station service demands simultaneously. A modified branch-and-price algorithm is used for the computation. This algorithm solves the linear relaxation of all nodes in a branch-and-bound tree using a column generation algorithm to derive a lower-bound value (LB) and derive an upper-bound value (UB) using a rapid branching strategy. The optimal solution is derived by iteratively updating the upper- and lower-bound values. Three acceleration strategies, namely, initial solution iteration, delayed constraints, and column removal, were designed to accelerate the computation. The effectiveness and efficiency of the proposed model and algorithm were tested using Wuhan-Guangzhou high-speed railroad data. The results show that the proposed model and algorithm can quickly reduce the defined cost function by 38.2% and improve the average travel speed by 10.7 km/h, which indicates that our proposed model and algorithm can effectively improve the quality of a constructed train timetable and the travel efficiency for passengers.

1. Introduction

A train timetable is a technical document that describes the operation of trains on certain railroad sections and the trains' departure and arrival times or passing times at stations. With the rapid development of high-speed railroads in China and various other countries during the last decade, the train timetabling problem (TTP) for high-speed railroads has become a new challenge for railway operators. The TTP for high-speed railroads differs from the traditional TTP in several ways:

(1) The mismatch between transportation demand and capacity causes the full use of the transportation capacity of railroads and the minimization of travel times of trains to be the main goals for the traditional TTP. In general, the TTP for high-speed railroads focuses on the maximal satisfaction of passenger travel demands and the improvement in customer service quality. Due to the high speed and high density of the high-speed railroad, passengers may have more choices of when to travel. Thus, the passenger flow of high-speed railroads shows distinct volatility in different periods. To satisfy the volatility characteristics of high-speed railroad passengers, we set a feasible departure time range for each train at its starting station according to the train service plan, as mentioned in Section 2.3.

(2) Passenger trains and freight trains run on traditional railroads, where the passenger trains have higher priority than freight trains. Thus, passenger trains may overtake freight trains at some stations. Medium- and high-speed passenger trains run on high-speed railroads, where medium-speed trains and high-speed trains have the same priority. Thus, the number of times overtaking occurs on high-speed railroads may be less than that of traditional railroads. We propose a punishment coefficient to control the number of occurrences on high-speed railroads in Section 3.1.

(3) The maintenance window for high-speed railroads is generally set at night, and maintenance requires a considerable length of time. The skylight for traditional railroads is more flexible and requires a shorter length of time.

The TTP has received considerable attention in recent decades as a fundamental aspect of railroad transportation management. Many researchers have explored train scheduling. Frank [1] was one of the earliest researchers to conduct a mathematical analysis of train timetables. He began to explore train scheduling for two-way railroad operations on a single track in 1966. Szpigel [2] converted the single-track TTP into a job scheduling problem and proposed a mixed-integer programming model, in which minimizing the total travel time was defined as the goal, which was subject to interval time constraints related to train overtaking or crossing. A branch-and-bound algorithm was employed for the computation of this model; however, it could be applied to only small-scale computations. Serafini and Ukovich [3] proposed the periodic event scheduling problem (PESP) in 1989, after which Schrijver and Steenbeek [4] applied the PESP to solve the TTP and pioneered research on periodic timetable optimization problems [5–11]. Periodic timetables help passengers memorize accurate departure times and effectively reduce the scale of the computation. Currently, periodic timetabling is the preferred choice of many railroad enterprises. Brännlund et al. [12] proposed a 0-1 integer programming model based on the discretization of time, the goal of maximizing the benefits for all trains, and the need to satisfy track capacity constraints. The Lagrange relaxation algorithm was employed for the computation of this model. Caprara et al. [13] explored periodic timetables based on graph theory and constructed an integer programming model for which the Lagrange relaxation algorithm was also used for computation. This model can be effectively applied to large-scale computations. Zhou and Zhong [14] explored the single-track TTP for conditions of limited track resources with the goal of minimizing the total travel time with safety headway constraints. Cacchiani et al. [15] proposed a time-space network-based optimization model for the TTP using a full timetable for a train as a variable. A column generation algorithm was utilized for the computation of this model. He et al. [16] explored the high-speed railroad TTP based on a time-space network model and the branch-and-price algorithm. They implemented a punishment value to reduce the deviations of train start times and the number of occurrences of train overtaking. They also implemented a strategy of increasing the amount of stopping at stations to increase the possibility of train overtaking but they did not consider the strategy of reducing the amount of stopping at stations to shorten passenger waiting times. These studies have yielded useful methods for the optimization of the TTP. However, due to the hierarchical optimization of traditional transportation management processes, passenger demands were not directly considered in these studies.

Another subset of previous studies has focused on passenger demands. Ceder [17] proposed an optimization framework for constructing bus timetables using passenger loading data. This framework can be used to synchronize the departure times of vehicles under dynamic passenger demands. Peeters and Kroon [18] employed the branch-and-price algorithm to solve the rolling stock scheduling problem for a given train timetable and passenger seat demands. Zhou et al. [19] constructed a bilevel programming model to optimize a passenger train operation plan and diagram with the goal of maximizing the profits of the railroad enterprise. Passengers can determine whether to travel by train and can select their transfer schemes based on ticket prices, departure and arrival times, switching times, and congestion charges. The railroad enterprise can improve its train operation plan and diagram based on passenger demands. Kaspi and Raviv [20] proposed a model for comprehensively optimizing a train operation plan and diagram with the goal of minimizing the train operation costs and passenger travel times based on time-varying passenger demands. This model did not have any train capacity constraints; this assumption reduced the computational complexity but yielded results that were inconsistent with the actual situation. Canca et al. [21] proposed a nonlinear integer programming model with the goal of minimizing passenger waiting times and the operational costs for the railroad enterprise. In this model, the departure and arrival times were determined based on dynamic passenger demands. Wang et al. [22] proposed an event-driven model that included departure events, arrival events, and events that correspond to changes in passenger arrival rates. They also constructed a nonlinear programming model with the multiple goals of minimizing the total passenger travel time and optimizing the energy consumption based on passenger transfer behavior in a railroad network. High-speed railroad operations enable highly dense train schedules with the ability to satisfy time-varying passenger demands. Niu et al. [23] proposed a nonlinear mixed-integer programming model based on time-varying origin-destination (OD) passenger demand matrices and explored the high-speed railroad TTP with the goal of minimizing passenger waiting times. Yue et al. [24] proposed a mathematical model for optimizing high-speed railroad timetables to simultaneously consider both passenger service demands and train scheduling. Lagrangian relaxation and a column generation algorithm were employed for computation. They verified the effectiveness of their model and algorithm via a case study of the Beijing-Shanghai high-speed railroad.

Stations are important nodes in a high-speed railroad transportation network; they are the locations of events, such as passenger boarding, alighting, and transfer. Therefore, some researchers have studied the simultaneous optimization of passenger and station service demands with the main goal of optimizing the train stop schedule plan. Suh et al. [25] implemented a skip-stop strategy for subway service planning based on OD matrix information, distances between stations, headways, and maximum link speeds. Deng et al. [26] proposed a bilevel programming model for train stop schedule planning based on the passenger transfer demands, the level of stations at which trains stop, the operational capacities of the train stations, and other factors. The objective function of the upper programming model included the generalized minimum travel cost and the number of train stops, whereas the lower programming model was a passenger flow assignment model with multiclass user equilibrium based on the train stop schedule. Zheng et al. [27] constructed a 0-1 integer programming model that considered skip-stop patterns, with the goal of minimizing the total passenger travel time. The tabu search algorithm was employed for computation. Li et

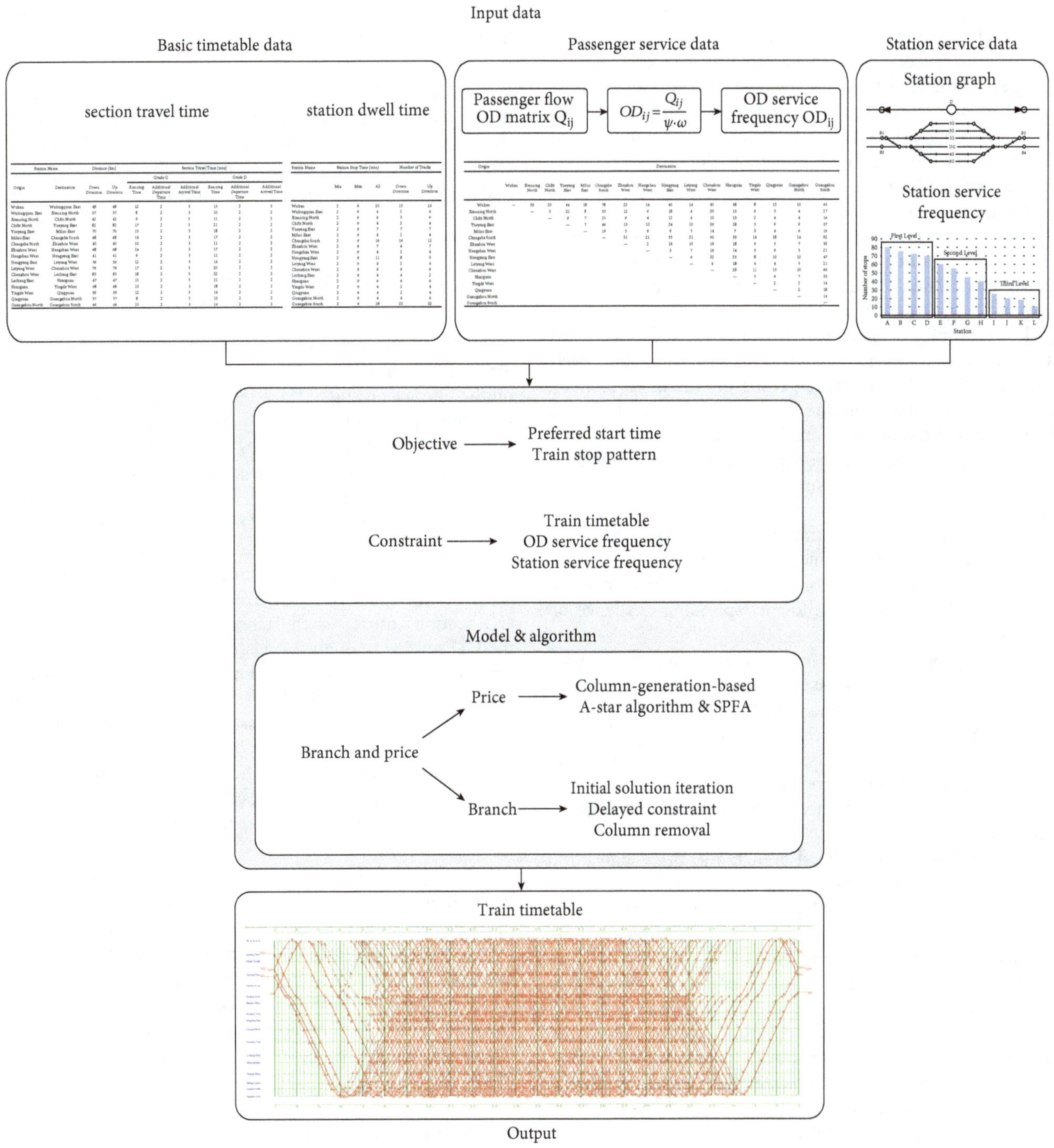

FIGURE 1: Overview of the proposed high-speed railroad train timetable optimization method.

al. [28] analyzed problems encountered with conventional train stop schedules and proposed a high-speed railroad stop schedule optimization model, in which the main constraints were related to the node service frequency, the interstation service accessibility, and the number of one-train stops, and the goal of optimization was to minimize the total number of stops at stations. They also quantitatively analyzed the effects of various factors on the construction of train stop schedule plans. Current research is primarily focused on the station stop schedules and does not consider the simultaneous optimization of train timetables.

We propose a new high-speed railroad train timetable optimization method using a time-space network-based approach and a modified branch-and-price algorithm to simultaneously consider passenger service demands, train scheduling, and station service demands. Figure 1 presents an overview of the method. The input data include the timetable parameters, passenger service data, and station service data.

First, the timetable parameters, such as the section travel times and dwell times at stations, are acquired from the existing diagram. Second, the service frequency of each station OD pair is inferred from the average daily passenger flow data provided by the passenger ticketing department. Last, the number of sidetracks at each station is determined based on the station graph. The stations are divided into nodes of different grades in accordance with the size, geographic location, and passenger flow data of each station. The service frequency of each station is determined according to its node grade. We propose a new integer programming model with constraints related to OD service frequency, station service frequency, and various operation time standards, with the goal of minimizing the deviation from the predefined train start times, the total number of train stops, and the dwell time at stations. The details regarding the modified branch-and-price algorithm and the three acceleration strategies developed for this algorithm are discussed in Section 4 of this paper. The output is a near-global-optimal train timetable.

The four main contributions of our study can be summarized as follows:

(1) We propose a new time-space network method to represent a train stop schedule plan and the comprehensive optimization of the train timetable. We use an associated labeling method to address the corresponding challenge of unconnected arc segments.

(2) We propose a new integer programming model that simultaneously considers passenger service demands, train scheduling, and station service demands.

(3) To calculate the pricing problem (PP), we employ a new method based on the A-star (A$*$) algorithm and the Shortest-Path Faster Algorithm (SPFA) to search for the shortest path in the unconnected and dynamic time-space network.

(4) We propose a modified branch-and-price algorithm and design three acceleration strategies for this algorithm: initial solution iteration, delayed constraints, and column removal. These algorithms and strategies can be used to quickly solve a very-large-scale TTP.

The remainder of this paper is organized as follows: In Section 2, we elaborate on the research problem and introduce the construction of the time-space network. In Section 3, we describe the new integer programming model. In Section 4, we discuss the extended branch-and-price algorithm and the three acceleration strategies. Section 5 presents a verification of the effectiveness of the proposed algorithm and model using data from the Wuhan-Guangzhou high-speed railroad. In Section 6, we discuss our conclusions and offer suggestions for future studies.

2. Problem Statement

This paper considers the high-speed TTP along one train direction of a two-direction railroad line under passenger demands, where the stations are sequentially numbered as $1, 2, \cdots, n_S$. Let $S = \{1, 2, \cdots, n_S\}$ be the set of stations. The set $N = \{1, 2, \cdots, n_N\}$ of trains travel along the line from the predetermined starting stations to the predetermined ending stations according to the train service plan. Let $t \in T$ be the index of times, where $T = \{0, 1, \cdots, n_T\}$ is the planning time horizon with equal one-minute intervals.

2.1. OD Service Frequency. The OD service frequency, which is denoted by $OD_{i,i'}$, is the minimum number of direct trains that serve passengers who travel between the stations of an OD pair. The relationship between the OD service frequency and the passenger flow data can be expressed by the following equation:

$$OD_{i,i'} = \frac{Q_{i,i'}}{\psi_j \cdot \omega_{i,i'}}, \quad (1)$$

where $Q_{i,i'}$ represents the average number of passengers who arrive at station $i \in S$ to travel to station $i' \in S : i' \neq i$ daily on the line, according to the passenger ticketing department; ψ_j is the capacity of train $j \in N$, i.e., the maximum number of passengers accommodated by train j; and $\omega_{i,i'}$ is the train seat occupation rate between station i and station i'.

Figure 2 lists the OD service frequencies of a line. Light red shading indicates OD sections with lower service frequencies. These sections have poor travel access, and passengers have few travel options. Yellow shading indicates OD sections with zero service frequency. Passengers on these OD sections must switch stations. As shown in Figure 2, the OD service frequencies related to station K are low; this situation is attributed to the low grade of the station and its low passenger traffic. A low OD service frequency also causes low attractiveness to passengers from sections related to station K, which inhibits an increase in passenger traffic. Therefore, a reasonable OD service frequency is an important condition for guaranteeing accessible services among stations and the satisfaction of passenger travel demands.

2.2. Station Service Frequency. Station service frequency is the number of trains that stop at and start from a station during a certain period [28]. In actual calculations, the grade of a station is determined based on various factors, such as political, economic, and cultural factors of the city in which the station is located and its attractiveness to passenger traffic, infrastructure, and operational capability. The service frequency of a station is determined in accordance with its grade.

2.3. Predefined Start Times. In theory, the start time of a train can have any value within a possible start time range. In practice, reasonable start time ranges can ensure that the train timetable exhibits reasonable performance in terms of the evenness of the train distribution and the satisfaction of the time-varying passenger demands. Suitable start time ranges can also tighten the feasible space of the resulting optimization models and reduce the search time. The start time range for a train can be determined based on the experience of the planners or acquired from existing timetables [23].

2.4. Description of a Train Timetable in the Form of a Time-Space Network. A description of a train timetable in the form of a time-space network can accurately reflect various

	A	B	C	D	E	F	G	H	I	J	K	L	M
A	-	26	46	9	7	10	10	25	6	13	2	4	26
B	10	-	58	8	12	10	11	30	6	11	2	6	32
C	33	46	-	25	31	18	24	64	14	29	3	11	69
D	10	6	28	-	12	11	15	33	7	18	1	3	26
E	6	9	29	11	-	7	13	29	6	14	0	4	26
F	6	10	21	8	7	-	9	20	4	9	3	2	17
G	10	13	30	12	13	7	-	26	6	13	1	5	22
H	16	24	65	34	29	18	29	-	18	35	5	14	69
I	6	8	20	8	8	8	8	20	-	11	3	4	16
J	7	11	31	20	15	4	14	39	10	-	3	6	25
K	3	0	7	3	5	1	3	10	2	5	-	4	6
L	4	6	17	8	7	6	6	19	6	6	5	-	17
M	16	24	70	28	21	19	24	68	20	27	10	23	-

FIGURE 2: OD service frequencies of a train line.

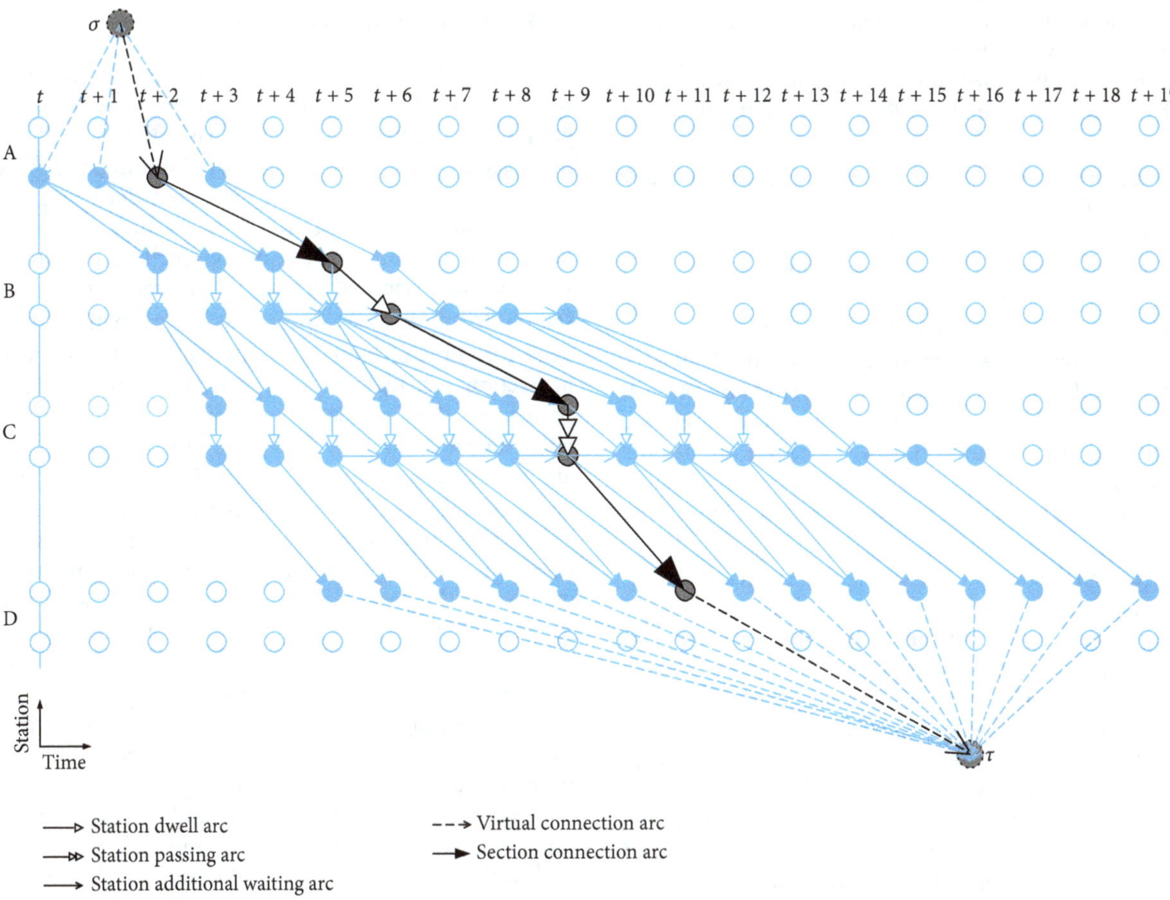

FIGURE 3: Illustration of a time-space network.

events and their occurrence times during train operations [16]. A time-space network is the combination of the physical network and the time horizon.

We use Figure 3 to demonstrate the time-space network formulation, in which the physical route of train j consists of four nodes/stations and three links/sections. The horizontal axis corresponds to discrete time, where the time horizon is represented as $[t, t + 19]$ with equal one-minute intervals. The vertical axis corresponds to space and represents stations and sections. The time-space vertex represents the arrival or departure state of train j at a physical node at the current time, and the time-space arc represents the movement of train j in stations or sections with the entering and leaving times.

The time-space network starts from the virtual departure node σ. Different virtual departure arcs are generated based on the feasible start time range. We define the section connection arcs to imply the section travel times of train j. When train j passes by a station, we define the station passing arcs to imply it. When train j stops at a station, additional starting and stopping times and dwell times are incurred.

Additional starting and stopping times are calculated in accordance with the stop statuses of train j at the two adjacent stations and added to the section connection arcs. The dwell times at stations are required to satisfy certain maximum and minimum dwell time requirements. The purpose of imposing a maximum dwell time is to tighten the feasible space of the optimization model and produce a more reasonable train timetable. The minimum dwell time is the minimum time required for passenger boarding, alighting, transfer, and other operations. We define the station dwell arcs to imply the minimum dwell times of train j at stations. Any additional dwell time beyond the minimum dwell time is represented by one or more additional waiting arcs. The time duration of each additional waiting arc is fixed to one minute in this paper. The network ends at the virtual arrival node τ. Thus, the path from σ to τ in the time-space network corresponds to a feasible timetable for train j. Each train j has multiple paths that constitute the path set $P_j \subset P$, where P is the set of all paths.

For example, the path $p \in P_j$, as shown by the black solid lines in Figure 3, starts from station A at time $t + 2$. Traveling with the section connection arc, which is subject to the section pure travel time and additional starting and stopping times, train j arrives at station B at time $t + 5$. After dwelling for one time interval, train j departs from station B at time $t + 6$. Train j passes by station C at time $t + 9$ after traveling with the section connection arc between station B and station C, which is subject to the section pure travel time and additional starting times. Likewise, train j reaches the destination at time $t+11$. In this manner, the train scheduling problem can be transformed into a routing problem in the time-space network.

We redesign the train stop schedule plan when optimizing the train timetable, unlike in the traditional TTP, in which the train stop pattern is predetermined. Thus, both station passing arcs and station dwell arcs are generated for the train path at every station. During the optimization process, the train can choose to either stop at or pass by a station, depending on demand. For stations at which trains must stop for technical reasons, only station dwell arcs and additional waiting arcs are generated. The existence of additional starting and stopping times and the variable stop pattern produce unconnected arcs in the time-space network. To address this issue, we add an associated label $(\beta_\alpha^{sto}, \beta_\alpha^{pst})$ to each arc $\alpha \in A$, where A is the set of arcs. When α is a section arc, β_α^{sto} indicates whether α stops at the ending station of this section, where a value of 1 represents yes and a value of 0 represents no, and β_α^{pst} indicates whether α stops at the starting station of that section. When α is a station arc, β_α^{sto} indicates whether α stops at this station, and β_α^{pst} indicates whether α stops at the preceding adjacent station.

As shown in Figure 4, neither C nor D is a must-stop station. The red path passes by station C. If the red path chooses to pass by station D (as on path p), the section operation arc is calculated based on the pure section travel time. The associated label is (0,0). If the red path chooses to stop at station D (as on path p'), the section operation arc is calculated using the pure section travel time and the additional stopping time at station D. The associated label is (1,0). The blue path stops at station C. If the blue path chooses to pass by station D (as on path q), the section operation arc is calculated using the pure section travel time and the additional starting time at station C. The associated label is (0,1). If the blue path chooses to stop at station D (as on path q'), the section operation arc is calculated using the pure section travel time, the additional starting time at station C, and the additional stopping time at station D. The associated label is (1,1). When selecting a time-space path, we choose the next arc from the subsequent connected arcs for which the β_α^{pst} label is equal to the β_α^{sto} label of the current arc.

3. Model Construction

3.1. Objective Function. We transfer the costs of the time-space nodes to the adjacent arcs. Thus, the cost of a time-space path is equal to the total cost of the included arcs. We define the objective function with the goal of minimizing this cost.

Let td_j be the predefined preferred departure time of train j from its starting station. tdr_p is the actual departure time from the starting station on path p. δ is the half-width of the permitted departure time range at the starting station. td_j^E and td_j^L are the earliest departure time and latest departure time of train j from its starting station. As noted in Section 2.3, the preferred start times of the trains are predefined. The permitted variations of the departure time fall in the range $[td_j-\delta, td_j+\delta]$. The cost of a virtual departure arc in this range is zero. Paths that lie outside this range are allowed but the punishment coefficient, c_1^{pen}/min, is added to achieve maximum control of the deviation of the departure time. The cost of the deviation is calculated and added to the cost of the virtual departure arc. In the example shown in Figure 5, the actual departure time of path $p \in P_j$ falls outside the allowed fluctuation range, and the cost of the corresponding virtual departure arc is $(tdr_p - td_j - \delta) \cdot c_1^{pen}$.

The stopping of a train at a station will cause an increase in the passenger travel time because it generates a dwell time and additional starting and stopping times. Our goal is to reduce the number of stops and the dwell time. The number of stops is associated with the punishment coefficient c_2^{pen}/num. The punishment coefficient for the dwell time c_3^{pen}/min includes the punishment for both the station dwell arc and any corresponding additional waiting arcs.

A punishment of c_4^{pen}/min is applied for additional waiting arcs to reduce but not preclude the occurrence of train overtaking.

The costs of section arcs, station passing arcs, and virtual arrival arcs are 0.

The cost c^σ of a virtual departure arc is

$$c^\sigma = \begin{cases} \left(\left|tdr_p - td_j\right| - \delta\right) \cdot c_1^{pen} & \left|tdr_p - td_j\right| > \delta \\ 0 & else \end{cases} \qquad (2)$$

$$\forall j, p \in P_j.$$

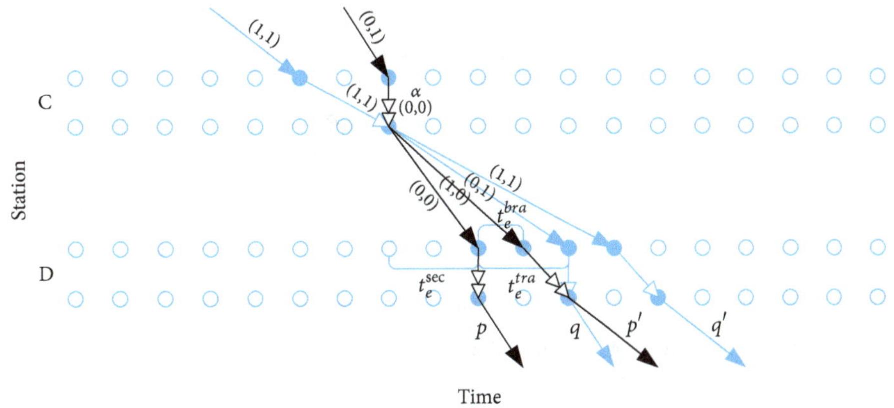

FIGURE 4: Illustration of unconnected arcs.

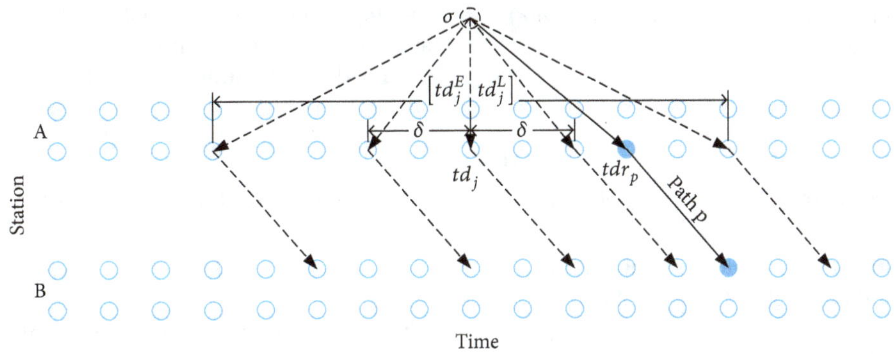

FIGURE 5: Illustration of the cost calculation for virtual departure arcs.

A^{sto} is the station dwell arc set. The cost c_{α}^{sto} of a station dwell arc $\alpha(\forall \alpha \in A^{sto})$ is

$$c_{\alpha}^{sto} = c_2^{pen} + \left(t_{\alpha}^{end} - t_{\alpha}^{beg}\right) \cdot c_3^{pen}, \tag{3}$$

where t_{α}^{beg} and t_{α}^{end} are the start time and end time of arc α.

A^{wai} is the station additional waiting arc set. The cost c_{α}^{wai} of an additional waiting arc $\alpha(\forall \alpha \in A^{wai})$ is

$$c_{\alpha}^{wai} = \left(t_{\alpha}^{end} - t_{\alpha}^{beg}\right) \cdot \left(c_3^{pen} + c_4^{pen}\right). \tag{4}$$

The cost c_p of a time-space path p is

$$c_p = c^{\sigma} + \sum_{\alpha \in A_p^{sto}} c_{\alpha}^{sto} + \sum_{\alpha \in A_p^{wai}} c_{\alpha}^{wai}. \tag{5}$$

The objective function is represented as

$$\min \sum_{j \in N} \sum_{p \in P_j} c_p x_p, \tag{6}$$

where the decision variable x_p indicates whether a certain path p appears in the optimal solution. It is equal to 1 if and only if path p is selected in the optimal solution and is equal to 0 otherwise.

3.2. Constraints. The network flow constraint,

$$\sum_{p \in P_j} x_p = 1 \quad \forall j, \tag{7}$$

represents the selection of only one path by each train.

The arrival safety headway constraint,

$$\sum_{p:\beta_{\alpha}^{p}=1} x_p + \sum_{p':\beta_{\alpha'}^{p'}=1} x_{p'} \leq 1 \qquad \forall e, \forall \alpha, \alpha' \in A_e^{\sec} : 0 \leq t_{\alpha'}^{end} - t_{\alpha}^{end} < t_{d(e)}^{arr}, \tag{8}$$

represents a constraint on the safety headway for consecutive arrivals of trains that travel in the same direction. $e \in E$ is the index of sections, where E is the set of sections. $\alpha' \in A_e^{\sec}$ is the index of arcs, where $A_e^{\sec}$ is the section arc set. $o(e)$ and $d(e)$ are the starting station and ending station of section e. β_{α}^{p} is the binary parameter, which is equal to 1 if arc α belongs to path p and is equal to 0 otherwise. $t_{d(e)}^{arr}$ is the safety headway for trains that arrive at station $d(e)$.

The departure safety headway constraint,

$$\sum_{p:\beta_{\alpha}^{p}=1} x_p + \sum_{p':\beta_{\alpha'}^{p'}=1} x_{p'} \leq 1 \qquad \forall e, \forall \alpha, \alpha' \in A_e^{\sec} : 0 \leq t_{\alpha'}^{beg} - t_{\alpha}^{beg} < t_{o(e)}^{dep}, \tag{9}$$

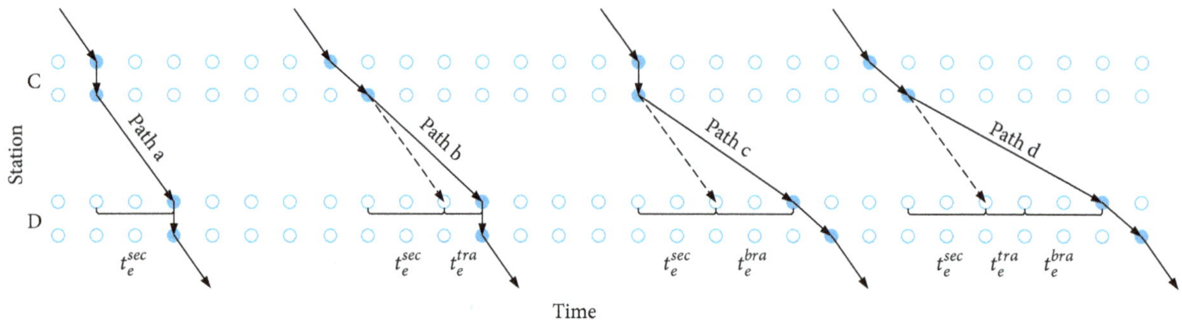

FIGURE 6: Illustration of the calculation of additional starting and stopping times.

represents a constraint on the safety headway for consecutive departures of trains that travel in the same direction. $t_{o(e)}^{dep}$ is the safety headway for trains that depart from station $o(e)$.

The overtaking constraint,

$$\sum_{p:\beta_\alpha^p=1} x_p + \sum_{p':\beta_{\alpha'}^{p'}=1} x_{p'} \le 1 \tag{10}$$
$$\forall e, \forall \alpha, \alpha' \in A_e^{\text{sec}} : \left(t_{\alpha'}^{beg} - t_\alpha^{beg}\right) \cdot \left(t_{\alpha'}^{end} - t_\alpha^{end}\right) \le 0,$$

represents the constraint that trains are not allowed to overtake other trains that travel in the same direction on a section.

The station capacity constraint is

$$\sum_{\alpha \in A_i^{dum}:t_\alpha^{beg} \le t, t_\alpha^{end} \ge t} \sum_{p:\beta_\alpha^p=1} x_p \le num_i \quad \forall i, t, \tag{11}$$

where num_i is the number of sidetracks at station i and A_i^{dum} is a set of virtual station arcs, including station passing arcs, station dwell arcs, and additional waiting arcs. A_i^{dum} is used to represent all possible operations of a train with respect to station i. The start time of an arc is the time at which the train enters the station from the previous section, and the end time of the arc is the time at which the train leaves the station and enters the next section.

The OD service frequency constraint,

$$\sum_{p:\tau_p^{i,i'}=1} x_p \ge OD_{i,i'} \quad \forall i, i', \tag{12}$$

represents the constraint that the number of trains that serve a given station OD pair should satisfy the passenger service demand of this pair. $\tau_p^{i,i'}$ is the coupled-stop index, which is equal to 1 if path p stops at both station i and station i' and is equal to 0 otherwise.

The station service frequency constraint,

$$\sum_{p:\mu_p^i=1} x_p \ge SF_i \quad \forall i, \tag{13}$$

represents the constraint that the number of trains that stop and start at a given station during a certain period should satisfy the service frequency requirement of the node to which this station belongs. SF_i is the station service frequency of station i, which corresponds to the station grade, and μ_p^i is the stop index, which is equal to 1 if path p stops at station i and is equal to 0 otherwise.

The departure time constraint is

$$td_j^E \le tdr_p \le td_j^L \quad \forall j, p \in P_j. \tag{14}$$

The station dwell time constraint is

$$t_i^{\min} \le t_i^{sto} \le t_i^{\max} \quad \forall i, p : \mu_p^i = 1, \tag{15}$$

where $t_i^{\max}$ and $t_i^{\min}$ are the maximum dwell time and minimum dwell time of trains at station i, and t_i^{sto} is the dwell time at station i on path p.

The section travel time constraint is

$$t_\alpha^{end} = t_\alpha^{beg} + t_e^{\text{sec}} + \mu_{p:\beta_\alpha^p=1}^{o(e)} t_e^{tra} + \mu_{p:\beta_\alpha^p=1}^{d(e)} t_e^{bra} \quad \forall e, \forall \alpha \in A_e^{\text{sec}}, \tag{16}$$

where t_e^{sec} is the pure section travel time of section e, and t_e^{tra} and t_e^{bra} are the additional starting time and stopping time of section e.

As shown in Figure 6, during the generation of the section arcs, the additional starting time and stopping time depend on the stopping statuses of the train at the two adjacent stations. These times are determined based on the results of the traction calculations.

The station travel time constraint is

$$t_\alpha^{end} = \begin{cases} t_\alpha^{beg} + t_i^{sto} & \mu_p^i = 1 \\ t_\alpha^{beg} & \mu_p^i = 0 \end{cases} \quad \forall i, p. \tag{17}$$

In the time-space network, the start time and end time of a station passing arc are equivalent. The end time of a station dwell arc is the sum of the start time and the minimum dwell time at this station. Any additional dwell time beyond the minimum dwell time is represented by one or more additional waiting arcs.

The decision variables are

$$x_p = \{0, 1\} \quad \forall p. \tag{18}$$

Model M1 (a generic model with passenger service demands and station service demands) consists of the objective function (6), which is subject to constraints (7)-(18). Constraints (14)-(17) are used to construct the time-space network. By limiting the departure time range at the starting station [(14)] and the maximum and minimum station dwell times [(15)], redundant time-space nodes and arcs can be effectively reduced. By defining the station dwell arcs and additional waiting arcs to imply the station dwell times [(17)], the number of station arcs can be effectively reduced. Thus, the complexity of the model and the scale of the time-space network are manageable. Model M1 is an integer programming model with a linear objective function and linear constraints. We apply an extended branch-and-price algorithm to solve this model.

4. Extended Branch-and-Price Algorithm

As the numbers of stations and trains increase, the computational scale of model M1 also increases. The problem becomes an integer linear programming problem on an extremely large scale. The number of generated feasible train paths, i.e., the number of decision variables, exceeds the number of constraints. Therefore, we propose a modified branch-and-price algorithm to solve this problem.

The branch-and-price algorithm employs a column generation algorithm to solve the linear relaxation of the nodes in the entire branch-and-bound tree, derives a reasonable lower-bound value for the original problem, and uses the branch-and-bound algorithm to solve the integer programming problem. The underlying idea of the column generation algorithm is to reduce the enumeration of all columns to only one set of feasible paths for the original problem in terms of a large-scale linear problem whose scale is considerably smaller than that of the original problem. A restricted master problem (RMP) is generated. Dual variables are derived by solving the RMP. Better paths are generated based on the PP and are added to the RMP. The algorithm iteratively alternates between the PP and the RMP until the best solution to the linear programming problem has been derived.

4.1. Restricted Master Problem. Model M1 is an integer linear programming model. In particular, (14)-(17) are used to construct the time-space network and are not applied to the RMP. The integer decision variables in (18) are relaxed to continuous variables, and the subset P_0 is selected from the path set P.

$$0 \leq x_p \leq 1 \quad \forall p \in P_0 \subset P \tag{19}$$

The RMP is defined as follows:

$$\text{RMP: Equations } (6)\text{-}(13) \text{ and } (19). \tag{20}$$

4.2. Pricing Problem. Solving the PP is a key step of the branch-and-price algorithm. The purpose is to seek the feasible solution that can reduce the objective function, by adding the column with the negative reduced cost to the RMP. The reduced cost of any column can be regarded as the price of this column.

γ_j, $\pi_{e,o,\alpha}$, $\beta_{e,d,\alpha}$, $\xi_{e,\alpha}$, $\lambda_{i,t}$, $\eta_{i,i'}$, and θ_i are dual variables that correspond to (7), (8), (9), (10), (11), (12), and (13), respectively. They represent shadow prices that are associated with time-space nodes, arcs, and other resources. For a certain train path $p \in P_j$, the reduced cost is

$$\begin{aligned}\varphi_p = c_p - \gamma_j - \sum_e \sum_{\alpha\in A_e^{sec}} \pi_{e,o,\alpha} - \sum_e \sum_{\alpha\in A_e^{sec}} \beta_{e,d,\alpha} \\ - \sum_e \sum_{\alpha\in A_e^{sec}} \xi_{e,\alpha} - \sum_i \sum_t \lambda_{i,t} - \sum_i \sum_{i'} \eta_{i,i'} - \sum_i \theta_i.\end{aligned} \tag{21}$$

If $\varphi_p \geq 0$, then all train paths optimally satisfy the original problem, and the optimal solution to the linear programming problem is derived. If $\varphi_p < 0$, further optimization is required. Considering the characteristics of the time-space network, the PP can be transformed into a shortest-path problem. The reduced cost of an entire time-space path can be calculated based on the costs and weights of the arcs that belong to this path. The weights of the arcs are updated using the dual variables derived from the RMP. The path with the negative reduced cost constitutes a new column to be added to the RMP [16]. As shown in (21), the weights of arcs can be negative; in general, the problem can be solved using the Bellman-Ford algorithm or the SPFA.

4.3. Calculation of the Shortest Path. In general, once the weights of all time-space arcs have been determined, the PP can be solved using a shortest-path algorithm. Due to (12), a conventional shortest-path algorithm is not applicable in this case. Equation (12) is an OD service frequency constraint that is related to any pair of different stop stations on a train path. Dual variables cannot be directly employed to update the arc weights; instead, we identify the stopping statuses along the entire path and then update the weights of the path using the dual variables that correspond to each pair of different stop stations. Due to the large scale of the time-space network, a search of all paths would be excessively time consuming and significantly affect the efficiency of the algorithm. Therefore, we propose a new method based on the A∗ algorithm and the reversed SPFA.

The A∗ algorithm can be regarded as a heuristic algorithm in which the estimation function $f^*(m) = g(m) + h^*(m)$ is used to conduct a network search. This algorithm has been widely applied in searches for an optimal path. In the time-space network, $f^*(m)$ is the estimated cost of a path that starts from the virtual departure node σ, passes through the node m, and arrives at the virtual arrival node τ. $g(m)$ is the actual cost of the path from σ to m. $h^*(m)$ is the estimated cost of the best path from m to τ. $h(m)$ is the actual cost of the best path from m to τ. The key to the A∗ algorithm is to select an appropriate estimation function. Since $g(m)$ is the actual cost of the path from the virtual departure node to the node m along the selected path, the problem involves the selection of an appropriate heuristic function $h^*(m)$ to achieve a quick and exhaustive search.

If the estimation function satisfies the admissibility condition [i.e., if the number of nodes subsequent to each node in the network is limited and $h^*(m) \leq h(m)$ is satisfied for each node m], then the best path for the original problem can be derived via the A$*$ algorithm [29]. Therefore, we reverse the original time-space network. Then, $h^*(m)$ can be derived as the cost of the shortest path from the virtual arrival node τ to a random node m, as calculated via the SPFA. If (12) is not considered, then $h^*(m) = h(m)$, and the search for the best path will be conducted exactly along the shortest path. All trains are assumed to stop at all stations after node m. We construct pairs from each station i' after node m and from each stop station i before node m and add the corresponding dual variables $\eta_{i,i'}$ to $h^*(m)$. The $h^*(m)$ derived in this manner must satisfy $h^*(m) \leq h(m)$. As noted in Section 2.4, the number of nodes subsequent to each node in the time-space network is limited, and, thus, the admissibility condition is satisfied. Therefore, the A$*$ algorithm can be used to derive the shortest path for the original problem.

The main procedure for solving the PP is as follows:

Step 1. The weights of the time-space arcs are updated using the dual variables derived from the RMP. The set $G^{\min}$ is used to store the calculated shortest path for each train. G_j represents the set of time-space paths for train j. A status priority queue $\{(m, g(m), f^*(m))\}$ is constructed. $cnt(m)$ denotes the number of times node m leaves the queue. $S(m)$ denotes the set of stop stations prior to node m. $G^{\min} = \varnothing$ and $j = 1$.

Step 2. $G_j{}'$ is derived through a reversal of G_j; in other words, all time-space arcs in G_j are reversed. In addition, $\sigma' = \tau$ and $\tau' = \sigma$. The shortest path from node σ' in $G_j{}'$ to any node $m' = m$ is calculated using the SPFA. The result can also be regarded as the cost of the shortest path from node m in G_j to τ. This cost is recorded as $h'(m)$. The calculation proceeds to Step 3.

Step 3. All items are cleared from the status priority queue. $cnt(\sigma) = 0$, $cnt(\tau) = 0$, $S(\sigma) = \varnothing$, and $S(\tau) = \varnothing$. The initial status $(\sigma, 0, h^*(\sigma))$ is added to the queue.

Step 4. The status priority queue is sorted in increasing order of the estimated cost $f^*(m)$. The lowest status $(u, g(u), f^*(u))$ is removed from the queue. $cnt(u) = cnt(u)+1$. If the node u is equal to τ and $cnt(u) = 1$, then the shortest path for train j is the output, and the calculation proceeds to Step 6; otherwise, it proceeds to Step 5.

Step 5. The statuses of all nodes subsequent to node u are calculated using a status transfer function and are added to the status priority queue. v represents an adjacent node subsequent to node u. $w[u][v]$ represents the weights of the arcs of the adjacent nodes (u, v). The status transfer function is

$$\begin{aligned}&(u, g(u), f*(u))\\&\longrightarrow \big(v, g(u) + w[u][v] + g'(v), g(u) + w[u][v]\\&\quad + g'(v) + h^*(v)\big).\end{aligned} \tag{22}$$

$g'(v)$ is derived from the set $S(u)$ and the stop status of the current arc (u, v). If the current arc is a section arc and the associated train stops at the ending station of the section, then we select each station k from $S(u)$ to form an OD pair with v. The corresponding dual variable $\eta_{k,v}$ is added to $g'(v)$, and $S(v) = S(u)+v$; otherwise, $g'(v) = 0$ and $S(v) = S(u)$. The train is assumed to stop at all stations after node v. Each station i from the set $S(v)$ and each station i' from the stations after node v form an OD pair, and the dual variable $\eta_{i,i'}$, which corresponds to this pair, is added to the $h'(v)$ calculated in Step 2. The result $[h^*(v)]$ is the estimated cost of the shortest path.

If $cnt(v) > 1$, then the calculated status is not added to the status priority queue. The calculation proceeds to Step 4.

Step 6. The shortest path for train j is added to $G^{\min}$. $j = j + 1$. If $j > n$, the calculation proceeds to Step 7; otherwise, it proceeds to Step 2.

Step 7. All train paths in $G^{\min}$ are ordered according to their costs. The path with the lowest cost is added to the RMP. The calculation ends.

4.4. Rapid Branching Strategy. We use a rapid branching strategy, in which only one branch is taken. The selected path, which is represented by a fractional variable, will directly appear in the next iteration. The corresponding train will choose and is allowed only to choose this path. The beam search algorithm is used for branching. The derived fractional variables are ordered from greatest to least according to their corresponding coefficients in the objective function. At each branching, only a certain number of promising variables are selected; this number is referred to as the beam width. This method has been demonstrated to be effective for a similar integer programming problem [30].

Figure 7 illustrates the workflow of the extended branch-and-price algorithm, where ε is the termination parameter.

4.5. Acceleration Strategies

4.5.1. Initial Solution Iteration. In the branch-and-price algorithm, new columns generated by solving the PP are iteratively added to the path set considered in the RMP to improve the objective value and to achieve the optimal result. As the iterations proceed, the scale of the path set considered in the RMP increases. An excessively large problem scale will affect the efficiency of the algorithm. Thus, strategies are necessary to limit the increase in the scale of the path set. In general, a reasonable initial feasible solution facilitates rapid convergence of the algorithm.

For these reasons, we designed an initial solution iteration strategy. As shown by the dotted box in Figure 7, once an integer solution is derived in the calculation process and the numerical difference between the corresponding objective function value and the current UB exceeds a certain threshold, the integer solution will be used as a new initial

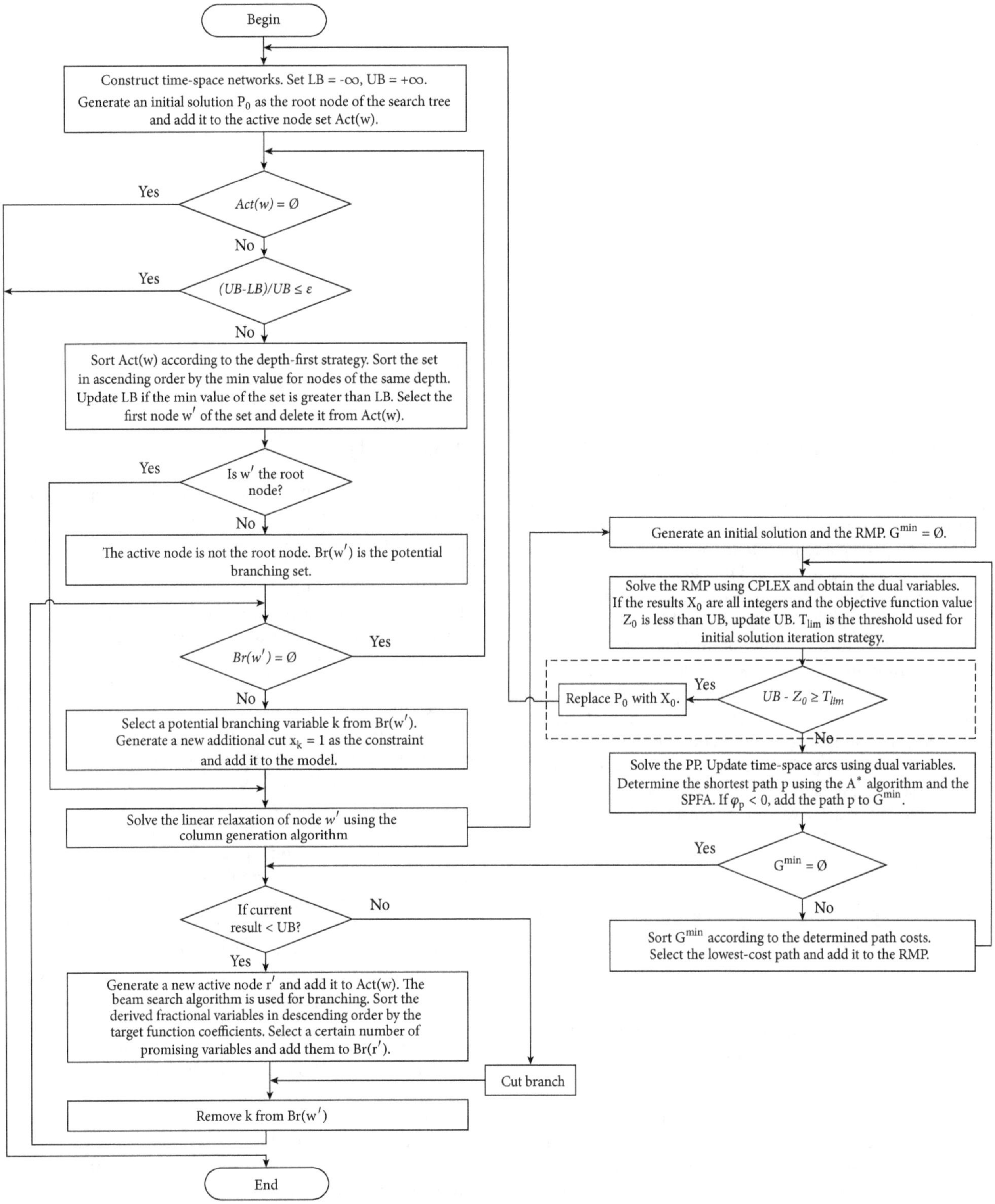

FIGURE 7: Workflow of the extended branch-and-price algorithm.

feasible solution to restart the execution of the entire branch-and-price algorithm. The purpose of specifying this threshold is to avoid consuming excessive time when frequently solving the linear relaxation of the root node. This initial solution iteration strategy improves the calculation efficiency and effectively reduces the processing scale.

4.5.2. Delayed Constraints. The delayed constraint strategy is specifically related to (8)-(11). In (8), as shown in Figure 8, (α, p) represents the section arc a and its corresponding path p. If path p exists in the current path set of the RMP, the time intervals between arriving at station C on path p and all adjacent paths must be greater than or equal to t_C^{arr}. The

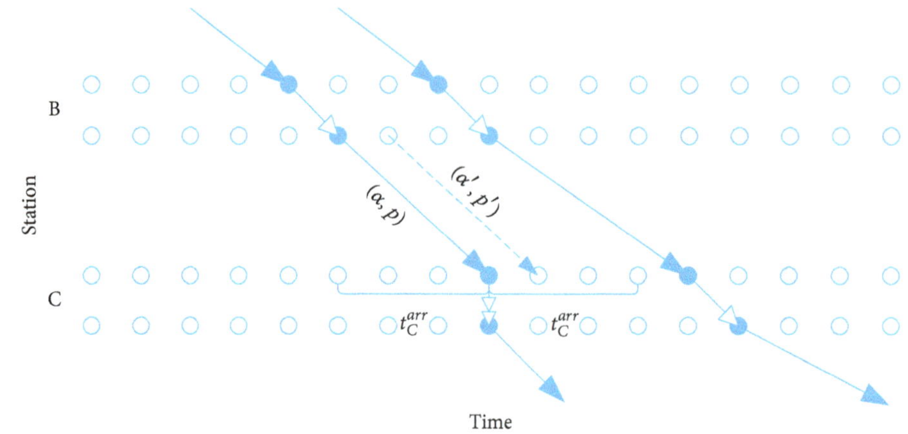

FIGURE 8: Illustration of the delayed constraint strategy.

dual variable of path p related to (8) is $\pi_{e,o,\alpha} = 0$. However, the weight of another arc (such as α') at station C in the time range $[t_\alpha^{arr}, t_\alpha^{arr} + t_C^{arr})$ is not affected by path p. If path p' is the shortest path selected by solving the PP and is added to the RMP, then path p and path p' must satisfy the constraint $x_p + x_{p'} \leq 1$ according to (8), and the dual variable of path p related to (8) may not be zero. In the calculation, the value of path p' may be less than 1 or even 0. In this case, the added path does not serve its ideal function in reducing the value of the objective function. Similarly, a time-space arc at station C in the time range $(t_\alpha^{arr} - t_C^{arr}, t_\alpha^{arr}]$ that is selected in the next iteration can conflict with path p. This constraint is referred to as a delayed constraint; the effect of a constraint in the current iteration can be shifted to the next iteration. To address these delayed constraints, we add a delay weight to each time-space arc. The value of the delay weight is determined based on the average value of the paths over several recent iterations (e.g., seven iterations) in the RMP. When the time-space network is updated using dual variables, the delay weight attributes of the paths are added to all selected arcs at station C in the time range $(t_\alpha^{arr} - t_C^{arr}, t_\alpha^{arr} + t_C^{arr})$. Delayed constraints related to (9)-(11) are handled in a manner similar to that applied for (8).

Applying the delayed constraint strategy to (8)-(11) effectively improves the efficiency of the branch-and-price algorithm in solving the linear relaxation of the nodes.

4.5.3. Column Removal. Due to the rapid branching strategy, in which a selected branching path is treated as a constraint that is added to the model, the train associated with a selected branching path must select only this path. All other paths related to this train in the current path set can be deleted. Any paths in the path set that have operational conflicts with the branching path [(8)-(11)] can be deleted. This train can be skipped in the calculation of the PP. When the arc weights are updated, the weights of all arcs that conflict with the branching path are set to $+\infty$. Thus, these conflicting arcs are avoided in the calculation of the shortest path. This column removal strategy can effectively control the processing scale and improve the efficiency of the algorithm.

5. Case Study

The effectiveness of the proposed model and algorithm was demonstrated using train data from the Wuhan-Guangzhou railroad. The algorithm was implemented using a Windows 7 computer with an Intel Xeon E3 3.30-GHz CPU and 8 GB of RAM. The programming tools were Microsoft Visual Studio 2010 with the C# language and ILOG CPLEX 12.7.1.0.

The Wuhan-Guangzhou high-speed railroad is an abbreviated name for the Wuhan-Guangzhou section of the Beijing-Guangzhou-Shenzhen-Hong Kong high-speed railroad. The length of the railroad is 968 km (601 miles), and the route has a total of 18 stations. The Wulongquan East station is an overtaking station that does not provide passenger transportation. The Lechang East station officially opened in May 2017 and was treated as a "passed-by" station in this experiment. All trains that run on this route are classified as Grade G and operate at a speed of 300 km/h. The operation of Grade D trains, with a speed of 250 km/h, was officially suspended on January 10, 2016. Due to the time designated for maintenance and testing, the actual passenger service time ranges from 6:30 AM to 0:00 AM. Additional infrastructure information is available on the website: https://en.wikipedia.org/wiki/Wuhan-Guangzhou_high-speed_railway.

Table 1 lists the station names and the number of tracks at each station on the Wuhan-Guangzhou railroad. Table 2 lists the section lengths, pure section travel times, and additional starting and stopping times.

To explicitly show the train timetabling results, the calculation was performed using trains that travel in the downward direction as an example.

Table 3 lists the predefined earliest and latest departure times for the trains at their starting stations. Table 4 lists the OD service frequencies inferred from the daily average passenger traffic data provided by the ticketing department. The grades of the station nodes and the corresponding service frequencies determined based on the station size, properties, infrastructure, railroad hub position, city population, economic data, and traffic data are shown in Table 5. All data are provided by the State Key Laboratory of Rail Traffic Control

TABLE 1: Station names and number of tracks.

Station Name	Station ID	Number of Tracks		
		Downward direction	Upward direction	All
Wuhan	1	15	15	20
Wulongquan East	2	2	4	4
Xianning North	3	3	6	6
Chibi North	4	2	4	4
Yueyang East	5	7	7	7
Miluo East	6	2	4	4
Changsha South	7	16	12	16
Zhuzhou West	8	4	7	7
Hengshan West	9	2	4	4
Hengyang East	10	8	9	11
Leiyang West	11	2	4	4
Chenzhou West	12	6	6	6
Lechang East	13	4	2	4
Shaoguan	14	6	6	6
Yingde West	15	2	4	4
Qingyuan	16	2	4	4
Guangzhou North	17	4	4	4
Guangzhou South	18	22	22	28

TABLE 2: Section lengths and section travel times.

Adjacent Station Pair	Distance (km)	Section Travel Time (min)		
		Running Time	Additional Starting Time	Additional Stopping Time
1-2	48	12	2	3
2-3	37	8	2	3
3-4	42	9	2	3
4-5	82	17	2	3
5-6	70	15	2	3
6-7	68	14	2	3
7-8	40	10	2	3
8-9	68	14	2	3
9-10	41	9	2	3
10-11	56	12	2	3
11-12	76	17	2	3
12-13	83	18	2	3
13-14	47	10	2	3
14-15	68	15	2	3
15-16	59	12	2	3
16-17	37	8	2	3
17-18	46	13	2	3

and Safety (Beijing Jiaotong University, Beijing, China). The values of the parameters n_S, n_N, n_T, δ, ε, c_1^{pen}, c_2^{pen}, c_3^{pen}, and c_4^{pen} were set to 18, 107, 1440, 12 min, 0.055, 1.5, 1.5, 1.2, and 0.5, respectively. The values of $t_i^{\min}$, $t_i^{\max}$, t_i^{dep}, and t_i^{arr} for trains at all stations were set to 2 min, 7 min, 2 min, and 3 min, respectively.

Based on these data and parameters, calculations were performed using model M1 and the extended branch-and-price algorithm. To achieve a balance between efficiency and accuracy, we applied a termination rule in the form of (23). When the difference between the UB and the LB was within the allowable range, the calculation was terminated. An approximately globally optimal solution was achieved. We performed 30 sets of experiments, and the average computation time of the extended branch-and-price algorithm was 199 s.

$$\frac{(UB - LB)}{UB} \leq \varepsilon. \tag{23}$$

Table 3: Earliest and latest departure times at starting stations.

Train	Starting Station	Earliest Time	Latest Time	Train	Starting Station	Earliest Time	Latest Time	Train	Starting Station	Earliest Time	Latest Time	Train	Starting Station	Earliest Time	Latest Time
1	1	15:27	16:27	28	1	15:58	16:58	55	1	07:44	08:44	82	7	07:00	08:00
2	1	17:33	18:33	29	1	13:05	14:05	56	1	07:55	08:55	83	7	07:25	08:25
3	1	17:45	18:45	30	1	10:50	11:50	57	1	08:08	09:08	84	7	09:37	10:37
4	1	12:39	13:39	31	1	16:11	17:11	58	1	08:23	09:23	85	7	10:28	11:28
5	1	13:57	14:57	32	1	12:21	13:21	59	1	08:45	09:45	86	7	11:25	12:25
6	1	16:23	17:23	33	1	15:02	16:02	60	1	09:14	10:14	87	7	15:40	16:40
7	1	08:30	09:30	34	1	16:54	17:54	61	1	09:48	10:48	88	7	18:35	19:35
8	1	13:50	14:50	35	1	15:39	16:39	62	1	10:30	11:30	89	7	18:57	19:57
9	1	12:52	13:52	36	1	16:40	17:40	63	1	10:39	11:39	90	7	07:00	08:00
10	1	09:29	10:29	37	1	17:17	18:17	64	1	11:31	12:31	91	7	07:00	08:00
11	1	13:22	14:22	38	1	18:16	19:16	65	1	11:55	12:55	92	7	07:10	08:10
12	1	14:13	15:13	39	1	11:05	12:05	66	1	12:14	13:14	93	7	07:40	08:40
13	1	11:49	12:49	40	1	19:53	20:53	67	1	13:29	14:29	94	7	09:00	10:00
14	1	09:43	10:43	41	1	07:00	08:00	68	1	14:37	15:37	95	7	11:49	12:49
15	1	07:00	08:00	42	1	09:00	10:00	69	1	14:50	15:50	96	7	12:28	13:28
16	1	14:55	15:55	43	1	10:23	11:23	70	1	15:18	16:18	97	7	14:49	15:49
17	1	12:08	13:08	44	1	11:38	12:38	71	1	16:47	17:47	98	7	17:24	18:24
18	1	19:20	20:20	45	1	13:12	14:12	72	1	17:04	18:04	99	7	19:23	20:23
19	1	20:30	21:30	46	1	14:07	15:07	73	1	17:26	18:26	100	7	20:30	21:30
20	1	12:32	13:32	47	1	15:06	16:06	74	1	17:54	18:54	101	7	20:36	21:36
21	1	12:59	13:59	48	1	16:28	17:28	75	1	18:05	19:05	102	7	20:36	21:36
22	7	07:00	08:00	49	1	11:12	12:12	76	1	18:55	19:55	103	7	20:36	21:36
23	7	14:00	15:00	50	1	07:00	08:00	77	1	18:43	19:43	104	5	10:35	11:35
24	1	10:10	11:10	51	1	07:00	08:00	78	1	18:30	19:30	105	5	16:45	17:45
25	1	15:48	16:48	52	1	07:00	08:00	79	1	19:02	20:02	106	1	07:00	08:00
26	1	18:24	19:24	53	1	07:12	08:12	80	7	14:30	15:30	107	7	12:53	13:53
27	1	09:36	10:36	54	1	07:26	08:26	81	7	07:00	08:00				

Table 4: OD service frequencies of sections that operate in the downward direction.

Stations	1	3	4	5	6	7	8	9	10	11	12	14	15	16	17	18
1	—	14	11	18	9	36	14	7	17	8	21	20	5	6	10	32
3		—	4	7	4	14	7	2	9	3	10	9	3	3	4	12
4			—	4	5	11	6	2	7	4	7	7	2	3	4	10
5				—	4	18	7	5	9	4	13	11	5	5	6	16
6					—	8	3	3	3	3	7	7	2	2	3	9
7						—	16	14	21	13	28	28	7	9	14	43
8							—	2	9	8	11	13	4	5	7	16
9								—	3	7	9	12	4	5	3	14
10									—	5	16	12	7	7	8	21
11										—	4	12	4	4	5	13
12											—	16	7	8	10	28
14												—	4	6	7	28
15													—	2	2	7
16														—	2	9
17															—	14
18																—

Table 5: Classification of node grades and corresponding station service frequencies.

Level	Station ID									Service Frequency
1	7	18	1	12						60
2	14	10	5	3	8					30
3	9	11	4	6	16	15	17	2	13	14

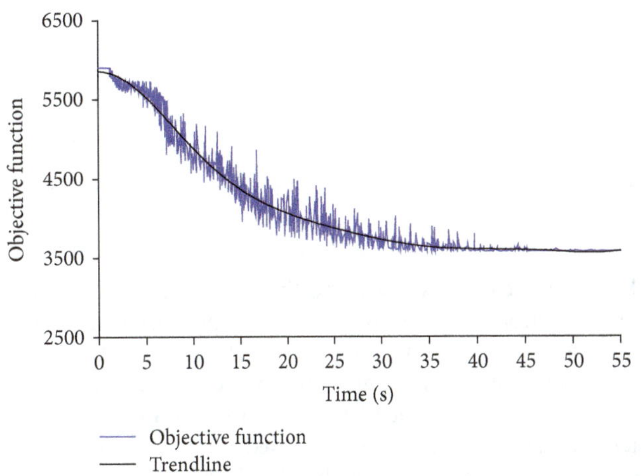

Figure 9: Variation in the objective function value of the root node over time.

As shown in Figure 9, during the process of solving the linear relaxation of the root node using the column generation algorithm, the value of the objective function initially decreases rapidly and then gradually tends toward stability as the calculation time increases.

As Figure 10 similarly illustrates, as the calculation time increases, both the average number of stops and the average dwell time decrease and gradually tend toward stability.

Figure 11 shows a comparison of the predefined and actual numbers of train departures during different hour-long periods. As shown in the figure, a reasonable selection of values for the parameters δ and c_1^{pen} can ensure that the majority of trains depart within the predefined departure time ranges.

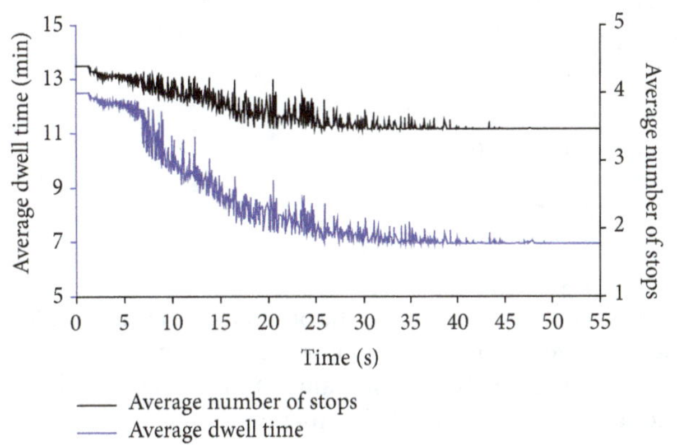

Figure 10: Variation in the parameters of the root node over time.

Table 6 presents a comparison between the real train timetable and the optimized train timetable in terms of the average travel speed, the average number of stops at stations, and the average station dwell time. The real case is the real train timetable for the Wuhan-Guangzhou high-speed railroad in February 2016 (refer to Dataset S1 in the Supplementary Materials), which was also employed as the initial solution for the optimal case. The timetable specified the operations of 107 downward-direction trains (from 6:30 AM to 0:00 AM). The minimum station dwell time was 2 min, and the maximum dwell time was 28 min. The average

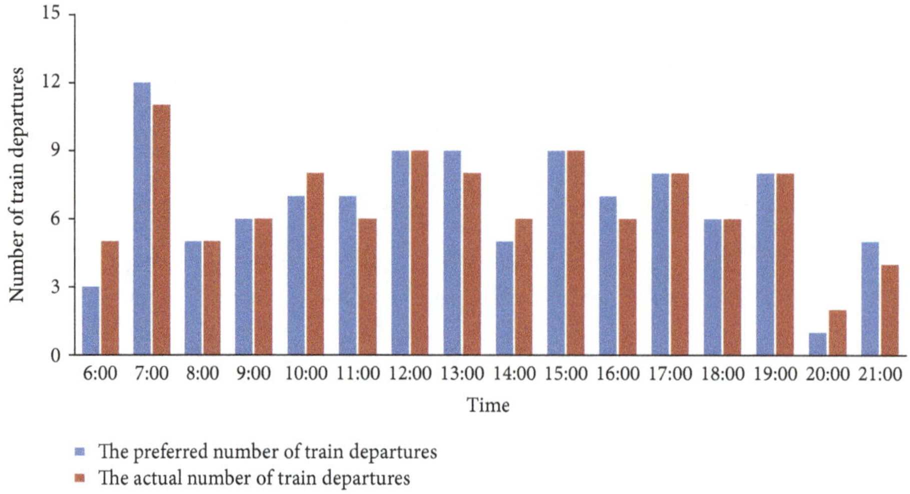

FIGURE 11: Preferred and actual numbers of train departures for each hour.

TABLE 6: Comparison between the real train timetable and the optimized train timetable.

Parameters	Real Case	Optimal Case	Change	Percent Change (%)
Number of trains	107	107	0	0
Average travel speed (km/h)	223.4	234.1	10.7	4.8
Average number of stops	4.4	3.5	-0.9	-20.3
Average dwell time (min)	12.5	7.1	-5	-43.2
Overtaking	40	0	-40	-100.0
Max dwell time (min)	28	6	-22	-78.6
Total cost of trains	5902.5	3645.7	-2257	-38.2
Average cost of trains	55.2	34.1	-21	-38.2
UB	—	3645.7	—	—
LB	—	3576	—	—
Computation time (s)	—	199.0	—	—
Gap	—	1.9%	—	—

travel speed was 223.4 km/h. The optimal case corresponds to the optimized train timetable, in which the number of trains, train types, and paths are the same as those in the real case. The gap value was calculated as follows:

$$\text{Gap} = \frac{(UB - LB)}{UB} \times 100\%. \tag{24}$$

As shown in Table 6, the average number of stops at stations and the average dwell time are reduced by 20.3% and 43.2%, respectively, in the optimized train timetable. The average travel speed is also significantly increased, and the total defined cost function of the trains is reduced by 38.2%. These results indicate that our proposed model and algorithm can effectively improve the quality of a constructed train timetable and the travel efficiency for passengers. Due to the more reasonable design of the train stop schedule plan, the number of occurrences of overtaking is reduced from 40 to 0, and the maximum dwell time is reduced by 78.6%. These improvements effectively reduce the workload for station operations and improve the operational capability of the station infrastructure. The total computation time of the optimal case was 199 s, and the gap value was 1.9%. The results demonstrate the efficiency of the proposed algorithm. We also tested the efficiency of the algorithm without the three acceleration strategies. In this case, the gap value was 50% when the calculation ran for 6 min. This result demonstrates that the three proposed acceleration strategies are effective in accelerating the algorithm. Figure 12 shows the train timetable from 6:30 AM to 0:00 AM based on the optimal case.

6. Conclusions

In this study, we developed a high-speed railroad train timetable optimization model based on a time-space network representation. The goal of the model is to minimize the deviations of the train start times, the total number of train stops, and the dwell time at stations, which are subject to the constraints related to the OD service frequency, train scheduling, and station service frequency. The train stop schedule plan is redesigned during the process of optimizing the train timetable. An experiment using data from the

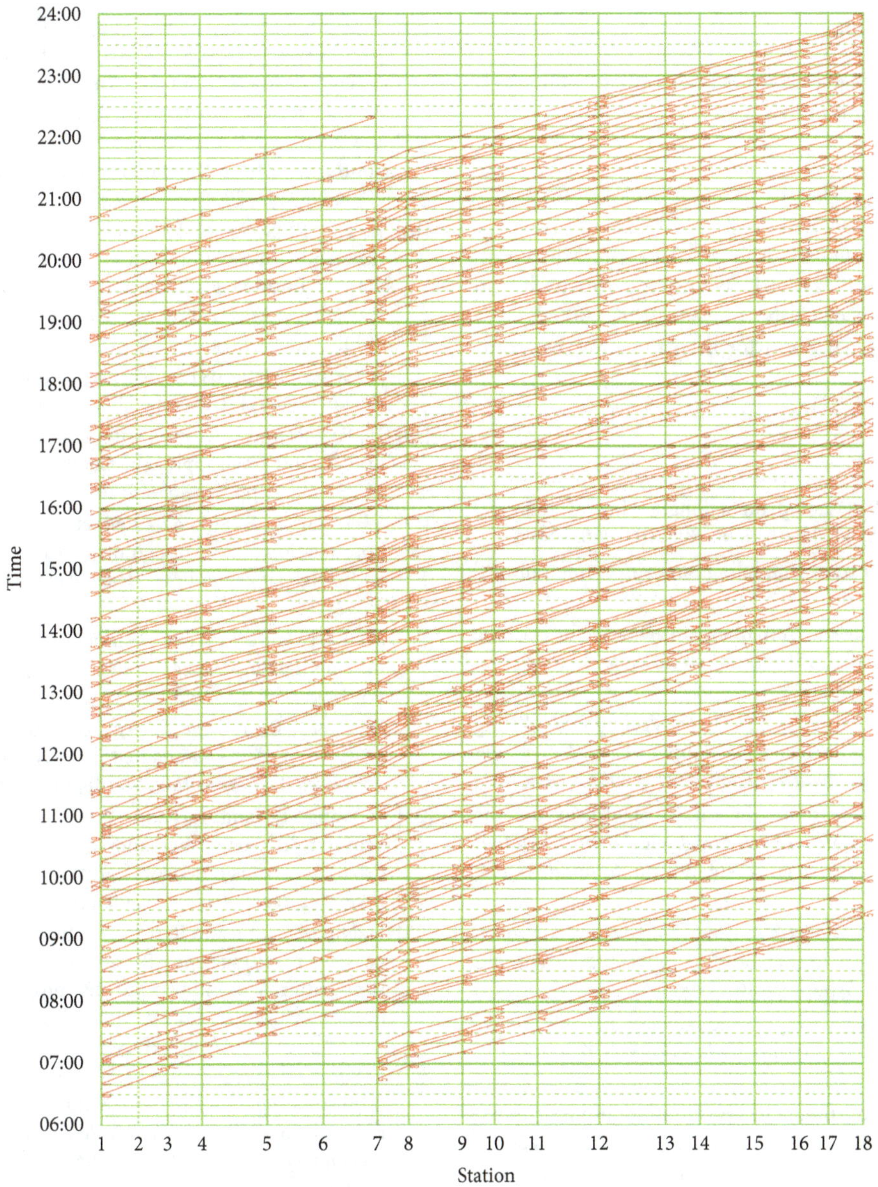

FIGURE 12: Optimized train timetable for the optimal case.

Wuhan-Guangzhou high-speed railroad demonstrated that the proposed model and algorithm can effectively reduce the number of stops at stations and the station dwell time to improve the quality of the constructed train timetables and the travel efficiency for passengers. First, we discussed the construction of a time-space network to represent a high-speed railroad environment. We used an associated labeling method to handle unconnected arcs. Second, we employed an extended branch-and-price algorithm to calculate the model. We also proposed a new method based on the A∗ algorithm and the SPFA to search for the shortest path in the unconnected and dynamic time-space network when solving the PP. Last, we designed three acceleration strategies: initial solution iteration, delayed constraints, and column removal. Our experiment demonstrated that these three strategies are effective in accelerating the algorithm.

Our future study will pursue three research directions. First, only constraints concerning the limited number of sidetracks at a station were considered when optimizing the train timetable; constraints related to routing schedules and track use at stations were not considered. Thus, addition work is needed to achieve the simultaneous optimization of the high-speed railroad train timetable and the station operation plan. Second, we plan to explore the impact of the OD/station service frequency on passenger demands. Last, we will perform sensitivity analyses to assess the effects of the punishment coefficients on the objective function.

Conflicts of Interest

The authors declare that there are no conflicts of interest regarding the publication of this paper.

Acknowledgments

This study was supported by National Key R&D Program of China (2018YFB1201402) and Technology Research and Development Program of China railway (2017X009-E). We thank Mingkai Bi and Yu Zhou for their constructive comments.

References

[1] O. Frank, "Two-Way Traffic on a Single Line of Railway," *Operations Research*, vol. 14, no. 5, pp. 801–811, 1966.

[2] B. Szpigel, "Optimal train scheduling on a single line railway," 1973.

[3] P. Serafini and W. Ukovich, "A mathematical model for periodic scheduling problems," *SIAM Journal on Discrete Mathematics*, vol. 2, no. 4, pp. 550–581, 1989.

[4] A. Schrijver and A. Steenbeek, "Dienstregelingontwikkeling voor Nederlandse Spoorwegen NS Rapport Fase 1," *Centrum voor Wiskunde en Informatica (Oktober 1993)*, 1993.

[5] M. A. Odijk, "A constraint generation algorithm for the construction of periodic railway timetables," *Transportation Research Part B: Methodological*, vol. 30, no. 6, pp. 455–464, 1996.

[6] K. Nachtigall and S. Voget, "A genetic algorithm approach to periodic railway synchronization," *Computers & Operations Research*, vol. 23, no. 5, pp. 453–463, 1996.

[7] C. Liebchen, "Symmetry for periodic railway timetables," *Electronic Notes in Theoretical Computer Science*, vol. 92, pp. 34–51, 2004.

[8] C. Liebchen, "Periodic timetable optimization in public transport," *Operations Research Proceedings 2006*, pp. 29–36, 2007.

[9] G. Caimi, M. Fuchsberger, M. Laumanns, and K. Schüpbach, "Periodic railway timetabling with event flexibility," *Networks*, vol. 57, no. 1, pp. 3–18, 2011.

[10] J.-H. Zhong, M. Shen, J. Zhang, H. S.-H. Chung, Y.-H. Shi, and Y. Li, "A differential evolution algorithm with dual populations for solving periodic railway timetable scheduling problem," *IEEE Transactions on Evolutionary Computation*, vol. 17, no. 4, pp. 512–527, 2013.

[11] D. Sparing and R. M. P. Goverde, "A cycle time optimization model for generating stable periodic railway timetables," *Transportation Research Part B: Methodological*, vol. 98, pp. 198–223, 2017.

[12] U. Brännlund, P. O. Lindberg, A. Nõu, and J.-E. Nilsson, "Railway timetabling using Lagrangian relaxation," *Transportation Science*, vol. 32, no. 4, pp. 358–369, 1998.

[13] A. Caprara, M. Fischetti, and P. Toth, "Modeling and solving the train timetabling problem," *Operations Research*, vol. 50, no. 5, pp. 851–861, 2002.

[14] X. Zhou and M. Zhong, "Single-track train timetabling with guaranteed optimality: branch-and-bound algorithms with enhanced lower bounds," *Transportation Research Part B: Methodological*, vol. 41, no. 3, pp. 320–341, 2007.

[15] V. Cacchiani, A. Caprara, and P. Toth, "A column generation approach to train timetabling on a corridor," *4OR*, vol. 6, no. 2, pp. 125–142, 2008.

[16] B. He, R. Song, S. He, and Y. Xu, "High-speed rail train timetabling problem: A time-space network based method with an improved branch-and-price algorithm," *Mathematical Problems in Engineering*, vol. 2014, Article ID 641562, 15 pages, 2014.

[17] A. Ceder, "Methods for creating bus timetables," *Transportation Research Part A: General*, vol. 21, no. 1, pp. 59–83, 1987.

[18] M. Peeters and L. Kroon, "Circulation of railway rolling stock: a branch-and-price approach," *Computers & Operations Research*, vol. 35, no. 2, pp. 538–556, 2008.

[19] W.-I. Zhou, F. Shi, Y. Chen et al., "Method of integrated optimization of train operation plan and diagram for network of dedicated passenger lines," *Journal of the China Railway Society*, vol. 2, 2011.

[20] M. Kaspi and T. Raviv, "Service-oriented line planning and timetabling for passenger trains," *Transportation Science*, vol. 47, no. 3, pp. 295–311, 2013.

[21] D. Canca, E. Barrena, E. Algaba, and A. Zarzo, "Design and analysis of demand-adapted railway timetables," *Journal of Advanced Transportation*, vol. 48, no. 2, pp. 119–137, 2014.

[22] Y. Wang, T. Tang, B. Ning, T. J. J. van den Boom, and B. De Schutter, "Passenger-demands-oriented train scheduling for an urban rail transit network," *Transportation Research Part C: Emerging Technologies*, vol. 60, pp. 1–23, 2015.

[23] H. Niu, X. Zhou, and R. Gao, "Train scheduling for minimizing passenger waiting time with time-dependent demand and skip-stop patterns: nonlinear integer programming models with linear constraints," *Transportation Research Part B: Methodological*, vol. 76, pp. 117–135, 2015.

[24] Y. Yue, S. Wang, L. Zhou, L. Tong, and M. R. Saat, "Optimizing train stopping patterns and schedules for high-speed passenger rail corridors," *Transportation Research Part C: Emerging Technologies*, vol. 63, pp. 126–146, 2016.

[25] W. Suh, K.-S. Chon, and S.-M. Rhee, "Effect of Skip-Stop Policy on a Korean Subway System," *Transportation Research Record Journal of the Transportation Research Board*, vol. 1793, no. 1, pp. 33–39, 2002.

[26] L.-B. Deng, F. Shi, and W.-I. Zhou, "Stop Schedule Plan Optimization for Passenger Train," *China Railway Science*, vol. 30, no. 4, pp. 102–107, 2009.

[27] L. Zheng, R. Song, S. W. He et al., "Optimization model and algorithm of skip-stop strategy for urban rail transit," *Journal of the China Railway Society*, vol. 31, no. 6, pp. 1–8, 2009.

[28] D. Li, B. Han, X. Li et al., "High-speed railway stopping schedule optimization model based on node service," *Journal of the China Railway Society*, vol. 35, no. 6, pp. 1–5, 2013.

[29] N. J. Nilsson, "Principles of artificial intelligence," *Pattern Analysis & Machine Intelligence IEEE Transactions on*, vol. PAMI-3, no. 1, 112 pages, 1980.

[30] T. Schlechte, "Railway Track Allocation," *Operations Research Proceedings*, vol. 203, no. 2, pp. 15–20, 2012.

5

Application of Data Clustering to Railway Delay Pattern Recognition

Fabrizio Cerreto,[1] **Bo Friis Nielsen**,[2] **Otto Anker Nielsen**,[1] **and Steven S. Harrod**[1]

[1]*Department of Management Engineering, Technical University of Denmark, 2800 Kongens Lyngby, Denmark*
[2]*Department of Applied Mathematics and Computer Science, Technical University of Denmark, 2800 Kongens Lyngby, Denmark*

Correspondence should be addressed to Fabrizio Cerreto; facer@dtu.dk

Academic Editor: Andrea D'Ariano

K-means clustering is employed to identify recurrent delay patterns on a high traffic railway line north of Copenhagen, Denmark. The clusters identify behavioral patterns in the very large ("big data") datasets generated automatically and continuously by the railway signal system. The results reveal the conditions where corrective actions are necessary, showing the cases where recurrent delay patterns take place. Delay profiles and delay change profiles are generated from timestamps to compare different train runs and to partition the set of observations into groups of similar elements. K-means clustering can identify and discriminate different patterns affecting the same stations, which is otherwise difficult in previous approaches based on visual inspection. Classical methods of univariate analysis do not reveal these patterns. The demonstrated methodology is scalable and can be applied to any system of transport.

1. Introduction

Operations analysis is the collection and review of performance data, such as punctuality and process cycle time. It is a key step in the continuous improvement of transport services, and several methods exist to collect and analyze data from operations. The increasing availability of automated data sources is offering new ways to analyze operations, providing deeper insight and more reliable information. Railway management is very accepting of these new possibilities, and considerable effort is made by operators and institutions to use operations analysis in feedback loops for improving the timetabling process [1–4]. A better understanding of the development of delays in railways, and in transportation in general, provides the opportunity to improve the processes and identify the factors affecting reliability. For example, causes of delays might be identified in misallocation of supplements and buffers in timetables, structural conflicts that require mitigation actions, suboptimal design of station processes, and inefficient procedures for preparing a train for departure. This paper demonstrates a data mining technique based on k-means clustering to identify recurrent delay patterns in transportation, identify the main reason for cluster membership, and provide managerial insight to improve timetables and processes.

Prior studies propose several methods that are currently in use for operation analysis, deploying sources of automatic data collection. These approaches can be divided into traditional statistical methods and big data techniques, which differ in both the use of data and in the output provided. Traditional methods tend to aggregate and summarize information, so these can provide a general picture or detailed information on specific stations or trains. These are typically proposed in the form of multiple univariate distribution analysis, where the occurrence of different delay patterns at the same station is not visible. Big data techniques can be used to investigate recurring patterns or internal structures in operations. These approaches are expanding, thanks to the growing availability of large amounts of data, and several techniques have been deployed to identify recurrences of delays and describe or predict delays. Advanced techniques such as neural networks, succession rules, Bayesian networks, and various methods of regression have been developed mainly to predict real time delays in railways, as described in Section 2. However, train delays are necessarily correlated over the progression of a complete journey, and these data relations both along the journey of a train and among adjacent train paths have not received as much attention in the literature.

This paper presents a big data technique to identify recurring delay patterns in railway operations. Big data refer to information assets characterized by high volume, velocity, and variety, whose value is extrapolated by analytical methods [5]. In this application, the absolute delay and delay change are tracked for individual train paths along a railway line, resulting in absolute delay and delay change profiles. In the papers based on univariate statistics, systematic delays in these profiles are identified through visual inspection. The manual search for similarities suffers from subjective interpretation from the operator and is easily biased by common artefacts of the representation. The technique presented in this paper applies k-means clustering to find recurrent patterns in train delay progression, so that management may identify processes for improvement or correction. In this way it is possible to support continuous quality improvement.

In Section 2, a literature survey of contemporary data analysis methods is offered. Section 3 presents the k-means cluster method and the structure of the data to be studied. Section 4 presents results from the study of a high density Danish railway line. The effectiveness of k-means clustering for this application is discussed in Section 5, particularly with regard to its novelty compared to existing literature, while conclusions of this paper are presented in Section 6.

2. Literature Survey

Operations analysis is fundamental in the continuous improvement process to manage and modify railway operations. Data collected from real operations, or from simulation models, has been used in the feedback loop to design and improve railway timetables for decades. Typically, even if timetables may change over time, some of the fundamental infrastructure and service behaviors will not be modified. Timetables are often the result of only minor modifications to the previous editions and need to consider problems discovered in earlier timetables. For example, after a structural change in the Danish railway timetable in 1998, after the opening of the Great Belt fixed link, the service structure remained largely unchanged until 2016 [6].

Data collection systems have proliferated in railway networks since 2000, and very large amounts of data are available today. Widespread systems to collect data increased both volume and variety of data, which are often collected by different systems at the same time. The methods to elaborate and interpret information from past operations evolved together with the amount and quality of data, starting from descriptive and inferential statistic and moving towards big data techniques. For example, delay probability density functions can be extrapolated from historical data and integrated in analytical models to estimate service reliability before operation [7]. Goverde et al. [8] perform extensive statistical analysis and distribution fitting of data from the Dutch railway network. Goverde et al. fit different distributions for arrival and departure delays and find that no general distribution fits all groups of recorded arrival delays.

Primary delay distributions derived from operational data are also often employed as input in simulation models to evaluate the propagation of delays. Sipilä [9] explores the effect of modified running time supplements in railway schedules through microsimulation of a Swedish railway line. The author identifies different strategies for running time supplement allocation by verifying the significance of the change in punctuality recorded in 1600 simulations of selected scenarios. Lindfeldt [10] describes a method to aggregate delay data from real records and isolate distributions of primary delays. These distributions are then used to formulate microsimulation models. The data consists of manual records from dispatchers that assign a delay cause code to every record greater than 4 minutes of delay on the Swedish railways. In absence of other sources of data, the reliability of manual record cannot be validated, although the whole simulation model and its results rely on the derived distributions. Studies from other countries show that manual input can be indeed unreliable [11, 12]. The same method to extract primary delay distributions is later used by Lindfeldt and Sipilä [13] in a simulation model to assess the effect of allowing freight trains to travel outside their assigned path. The authors demonstrate that the realized travel times of freight trains could be shortened considerably without affecting the performance of other trains. The reduction of unnecessary waits for traffic management and the permission to depart before schedule reduced the average travel time on one side but increased its variability on the other.

Historical data also provides insight into the factors that influence service reliability. Olsson and Haugland [14] apply regression analysis on the Norwegian railway network and identify the most relevant factors for punctuality, such as absolute passenger flow and passenger occupation ratio. Gorman [15] uses regression analysis on data from American single-tracked freight railways to identify the factors that contribute the most to prolongation of railway running times. Gorman predicts congestion delay based on meets and passes scheduled as a consequence of speed heterogeneity. Again, in simulation, Shih et al. [16] apply an approach similar to Gorman's to determine the best capacity expansion strategy in terms of reduction of average prolongation of running time for freight trains. Shih et al. identify functional relationships, through regression of simulation results, between average delay per train-mile and several factors, such as the relative length of the doubled-tracked section of a railway line. Lindfeldt [17] applies multilinear regression with special focus on F-statistics to investigate factors generating delays on the Swedish railway network. Lindfeldt measures delay changes over selected routes and analyzes their distributions. In particular, the response variables are the share of trains with a delay increase, the median change in delay, and its standard deviation on the route. Statistically significant explanatory variables are found in the traffic volume for both passenger and freight trains. Among passenger trains the most significant variables are average speed and traffic heterogeneity, and for freight trains it is the number of stations on the route with at least three tracks.

Timestamps and recorded deviations from schedule can be integrated with information from other sources. For example, incident reports may be compiled in case of larger disruptions. Such reports include information about the typology of the incident, the train affected by the primary delay, other trains involved, the secondary delays generated, and the recovery plans taken by the dispatchers. Schittenhelm and Richter [1] describe the reporting system in the Danish railways (the same system in service at the time of this study) and introduce a quantile-based approach to depict the development of train delays en route. The plots confirm the general understanding of delays from experienced operators and can be used to quantify the magnitude of expected disruption. The quantile-based approach, though, describes operations as a whole, and it is not able to distinguish systematic delays occurring at individual stations, but with different origins, so analysis of individual train services is necessary to identify peculiar delay patterns. Richter [18] introduces new metrics to identify improvement actions, based on data from automatic detection systems. Richter sorts the trains according to recorded delay and identifies the worst one in a percentile approach, associated with recorded delay causes. A similar approach is adopted with regard to change in deviation, or delay jump, recorded on line sections, so that most critical geographical areas are identified. Lastly, Richter proposes a tabular representation of the median delay of individual trains recorded at station, sorted by scheduled time and geographical location. In this way, the analyst can identify which specific trains typically suffer from primary delays, also characterized by geographical location, and which trains are typically affected.

Similarly, Peterson [4] studies the on-time performance along the path of specific train services, using the rolling average delay of the last three timing points. Such on-time performance is plotted for all the repetitions of a specific train service over a time period and compared to the average, standard deviation, and 75th percentile. Peterson identifies empty areas in the pool of plotted delay profiles and interprets these as recurrent delay patterns given by discrete dispatching choices along the train path. Peterson also interprets recurrent increases or decreases of vehicle delay as segments of insufficient or excess running time supplement, respectively. Reliability of service is described by the standard deviation of recorded delays. Peterson used the mentioned measures in a feedback loop to redistribute the running time supplement in train paths according to the recorded performance.

Andersson et al. [19] assess the effectiveness of running time supplement in railway schedules from empirical data collected on a Swedish railway line. The study plots the recorded delays over the train itinerary overlapped with scheduled running time supplement and compares pair-wise stacked plots from different railway services, stopping patterns, or directions. The identification of misallocation of running time supplement is based on visual search for recurrent delay patterns, and a few different dispatching tactics are identified in clusters of similar delay profiles. Andersson et al. highlight the existence of a threshold value of delay that triggers prioritization of other trains that are traveling on schedule. The observations are clustered in groups and show recurrent delay patterns, and the analysis is supported by detailed analysis of possible conflicts among individual train itineraries. Noticeably, the authors demonstrate that the measures of punctuality currently in use on the Swedish network hide the effects of running time supplement misallocation and delays developed en route. Even though the punctuality at the final destination is a measure of railway performance very common among railway operators, it does not express how trains increase or recover from delays along their journey. Schittenhelm [20] provides a sample of similar measuring approaches in the European railway industry. In a later study, Andersson et al. [21] underline the relevance of critical points for network robustness by plotting delay profiles and showing that the profiles cluster around critical points according to different dispatching strategies. Advanced clustering techniques may support the identification of different strategies to compute the effects on robustness.

Lastly, van Oort et al. [22] evaluate data collected automatically on public transport services with a combination of statistical methods and visual representation. The study represents delay data similarly to Peterson [4], Andersson et al. [19, 21], and Schittenhelm and Richter [1], plotting the recorded delay over individual repetitions of the same service path, and adds the plot of relevant delay percentiles over the stations. The shape of the percentile-based delay profiles highlights recurrent patterns in the deviation from schedule. The representative delay profiles appear different depending on the percentile they represent. Patterns found included the presence of typical early arrival at stations in bus services, followed by waiting time until the scheduled departure time, or recurrent delay drops or increases at specific stations. The delay plots are combined with the measured headway from the previous vehicle. While the delay plots would suggest allocating more running time supplements at systematic increases of delay, structural delays that cannot be compensated by timetable slack are highlighted in the plots of headways, where service unreliability corresponds to scattered recorded headways. A percentile approach was also presented by van Oort et al. to characterize and sort the stations according to performance, similarly to previous literature.

The statistical analyses presented above are suitable for general description of the system performance but lack specific insight into recurrent delay patterns that occur in operation and into the relationships between delays at different locations. The literature presented in this survey focuses on the univariate analysis of selected measures, such as delays at single stations. Traditional metrics common in the railway industry, such as punctuality, have also been found unrepresentative of the actual service reliability. The methods that include the multidimensional aspect of the problem mostly deal with delay profiles, the sequences of delays recorded on individual train itineraries. The quality of these analyses often relies on visual inspection of plotted data and the observer-operated search for matching delay profiles. This search lacks a standardized methodology and is influenced by the plotting layout and the subjective interpretation, which is based on personal experience.

Big data techniques have arisen recently and seek to make use of the very large amount of information that is provided by automatic data collection systems, overcoming the mentioned issues of traditional methods. The term big data is rather broad and includes different techniques that serve specific purpose. The common characteristics of these techniques are volume, velocity, and variety, meaning large amounts of data, generated at high speed, possibly by different sources with different or no structure [5]. As opposed to standard statistical analyses, where hypotheses are formulated and tested, big data techniques search for internal structures directly in the data. Data generated by automatic sources typically fit in the big data criteria. In railways, several data mining techniques were developed in the last years, following different approaches and searching for different types of information. The interest in these techniques is rising, together with the increasing availability of structured data. Industrial applications of these techniques are spreading, and new approaches are being studied also among public institutions [2].

Event mining is a technique based on time sorted logs, where relations between different events are found based on their coincidences. Hansen et al. [23] combine an event mining tool and standard statistics to predict the actual running times of trains to the next station, given all the recorded current delays. Dependencies between pairs of events are found or "mined" in timed event graphs created from the timestamps of individual trains, which correspond to events of occupation and release of blocking sections. The process times between events are inspected by standard statistics, resulting in conditional probabilities of process times, given the recorded delays of all relevant trains in the system. Such a model, though, relies considerably on very detailed knowledge about the infrastructure and requires data which is not commonly available from railway infrastructure managers.

Goverde and Meng [11] use the same information source and similar technique to identify and analyze route conflicts and identify delay chains. Infrastructure data and operation data are integrated so that it is possible to identify a train that is occupying a blocking section linked to a signal at danger for another train. Delay trees are built and traced backwards to identify the primary causes, so individual delays can be classified automatically into primary and secondary, and the correct attribution of delay causes can be verified. Interestingly, the authors verify that more than half of the delay cause records were assigned wrongly by the dispatchers, stating that, in the Netherlands, this type of manual input is not reliable and objective enough to be deployed in data analysis.

Kecman and Goverde [24] extend the model to include nonlogged line sections, where it is not possible to distinguish delays due to signaling impositions and delays due to primary causes. Delay chains are also traced in less detailed data by Sørensen et al. [12]. Based on the time sequences at stations experiencing disturbed operations, the authors identify the trains generating the conflicts and the trains suffering from the conflicts. The analysis is used to identify primary delays, describe single days of operation, identify frequent trains originating, or subject to, delay chains, and identify point stations where most of the primary or secondary delays are generated. In a comparison with manually recorded delay causes, the study finds relevant inconsistencies with the primary delays traced in the delay chains, in accordance with Goverde and Meng [11]. The method described, though, is only valid for single track lines and does not identify multiple primary delays.

Cule et al. [25] introduce association rules to identify delays recurring often together and set up an episode mining framework to highlight frequent delay patterns from train timestamps at stations. However, association rules can highlight common recurrences but cannot explain relations of causality between two events, so primary and secondary delays cannot be distinguished. Similarly, Wallander and Mäkitalo [26] identify delay chains according to the manual delay cause records from the dispatchers and based on timestamps at stations with granularity of 1 minute. The succession rules used are very similar to association rules but consider the time dependencies, so that events taking place earlier can be assumed to be the cause of events happening later under the same circumstances. Trains are characterized by the number and magnitude of conflicts they generate so that improvement actions can be concentrated. Association rules have also been adopted to evaluate the effectiveness of delay prevention actions on Japanese suburban networks by Yabuki et al. [27]. Yabuki et al. compare the association among occurrence of delays of different trains, change in delays, extension of running and dwelling times, and realized headway in before/after scenario comparison. The downside of such models is that association rules can be set between binary variables, so the development of delays depicted does not include its magnitude. Further, the number of associations to be analyzed grows exponentially with the number of potential pairs of events, so the analyses must be limited to short time frames of operation.

Neural networks are a big data method that learns from historical records and uses the relations identified among variables to predict an output, given unseen values of the input variables. This technique is particularly suited to delay prediction and has been deployed in multiple studies. Neural networks look for dependencies in the data, as opposed to simulation models, which are based on interaction rules between objects defined initially. Malavasi and Ricci [28] use neural networks to predict the total experienced delay on a railway line, given its geometrical and technological characteristics, and its scheduled utilization over time. In comparison to simulation, Malavasi and Ricci find neural networks more robust against extreme-valued input, which implicates more likely case-overfitting with simulation. Kecman et al. [29] propose a Bayesian network delay prediction model. In this case, input includes the timetable and recorded delays at all stations. Each delay is assumed to depend only on direct connections in a timed event graph, meaning the recorded delay for the same train at a previous station and for the previous train at the same station. Conditional delay distributions are assumed to be Gaussian, and the parameters are derived through recursive Generalized Linear Models. Chapuis [30] deploys the same assumed delay dependency in a neural network model, where input includes delay of the previous train and at the previous station, and distance to the next station. Such a model can predict the delay of a train at the next station. Independent of the actual infrastructure, this model is generic and can be applied at any station of the railway network. The downside of neural networks, though, is the risk of data overfitting, reducing the prediction capability, although this risk is lower in neural networks than in simulation models.

In response, Marković et al. [31] introduce Support Vector Regression (SVR) to establish a functional relationship between characteristics of the railway system and train delays. Train category, scheduled time, infrastructure, and share of journey completed are identified as most influencing factors to predict the train delay at one station. The authors show that SVR generalizes better than an artificial neural network, which seeks to minimize error of prediction in the historical dataset. Interestingly, the authors assume that the performance of delay prediction can be improved by grouping delays by magnitude, as factors generating smaller delays differ from factors that generate larger disturbances.

Kecman and Goverde [32] apply big data techniques to predict running and dwelling times from actual operation data, based on records from block sections occupations. The study uses random forests of tree-based models, to predict nonlinear relations between input variables and process times, with sufficient robustness to outliers in the data, lowered risk of overfitting, and focus on real time application. Running time predictors are calculated for every block section, and dwelling time predictors are calculated for every station platform. Among the interesting findings, the running times are longer if the headway to the preceding train is short, meaning that the succeeding trains tend to slow down to smoothen the trip and reduce the risk of encountering a yellow signal. Moreover, the authors find no evidence to support the hypothesis that trains run faster when delayed. All the trains were found to run at approximately the maximum performance in any condition. The authors suggest that, in case of insufficient prediction accuracy, new variables might be included in the model, such as platform shape for dwelling times.

Big data techniques focus mainly on the prediction of delays and running times or on the identification of delay chains and realized delay propagation among trains. New applications of these techniques would support the analysis of the realized development of delays along the path of individual train delays. As shown by statistical analysis and visual search for patterns presented by Schittenhelm and Richter [1, 18], Peterson [4], Andersson et al. [19, 21], and van Oort et al. [22], this type of data contains a great deal of information yet to be explored, which would provide insight into the effectiveness of running time supplements and into the presence of structural issues that generate delay in transport operation. In this paper we present a clustering technique to identify recurrent delay patterns among train services, based on readily available data, which leaves room for inference on the factors that generate specific delay patterns. The result shows that, within comparable train trajectories and stopping patterns, different train services accumulate delay at different stations and that recovery shapes differently according to the route direction. Inferences on the cluster composition show the most frequent service characteristics in each cluster. Such information could inform the allocation of correction measures to improve timetables. Tables 1 and 2 summarize the literature just reviewed.

3. Identification of Recurrent Delay Patterns Using Big Data Techniques

In this paper, a delay profile of a train run is defined as the set of recorded deviations throughout its path or a part of it, on a specific date. Note that deviation is reported as the time difference between a scheduled and a realized event, such as arrival, departure, or nonstop timing point. Even though delay is often used to refer to positive deviations, a delay profile can include null and negative values. A delay profile is a powerful representation of operation and the comparison of several delay profiles along the same service path allows the identification of recurrent delay patterns and such a representation method has already been presented in the literature [1, 4, 18, 19, 21, 22]. Delay change, also called delay jump, is the difference in deviation between two consecutive stations and represents the delay recovery or increase. Schittenhelm and Richter [1] use this measure to assess delay increases or time gains between stations, and Goverde and Meng [11] use it to identify delay chains in railway operation. We define a delay change profile of a train as the set of recorded delay changes along its path or a part of it.

A dataset of delay profiles consists of all the delay profiles recorded in a defined period, stacked together. Fields, or variables, of the dataset are the events at every station, whereas observations are individual train runs from a selected service. Such a dataset can refer to a specific train service or to several services following the same stopping pattern, so that the fields can be aggregated. The first case is intended for infrequent services, typically long-distance trains, where every single service may have its own characteristics in terms of planned demand, scheduled rolling stock, or the time of crossing congested nodes. Suburban and regional railway services are often scheduled in constant stopping patterns at high frequency and could, thus, be analyzed together, expecting characteristics of operation to be more homogenous across services. A dataset of delay change profiles is defined analogously to delay profile datasets, where the fields contain the change in deviation in place of the absolute deviation.

Previous researches presented on delay and delay change profiles interpret recurrent patterns by visual search for similarities [1, 4, 19, 21]. The systematic analysis of these two types of datasets through clustering algorithms allows the identification of patterns that are not necessarily visible or that could be wrongly associated by subjective interpretation.

Clustering techniques partition a dataset into a collection of groups of similar observations. In this study, clustering is used to partition the datasets of delay profiles and

TABLE 1: Review of previous uses of univariate statistics in railway operation analysis.

	Environment		Technique					Purpose
	Real operation	Simulation	Distribution fitting	Test significance	Regression analysis	Percentile sorting	Visual inspection	
Goverde et al. (2001)	X		X					Distributions of primary and secondary delays
Sipilä (2010)		X		X				Comparison running time supplement strategies
Lindfeldt (2010)	X	X	X					Distributions of primary delays from real operation for simulation
Olsson and Haugland (2004)	X				X			Factors affecting punctuality
Gorman (2009)	X				X			Factors that generate delays on single track lines
Lindfeldt (2010)	X				X			Factors that increase delays on line segments
Lindfeldt and Sipilä (2014)		X		X				Travel times with different operation models, with/without free freight operation
Shih et al. (2014)		X			X			Factors affecting average delay per train-mile
Schittenhelm and Richter (2009)	X					X	X	Visual inspection of quantile-based representation of deviations and change in deviation
Richter (2010)	X					X		Delay tabular representation and sorting train service performance
Peterson (2012)	X					X		Rolling average delay for specific train services
Andresson et al. (2011)	X						X	Assessment of effectiveness of running time supplements
Andresson et al. (2013)	X						X	Identification of critical points for robustness
van Oort et al. (2015)	X						X	Delay profiles, headway profiles

TABLE 2: Review of previous uses of big data techniques in railway operation analysis.

	Technique								Level of detail			
	Event mining	Association rules	Succession rules	Neural networks	Bayesian networks	Random forests	Support Vector Regression	Clustering	Track sections	Station	Input	Purpose
Hansen et al. (2010)	X								X		Current delays of al trains	Prediction of running time to next station
Goverde and Meng (2011) Kecman and Goverde (2012)	X								X		Timestamps	Delay chains, actual primary delay causes
Sørensen et al. (2017)	X									X	Timestamps	Delay chains on single track lines, actual primary delay causes
Cule et al. (2011)		X								X	Timestamps	Delay patterns
Wallander and Mäkitalo (2012)			X							X	Timestamps, delay causes from dispatchers	Delay chains
Yabuki et al. (2015)		X								X	Timestamps	Comparison of real scenarios
Malavasi and Ricci (2001)				X					X		Physical infrastructure and utilization ratio	Prediction of total realized delay on a network
Kecman et al. (2015)a					X					X	Current train delay, last delay at station	Delay prediction at next stations
Chapuis, (2017)				X						X	Current train delay, last delay at station, distance	Delay prediction at next stations
Marković et al. (2015)							X			X	Infrastructure and train journey characteristics	Delay prediction at next stations
Kecman et al. (2015)b						X			X		Current traffic condition, actual train position, delays of the day	Running time and dwelling time prediction
Cerreto et al. *(2018) (this paper)*								*X*		*X*	*Timestamps*	*Recurrent delay patterns across stations*

identify train services that are candidates for identification of common causality. Inference on common factors appearing in observations clustered together facilitates the assessment of delay patterns in association with specific characteristics of a transport service, such as time of the day (peak/off-peak), day of the week, or equipment used. The clustering process is realized through measures of similarity between elements in the same cluster and dissimilarity between elements from different clusters. Several methods and metrics are available to accomplish the task, suitable for different uses. Hierarchical algorithms proceed by splitting or merging observations recursively and are preferred when a nested structure is assumed in the clusters. In contrast, partitional algorithms do not impose a hierarchical structure and find all the clusters at the same time. K-means clustering is a partitional algorithm and was chosen due to its simplicity and frequent appearance in the literature [33].

K-means clustering is an iterative clustering process based on the identification of the mean element in each cluster. Every cluster is represented by its centroid, calculated as the average of the elements of the cluster, and every observation is assigned to the cluster corresponding to the closest centroid. Given a number k of initial centroids, the algorithm executes the following steps:

(1) Assigning every element to the cluster with the closest centroid

(2) Calculating the new centroids of all the clusters as the mean of the elements

(3) Repeating until convergence, which is met when no element changes cluster between consecutive iterations.

This simple method requires three user-specified parameters, which might be hard to determine beforehand: the distance metric, the number of clusters k, and the cluster initialization. Euclidean distance is often used to determine the difference between observations, but other metrics are available, such as the L_1 distance [34]. The number of clusters k is the most difficult parameter to estimate, as there is no perfect mathematical criterion. The parameter k is typically determined according to available knowledge about the data or interpreting and evaluating the meaning of several independent partitions realized for different values of k. The initial centroids might influence the resulting clusters, so the initialization is often chosen among several independent partitions that result from sampling k initial centroids among the observations. The influence of initialization, however, generally diminishes with the dimensionality of the dataset [33].

A substantial contribution to the simplicity of the method is given by the required structure of the data. Contrary to observer-operated search, clustering methods rely on the numerical relations between variable values recorded across single observation. It is, thus, unnecessary for the clustering algorithm to preprocess the data and sort the recorded delays for every train/observation. In the method proposed in this paper, k-means clustering is applied to observations of a multidimensional variable, whose size corresponds to the number of timing points of a fixed stopping pattern, where

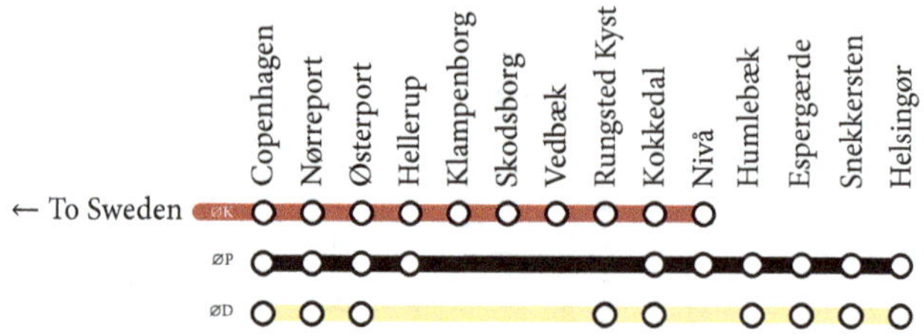

FIGURE 1: DSB services and stopping patterns on the Kystbanen.

the fields contain the delays, or delay changes, respectively, recorded at the individual timing points. Every observation of this multidimensional variable is a vector and represents a single train run.

4. Case Study: The Kystbanen, Copenhagen

The *Kystbanen* (coastline) is a double-tracked railway in the Copenhagen region. It is one of the busiest railway lines in the network of Banedanmark, the Danish infrastructure manager, and it is operated with regional standards, with some international services. It is operated nearly entirely by DSB, the largest Danish railway undertaking, which runs three different service types. The timetable is cyclic, and the services operate different stopping patterns during the day, as illustrated in Figure 1.

(i) The Øresund trains ("ØK") run all day every 20 minutes on a limited section of the coastline, between Copenhagen and Nivå. These trains operate between Denmark and Sweden across the Øresund bridge and stop at every station in Danish territory.

(ii) The regional trains ("ØP") run all day every 20 minutes as well, but they only operate in Denmark and run the whole coastline. These trains skip selected stops between Copenhagen and Nivå.

(iii) Additional trains are operated in the morning and afternoon peak hours. The rush hour trains ("ØD") operate every 20 minutes between Copenhagen and Helsingør, skipping other selected stops.

Fewer trains with modified stopping patterns run at night, so only weekday operation between 4:30 and 20:00 is considered in this study. The sections between Copenhagen and Østerport and between Snekkersten and Helsingør are shared with other services and operators.

In the resulting charts, stations are identified by a code specified by the infrastructure manager. Station codes and names are reported in Table 3.

Banedanmark provided a set of timestamps that state the scheduled and realized times of the trains at every timing point from April to December 2014. The records include information about the operation and about the timing points, such as station name, train ID, train category, scheduled time, and recorded deviation. Banedanmark relies on automatic train detection systems, based on the signaling system components. Typically, the track circuit boundaries do not correspond exactly to the platforms, and an offset

TABLE 3: Station codes and names on the Kystbanen.

Station code	Station name	Distance from KH [km]
KH	København H (Copenhagen central station)	0,0
KN	Nørreport	1,5
KK	Østerport	3,1
HL	Hellerup	7,8
KL	Klampenborg	13,3
SÅ	Skodsborg	18,8
VB	Vedbæk	22,1
RU	Rungsted Kyst	26,1
OK	Kokkedal	29,1
NI	Nivå	32,5
HUM	Humlebæk	36,3
GÆ	Espergærde	40,0
SQ	Snekkersten	42,7
HG	Helsingør	46,2

is generated between the time recorded by the automatic system and the actual time a train arrives at the platform or departs. This is a rather common problem, and it is also reported in the Netherlands [32] and Norway [12]. For the Danish network, a correction factor was calculated by Banedanmark using statistical analyses of GPS positions of train trajectories in collaboration with the main rail operator, DSB. The method and results are described by Richter et al. [35, 36]. Nørreport station is the only underground station on the line, so GPS correction is not available, which is visible as a sawtooth pattern common to all train services in the delay profiles presented below, with a slightly underestimated delay for arrival records at Nørreport and overestimated one for departure records from the same station. Similarly, delay change records are shifted to negative values for arrivals at Nørreport and at Østerport, whereas higher positive values are recorded for delay changes at departures from Nørreport. The bias is systematic and has the same exact effect on all the trains; therefore its influence on clustering can be neglected.

The train timestamps were rearranged by an automatic algorithm to create datasets as described in Section 3, by means of the commercial software SAS 9.4 TS Level 1M4, by SAS Institute Inc., Cary, NC, USA. Observations corresponded to a realized train on a given date, and the fields contained the recorded delay at every station. Data from every station was divided into arrival, departure, and pass-through times, where trains did not stop. Each record is the delay profile or the delay change of a train on a date and represents one observation of the given train. Every variable identifies the station code and the type of timestamp, which can be entrance to the station, I ("Indkørsel"), exit from the station, U ("Udkørsel"), or pass-through station, G ("Gennemkørsel"), which is used where trains do not stop.

The analysis is intended to report delay patterns. Consequently, punctual trains are discarded from the dataset. In Denmark, punctuality measurements are based on a delay threshold of 5 minutes for regional and long-distance trains, such as the Kystbanen. However, for internal management purposes, the infrastructure manager Banedanmark creates a delay report every time a train reaches at least 3 minutes of delay, containing information on the delay cause and on possible other trains hindered. Consequently, only trains with at least one recorded delay greater than or equal to 3 minutes are considered relevant in the present case study. Delay distributions are known to include large shares of trains with short delays, with decreasing frequency for larger delays [7, 8]. Largely unbalanced clusters are a known issue in clustering algorithms and are an object of study to reduce the interference of large clusters [37]. In this case, punctual trains can therefore be considered as a compact cluster derived by prior knowledge, and they can be filtered out from the cluster analysis. The operation of filtering can be considered noise reduction and improves the quality of clustering, as the k-means procedure tends to generate spherical clusters of same radius [38]. According to Marković et al. [31], large delays are influenced by different factors other than smaller delays, which further supports the filtering choice. However, in different contexts, the filtering threshold might be set equal to a different value or not be applied at all.

Given the characteristic high frequency of train services on this line, clustering was operated by stopping patterns rather than by train numbers, so trains were grouped together by direction and service category. Grouping trains with similar characteristics and same stopping patterns increases data availability in the comparison and does not disqualify the result. In fact, such grouping was already proposed by Schittenhelm and Richter [1].

As explained in Section 3, k-means clustering requires choosing the number of clusters k in advance. To set the number of clusters, the k-means algorithm was repeated with different values of k, and the best result was selected using criteria from Jain [33]. The number of clusters k should be large enough to represent different patterns. At the same time, as k increases, the same delay patterns tend to split into more clusters, and k should remain small enough to prevent the generation of duplicate clusters. In detail, for every combination of train category, direction, and clustering variable (delay or delay change), k was set as the highest integer that did not generate duplicate clusters. That is, the univariate distributions of delays, or delay changes, in every cluster should be different from all the other clusters for at least one station. Since k is selected independently for all the mentioned cases, the same set of trains might best be represented by a different number of clusters when the algorithm operates on the delay variables or on the delay change variables. The $L1$ distance was used as a clustering metric between observations, as suggested by Kashima et al. [34].

K-means clustering was performed on the described dataset by the commercial software MATLAB R2017a, by The MathWorks, Inc. Selected results of the method application are reported in Section 4.1, clustering on either delay profiles or delay change profiles.

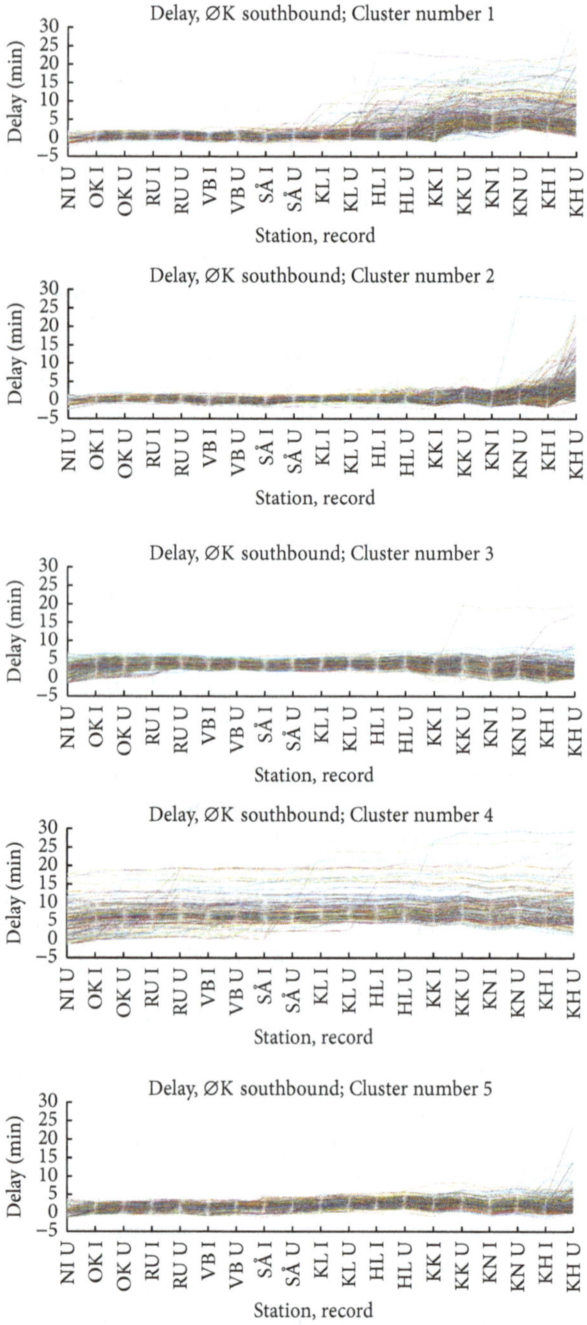

FIGURE 2: Resulting clusters in southbound ØK trains, Nivå–Copenhagen.

4.1. Clustering Results. Figure 2 illustrates the effectiveness of delay profiles clustering on ØK southbound trains, on the delay variables. Note, after a stop at Copenhagen central station, these trains proceed to Sweden. The charts show that similar delay profiles are grouped together with low variance around the average centroid of each cluster, highlighting recurrent patterns. The resulting clusters can be interpreted as follows:

(1) Cluster 1: it includes trains that are punctual on the first section of the line but suffer delays approaching the most congested area of Copenhagen, mainly from Klampenborg and from Østerport.

(2) Cluster 2: it includes trains that are punctual throughout the complete journey, which receive delays leaving from Copenhagen.

(3) Cluster 3: it includes trains that are nearly punctual or anyway within 5 minutes of delay through the complete journey and across Copenhagen central station; from Hellerup, a marginal delay recovery is visible for these trains.

(4) Cluster 4: it includes the most delayed trains, being delayed throughout the whole itinerary or it largest part.

(5) Cluster 5: it includes punctual trains with slightly, but steadily, increasing delay across stations.

Some clusters present outliers, such as Clusters 2 and 3. Even though some delay profiles may appear considerably different from other profiles in the same clusters, these observations were assigned to the cluster with the closest centroid. This means that, in selected cases, the delay profiles are the representation of rather unique events, which may be neglected after more detailed analysis in the composition of the individual clusters.

Individual clusters are characterized through the mean values of the aforementioned measures. The following measures were computed for each train run to characterize the individual clusters:

(i) Average, minimum, and maximum delay across stations

(ii) Range of delays across stations

(iii) Standard deviation of delays recorded across stations

(iv) Initial delay, the delay at first station

(v) Final delay, the delay at the last station

(vi) Overall delay change, difference between final and initial delay—positive values mean the delay has increased from first to last station

(vii) Maximum delay change across stations.

Cluster characteristics are summarized in Table 4.

4.2. Comparison with Percentile-Based Approaches on Delay Profiles. In this section, a comparison is provided between the pooled data and the clusters on the dataset of delay profiles. The same percentile representation of delay profiles is shown, as proposed by Schittenhelm and Richter [1], Peterson [4], and van Oort et al. [22]. These authors represented different percentiles. For the sake of clarity, only the 15th, 50th, and 85th percentiles and the average are displayed in the following diagrams.

Figure 3 shows the distribution of delays of the entire dataset of ØK southbound trains. The only pattern visible is a slight increase in delay towards Copenhagen, more evident for the more delayed trains, represented by the 85th percentile. Even though a large portion of punctual trains

TABLE 4: Characterization of delay profile clusters and southbound ØK trains, Nivå–Copenhagen.

Cluster	Number of observations	Mean average delay [min]	Mean STD of delays [min]	Mean initial delay [min]	Mean final delay [min]	Mean min delay [min]	Mean max delay [min]	Mean delay range [min]	Mean max delay change [min]	Mean overall delay change [min]
(1)	270	2,26	2,78	−0,95	6,14	−1,06	7,72	8,78	4,88	7,09
(2)	418	0,55	1,47	−1,05	4,71	−1,24	5,27	6,52	4,69	5,76
(3)	381	3,09	1,12	1,70	1,80	0,53	4,64	4,11	2,69	0,11
(4)	159	7,65	1,92	4,59	8,03	3,73	10,21	6,47	6,79	3,44
(5)	395	1,92	1,14	−0,28	2,23	−0,47	4,10	4,57	2,25	2,51
Total	1623	2,46	1,57	0,35	3,99	−0,12	5,73	5,85	3,87	3,64

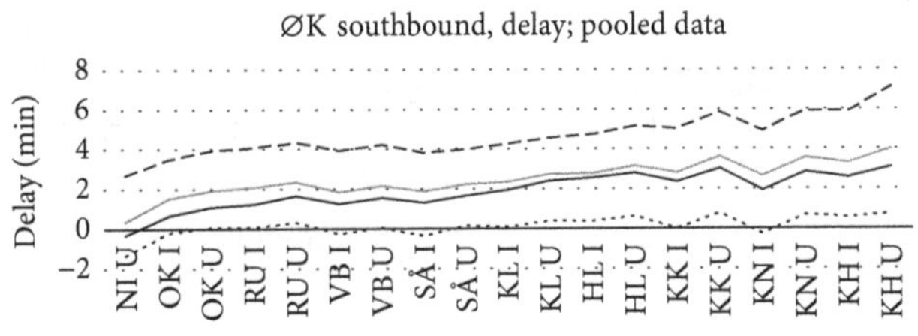

FIGURE 3: Delays recorded for ØK southbound trains. 15th percentile: dotted; median: solid black; 85th percentile: dashed; average: solid gray.

was discarded from the dataset, the residual distribution of delays remains positively skewed, as shown by the average constantly higher than the median value.

The new information revealed by the clustering algorithm is provided in Figure 4. In this figure, the individual internal distributions of delays are compared to the pooled delay distribution from Figure 3. Figure 4 shows, for each cluster, the difference between the cluster statistic at each station and the equivalent pooled statistic from Figure 3.

In Figure 4, the 15th and 85th percentiles and the median line of the internal cluster delay profiles distributions are compared to the distribution of pooled delay profiles. The clusters where the difference of 85th percentile from the pooled dataset is lower than the difference of the 15th percentile have tighter distributions of delay profiles compared to the pooled dataset, increasing the significance of the identified pattern. The local deviation present in the clusters represents the information hidden in the pooled dataset, which is instead brought to light by the clustering.

4.3. Comparison with Percentile-Based Approaches on Delay Change Profiles. In this section, a comparison is provided between the pooled data and the clusters on the dataset of delay change profiles. The same representation of delay change profiles based on the median is shown, as proposed by Schittenhelm and Richter [1, 18], supplemented with the average, that is, the cluster centroid.

Figure 5 shows the delay change profiles of the entire dataset of ØD northbound trains. A generalized positive delay change is visible at the last station. The large changes in delay from location KN I to KK I are linked to the known deviation in the timestamps at Nørreport.

The differences between the pooled median and average delay change profile and the same profiles from individual clusters are represented in Figure 6. In this case, the information gained by clustering is more evident. All the clusters remain similar to the pooled data at most stations, except few stations, where a large difference is recorded in the delay change.

Every cluster is characterized by at least one larger delay change at one station, which would be hidden in the pooled distribution of delay change profiles. Noticeably, the negative effect of different delay patterns overlapping is evident for *KN I* records. All the clusters deviate negatively from the pooled data by around 0,5 minutes, except Cluster 2, which deviates positively by around 1,5 minutes from the pooled profile. This means that the pooled profile was shifted by one single cluster to a central value, hiding both the frequent delay recovery and the specific delay increase from Cluster 2.

4.4. Inference on the Clusters. In this section, results from clustering of delay profiles and delay change profiles are investigated to identify relations with cluster characteristics, using heuristic classification. For the sake of conciseness, only cluster centroids are reported in Figures 7 and 10, and only a sample of the results is reported, which is ØD northbound trains and ØK southbound trains. Figure 7 shows results from clustering delay change profiles for ØD trains to Helsingør.

Clusters can be interpreted as follows:

(1) Cluster 1: regular delay increases at the last three stations, where trains become unpunctual.

(2) Cluster 2: delay increase arriving at the first stop, Nørreport.

(3) Cluster 3: trains are considerably delayed arriving at the final destination, Helsingør.

(4) Cluster 4: trains are without remarkable delay changes: these trains tend to keep the same delay throughout the whole journey.

(5) Cluster 5: specific delay increases at Humlebæk arrival; trains in this cluster show also smaller recovery at Skodsborg arrival, compared to other clusters.

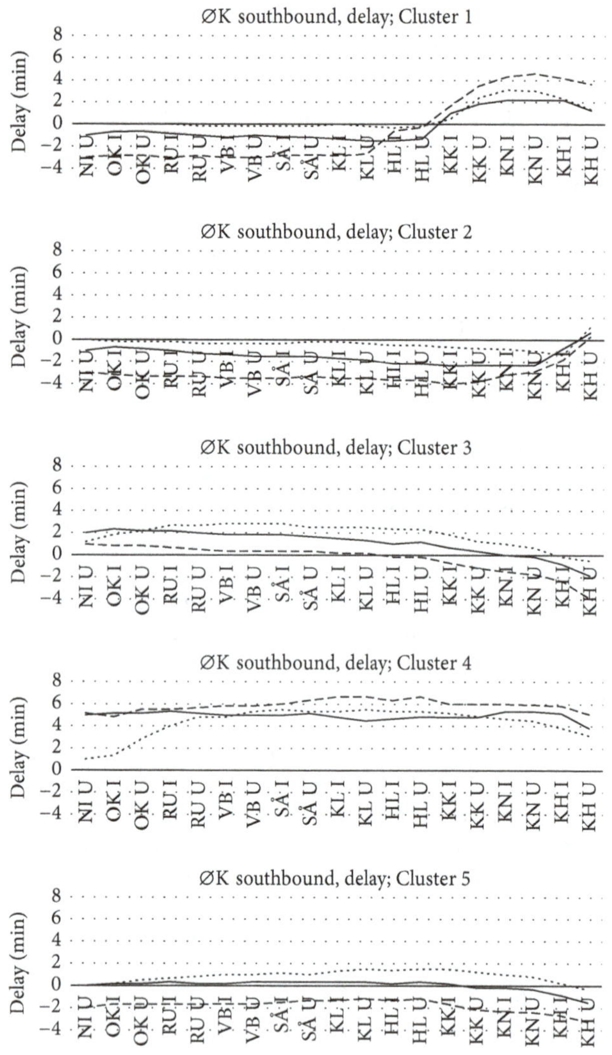

Figure 4: Differences in the distributions of delays recorded for ØK southbound trains. Each cluster's internal distribution is compared to the pooled distribution. 15th percentile: dotted; median: solid; and 85th percentile: dashed.

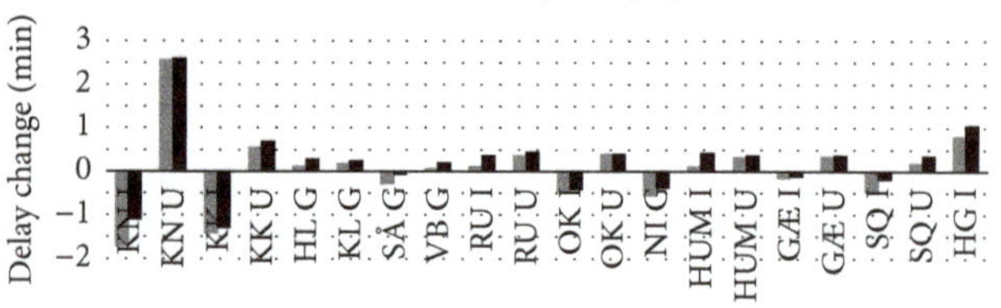

Figure 5: Delay changes recorded for ØD northbound trains. Median in bright shade and average in dark shade.

(6) Cluster 6: these trains accumulate delays passing the stations of Hellerup and Klampenborg; on the other side, compared to other clusters, the average delay increase at final destination Helsingør is smaller.

Inference on the cluster population shows that some patterns are specific for selected train services, identified by their

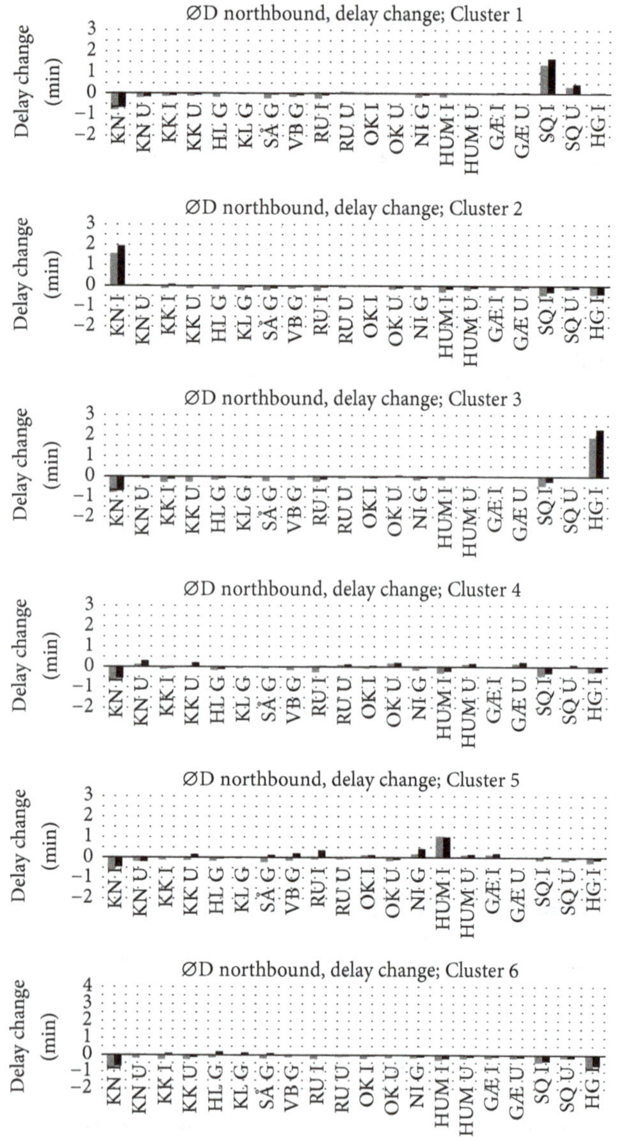

Figure 6: Delay changes recorded for ØD northbound train, by clusters. Median in bright shade and average in dark shade.

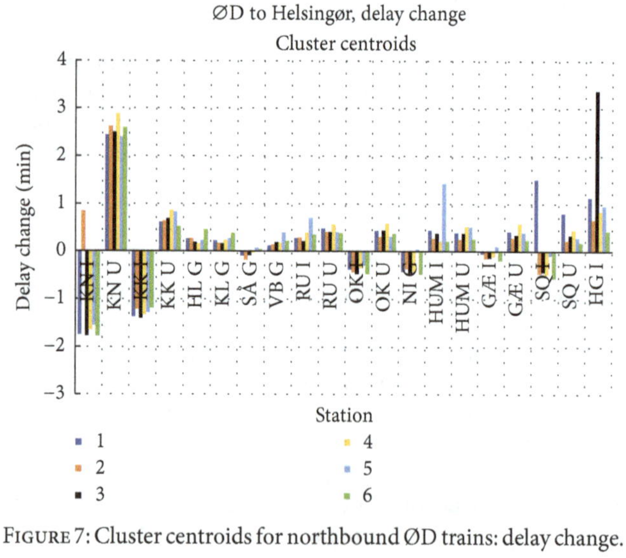

Figure 7: Cluster centroids for northbound ØD trains: delay change.

Time band	Departure time from KH	Train number	Cluster 2	6	4	1	3	5
2 AM Peak	06:18	4413	18%	25%	14%	14%	18%	11%
	06:38	4415	4%	36%	4%	16%	32%	8%
	06:58	4417	26%	33%	7%	19%	4%	11%
	07:18	4419	6%	22%	8%	31%	8%	25%
4 PM Peak	15:18	4467	21%	25%	17%	4%	21%	13%
	15:38	4469	19%	30%	7%	12%	23%	9%
	15:58	4471	44%	16%	16%	4%	8%	12%
	16:18	4473	5%	15%	28%	18%	20%	15%
	16:38	4475	43%	13%	21%	13%	10%	2%
	16:58	4477	20%	15%	39%	9%	2%	15%
	17:18	4479	16%	32%	15%	12%	9%	16%
	17:38	4481	31%	14%	19%	19%	5%	12%
	17:58	4483	46%	14%	14%	6%	10%	10%
		Cluster share	*24%*	*22%*	*17%*	*14%*	*12%*	*12%*

FIGURE 8: Northbound ØD trains. Cluster share by train service ID. The color code compares the individual row's distributions among clusters to the overall distribution among clusters reported in the last row. Clusters sorted by size.

Time	Type	Time band	Cluster 2	6	4	1	3	5
6:20–8:20	Peak AM	2	13%	28%	9%	21%	15%	15%
15:20–18:00	Peak PM	4	27%	20%	19%	12%	11%	11%
		Cluster share	*24%*	*22%*	*17%*	*14%*	*12%*	*12%*

FIGURE 9: Northbound ØD trains. Cluster share by time band. The color code compares the individual row's distributions among clusters to the overall distribution among clusters reported in the last row. Clusters sorted by size.

train number. Figure 8 shows how every train service ID is spread across clusters. In each column, the shade represents the difference between individual percentages and the cluster share, where the brightest colors are associated with the values furthest from the cluster share. Green is positive difference, that is, larger percentages than the cluster share; red is negative difference, that is, smaller percentages than the cluster share.

Delay change profiles in Clusters 1 and 5 represent typical behavior of service 4419, whereas Cluster 2 is considerably more frequent in services 4471, 4473, and 4483. Cluster 3 is more common among services 4415, three times more frequent than the whole population distribution across clusters, 4467, 4469, and 4473, which double the average frequencies. Cluster 4 is typical for service 4477, and, lastly, Cluster 6 represents a large share of services 4417 and, again, 4415. Further investigation on other factors may reveal the causes that rule the train services' cluster membership.

The analysis of Figure 8 shows the existence of a relation between train IDs in a specific time band and cluster membership. This is shown in detail in Figure 9, where cluster membership is aggregated in time bands. The same color coding as Figure 8 is applied.

The timetable is divided into time bands according to the overall service frequency on the line, so that time bands 2 and 4 are the AM and PM peak periods, respectively, when 9 trains/h per direction are operated. Time bands 1, 3, and 5 are the remaining off-peak periods, when ØD trains are not operated, so only 6 trains/h occupy the line in each direction, allowing for larger headway buffers between trains. At the same time, smaller congestion is expected, in off-peak periods, both on the train traffic and on the number of passengers to board or alight at the stations.

In this case, morning peak shows recurrent delay patterns presented by Clusters 1 and 6, whereas patterns represented by Clusters 2 and 4 are rare in this time band. Oppositely, the distribution of trains in the PM peak hour is similar to the overall distribution.

Further inference on the clusters of ØD northbound trains might highlight interferences from other trains. Lokaltog trains run mostly on a network independent of Banedanmark's and share with ØD and ØP trains the line section between Snekkersten and Helsingør. ØD northbound trains are scheduled at a short headway after Lokaltog trains from Snekkersten to Helsingør. The analysis of timestamps from Lokaltog trains on this section and of the realized headways between Lokaltog and ØD northbound trains might suggest that Clusters 1 and 3, which increase the delay near Helsingør, are actually the result of delay propagation from Lokaltog trains to ØD trains.

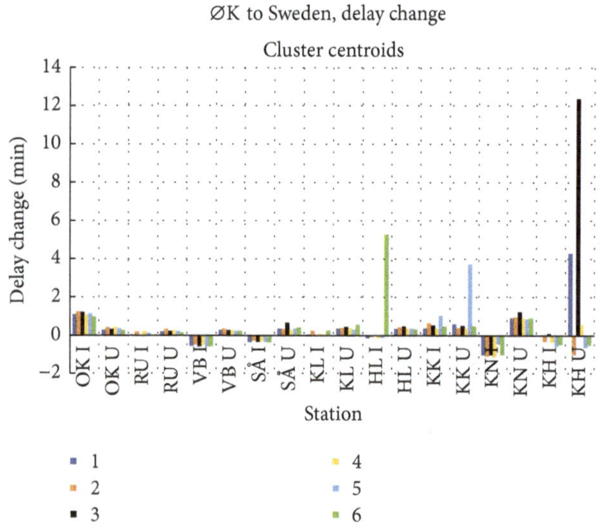

Figure 10: Cluster centroids for southbound ØK trains: delay change.

The clustering results from other service categories, with different stopping patterns, can be related to the time periods of the day. For example, ØK southbound trains are reported in Figures 10 and 11.

Figure 10 represents the centroids of resulting clusters in train category ØK southbound, according to delay change. Besides, the distribution of trains across clusters is summarized in Figure 11, disaggregated by time bands, highlighted in the same color code as Figures 8 and 9. Note that the number of clusters in the delay change profiles of ØK southbound trains is different from the number of clusters in delay profiles. This is not necessarily inconsistent, as the two variables express different aspects of the development of delays. In this case, the cluster share by time band explains the nature of two clusters. In particular, Cluster 1 is considerably more frequent in the PM peak hour, whereas Cluster 2 is more typical of trains in the AM peak hour. This result can be reasonably interpreted as delays generated by passenger congestion. In fact, delay increases in the PM peak hour appear at departures from Copenhagen, where a large number of passengers leave towards Sweden. On the contrary, Cluster 2 represents delays increases collected across stations towards Copenhagen and a delay recovery departing from Copenhagen, where fewer passengers are expected to board. The cluster share for Clusters 3 to 6 is comparable with the overall distribution across different time bands, so these delay patterns cannot directly be associated with time of the day. Further research may reveal factors that rule the cluster membership for these clusters.

More disaggregated analysis of cluster composition according to train number, or service ID, is in accordance with aggregated time bands. This is valuable especially for time band 3, which is the most populated time band according to the timetable. Figure 12 shows that even if the overall distribution of trains in time band 3 across clusters is very similar to the overall distribution, specific train services present different typical delay patterns. In this case, further analysis of train service characteristics should indicate a better disaggregation of train services in a specific time band. The same color code as in tables in Figures 8, 9, and 11 is applied in Figure 12.

Even though recurrent patterns are also clear in the delay profiles dataset, the results could not be explained by the available variables. Further research might identify relations that guide the clustering of delay profiles on this line, such as realized headways, weather conditions, passenger counts, and recorded delay causes.

5. Discussion

The clustering method proposed in this paper finds it strengths in being automatic, unbiased, flexible, and simple. A comparison to methods presented in the literature is provided in this section. Previous approaches [1, 4, 18, 19, 21, 22] extracted information from delay profiles mainly through observation, occasionally combined with multiple univariate statistical analyses and observation ranking. In most studies, the complete dataset was plotted in the form of delay profiles, and the identification of frequent patterns among the observations relied on the observer's ability. Visual inspection is typically affected by subjective interpretation, which can differ across analysts, and suffers from low effectiveness of naked eye to average data represented graphically. In some studies, supporting measures were plotted with the full dataset, such as average profile, median, and selected percentiles to represent the distributions.

The application of these measures as multiple univariate distributions, though, does not catch the interdependencies of delays at different stations and does not provide information about the development of delays along the train journey. The method proposed in this paper allows automatic identification of delay patterns, removing, thus, the influence of subjective interpretation of delay profiles. Furthermore, profile clustering allows the identification of similar delay profiles in the entire pool of records. Note that even though the clustered delay profiles were plotted in this paper, the observation of the profiles did not play a role in the identification of similarities. This exact process is indeed performed by the clustering algorithm, and the results are then plotted for an easier comprehension of the development of delays in the individual clusters. The metrics provided as 15th, 50th, and 85th percentile would be sufficient to describe the distributions within individual clusters and might be used in replacement of the cluster plots.

Compared to big data techniques proposed in the literature for other purposes in analysis of transport operation [11, 23, 24, 32], this method relies on readily available data and does not need detailed knowledge on the infrastructure and occupation of individual blocking sections. It can therefore be scaled to different levels of detail or transferred to other modes of transportation where delay can be measured at fixed points on a given path, such as bus networks or air traffic. It is a very common practice of transport operators to provide live data on delays recorded on their own network, which can be recorded accessing public websites.

Start time	Type	Time band	Cluster 2	Cluster 4	Cluster 1	Cluster 5	Cluster 3	Cluster 6
04:30	Off-peak	1	35%	30%	15%	12%	3%	5%
06:20	Peak AM	2	49%	26%	10%	11%	1%	3%
08:20	Off-peak	3	39%	29%	13%	13%	3%	3%
15:20	Peak PM	4	30%	26%	24%	9%	9%	2%
18:00	Off-peak	5	40%	29%	18%	9%	3%	1%
		Cluster share	*40%*	*28%*	*15%*	*12%*	*3%*	*2%*

FIGURE 11: Southbound ØK trains. Cluster share by time band. The color code compares the individual row's distributions among clusters to the overall distribution among clusters reported in the last row. Clusters sorted by size.

Time band	Departure time from NI	Train number	Cluster 2	Cluster 4	Cluster 1	Cluster 5	Cluster 3	Cluster 6
1	06:01	1314	38%	20%	23%	13%	3%	5%
	06:21	1316	32%	44%	8%	8%	4%	4%
	06:41	1318	33%	33%	10%	14%	5%	5%
2	07:01	1320	59%	12%	12%	15%	0%	2%
	07:21	1322	50%	29%	13%	4%	0%	4%
	07:41	1324	51%	31%	10%	5%	0%	3%
	08:01	1326	41%	32%	14%	11%	2%	2%
	08:21	1328	53%	18%	0%	22%	2%	5%
	08:41	1330	42%	33%	14%	3%	3%	6%
3	09:01	1332	43%	30%	4%	20%	0%	4%
	09:21	1334	39%	32%	16%	13%	0%	0%
	09:41	1336	41%	24%	15%	17%	0%	2%
	10:01	1338	24%	24%	27%	16%	5%	3%
	10:21	1340	31%	39%	19%	3%	0%	8%
	10:41	1342	57%	14%	11%	14%	0%	5%
	11:01	1344	45%	24%	12%	15%	0%	3%
	11:21	1346	40%	47%	10%	3%	0%	0%
	11:41	1348	53%	35%	7%	5%	0%	0%
	12:01	1350	26%	29%	21%	15%	3%	6%
	12:21	1352	47%	32%	8%	8%	0%	5%
	12:41	1354	38%	16%	28%	16%	0%	3%
	13:01	1356	46%	19%	15%	8%	8%	4%
	13:21	1358	50%	31%	15%	0%	4%	0%
	13:41	1360	59%	24%	7%	3%	7%	0%
	14:01	1362	29%	34%	17%	17%	3%	0%
	14:21	1364	44%	32%	12%	4%	8%	0%
	14:41	1366	37%	37%	10%	15%	2%	0%
	15:01	1368	24%	24%	16%	30%	4%	1%
	15:21	1370	43%	25%	6%	14%	10%	2%
	15:41	1372	35%	41%	5%	11%	5%	3%
	16:01	1374	36%	21%	17%	17%	8%	2%
	16:21	1376	36%	33%	14%	6%	6%	6%
4	16:41	1378	37%	27%	22%	7%	5%	2%
	17:01	1380	19%	29%	29%	13%	6%	3%
	17:21	1382	39%	18%	18%	5%	18%	0%
5	17:41	1384	45%	29%	23%	3%	0%	0%
	18:01	1386	18%	43%	30%	8%	3%	0%
	18:21	1388	47%	28%	16%	9%	0%	0%
	18:41	1390	40%	24%	16%	12%	4%	4%
	19:01	1392	58%	15%	13%	15%	0%	0%
	19:21	1394	34%	31%	17%	10%	7%	0%
	19:41	1396	48%	26%	7%	7%	11%	0%
	20:01	1398	31%	38%	22%	6%	0%	3%
		Cluster share	*40%*	*28%*	*15%*	*12%*	*3%*	*2%*

FIGURE 12: Southbound ØK trains. Cluster share by service ID. The color code compares the individual row's distributions among clusters to the overall distribution among clusters reported in the last row. Clusters sorted by size.

Furthermore, the partition of operation into recurrent delay patterns allows inference on individual clusters, which is not possible with association or succession rules [25–27]. These methods do not provide causality connection and can only be used to compare scenarios, for example, before and after delay mitigation countermeasures have been implemented. Results from clustering can be inferred with other mining techniques to identify further connections between specific system factors and delay membership, so that the causes of specific delays can be identified, and the effects of corrective actions can be estimated beforehand.

Alongside flexibility, the strength of this method resides in its simplicity. Unsupervised learning methods, such as clustering, aim at the identification of internal structures of the system. Supervised learning methods, in contrast, attempt to predict results, based on assumed connections in the input. For these reasons, neural networks [28, 30], Bayesian networks [29], and supporting vector regression methods [31] require initial assumptions on the factors that have direct effect on the desired output, which can be cumbersome to identify and could be hidden. The clustering method proposed here does not require initial assumptions, so any recurrent delay pattern can be identified. In particular, the k-means algorithm was selected, being the most common algorithm for partitional clustering. Even though several clustering methods and algorithms exist in the literature, none of them is clearly preferred to the others [33]. It is important to stress the fact that the output of clustering algorithms only suggests hypotheses and that the interpretation of results plays a more relevant role than seeking the best clustering principle. However, further research might improve the method. For example, a different choice of the clustering statistic between observations might be explored. In addition, the choice of the parameter k might be supported by advanced techniques and metrics. In this paper, k was set through statistical analysis of the associated clusters, but further studies might reveal more efficient methods integrated in the clustering algorithm itself. Lastly, the clustering results might depend on the punctuality threshold selected to filter out punctual trains, if applied.

The relations found in inference from resulting clusters can, eventually, be considered and implemented in the mentioned supervised data mining methods. The use of other sources of information can be further investigated, for example, the rolling stock equipment deployed or information on delay causes collected by train dispatchers. The clustering algorithm itself cannot provide information on the causes of delays, but relevant relationships with external variables might be found through the inference on the clusters. The implementation of information recorded by the dispatchers on primary and secondary delays could support the identification of delay propagation. However, previous studies in Europe highlighted the unreliability of such manually recorded data [11, 12]. These procedures are different for each infrastructure manager and should comply with different national regulations. This input should be analyzed in detail before being implemented in the inference on clusters. The timestamps might be integrated with data from other railway undertakings, so that the realized headways could be investigated and included in the cluster inferences. The effects of delay propagation might be thus investigated and the dispatching strategies possibly improved. Passenger counts or boarding/alighting timings could also reveal that specific localized delay increases are linked to passenger exchange and might suggest modifications in the scheduled dwelling times. Useful information from the railway undertakings might include differences between planned and realized train compositions or the use of energy saving strategies. Driving support systems are spreading among train operators to reduce the energy consumption and thus the operating cost, especially for diesel-powered railways. The effects of such systematic patterns in the driving style are, in any event, expected to emerge in the clustering algorithm, especially with more detailed data in the positioning. Further development of this method might expand its application to other industrial processes or other transportation modes. The service timekeeping could be measured at designated checkpoints, to build standard delay profiles and delay change profiles.

6. Conclusions

In this paper, a new method is presented to analyze railway operations, based on big data techniques. Previous studies highlighted the need for tools to analyze railway operation based on data from automatic data collection sources and to automatically detect delay patterns [1]. K-means clustering is here applied to train delay records from automatic train detection systems to identify systematic delays, rearranged in delay profiles and delay change profiles. This method is automatic, unbiased, flexible, and simple.

Both institutions and industry are showing great interest in big data applications [2]. The method described in this paper provides a managerial tool to identify recurrent delay patterns that affect the service reliability. A localized analysis with additional information supports the identification of the causes of individual patterns, so that specific countermeasures can be designed. For example, dispatching strategies might be modified when a structural conflict is detected; the boarding and alighting process might be improved at stations where delay increases recurrently. If the causes of recurrent delays are identified in frequent conflicts, small modifications to the timetable slack might be a solution to reduce delay propagation.

The effectiveness of this approach is demonstrated in an application on a Danish regional railway line. The application shows that it is possible to identify systematic delays at specific stations in a congested area and to identify different delay patterns. Furthermore, delay patterns can be conveniently associated with specific time periods of the day, showing time dependency, reasonably explained by the prevailing passenger flow direction. Specific delay patterns are demonstrated to be characteristic of individual train service IDs, which could depend on other service characteristics, such as structural conflicts with other trains in specific sections of the line, use of specific rolling stock equipment, or connections to other transport services. The implementation of other sources of information might improve the inference on the clusters, such as weather conditions, passenger counts, information from the dispatchers, or rolling stock characteristics.

Further development of this method might improve the selection of the number of clusters, identify new clustering metrics between observations, or integrate additional sources of information to improve the inference on clusters.

Conflicts of Interest

The authors declare that there are no conflicts of interest regarding the publication of this paper.

Acknowledgments

This work was funded by a Dean Grant from the Technical University of Denmark (DTU) and by the Danish Innovation Fund through the IPTOP (Integrated Public Transport Optimisation and Planning) project.

References

[1] B. Schittenhelm and T. Richter, "Railway Timetabling Based on Systematic Follow-up on Realized Railway Operations," in *Proceedings of the Proceedings from the Annual Transport Conference at Aalborg University*, H. Lahrmann, Ed., Aalborg: Traffic Research Group, Aalborg University, 2009.

[2] A. D'agostino, *Big Data in Railways - Common Occurrence Reporting Programme*, 2016, Big Data in Railways - Common Occurrence Reporting Programme.

[3] T. Richter, "En bedre jernbane gennem højere datakvalitet," in *Trafikdage på Aalborgs Universitet. Aalborg: Traffic Research Group*, H. Lahrmann, Ed., Aalborg University, 2008.

[4] A. Peterson, "Towards a robust traffic timetable for the swedish southern mainline," in *Proceedings of the Computers in Railways XIII*, pp. 473–484, WIT Transactions on The Built Environment, New Forest, UK.

[5] A. De Mauro, M. Greco, and M. Grimaldi, "A formal definition of Big Data based on its essential features," *Library Review*, vol. 65, no. 3, pp. 122–135, 2016.

[6] J. Hansen, "Future Railway Development and Performance," in *Proceedings of the Danish Rail Conference*, The Danish Rail Sector Association, Copenhagen, 2015.

[7] M. Carey, "Ex ante heuristic measures of schedule reliability," *Transportation Research Part B: Methodological*, vol. 33, no. 7, pp. 473–494, 1999.

[8] R. M. P. Goverde, G. Hooghiemstra, and H. P. Lopuhaä, *Statistical Analysis of Train Traffic: The Eindhoven Case*, IOS Press, Incorporated, 2001.

[9] H. Sipilä, "Simulation of modified timetables for high speed trains Stockholm – Göteborg," in *Proceedings of the First International Conference on Road and Rail Infrastructure - CETRA2010*, S. Lakušić, Ed., p. 1, University of Zagreb, Opatija, Croatia, 2010.

[10] O. Lindfeldt, "Evaluation of punctuality on a heavily utilised railway line with mixed traffic," in *Proceedings of the 11th International Conference on Computer System Design and Operation in the Railway and Other Transit Systems, COMPRAIL 2008*, pp. 545–553, esp, September 2008.

[11] R. M. P. Goverde and L. Meng, "Advanced monitoring and management information of railway operations," *Journal of Rail Transport Planning and Management*, vol. 1, no. 2, pp. 69–79, 2011.

[12] A. Ø. Sørensen, A. D. Landmark, N. O. E. Olsson, and A. A. Seim, "Method of analysis for delay propagation in a single-track network," *Journal of Rail Transport Planning and Management*, vol. 7, no. 1-2, pp. 77–97, 2017.

[13] A. Lindfeldt and H. Sipilä, "Simulation of freight train operations with departures ahead of schedule," in *Proceedings of the 14th International Conference on Railway Engineering Design and Optimization, COMPRAIL 2014*, pp. 239–249, Italy, June 2014.

[14] N. O. E. Olsson and H. Haugland, "Influencing factors on train punctuality - Results from some Norwegian studies," *Transport Policy*, vol. 11, no. 4, pp. 387–397, 2004.

[15] M. F. Gorman, "Statistical estimation of railroad congestion delay," *Transportation Research Part E: Logistics and Transportation Review*, vol. 45, no. 3, pp. 446–456, 2009.

[16] M.-C. Shih, C. T. Dick, S. L. Sogin, and C. R. L. Barkan, "Comparison of capacity expansion strategies for single-track railway lines with sparse sidings," *Transportation Research Record*, vol. 2448, pp. 53–61, 2014.

[17] A. Lindfeldt, "A study of the performance and utilization of the Swedish railway network," in *Proceedings of the First International Conference on Road and Rail Infrastructure - CETRA2010*, S. Lakušić, Ed., University of Zagreb, Opatija, Croatia, 2010.

[18] T. Richter, "Systematic analyses of train run deviations from the timetable," in *Proceedings of the 12th International Conference on Computer System Design and Operation in the Railways and other Transit Systems, COMPRAIL 2010*, pp. 651–662, China, September 2010.

[19] E. Andersson, A. Peterson, and J. Törnquist Krasemann, "Robustness in Swedish Railway Traffic Timetables," in *Proceedings of the 4th International Seminar on Railway Operations Modelling and Analysis*, S. Ricci, I. A. Hansen, G. L. Longo, D. Pacciarelli, J. Rodriguez, and E. Wendler, Eds., pp. 1–18, Rome, 2011.

[20] B. H. Schittenhelm, "lanning With Timetable Supplements in Railway Timetables," in *Proceedings of the Annual Transport Conference at Aalborg University*, Aalborg, DK: trafikdage, 2011.

[21] E. Andersson, A. Peterson, and J. Törnquist Krasemann, "Introducing a New Quantitative Measure of Railway Timetable Robustness Based on Critical Points," in *Proceedings of the 5th International Seminar on Railway Operations Modelling and Analysis (IAROR): RailCopenhagen2013*, vol. 3, pp. 1–19, Copenhagen, 2013.

[22] N. van Oort, D. Sparing, T. Brands, and R. M. P. Goverde, "Data driven improvements in public transport: the Dutch example," *Public Transport*, vol. 7, no. 3, pp. 369–389, 2015.

[23] I. A. Hansen, R. M. P. Goverde, and D. J. Van Der Meer, "Online train delay recognition and running time prediction," in *Proceedings of the 13th International IEEE Conference on Intelligent Transportation Systems, ITSC 2010*, pp. 1783–1788, Portugal, September 2010.

[24] P. Kecman and R. M. P. Goverde, "Process mining of train describer event data and automatic conflict identification," in *Computers in Railways XIII, WIT Transactions on the Built Environment*, C. A. Brebbia, N. Tomii, P. Tzieropoulos, and J. M. Mera, Eds., vol. 127, pp. 227–238, 2012.

[25] B. Cule, B. Goethals, S. Tassenoy, and S. Verboven, "Mining train delays," *Lecture Notes in Computer Science (including*

subseries Lecture Notes in Artificial Intelligence and Lecture Notes in Bioinformatics): Preface, vol. 7014, pp. 113–124, 2011.

[26] J. Wallander and M. Mäkitalo, "Data mining in rail transport delay chain analysis," *International Journal of Shipping and Transport Logistics*, vol. 4, no. 3, pp. 269–285, 2012.

[27] H. Yabuki, T. Ageishi, and N. Tomii, "Mining the Cause of Delays in Urban Railways based on Association Rules," in *CASPT2015*, pp. 1–16, 2015.

[28] G. Malavasi and S. Ricci, "Simulation of stochastic elements in railway systems using self-learning processes," *European Journal of Operational Research*, vol. 131, no. 2, pp. 262–272, 2001.

[29] P. Kecman, F. Corman, A. Peterson, and M. Joborn, "Stochastic prediction of train delays in real-time using Bayesian networks," in *Proceedings of the in. Proceedings of the 13th Conference on Advanced Systems in Public Transport (CASPT*, p. 18, Rotterdam (Netherlands), 2015.

[30] X. Chapuis, "Arrival Time Prediction Using Neural Networks," in *7th International Conference on Railway Operations Modelling and Analysis*, N. Tomii, I. A. Hansen, J. Rodriguez, P. Pellegrini, S. Dauzère-Pérès, and D. De Almeida, Eds., pp. 1500–1510, International Association of Railway Operations Research, Lille, Fra, 2017.

[31] N. Marković, S. Milinković, K. S. Tikhonov, and P. Schonfeld, "Analyzing passenger train arrival delays with support vector regression," *Transportation Research Part C: Emerging Technologies*, vol. 56, pp. 251–262, 2015.

[32] P. Kecman and R. M. P. Goverde, "Predictive modelling of running and dwell times in railway traffic," *Public Transport*, vol. 7, no. 3, pp. 295–319, 2015.

[33] A. K. Jain, "Data clustering: 50 years beyond K-means," *Pattern Recognition Letters*, vol. 31, no. 8, pp. 651–666, 2010.

[34] H. Kashima, J. Hu, B. Ray, and M. Singh, "K-means clustering of proportional data using L1 distance," in *Proceedings of the 2008 19th International Conference on Pattern Recognition (ICPR)*, pp. 1–4, Tampa, FL, USA, December 2008.

[35] T. Richter, A. Landex, and J. L. E. Andersen, "Precise and accurate train run data: Approximation of actual arrival and departure times," in *Proceedings of the Precise and accurate train run data: Approximation of actual arrival and departure times*, International Association of Railways, Sydney, Australia, 2013.

[36] T. Richter, *Data aggregation for detailed analysis of train delays*, C. A. Brebbia, N. Tomii, J. M. Mera, B. Ning, and P. Tzieropoulos, Eds., WIT Transactions on the Built Environment. WIT-Press, 2012.

[37] J. Wu, *Advances in K-means Clustering*, Springer, Berlin, Heidelberg, 2012.

[38] T. Hastie, R. Tibshirani, and J. Friedman, *The Elements of Statistical Learning*, Springer, New York, NY, USA, 2008.

6

Modelling Rail-Based Park and Ride with Environmental Constraints in a Multimodal Transport Network

Xinyuan Chen and Inhi Kim

Monash Institute of Transport Studies, Australia

Correspondence should be addressed to Inhi Kim; inhi.kim@monash.edu

Guest Editor: David Z. W. Wang

Environmental sustainability is a significant aspect in the sustainable development of modern urban cities, especially in the road transport system. As traffic demands increase, public transport requires more promotion to accommodate the increasing travel demands while maintaining the environmental quality. Public transport, however, is less attractive in vast suburb areas mainly due to its longer travel distance and waiting time. Therefore, this paper proposes a rail-based Park-and-Ride (RPR) scheme to promote public transport in the multimodal transport network. To remedy the heterogeneous distribution of vehicle pollutants in the network, regulations in environmental sensitive districts are required and studied in this paper. To quantitatively evaluate and analyse this joint RPR and environmental regulation strategy in multimodal transport systems, this paper develops an environmental constrained combined modal split and traffic assignment (EC-CMSTA) model. The proposed formulation adopts the concept of fix-point to reformulate the nonlinear complementarity conditions associated with the combined modal split and user equilibrium conditions, which is subsequently incorporated into a VI formulated nonlinear complementarity conditions associated with environmental constraints. The proposed VI formulation can handle a general constraint structure, which enhances the modelling adaptability and flexibility. The strictly monotone and Lipschitz continuity properties of this model are rigorously proved, giving rise to efficient algorithms for the model. A customized projection based self-adaptive gradient projection (SAGP) algorithm is then developed. Numerical studies demonstrate that the EC-MSTA model could enhance the behavioural modelling of network users' travel decisions and assist in quantitatively evaluating the effectiveness of RPR schemes and environmental regulations.

1. Introduction

Environmental protection is an important component in the sustainable development of modern urban cities and closely related to the transport system. According to Hong Kong Air Pollutant Emission Inventory (2016), the amount of CO generated in the year of 2016 in Hong Kong is 58600 tones and among them, road transport contributes to 31500 tones and shares 54% of the CO emissions. Also, the US's environmental protection agency reported motorized vehicles in the United States contribute to 86% of CO emissions, 62% of NO_x emissions, 12% of PM2.5 emissions, and 7% of SO_2 emissions [1]. Kitthamkesorn et al. [2] pointed out that only about 20% of the world's town residents enjoy good enough air quality as measured by the levels of emissions and over 600 million people in urban areas worldwide were exposed to dangerous levels of traffic-generated emissions from congestion. In addition, these emissions have not only adverse effects on human health but also a source of greenhouse gas that contributes to global warming [3].

With these adverse effects, the promotion of alternative green transport modes is thus pressing for the governments and transportation authorities to accommodate the growing travel demand while maintaining the environmental quality. Public transport aiming to satisfy the travel demand of urban commuters safely, quickly, and affordably is therefore a significant alternative to private transport. However, in vast suburb areas which are far away from an urban centre, public transport becomes less attractive, mainly due to the unreliable time table, tortuous bus itinerary, much lower travel speed, and lack of door to door services [4]. Large-scale deployments of public transport system in the remote areas are hardly seen as an economical available strategy. Park and Ride (P&R), therefore, becomes an important strategy to promote public transport since it enables travellers to drive when the road is not congested and to change a mode to transit accessing to CBD without traffic congestion. Since its first implementation in 1930s, P&R has been widely recognized as a significant traffic management strategy and becomes an indispensable part of majority modern urban cities.

P&R can be categorized into bus-based P&R (BPR) and rail-based P&R (RPR) based on its ride component [5]. In a metropolis with advanced rail system, rail-based P&R is superior to bus-based P&R, due to the faster speed, higher level of safety, comfortability, and its advantages in environmental protection (zero-emission).

Despite the fact that environmental deterioration could somehow be alleviated by promoting public transport through RPR schemes, the environmental requirements are usually not strictly met in certain districts mainly due to the heterogeneous distribution of pollution in the network [3]. To remedy this limitation, specific environmental regulations, therefore, are required to certain district, e.g., the city business district (CBD) with many tall buildings, to protect the environmental quality in this area.

Therefore, we aim to use RPR services to assist mode shift to reduce the overall vehicle travel demands in the road network and then apply area-based regulation scheme to deal with the environmental deterioration caused by heterogeneity of traffic flows. The environmental protection approaches are integrated as a combined strategy in this paper. Previous studies revealed that a systematic study is necessary to properly understand and evaluate the effectiveness of the environmental regulations, as well as the RPR scheme. Thus, this paper focuses on the modelling and solution algorithm development of the proposed environmental protection strategy.

2. Literature Review

Road network equilibrium approaches are commonly preferred for estimating vehicle emissions and noises [3], because the distribution of vehicle pollution heavily relies on the states of traffic flows [6]. Other than the overall pollution level, the spatial distribution level should be controlled since it is heterogeneous.

One line of the research developed models by setting environmental protection as objectives [3], modelling environmental considerations as an objective function. Rilett and Benedek [7] proposed the concept of emission optimum (EO) and emission-oriented user equilibrium (EOUE), in which the former one assumes travellers make route choices with the objective of minimizing the total emissions or noises, and the latter includes the toll charges to represent the impact on the environment. Later, Benedek and Rilett [8] made extension by modelling the assignment of vehicles based on equity which considers the objectives of neighbouring residents, rather than the system operator's or drivers' objectives. Sugawara and Niemeier [9] used the emissions-optimized trip assignment model to estimate the maximum carbon monoxide emission reduction under varying congestion levels.

Although the overall environmental quality could somehow be improved by taking the environmental considerations as the objective, the pollution level is not always strictly met on each link and zone in the road network. Therefore, another line of the research focuses on modelling the emission as environmental constraints. Chen, Zhou and Ryu [10] developed a traffic assignment model by considering various physical and environment restrictions as side constraints, which is formulated as a VI problem and solved by a predictor-corrector (PC) algorithm. Li et al. [11] developed a toll design model by simultaneously considering congestion and environmental externalities. Recently, Xu, Chen, and Cheng [1] reformulated the environmental constrained traffic assignment model through a smooth gap function to enhance the modelling adaptability and flexibility.

As a rational extension of traffic assignment problem, Nagurney, Ramanujam, and Dhanda [12] extended the environmental constrained traffic assignment problem to the case of multimodal traffic network. However, the model presented in this paper simply follow the deterministic user equilibrium (DUE) principle for both modal split and traffic assignment. In order to evaluate the environmental regulation strategy in the multimodal transport network, a better behavioural model is needed. Therefore, the combined modal split and traffic assignment model (CMSTA) is adopted in this paper. In the literature, the CMSTA problems are widely adopted to evaluate the P&R schemes. Fernandez et al. [13] conducted an initial work to model P&R in deterministic case. Later, this work was extended to the case of asymmetric link travel time functions by García and Marín [14] and to the logit-based stochastic model by Li et al. [15] and Lam et al. [16]. Recently, Liu and Meng [5] provided a formulation for the bus-based P&R in a network with the congestion charge. Pineda et al. [17] and Song, He, and Zhang [18] developed an integrated P&R facilities and transit service optimization model. Kitthamkesorn et al. [2] and Kitthamkesorn and Chen [19] considered the influence of "go green" transport modes on environmental protection by taking use of recent advances in logit model. Liu et al. [20] proposed a new concept of P&R, remote park and ride, and developed an exact solution method for the optimal design of parking facilities, following the nonlinear valid inequality technique [21, 22]. Some other studies focus on the P&R schemes in a linear corridor, e.g., Liu et al. [23], Wang and Du [24], and Du and Wang [25]. However, these studies mainly focus on congestion alleviation. As abovementioned, the environmental requirements in certain areas are not considered. Therefore, a more general modelling framework that incorporates environmental requirements is needed. To this end, this paper aims to build an environmental constrained combined modal split and traffic assignment (EC-CMSTA) model to better evaluate and analyse environmental sustainable transportation development strategies.

This paper distinguishes itself from previous studies in threefold aspects. First, we propose EC-CMSTA, CMSTA model with side constraints, which could handle the general environmental constraint structures, including linear or nonlinear, link based, or area-based constraints. Also, the general link/route cost structure is considered here. Second, by incorporating general side constraints into CMSTA, solving is much more complex compared to the traditional side constrained unimodal traffic equilibrium problems. To explicitly consider the general environmental constraints and make use of existing efficient algorithms, we propose a VI formulation by adapting the concept of fix-point. This VI formulation is strictly monotone and Lipschitz continuous, permitting a number of existing efficient algorithms for its solution [10, 26]. Third, even though the solution methods for the traffic assignment problems with side constraints have been well studied with numerous achievements [1, 10, 12, 27], solutions for the multimodal traffic equilibrium with side constraints are relatively limited. Recently, Meng, Liu and Wang [28] proposed a CA incorporated self-adaptive PC algorithms to solve asymmetric stochastic user equilibrium (SUE) problem with elastic demand and link capacity constraints. Inspired by Meng, Liu, and Wang [28], we extend this approach to multimodal traffic equilibrium problem and develop a customized projection based algorithm with a self-adaptive step size scheme which permits the convergence under a much milder assumption of the underlying mapping.

The remainder of this paper is organized as follows: we present assumptions and an EC-CMSTA model for the environmental regulations in the multimodal transport system with RPR scheme in the next section. A general VI formulation and a projection based solution algorithm are provided in Sections 3 and 4. Section 5 provides two numerical studies to validate the proposed methodology. Finally, Section 6 concludes this study.

3. Problem Description

Considering a specific environmental regulation strategy, to analyse its impact in the multimodal transport system, we build a model for the equilibrium flows following environmental requirements.

Given a multimodal transportation network represented by a strongly connected topology graph $G(N, A)$ where N is the set of nodes and A is the set of links. The origins, destinations, and RPR facilities are scattered in the network and represented by certain nodes, $n \in N$. The set of links A can be further categorized into three subsets: set of autolinks A^a; set of rail links A^r; and set of P&R links A^p; $A = A^a \cup A^r \cup A^p$.

3.1. Modelling Environmental Requirements. As abovementioned, the environmental requirements are not strictly met in the road network. Thus, it is necessary to explicit considering environmental protection thresholds to prevent environmental deterioration in certain zones. In the literature, the environmental constraints can be expressed into a general form:

$$g_e(\mathbf{v}) \leq 0, \quad \forall e \in E \tag{1}$$

where e is a component of E and E represents a set of environmental sensitivity elements. In the case of link based environmental protection strategy, e represents a link which is assumed in this paper. $\mathbf{v}$ is the vector of link flows; $g_e(\mathbf{v})$ is a function to measure the amount of vehicle pollution and environmental thresholds. Equations (1) ensure the amount of pollution no larger than predetermined thresholds. Note that (1) are general formulation that could measure different types of vehicle pollutants [1]. Herein, we make assumption as follows.

Assumption 1. The environmental threshold measure function $g_e(\mathbf{v}), \forall e \in E$, is a continuously differentiable and strictly monotone increasing function of the link flows $\mathbf{v}$.

Suppose the total number of environmentally sensitive areas in the network is I. The inbound flows of these areas should be restricted according to environmental protection strategies. All the links in the area $i \in I$ are grouped into set E_i, where $E_i \in A$. Then, the specific link based environmental constraints in these areas could be expressed as

$$g_a(\mathbf{v}) \cdot v_a \leq \overline{g}_a, \quad \forall a \in E_i, \ i \in I \tag{2}$$

where $g_a(\mathbf{v})$ is the amount of vehicle pollutants per vehicle on link a and $\overline{g}_a$ is the environmental protection threshold on link a imposed by the operators.

3.2. Network Assessment. Due to the fact that the travel decisions of a network user consist of modal choice and route choice, a combined modal split and traffic assignment model is needed. The travel impedance on an auto or rail link is clearly influenced by its flows. For the RPR links, each of them always connects an autolink and a rail link, so that the RPR flows could transfer. The total travel impedance on a RPR link includes the time spent on parking, waiting/boarding the dedicated bus as well as the bus in-vehicle travel time, which is clearly flow-dependent. We then make the following assumption.

Assumption 2. The link travel time $t_a(\mathbf{v}), a \in A$ is a continuously differentiable, monotone increasing of the link flow $\mathbf{v}$.

Let K_m^{od} denote the set of paths under m; $f_k^{m,od}$ denotes the flow on path $k \in K_m^{od}$ for mode m; q_m^{od} denotes the demand of mode m; and $\overline{q}^{od}$ denotes the total travel demand between OD pair (o, d). We have the following flow conservation conditions:

$$\sum_{m \in M_{od}} q_m^{od} = \overline{q}^{od}, \quad \forall od \in W \tag{3}$$

$$q_m^{od} \geq 0, \quad \forall m \in M_{od}, \ od \in W \tag{4}$$

$$\sum_{k \in K_m^{od}} f_k^{m,od} = q_m^{od}, \quad \forall m \in M_{od}, \ od \in W \tag{5}$$

$$f_k^{m,od} \geq 0, \quad \forall k \in K_m^{od}, \ m \in M_{od}, \ od \in W \tag{6}$$

$$v_a = \sum_{od \in W} \sum_{m \in M_{od}} \sum_{k \in K_m^{od}} f_k^{m,od} \delta_{a,k}^{m,od}, \quad \forall a \in A \tag{7}$$

where (3) and (5) define demand/flow conservation conditions. Equations (4) and (6) are the nonnegativities. Equations (7) define the link-path flow relationship. Constraints (1), (3)-(7) define the feasible set of Ω of $(\mathbf{v}, \mathbf{f}, \mathbf{q})$ where $\mathbf{v} := (v_a, a \in A)$ is the vector of link flows; $\mathbf{f} := (f_k^{em,od}, k \in K_m^{od}, m \in M^{od}, od \in W)$ is the vector of path flows; and $\mathbf{q} := (q_m^{od}, m \in M^{od}, od \in W)$ is the vector of modal demands.

Let W denote the set of OD pairs; $\overline{q}^{od}$ denote the total travel demand for $(o,d) \in W$; and M_{od} denote the set of modes between OD pair (o,d). The multinomial logit (MNL) model is adopted to analyse the modal split pattern. The probability of choosing any mode m between an OD pair (o,d) equals

$$p_m^{od} = \frac{\exp\left(-\theta \cdot \tau_m^{od}\right)}{\sum_{i \in M_{od}} \exp\left(-\theta \cdot \tau_i^{od}\right)}, \quad \forall m \in M,\ od \in W \tag{8}$$

where $\theta \in [0, +\infty)$ is the dispersion parameter and τ_m^{od} is the deterministic component of the utility of mode m between OD pair (o,d), which can be taken as the equilibrium path travel time of mode m. The satisfaction of MNL, $S_{od}(\tau_m^{od})$, equals $\tau_m^{od} + (1/\theta)\ln(q_m^{od})$, which is concave function [28].

The user equilibrium path flow solution is then formulated by the following conditions:

$$f_k^{m,od}\left(c_k^{m,od} - \varphi_m^{od}\right) = 0, \quad \forall k \in K_m^{od},\ m \in M_{od},\ od \in W \tag{9}$$

$$f_k^{m,od} \geq 0, \quad \left(c_k^{m,od} - \varphi_m^{od}\right) \geq 0, \quad \forall k \in K_m^{od},\ m \in M_{od},\ od \in W \tag{10}$$

where φ_m^{od} is the Lagrange multiplier associated with the constraints (5), which represents the travel impedance of using mode m. At equilibrium, the travel time $c_k^{m,od}$ of a used path $k \in K_m^{od}$ equals φ_m^{od}. All commuters under mode m have identical path travel time, and no traveller can unilaterally reduce his/her travel time by changing routes.

4. Variational Inequality Formulation

In this section, we propose two equivalent VI formulations for the proposed EC-CMSTA model. In order to strengthen the flexibility and adaptability of the proposed model, we further make the following two assumptions.

Assumption 3. The link travel time function is asymmetric, i.e., t_a is not only affected by its own flow v_a, but also flows on some other links, which is expressed as $t_a = t_a(\mathbf{v})$.

Assumption 4. The path travel time c_k is nonadditive to its links, i.e., $c_k \neq \sum_a t_a \delta_{a,k}$, where $\delta_{a,k} = 1$ if path k uses link $a \in A$ and 0 otherwise.

Then, the VI mode for EC-CMSTA is formulated as follows:

$$\sum_{od \in W} \sum_{m \in M_{od}} \sum_{k \in K_m^{od}} c_k^{m,od}\left(\mathbf{f}^*, \mathbf{q}^*\right)\left(f_k^{m,od} - f_k^{m,od*}\right) + \sum_{od \in W} \sum_{m \in M_{od}} \tilde{c}_m^{od}\left(\mathbf{f}^*, \mathbf{q}^*\right)\left(q_m^{od} - q_m^{od*}\right) \geq 0, \quad \forall\left(\mathbf{f}^*, \mathbf{q}^*\right) \in \Omega \tag{11}$$

where $c_k^{m,od}$ is the travel time on route k, where $k \in K_m^{od}$. $\tilde{c}_m^{od} = (1/\theta)\ln(q_m^{od})$ represents the entropy term in travel cost function of mode m, $m \in M_{od}$; $f_k^{m,od*}$ and q_m^{od*} are the optimal solution of this problem; $\mathbf{f}^*$ and $\mathbf{q}^*$ are the vector of optimal solution.

Let $\mathbf{c}$ be the vector of $c_k^{m,od}$ and $\tilde{\boldsymbol{\tau}}$ be the vector of $\tilde{c}_m^{od}$, i.e., $\mathbf{c} := (c_k^{m,od},\ k \in K_m^{od},\ m \in M_{od},\ od \in W)^T$ and $\tilde{\boldsymbol{\tau}} := (\tilde{c}_m^{od},\ m \in M_{od},\ od \in W)^T$. The VI formulation can be simplified as

$$\mathbf{c}\left(\mathbf{f}^*, \mathbf{q}^*\right)^T\left(\mathbf{f} - \mathbf{f}^*\right) + \tilde{\boldsymbol{\tau}}\left(\mathbf{f}^*, \mathbf{q}^*\right)^T\left(\mathbf{q} - \mathbf{q}^*\right) \geq 0, \quad \forall\left(\mathbf{f}^*, \mathbf{q}^*\right) \in \Omega \tag{12}$$

Note that the inclusion of nonlinear environmental constraints makes the feasible set more complex in general. Existing solution algorithms for CMSTA cannot be directly adopted to solve it. Most of existing studies focus on solving the simplified linear link capacity constrained traffic equilibrium problems by a penalty based algorithm. However, due to the complexity of EC-CMSTA (simultaneous achieving equilibrium for both route flow and modal demand), reformulating it into an integrated formulation or directly solving it by a penalty/gradient based algorithm is less effective. Therefore, this paper develops an alternative approach to reformulate the EC-CMSTA which permits existing solution algorithms for its solution.

4.1. Alternative VI Formulation. Daganzo [29] built a fix-point model in terms of link flows for the asymmetric SUE problem with fixed demand and showed availability of the MSA, which was extended by Cantarella [30] to the case of asymmetric SUE problem with elastic demand. Recently, Meng, Liu and Wang [28] further made extension by incorporating link capacity constraints. This problem was formulated as a VI problem by incorporating the fix-point formulated asymmetric SUE problem with elastic demand. Following the same technique, we propose a fixed point reformulated CMSTA model which is expressed as follows:

$$v_a = \sum_{od \in W} \overline{q}^{od} \sum_{m \in M_{od}} p_m^{od}\left(\mathbf{f}(\mathbf{v})\right) \times P_a^{m,od}\left(\mathbf{f}(\mathbf{v})\right), \quad \forall a \in A \tag{13}$$

where $\mathbf{v}$ is the equilibrium link flow solution, p_m^{od} is the probability of mode choice defined in (8), and $P_a^{m,od}$ is the link usage probability defined by

$$P_a^{m,od}\left(\mathbf{f}(\mathbf{v})\right) = \sum_{k \in K_m^{od}} p_k^{m,od}\left(\mathbf{f}(\mathbf{v})\right) \delta_{k,a}^{m,od}, \quad \forall a \in A \tag{14}$$

and $p_k^{m,od}$ is the probability of path k perceived as the shortest one among the path set K_m^{od}. In the deterministic case, $p_k^{m,od}$ equals one if path k is the shortest path among alternatives and zero otherwise.

A set of generalized SUE conditions with elastic demand and capacity constraints was proposed by Meng, Liu, and Wang [28] for the optimal solution of SUE problem with elastic demand, asymmetric link travel time function and link capacity constraints. These conditions can be further extended to the multimodal transport system which is the problem addressed in this paper. Let $\mathbf{v}$ denote the equilibrium link flow pattern of the addressed problem if and only if there exists a vector of Lagrangian multipliers corresponding to environmental constraints, denoted by $\mathbf{u}^* = (u_e^*, e \in E)^T$.

$$v_a(\mathbf{u}) = \sum_{od \in W} \sum_{m \in M_{od}} q_m^{od}(\mathbf{f}(\mathbf{v}), \boldsymbol{\lambda}(\mathbf{u}^*)) \times P_a^{m,od}(\mathbf{f}(\mathbf{v}), \boldsymbol{\lambda}(\mathbf{u}^*)), \quad \forall a \in A \tag{15}$$

$$-g_e(\mathbf{v}) \geq 0, \quad \forall e \in E \tag{16}$$

$$u_e^* \cdot (-g_e(\mathbf{v})) = 0, \quad \forall e \in E \tag{17}$$

$$u_e^* \geq 0, \quad \forall e \in E \tag{18}$$

where $\boldsymbol{\lambda}(\mathbf{u}^*)$ is defined by

$$\boldsymbol{\lambda}(\mathbf{u}^*) = \left(\lambda_k^{m,od} = \sum_{e \in E} u_e^* \zeta_{e,k}^{m,od}, \ \forall k \in K_m^{od}, \ m \in M_{od}, \ od \in W \right)^T \tag{19}$$

and $\zeta_{e,k}^{m,od} = 1$ if path $k \in K_m^{od}$ goes through $e \in E$ and $\zeta_{e,k}^{m,od} = 0$, otherwise. Based on the optimal Lagrangian multiplier $\mathbf{u}^*$, the EC-CMSTA is identical to the CMSTA in terms of generalized link travel time functions.

$$\hat{t}_a(\mathbf{v}, \mathbf{u}^*) = \begin{cases} t_a(\mathbf{v}) + u_a^*, & \forall a \in E \\ t_a(\mathbf{v}), & \forall a \in A/E \end{cases} \tag{20}$$

Equations (16)-(18) imply that if there is no environmental constraint on any link $a \in E$, $u_a^* = 0$.

Note that the complementary slackness conditions (see (16)-(18)) could be regarded as a nonlinear complementary problem (NCP) of the vector of Lagrangian multipliers $\mathbf{u}^*$. Due to the feasible set of $\mathbf{u}$ is defined on the whole nonnegative orthant, this NCP problem can be equivalently formulated as a VI model. Hence, the alternative formulation of VI (11) is proposed as follows:

$$\mathbf{g}(\mathbf{u}^*)^T (\mathbf{u} - \mathbf{u}^*) \geq 0, \quad \forall \mathbf{u} \in \mathfrak{R}_+^{|E|} \tag{21}$$

where $\mathfrak{R}_+^{|E|} = (u_e \geq 0, e \in E)^T$ denotes the feasible set. $\mathbf{g}(\mathbf{u})$ is the $|E|$ dimensional vector functions, which measures vehicle pollution:

$$\mathbf{g}(\mathbf{u}) = (-g_e(\mathbf{v}(\mathbf{u})), \ \forall e \in E)^T : \mathfrak{R}_+^{|E|} \longrightarrow \mathfrak{R}^{|E|} \tag{22}$$

$\mathbf{g}(\mathbf{u}^*)$ is thus the optimal solution of the VI problem (21). Herein, $\mathbf{v}(\mathbf{u})$ is a link flow solution for the CMSTA with the generalized link travel time functions (20) (without environmental constraints). In other words, $\mathbf{v}(\mathbf{u})$ is a solution of the fix-point model with the Lagrangian multiplier $\mathbf{u}$:

$$v_a(\mathbf{u}) = \sum_{od \in W} \sum_{m \in M_{od}} q_m^{od}(\hat{\mathbf{t}}(\mathbf{v}, \mathbf{u})) \times P_a^{m,od}(\hat{\mathbf{t}}(\mathbf{v}, \mathbf{u})), \quad \forall a \in A \tag{23}$$

where $\hat{\mathbf{t}}(\mathbf{v}, \mathbf{u}) = (\hat{t}_a(\mathbf{v}, \mathbf{u}), \ \forall a \in A)$ denotes the vector of generalized link travel time functions.

Note that the feasible set $\mathfrak{R}_+^{|E|}$ of $VI(\mathbf{g}, \mathfrak{R}_+^{|E|})$ is unbounded, which makes it difficult to shown the existence of a solution to VI model $VI(\mathbf{g}, \mathfrak{R}_+^{|E|})$. Fortunately, according to Meng, Liu and Wang [28], one can always find an upper bound $\overline{M}$ for $u_a, \forall a \in E$. Therefore, we define a subset $\overline{\Omega}$ of $\mathfrak{R}_+^{|E|}$ which is bounded by $\overline{M}$. The restricted VI model is subsequently formulated as

$$\mathbf{g}(\mathbf{u}^*)^T (\mathbf{u} - \mathbf{u}^*) \geq 0, \quad \forall \mathbf{u} \in \overline{\Omega} \tag{24}$$

The feasible set $\overline{\Omega}$ is a nonempty, closed, and convex set. To show the existence of solution to $VI(\mathbf{g}, \overline{\Omega})$ as well as the convergence of its solution algorithm, the following properties of vector function $\mathbf{g}(\mathbf{u})$ are proposed. Rigorous proofs of Propositions 5 and 6 are given in the Appendices A and B, respectively.

Proposition 5. *Vector function* $\mathbf{g}(\mathbf{u})$ *is strictly monotone on* $\overline{\Omega}$*; namely,*

$$(\mathbf{g}(\mathbf{u}') - \mathbf{g}(\mathbf{u}''))^T (\mathbf{u}' - \mathbf{u}'') > 0, \quad \forall \mathbf{u}', \mathbf{u}'' \in \overline{\Omega} \tag{25}$$

Proposition 6. *Vector function* $\mathbf{g}(\mathbf{u})$ *is Lipschitz continuous on* $\overline{\Omega}$*; namely,*

$$\left\| \mathbf{g}(\mathbf{u}') - \mathbf{g}(\mathbf{u}'') \right\|_2 \leq L \left\| \mathbf{u}' - \mathbf{u}'' \right\|, \quad \forall \mathbf{u}', \mathbf{u}'' \in \overline{\Omega} \tag{26}$$

Though we have Propositions 5 and 6, the existence of a solution to VI model $VI(\mathbf{g}, \overline{\Omega})$ can be ensured. We proceed to prove the equivalence between the VI model $VI(\mathbf{g}, \overline{\Omega})$ and the generalized CMSTA conditions and summarized in Proposition 7. Rigorous proofs are given in Appendix C.

Proposition 7. $\mathbf{u}^* = (u_a^*, \ \forall a \in E)$ *is a solution of* $VI(\mathbf{g}, \overline{\Omega})$ *if and only if* $\mathbf{u}^*$ *and* $\mathbf{v}(\mathbf{u}^*)$ *fulfil the generalized CMSTA conditions (15)-(18).*

5. Solution Algorithm

As the vector function $\mathbf{g}(\mathbf{u})$ of the VI model $VI(\mathbf{g}, \overline{\Omega})$ is an implicit function, it is burdensome to directly evaluate its gradient. On the other hand, with any given Lagrangian multiplier vector $\mathbf{u}$, the value of vector function $\mathbf{g}(\mathbf{u})$ can be evaluated by solving the CMSTA with generalized travel time with no constraints, which can be efficiently solved by a number of existing solution algorithms.

Projection type algorithms have been thoroughly studied and proven to be efficient in solving traffic equilibrium problems [31–33]. Among them, the gradient projection (GP) is commonly adopted due to its concise form and good performance, for which a recent survey can be found in Ryu, Chen, and Choi [34]. The self-adaptive gradient projection (SAGP) algorithm was originally proposed by He et al. [35], where the gradient projection operation is adopted to establish the descent direction, coupled with a self-adaptive step size scheme to permit the convergence under a much milder assumption of the underlying mapping, and thus can handle general cost mapping structures and generalized side constraints. Later, Han and Sun [36] adopted this method to solve the asymmetric traffic assignment problems. Chen, Zhou, and Xu [32] further used this method to solve the asymmetric traffic assignment problem with nonadditive route cost functions. Inspired by previous studies, this paper presents the SAGP algorithm specifically designed for solving the CMSTA with generalized travel time functions.

As for the Lagrangian multipliers $\mathbf{u}$, the projection operation on Ω of the VI model $VI(\mathbf{g}, \overline{\Omega})$ is straightforward, as well. Propositions 5 and 6 imply the monotony and continuity of the VI model $VI(\mathbf{g}, \overline{\Omega})$. Therefore, the projection based algorithm is also suitable to evaluate the Lagrangian multipliers solutions. To this end, this section proposes a nested SAGP algorithm to solve the EC-CMSTA model which is presented as follows.

Step 0 (initiation). Set parameters of SAGP within the given range: $\delta \in (0, 1)$, $\mu \in [0.5, 1]$, $\varepsilon > 0$, $\alpha_{\max} > 0$, $\alpha^{(0)} > 0$, $\gamma^{(0)} = \alpha^{(0)}$. Set the iteration counter: $n = 0$, and the initial solution which should be a feasible solution: $\mathbf{u}^{(0)} = (u_a^{(0)} = 0, a \in E)^T$.

Step 1 (self-adaptive scaling procedure). This step aims to find the largest step size $\alpha^{(n+1)}$ that satisfies the given criteria in (27). To this end, let $\alpha^{(n+1)} = \mu^{l^{(n)}}\gamma^{(n)}$, and set nonnegative integer $l^{(n)} = 0$. We perform the following procedures, if the given criterion is satisfied, terminate and go to Step 2; otherwise, let $l^{(n)} = l^{(n)} + 1$, and perform the following procedures again.

Step 1.1. Given vector $\mathbf{u}^{(n)}$, evaluate the CMSTA with generalized travel time functions (23), using SAGP algorithm, and obtain the equilibrium flows $(\mathbf{v}^{(n)}, \mathbf{q}^{(n)})$. The solutions are subsequently used to evaluate the environmental measurement function $\mathbf{g}(\mathbf{u}^{(n)})$.

Step 1.2. Perform the projection operation for $\mathbf{u}^{(n)}$ to obtain a tentative vector $\tilde{\mathbf{u}}^{(n+1)}$, i.e., $\tilde{\mathbf{u}}^{(n+1)} = P_{\Omega}[\mathbf{u}^{(n)} - \alpha^{(n+1)} \cdot \mathbf{g}(\mathbf{u}^{(n)})]$.

Step 1.3. Let $\overline{\mathbf{u}}^{(n+1)} = \max[0, \mathbf{u}^{(n)} - \alpha^{(n)}\mathbf{g}(\mathbf{u}^{(n)})]$. If the following inequality is satisfied

$$\begin{aligned}(2-\delta)\,\alpha^{(n+1)}&\left(\mathbf{u}^{(n)}-\tilde{\mathbf{u}}^{(n+1)}\right)^T\left(\mathbf{g}\left(\mathbf{u}^{(n)}\right)-\mathbf{g}\left(\tilde{\mathbf{u}}^{(n+1)}\right)\right)\\&+\left(\alpha^{(n+1)}\right)^2\left\|\mathbf{g}\left(\mathbf{u}^{(n)}\right)-\mathbf{g}\left(\tilde{\mathbf{u}}^{(n+1)}\right)\right\|^2\\&\geq\max\left\{0,\frac{\left(\alpha^{(n+1)}\right)^2-\left(\alpha^{(n)}\right)^2}{\left(\alpha^{(n)}\right)^2}\left\|\mathbf{u}^{(n)}-\overline{\mathbf{u}}^{(n+1)}\right\|^2\right\}\end{aligned} \tag{27}$$

go Step 2. Otherwise, let $l^{(n)} = l^{(n)} + 1$, update step size $\alpha^{(n+1)}$, and go back to Step 1.1.

Step 2 (selection of $\gamma^{(n+1)}$). If the following condition is satisfied, let $\gamma^{(n+1)} = \max\{\alpha^{(n+1)}/\mu, \alpha_{\max}\}$.

$$\begin{aligned}0.5\alpha^{(n+1)}&\left(\mathbf{u}^{(n)}-\tilde{\mathbf{u}}^{(n+1)}\right)^T\left(\mathbf{g}\left(\mathbf{u}^{(n)}\right)-\mathbf{g}\left(\tilde{\mathbf{u}}^{(n+1)}\right)\right)\\&+\left(\alpha^{(n+1)}\right)^2\left\|\mathbf{g}\left(\mathbf{u}^{(n)}\right)-\mathbf{g}\left(\tilde{\mathbf{u}}^{(n+1)}\right)\right\|^2\\&\geq\max\left\{0,\frac{\left(\alpha^{(n+1)}\right)^2-\left(\alpha^{(n)}\right)^2}{\left(\alpha^{(n)}\right)^2}\left\|\mathbf{u}^{(n)}-\overline{\mathbf{u}}^{(n+1)}\right\|^2\right\}\end{aligned} \tag{28}$$

Step 3 (convergence test). If the convergence criterion is satisfied, terminate; otherwise, let iteration counter: $n = n+1$, go back to Step 1.

Note that the CMSTA with generalized travel time functions in Step 1.1 is also solved by the SAGP algorithm, which is not repeated here for conciseness.

6. Numerical Examples

In this section, two experiments based on a linear network and the Sioux-Falls network are provided to demonstrate the features of the proposed EC-CMSTA and validity of the proposed solution algorithm.

It is well known that carbon monoxide (CO), carbon dioxide (CO_2), sulphur dioxide (SO_2), nitrogen oxides (NO_x), respirable suspended particulates (RSP or PM_{10}), fine suspended particulates (FSP or $PM_{2.5}$), and volatile organic compounds (VOC) are the main types of vehicular emission pollutants that are critical for air quality protection. Among others, CO is taken as an important indicator of the level of vehicle atmosphere pollution, mainly because (1) vehicles contribute to the majority of CO emissions in the atmosphere; (2) CO is one of the most critical pollutants among the various types of vehicular emissions; (3) the emission functions of other pollutants are similar to that of CO [11]. In addition, this consideration has also been used in a number of recent studies, e.g., Yin and Lawphongpanich [37]; Nagurney, Qiang and Nagurney [38]; Chen, Zhou and Ryu [10]; Li et al. [11]; Xu, Chen and Cheng [1]; Yang, Ban and Ma [39]; Kitthamkesorn et al. [2]. Therefore, we use CO as an illustration to model vehicle emission. The nonlinear macroscopic formula of CO emission rate proposed by Wallace et al. [6] is used for analysis as shown below:

$$g_a(t_a) = 0.2038 \cdot t_a \cdot \exp\left(\frac{0.7962 \cdot L_a}{t_a}\right) \tag{29}$$

where $g_a(t_a)$ is the CO emission rate on link a measured in grams per hour. L_a is the length of link a measured in kilometers, and t_a is the link travel time measured in minutes. This nonlinear function can be shown to be convex and continuously differentiable. The Bureau of Public Road (BPR) type function is adopted for the link travel time, i.e., $t_a = t_a^0(1+0.15(v_a/C_a)^4)$ for road links, and $t_a = t_a^0(1+0.1(v_a/C_a)^4)$ for rail and RPR links. The dispersion parameter of MNL mode is set as $\theta = 1.0$.

6.1. A Linear Corridor. As shown in Figure 1, consider a linear corridor $G(N, A)$, with one rail line (dotted line) from suburb o to the urban city area d and another highway (solid line)

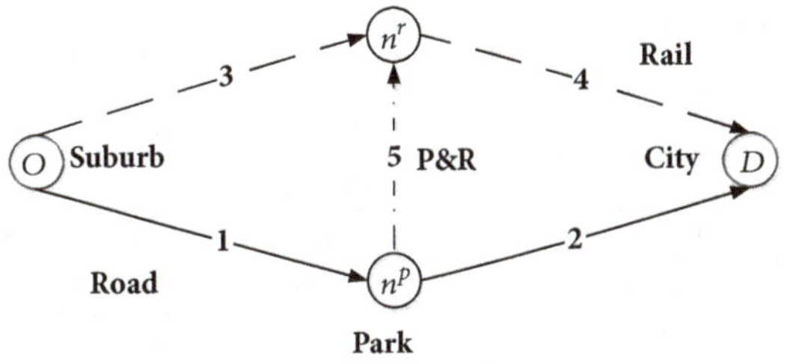

FIGURE 1: A linear corridor with RPR facility.

TABLE 1: Attributes of the linear corridor.

Link Type	Auto		Rail		RPR
Link No.	1	2	3	4	5
Free Flow Travel Time (t_a^0)	6	4	12	5	3
Capacity (C_a)	250	200	350	350	350

TABLE 2: Modal share and equilibrium travel time.

Modes	With RPR scheme	
	Demand	Travel time
Auto	250.45	16.66
Train	150.15	17.16
RPR	99.38	17.58

connecting (o, d). The OD demand is assumed to be 500 per hour. One RPR site is constructed at n^p, adjacent to the train station n^r. Travellers transfer from the RPR site n^p to the train station n^r through the RPR link (link 5). Link 2 passes into an environmentally sensitive urban area, which exerts restriction on the total CO emissions on link 2. Using the CO emission rate function defined by (29), the nonlinear environmental constraint on link a can be expressed as

$$0.2038 \cdot t_a \cdot \exp\left(\frac{0.7962 L_a}{t_a}\right) \cdot v_a - \overline{g}_a \leq 0, \quad \forall a \in A \quad (30)$$

where $\overline{g}_a$ is the CO emission threshold on link a. Attributes of the network are given in Table 1.

Setting the environmental constraints at 500, we first examine the correctness of the results. As expected, the results in Table 2 satisfy the modal demand conversation condition and the MNL based modal choice condition. The Lagrangian multiplier on link 2 equals 1.73, permitting the emission on link 2 subject to the requirements.

We proceed to explore the influence of the environmental constraint on the flow shift of modal demands, link flows, and total system travel time by varying the emission constraint $\overline{g}_2$ between 0 and 800. The corresponding modal demand shift pattern under environmental constraint is given in Figure 2. We can see that as $\overline{g}_2$ increases (relax the CO emission constraints), the auto demand first increases and then becomes stable; meanwhile, the demands of rail and RPR show the crosscurrent. It makes sense because relaxing $\overline{g}_2$ is similar to increase the road capacity. When the emission level is beyond 631.47, constraint (30) is not blinding anymore which means the PRP facilities are able to achieve the environmental

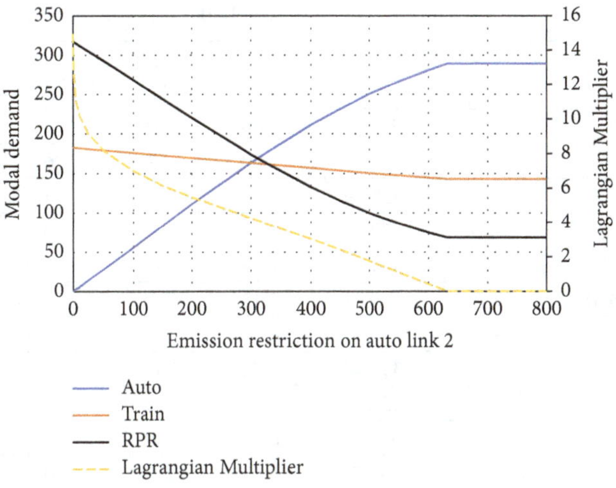

FIGURE 2: Modal split pattern with various environmental constraint level.

protection requirements without additional regulations or restrictions. The dotted line indicates the evolution of Lagrangian multiplier on link 2, u_2^*. As the environmental regulation can be strengthened, u_2^* keeps increasing and trends to be infinity, when $\overline{g}_2 = 0$.

Detailed results of link flows, link travel time, and Lagrangian multiplier are shown in Table 3. As marked by bold and italic text in Table 3 the minimal system travel cost is achieved when $\overline{g}_2$ is increased to around 631.47. In other words, setting $\overline{g}_2 \geq 631.47$ is equivalent to have no CO emission constraint on link 2, which reduces the problem into CMSTA. Meanwhile, the Lagrangian multiplier u_2^* reduces to zero as well indicating that there is no restrictions for vehicles which come into this area.

6.2. Sioux-Falls Network. Another numerical example built from the Sioux-Falls network (Figure 3) is used to further access the performance of the proposed model and solution algorithm for solving the environmental constrained CMSTA problem. This study network consists of 30 nodes and 76 links. The area highlighted by the red ellipse is the CBD district. It has four rail lines between the suburb areas and the city loop. Detail attributes of this network are tabulated in Appendix D. We focus on the travel demands in commuting hours, where the majority of travel demands origin from the suburbs and go into business districts. There are 24 OD pairs and the OD demand data are given in Appendix E. Four sites (nodes 3, 6, 19, and 14) which locate close to rail lines are selected as RPR car parks where RPR users park their car and transfer to rail mode.

The column generation scheme is adopted in this case and incorporated into our proposed solution algorithm to generate paths in evaluating the equilibrium flows. In the multimodal transport network, we first examine the effect of the proposed RPR scheme with the CMSTA model and access the emissions in the road subnetwork. 5 links in the CBD district are subsequently identified as the critical links, and be further grouped into the link set $\overline{A} = \{14, 22, 29, 46, 48\}$. As shown in Figure 4, the provision of RPR facilities significantly

TABLE 3: Link flows and link cost with varied restricted CO emission level.

$\overline{g}_2$	Link 1		Link 2			Link 3		Link 4		Link 5	
	Time	Flow	Time	Flow	u^*	Time	Flow	Time	Flow	Time	Flow
0	8.33	317.29	4.00	0.00	$+\infty$	12.09	182.71	7.08	499.94	3.20	317.23
100	8.55	324.24	4.00	55.32	6.99	12.08	175.76	6.30	444.68	3.10	268.92
200	8.76	330.69	4.06	110.34	5.45	12.07	169.31	5.77	389.66	3.05	220.35
300	8.97	337.09	4.27	163.54	4.24	12.06	162.91	5.43	336.46	3.02	173.55
400	9.21	343.56	4.75	211.36	3.02	12.05	156.44	5.23	288.64	3.01	132.20
500	9.45	349.85	5.48	250.47	1.73	12.04	150.15	5.13	249.53	3.00	99.38
600	9.69	355.77	6.34	281.02	0.41	12.03	144.23	5.08	218.98	3.00	74.75
631.47	***9.77***	***357.57***	***6.62***	***289.22***	***0.00***	***12.03***	***142.43***	***5.07***	***210.78***	***3.00***	***68.35***
700	9.77	357.57	6.62	289.22	0.00	12.03	142.43	5.07	210.78	3.00	68.35
800	9.77	357.57	6.62	289.22	0.00	12.03	142.43	5.07	210.78	3.00	68.35
900	9.77	357.57	6.62	289.22	0.00	12.03	142.43	5.07	210.78	3.00	68.35

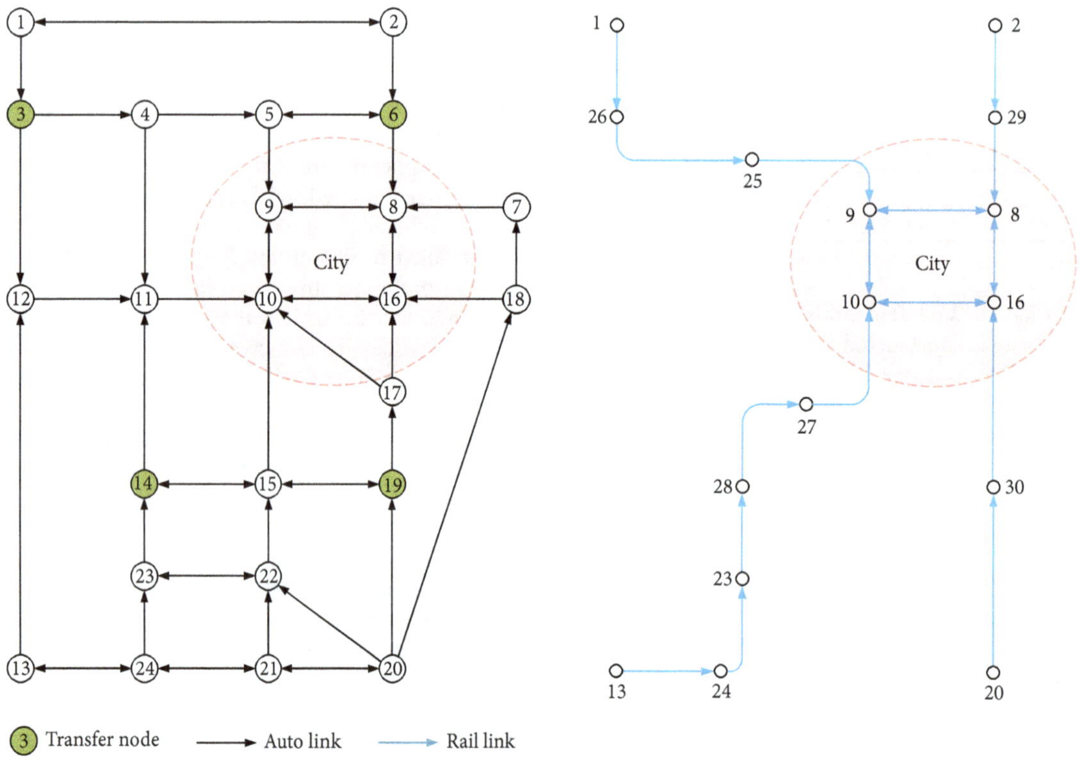

FIGURE 3: Modified Sioux-Falls network with RPR scheme.

influence travellers' mode choice as well as route choice behaviour. 14.03% of previous autodrivers and 5.51% of previous train riders shift to RPR mode, due to the utility provided by RPR services. The amount of emissions on critical links also significantly reduced with the provision of RPR schemes. This phenomenon indicates RPR schemes could alleviate the environmental deterioration by redistributing the flow and modal demand in the network and reflects the rationality of modelling users modal choice and route choice decisions in a combined model, i.e., CMSTA.

We proceed to exert environmental constraints on these critical links. Here, we consider five scenarios of the thresholds and summarize in Table 4. Parameters in the SAGP algorithm are set as follows: $\mu = 0.85$, $\delta = 0.8$, and $\alpha^{(0)} = 1.0$. In order to reflect the convergence trend of the SAGP algorithm apparently, we take the following logarithmic value of the maximum absolute error as a performance index, namely,

$$\varepsilon = \log_{10}\left(\max\left(\left\|T_a - g_a v_a\right\|_2, \forall a \in \overline{A}\right)\right) \tag{31}$$

The SAGP algorithm is then used to solve the environmental constrained CMSTA in terms of the five scenarios provided in Table 4. Figure 5 depicts the logarithmic value of the maximum absolute error versus the number of iterations used for solving scenarios 1, 2, 3, and 4. It clearly shows that the

TABLE 4: Five scenarios of environmental constraints.

Link ID	Scenario 1 Threshold T_a^1	Scenario 2 Threshold T_a^2	Scenario 3 Threshold T_a^3	Scenario 4 Threshold T_a^4	Scenario 5 Threshold T_a^5	Emission	Link flow
14	500	600	700	800	900	724.16	320.28
22	500	600	700	800	900	1007.05	370.93
29	500	600	700	800	900	796.13	319.51
46	500	600	700	800	900	866.88	351.1
48	500	600	700	800	900	735.61	531.39

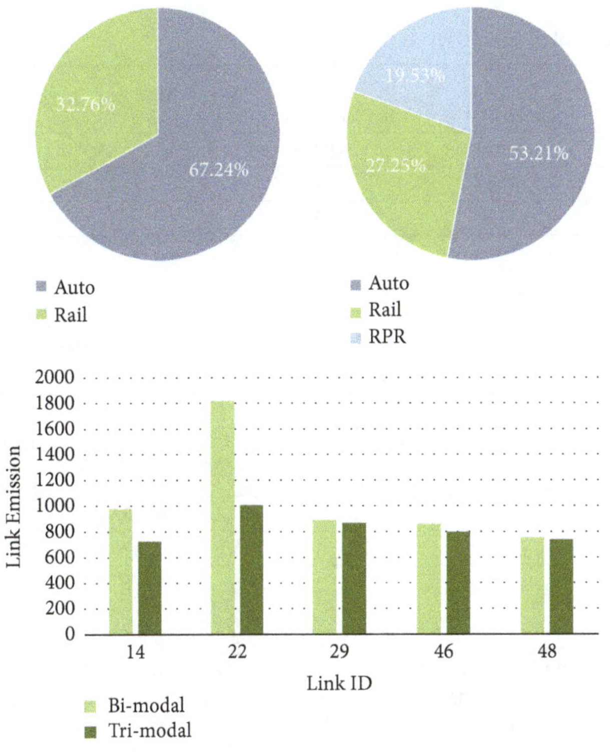

FIGURE 4: Mode share and emissions with/without RPR scheme.

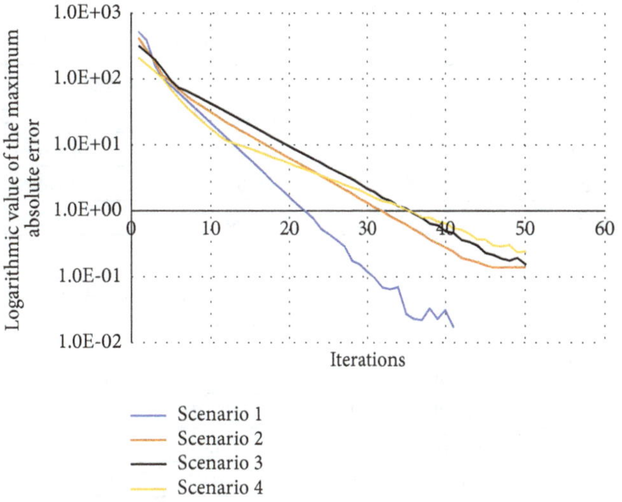

FIGURE 5: Convergence trend of the SAGP algorithm.

algorithm can converge to a solution of VI model $VI(\mathbf{g}, \overline{\Omega})$. In scenario 5, the solution converges much faster than other scenarios because only link 22 is constrained.

Table 5 gives the resultant link emissions and corresponding Lagrangian multiplier for each scenario. We can observe that the emissions on each link are bounded by the required thresholds (some link emissions slightly excess the requirements caused by the computation error). The corresponding multiplier takes positive value only on the links with emission bounded by the threshold. These two phenomena tally with (16)-(18), which numerically verify the effectiveness of the VI model $VI(\mathbf{g}, \overline{\Omega})$ and the SAGP algorithm. In addition, from scenario 1 to scenario 5, it can be seen that, by tightening the thresholds, the corresponding Lagrangian multipliers become larger, which indicates that it would be more costly to restrict the emission to a much lower level.

Table 6 further investigates the modal shift trend from scenario 1 to scenario 5. As the values of threshold getting smaller, travellers who use automode become less. The environmental regulations in road network could significantly promote the modal shift from auto to "green" travel modes (i.e., Train and RPR). Taking scenario 1, for example, the automode share is depressed about 11.11%, from 53.21% to 42.10%, where 5.38% of travel demand shifts to rail mode and 5.72% rest shifts to RPR mode. This phenomenon indicates that environmental regulations significantly influence not only travellers' route choice behaviour but also mode choice behaviour.

To further see the impacts of demand level on the Lagrangian multipliers, a sensitivity test is carried out on the dispersion parameter on the modal choice function (8). Without loss of generality, the first scenario of threshold setting is taken for this test, and then the value of θ is changed from 0.5 to 0.9 with an interval of 0.1. The Lagrangian multipliers obtained by the SAGP algorithm in these five cases are presented in Table 7. We noted that a larger value of θ results in a larger level of autodemand level and the road network congestion level is higher accordingly. Table 7 indicates the generalized CMSTA conditions can still be fulfilled. As the demand level ascends, the corresponding Lagrangian multipliers also increases, meaning that it is more difficult to maintain the threshold constraints at a higher demand level.

We further compare the network-level performance before and after implementing the set of environmental

Table 5: Link emission and corresponding Lagrangian multiplier.

Link ID	Scenario 1		Scenario 2		Scenario 3		Scenario 4		Scenario 5		Without
	Emission	LM	Emission	LM	Emission	LM	Emission	LM	Emission	LM	
14	499.99	1.64	600.06	0.93	700.07	0.22	728.78	0.00	726.51	0.00	724.16
22	500.00	2.29	600.03	1.74	700.10	1.20	800.17	0.64	900.00	0.38	1007.05
29	500.00	0.79	600.00	0.23	700.03	0.10	800.06	0.03	867.92	0.00	866.88
46	500.02	2.32	600.14	1.5	700.09	0.77	800.21	0.15	803.76	0.00	796.13
48	500.03	0.86	600.02	0.71	699.99	0.32	743.59	0.00	736.09	0.00	735.61

∗LM denotes Lagrangian multiplier.

Table 6: Mode share with different threshold levels.

Mode	With Threshold					Without Threshold
	Scenario 1	Scenario 2	Scenario 3	Scenario 4	Scenario 5	
Auto	42.10%	45.68%	48.96%	51.53%	52.35%	53.21%
Train	32.64%	30.64%	28.89%	27.61%	27.39%	27.25%
RPR	25.26%	23.68%	22.15%	20.86%	20.27%	19.53%

Table 7: Sensitivity test on the dispersion parameter.

link ID	$\theta = 0.5$		$\theta = 0.6$		$\theta = 0.7$		$\theta = 0.8$		$\theta = 0.9$	
	Emission	LM	Emission	LM	Emission	LM	Emission	LM	Emission	LM
14	499.99	1.64	500.00	1.70	499.98	1.73	500.01	1.74	500.02	1.75
22	500.00	2.29	500.00	2.02	500.00	1.82	500.02	1.66	500.01	1.54
29	500.00	0.79	500.01	0.68	500.01	0.59	500.00	0.52	500.01	0.48
46	500.02	2.32	500.03	2.38	500.03	2.41	500.00	2.42	500.06	2.43
48	500.03	0.86	500.01	0.85	499.94	0.84	500.02	0.84	499.96	0.83

∗LM denotes Lagrangian multiplier.

restraints. The first scenario of threshold setting is still taken for this test. Without loss of generality, the total travel time (TTT) and total vehicle distances (TVD) travelled are used in this study to measure the network-wide performance. From Table 8 we can observe that implementation of environmental constraints decreases both TTT and TVD (-13.89% decrease of TTT and -13.19% decrease of TVM), which implies that some travellers must be shifting their modes in order to benefit the air quality of the CBD area.

7. Conclusions

Environmental protection is an important consideration in planning a transportation system. Environmental constraint is a useful means to explicitly reflect environmental protection requirements. In this paper, we studied a combined environmental protection strategy in the multimodal transport network: (1) a rail-based park-and-ride scheme to assist mode shift to reduce the overall vehicle travel demands in the road network and (2) an area-based regulation scheme to deal with the environmental deterioration caused by heterogeneity of traffic flows. To quantitatively evaluate the effectiveness of the proposed scheme, we built an environmental constrained combined modal split and traffic assignment (EC-CMSTA) model. Specifically, we developed an equivalent VI reformulation for the EC-CMSTA. Different from the explicit formulation as a VI or NCP, the proposed reformulation adapts the concept of fix-point to reformulate the CMSTA conditions, which is, later, incorporated into the VI formulated nonlinear complementarity conditions associated with the nonlinear environmental constraints. The proposed VI reformulation can handle a general nonlinear environmental constraint structure and a general link and route cost structure, enhancing the modelling adaptability and flexibility, and in addition permits a number of efficient algorithms for its solution. A customized projection algorithm with self-adaptive step size scheme is adopted in this paper. Numerical examples demonstrated the EC-CMSTA model has the potential to enhance the behavioural modelling of network users' travel decisions and to assist in quantitatively evaluating the effectiveness of RPR schemes and environmental regulations.

Several directions are worthy of further investigation. First, in this paper, we use CO as an indicator to model vehicle emission. Further study will take other types of emission into consideration, e.g., the CO_2. Second, a number of efficient solution algorithms proposed by recent studies, e.g., Ryu, Chen, and Choi [34]; Ryu, Chen, and Choi [40], should be explored to further improve the solution efficiency. Third, we plan to test our model on larger networks with more real world applications. For very large-scale problems, existing algorithms may be not efficient and specific algorithms should be developed.

TABLE 8: Network performance with and without environmental regulations.

Network performance	Without environmental Regulation (a)	With environmental regulation (b)	Relative difference [(b-a)/a]
Total travel time	41442.50	35685.36	-13.89%
Total vehicle distance	38995.10	33851.32	-13.19%

Appendix

A. Proof of Proposition 5

Vector function $\mathbf{g}(\mathbf{u})$ is strictly monotone on $\overline{\Omega}$; namely,

$$\left(\mathbf{g}\left(\mathbf{u}'\right)-\mathbf{g}\left(\mathbf{u}''\right)\right)^T\left(\mathbf{u}'-\mathbf{u}''\right)>0, \quad \forall \mathbf{u}',\mathbf{u}''\in\overline{\Omega} \quad \text{(A.1)}$$

Proof. For any two distinct nonnegative Lagrangian multiplier vectors $\mathbf{u}'$ and $\mathbf{u}''$, let $\mathbf{v}(\mathbf{u}')$ and $\mathbf{v}(\mathbf{u}'')$ denote the corresponding CMSTA link flow solutions, and similarly, let $\mathbf{f}(\mathbf{u}')$, $\mathbf{f}(\mathbf{u}'')$, $\mathbf{q}(\mathbf{u}')$, and $\mathbf{q}(\mathbf{u}'')$ denote the corresponding CMSTA path flow and modal demand solutions. We have

$$\mathbf{v}\left(\mathbf{u}'\right)=\Delta\mathbf{f}\left(\mathbf{u}'\right) \quad \text{(A.2)}$$

$$\mathbf{v}\left(\mathbf{u}''\right)=\Delta\mathbf{f}\left(\mathbf{u}''\right) \quad \text{(A.3)}$$

$$\mathbf{f}\left(\mathbf{u}'\right)=\mathbf{q}\left(\boldsymbol{\tau}'\right)^T\cdot\mathbf{p}\left(\widehat{\mathbf{c}}'\right) \quad \text{(A.4)}$$

$$\mathbf{f}\left(\mathbf{u}''\right)=\mathbf{q}\left(\boldsymbol{\tau}''\right)^T\cdot\mathbf{p}\left(\widehat{\mathbf{c}}''\right) \quad \text{(A.5)}$$

$$\mathbf{q}\left(\widehat{\boldsymbol{\tau}}'\right)=\overline{\mathbf{q}}^T\cdot\overline{\mathbf{p}}\left(\widehat{\boldsymbol{\tau}}'\right) \quad \text{(A.6)}$$

$$\mathbf{q}\left(\widehat{\boldsymbol{\tau}}''\right)=\overline{\mathbf{q}}^T\cdot\overline{\mathbf{p}}\left(\widehat{\boldsymbol{\tau}}''\right) \quad \text{(A.7)}$$

where Δ defines the link-path incidence, $\widehat{\mathbf{c}}'=\Delta^T\widehat{\mathbf{t}}(\mathbf{v}(\mathbf{u}'),\mathbf{u}')$, $\widehat{\mathbf{c}}''=\Delta^T\widehat{\mathbf{t}}(\mathbf{v}(\mathbf{u}''),\mathbf{u}'')$, $\mathbf{p}(\widehat{\mathbf{c}}')=(p_k^{m,od}(\widehat{\mathbf{c}}'),\ \forall k\in K_m^{od},\ m\in M_{od},\ od\in W)^T$, $\mathbf{p}(\widehat{\mathbf{c}}'')=(p_k^{m,od}(\widehat{\mathbf{c}}''),\ \forall k\in K_m^{od},\ m\in M_{od},\ od\in W)^T$, $\widehat{\boldsymbol{\tau}}'=(\tau_m^{\prime od}(\widehat{\mathbf{c}}'),\ \forall m\in M_{od},\ od\in W)^T$, $\widehat{\boldsymbol{\tau}}''=(\widehat{\tau}_m^{\prime\prime od}(\widehat{\mathbf{c}}''),\ \forall m\in M_{od},\ od\in W)^T$, $\overline{\mathbf{p}}(\widehat{\boldsymbol{\tau}}')=(p_m^{od}(\widehat{\boldsymbol{\tau}}'),\ \forall m\in M_{od},\ od\in W)^T$, $\overline{\mathbf{p}}(\widehat{\boldsymbol{\tau}}'')=(p_m^{od}(\widehat{\boldsymbol{\tau}}''),\ \forall m\in M_{od},\ od\in W)^T$, and $\overline{\mathbf{q}}=(\overline{q}^{od},\ \forall od\in W)^T$.

According to the definition of CMSTA conditions, the modal travel time τ_m^{od} is taken as the equilibrium path travel time of mode m, it follows that

$$\tau_m^{od}\le c_k^{m,od},\quad \forall k\in K_m^{od}; \quad \text{(A.8)}$$

together with the flow conservation condition (5), we have

$$\mathbf{q}\left(\boldsymbol{\tau}'\right)^T\boldsymbol{\tau}'\le\mathbf{f}\left(\mathbf{u}'\right)^T\mathbf{c}' \quad \text{(A.9)}$$

$$\mathbf{q}\left(\boldsymbol{\tau}''\right)^T\boldsymbol{\tau}''\le\mathbf{f}\left(\mathbf{u}''\right)^T\mathbf{c}'' \quad \text{(A.10)}$$

Subtracting (A.9) and (A.10) yields that

$$\mathbf{q}\left(\boldsymbol{\tau}'\right)^T\left(\boldsymbol{\tau}'-\boldsymbol{\tau}''\right)\le\mathbf{f}\left(\mathbf{u}'\right)^T\left(\mathbf{c}'-\mathbf{c}''\right) \quad \text{(A.11)}$$

$$\mathbf{q}\left(\boldsymbol{\tau}''\right)^T\left(\boldsymbol{\tau}'-\boldsymbol{\tau}''\right)\le\mathbf{f}\left(\mathbf{u}''\right)^T\left(\mathbf{c}'-\mathbf{c}''\right); \quad \text{(A.12)}$$

namely,

$$\begin{aligned}&\left(\mathbf{q}\left(\boldsymbol{\tau}'\right)-\mathbf{q}\left(\boldsymbol{\tau}''\right)\right)^T\left(\boldsymbol{\tau}'-\boldsymbol{\tau}''\right)\\&\quad\le\left(\mathbf{f}\left(\mathbf{u}'\right)-\mathbf{f}\left(\mathbf{u}''\right)\right)^T\left(\mathbf{c}'-\mathbf{c}''\right)\end{aligned} \quad \text{(A.13)}$$

The monotonicity of function (8) implies that

$$\left(\mathbf{q}\left(\boldsymbol{\tau}'\right)-\mathbf{q}\left(\boldsymbol{\tau}''\right)\right)^T\left(\boldsymbol{\tau}'-\boldsymbol{\tau}''\right)>0 \quad \text{(A.14)}$$

According to (A.13) and (A.14), we thus obtain

$$\left(\mathbf{f}\left(\mathbf{u}'\right)-\mathbf{f}\left(\mathbf{u}''\right)\right)^T\left(\mathbf{c}'-\mathbf{c}''\right)>0 \quad \text{(A.15)}$$

As $\mathbf{c}'=\Delta^T\widehat{\mathbf{t}}(\mathbf{v}(\mathbf{u}'),\mathbf{u}')$ and $\widehat{\mathbf{c}}''=\Delta^T\widehat{\mathbf{t}}(\mathbf{v}(\mathbf{u}''),\mathbf{u}'')$, (A.15) can be equivalently written as

$$\begin{aligned}&\left(\mathbf{f}\left(\mathbf{u}'\right)-\mathbf{f}\left(\mathbf{u}''\right)\right)^T\\&\quad\cdot\left(\Delta^T\widehat{\mathbf{t}}\left(\mathbf{v}\left(\mathbf{u}'\right),\mathbf{u}'\right)-\Delta^T\widehat{\mathbf{t}}\left(\mathbf{v}\left(\mathbf{u}''\right),\mathbf{u}''\right)\right)>0\end{aligned} \quad \text{(A.16)}$$

Rearranging the left side of (A.16), it follows that

$$\begin{aligned}&\left(\mathbf{v}\left(\mathbf{u}'\right)-\mathbf{v}\left(\mathbf{u}''\right)\right)^T\\&\quad\cdot\left(\Delta^T\widehat{\mathbf{t}}\left(\mathbf{v}\left(\mathbf{u}'\right),\mathbf{u}'\right)-\Delta^T\widehat{\mathbf{t}}\left(\mathbf{v}\left(\mathbf{u}''\right),\mathbf{u}''\right)\right)>0\end{aligned} \quad \text{(A.17)}$$

According to the definition of generalized link travel time function, (20), we have

$$\left(\mathbf{v}\left(\mathbf{u}'\right)-\mathbf{v}\left(\mathbf{u}''\right)\right)^T\left(\mathbf{u}'-\mathbf{u}''\right)>0 \quad \text{(A.18)}$$

Together with Assumption 1, $\mathbf{g}(\mathbf{u})$ is a strictly monotone function with respect to link flow solutions $\mathbf{v}(\mathbf{u})$; it follows that

$$\left(\mathbf{g}\left(\mathbf{u}'\right)-\mathbf{g}\left(\mathbf{u}''\right)\right)^T\left(\mathbf{u}'-\mathbf{u}''\right)>0 \quad \text{(A.19)}$$

which implies $\mathbf{g}(\mathbf{u})$ is strictly monotone with respect to $\mathbf{u}$ on $\overline{\Omega}$. □

B. Proof of Proposition 6

Vector function $\mathbf{g}(\mathbf{u})$ is Lipschitz continuous on $\overline{\Omega}$; namely,

$$\left\|\mathbf{g}\left(\mathbf{u}'\right)-\mathbf{g}\left(\mathbf{u}''\right)\right\|_2\le L\left\|\mathbf{u}'-\mathbf{u}''\right\|,\quad \forall \mathbf{u}',\mathbf{u}''\in\overline{\Omega} \quad \text{(B.1)}$$

Proof. Following the approach presented by Meng, Liu, and Wang [28], it is easily to conclude that $\mathbf{v}(\mathbf{u})$ is continuously differentiable in a neighbourhood of $\mathbf{u}_0$ according to the implicit function theorem [41].

Since $\mathbf{v}(\mathbf{u})$ is continuously differentiable on $\overline{\Omega}$, its Jacobian matrix $\nabla_{\mathbf{u}}\mathbf{v}(\mathbf{u})$ is thus continuous on $\overline{\Omega}$. The two-norm of $\nabla_{\mathbf{u}}\mathbf{v}(\mathbf{u})$ is therefore bounded from the nonempty closed and convex set $\overline{\Omega}$; namely, there is a positive constant such that

$$\left\|\nabla_{\mathbf{u}}\mathbf{v}\left(\mathbf{u}\right)\right\|_2\le L_1,\quad \forall \mathbf{u}\in\overline{\Omega} \quad \text{(B.2)}$$

According to the Theorem 3.2.4 of Ortega and Rheinboldt [41] (mean-value theorem), it can be seen that

$$\left\|\mathbf{v}\left(\mathbf{u}'\right)-\mathbf{v}\left(\mathbf{u}''\right)\right\|_2 \leq L_1\left\|\mathbf{u}'-\mathbf{u}''\right\|_2, \quad \forall\mathbf{u}',\mathbf{u}''\in\overline{\Omega} \tag{B.3}$$

Similarly, according to **Assumption 1**, $\mathbf{g}(\mathbf{v})$ is a continuous differentiable function w.r.t. $\mathbf{v}$. The Jacobian matrix $\nabla_{\mathbf{v}}\mathbf{g}(\mathbf{v})$ is thus continuous on Ω which is also nonempty closed and convex. Thus, there is a positive constant such that

$$\left\|\nabla_{\mathbf{v}}\mathbf{g}(\mathbf{v})\right\|_2 \leq L_2, \quad \forall\mathbf{v}\in\Omega \tag{B.4}$$

and we have

$$\left\|\mathbf{g}\left(\mathbf{v}'\right)-\mathbf{g}\left(\mathbf{v}''\right)\right\|_2 \leq L_2\left\|\mathbf{v}'-\mathbf{v}''\right\|_2, \quad \forall\mathbf{v}',\mathbf{v}''\in\Omega \tag{B.5}$$

According to (B.3) and (B.5), we thus have

$$\begin{aligned}\left\|\mathbf{g}\left(\mathbf{u}'\right)-\mathbf{g}\left(\mathbf{u}''\right)\right\|_2 &\leq L_2\left\|\mathbf{v}'-\mathbf{v}''\right\|_2\\ &\leq L_2\cdot L_1\left\|\mathbf{u}'-\mathbf{u}''\right\|, \\ &\qquad \forall\mathbf{u}',\mathbf{u}''\in\overline{\Omega}\end{aligned} \tag{B.6}$$

This completes the proof. □

C. Proof of Proposition 7

$\mathbf{u}^* = (u_a^*, \forall a\in E)$ is a solution of $VI(\mathbf{g}, \mathfrak{R}_+^{|E|})$ if and only if $\mathbf{u}^*$ and $\mathbf{v}(\mathbf{u}^*)$ fulfil the generalized CMSTA conditions (15)-(18).

Proof.

Necessary Condition. Suppose $\mathbf{u}^*$ is a solution of $VI(\mathbf{g}, \mathfrak{R}_+^{|E|})$, and we now show that $\mathbf{u}^*$ fulfil the generalized CMSTA conditions.

We first demonstrate that $u_a^* < \overline{M}, \forall a\in E$. Assume that there is at least on element $b\in E$ with $u_b^* = \overline{M}$. Then, we define a feasible vector $\mathbf{u}'\in\mathfrak{R}_+^{|E|}$.

$$\mathbf{u}' = \left(u_a' = u_a^*, \ \forall a\in E\setminus\{b\}, \ u_b' = 0.5\overline{M}\right)^T. \tag{C.1}$$

Substituting $\mathbf{u}$ in the VI model $VI(\mathbf{g}, \mathfrak{R}_+^{|E|})$ with vector $\mathbf{u}'$, it follows that

$$\left(-g_b\left(v_b\left(\mathbf{u}^*\right)\right)\right)\cdot\left(0.5\overline{M}-\overline{M}\right)\geq 0; \tag{C.2}$$

namely,

$$g_b\left(v_b\left(\mathbf{u}^*\right)\right)\geq 0 \tag{C.3}$$

Now, we proceed to demonstrate $u_a^* < \overline{M}, \forall a\in E$ by using an apagogical approach. Note that, according to the monotony increasing of $\mathbf{g}(\mathbf{v})$, there exists a upper bound of link flow $\overline{v}_b$ that, for any flow $v_b < \overline{v}_b$, we have

$$g_b\left(v_b\left(\mathbf{u}^*\right)\right) < 0 \tag{C.4}$$

Then, we proceed to show $v_a(\mathbf{u}^*) < \overline{v}_a, \forall a\in E$. Suppose at least one flow loaded path on the road network uses link b. According to the DUE conditions, for any flow loaded paths, they have identical path travel time. Denote one of these paths as k_o and its $\lambda_{k_0}^{m,od}(\mathbf{u}^*)$ follows that

$$\lambda_{k_0}^{m,od}\left(\mathbf{u}^*\right) = \sum_{a\in E} u_a^*\zeta_{e,k_0}^{m,od} > \overline{M} \tag{C.5}$$

Therefore, its generalized path travel time and the corresponding modal travel time fulfil the condition:

$$c_{k_0}^{m,od}\left(\mathbf{u}^*\right)+\lambda_{k_0}^{m,od}\left(\mathbf{u}^*\right) > \overline{M} \tag{C.6}$$

$$c_m^{od}\left(\mathbf{u}^*\right)+\lambda_m^{od}\left(\mathbf{u}^*\right)\geq c_{k_0}^{m,od}\left(\mathbf{u}^*\right)+\lambda_{k_0}^{m,od}\left(\mathbf{u}^*\right) > \overline{M} \tag{C.7}$$

According to the monotonicity of function (8), we have

$$q_m^{od}\left(c_m^{od}\left(\mathbf{u}^*\right)+\lambda_m^{od}\left(\mathbf{u}^*\right)\right)\leq q_m^{od}\left(\overline{M}\right) < \frac{\overline{v}_b}{|W|\cdot|M|} \tag{C.8}$$

In other words,

$$v_b < \sum_{od\in W}\sum_{m\in M_{od}} q_m^{od} < \overline{v}_b \tag{C.9}$$

$$g_b\left(v_b\left(\mathbf{u}^*\right)\right) < 0 \tag{C.10}$$

However, (C.10) contradicts (C.3), then the hypothesis $u_b^* = \overline{M}$ is incorrect. In other words,

$$u_a^* < \overline{M}, \quad \forall a\in E \tag{C.11}$$

To further prove the necessary conditions, we construct the following two specific vectors for any link $a\in E$:

$$\mathbf{u}' = \left(u_a' = \left(1-\gamma_1\right)u_a^*+\gamma_1\overline{M}, \ \forall a\in E\right)^T \tag{C.12}$$

$$\mathbf{u}'' = \left(u_a'' = \gamma_1 u_a^*, \ \forall a\in E\right)^T \tag{C.13}$$

where $\gamma_1\in(0,1)$ is a given parameter. Substituting vector $\mathbf{u}$ of VI model $VI(\mathbf{g}, \mathfrak{R}_+^{|E|})$ with these two particular vectors, respectively, yields that

$$g_a(\mathbf{v})\left(\gamma_1\left(\overline{M}-u_a^*\right)\right)\leq 0 \tag{C.14}$$

$$g_a(\mathbf{v})\left(\left(\gamma_1-1\right)u_a^*\right)\leq 0 \tag{C.15}$$

Since $M - u_a^* > 0$, (C.14) implies that

$$g_a(\mathbf{v})\leq 0, \quad \forall a\in E \tag{C.16}$$

In accordance with the fact that $u_a^*\geq 0$, we thus have

$$\left(-g_a(\mathbf{v})\right)\cdot u_a^*\geq 0, \quad \forall a\in E \tag{C.17}$$

Hence, (C.17) in conjunction with (C.15) implies that

$$\left(-g_a(\mathbf{v})\right)\cdot u_a^* = 0, \quad \forall a\in E \tag{C.18}$$

In other words, $\mathbf{u}^*$ and $\mathbf{v}(\mathbf{u}^*)$ fulfil the generalized CMSTA conditions.

Sufficient Condition. Suppose $\mathbf{u}^*$ fulfil the generalized CMSTA conditions (15)-(18); we have the following conditions:

$$-\mathbf{g}\left(\mathbf{v}\left(\mathbf{u}^*\right)\right)\geq\mathbf{0} \tag{C.19}$$

$$\mathbf{g}\left(\mathbf{v}\left(\mathbf{u}^*\right)\right)^T\cdot\mathbf{u}^* = 0 \tag{C.20}$$

In addition, we have $\mathbf{u}\geq 0$. Therefore, according to (C.19) and (C.20), it follows that

$$\mathbf{g}\left(\mathbf{v}\left(\mathbf{u}^*\right)\right)^T\cdot\mathbf{u}\geq\mathbf{g}\left(\mathbf{v}\left(\mathbf{u}^*\right)\right)^T\cdot\mathbf{u}^* \tag{C.21}$$

$$\mathbf{g}\left(\mathbf{v}\left(\mathbf{u}^*\right)\right)^T\cdot\left(\mathbf{u}-\mathbf{u}^*\right)\geq 0 \tag{C.22}$$

This completes the proof. □

TABLE 9

ID	T	H	Fft	Capa	Type
1	8	9	3	400	Auto
2	9	8	3	400	Auto
3	9	10	3	400	Auto
4	10	9	3	400	Auto
5	10	16	3	400	Auto
6	16	10	3	400	Auto
7	8	16	3	400	Auto
8	16	8	3	400	Auto
9	1	3	3	800	Auto
10	3	4	4	1200	Auto
11	4	3	4	1200	Auto
12	4	5	5	800	Auto
13	5	4	5	800	Auto
14	5	9	5	800	Auto
15	1	2	8	600	Auto
16	3	12	6	1200	Auto
17	4	11	7	600	Auto
18	5	6	3	800	Auto
19	6	5	3	800	Auto
20	12	11	5	600	Auto
21	2	6	3	800	Auto
22	6	8	6	800	Auto
23	2	1	8	600	Auto
24	13	24	3	600	Auto
25	24	23	4	600	Auto
26	23	14	4	600	Auto
27	14	15	3	1200	Auto
28	15	14	3	800	Auto
29	15	10	6	800	Auto
30	13	12	7	800	Auto
31	24	21	3	600	Auto
32	21	24	3	600	Auto
33	23	22	3	600	Auto
34	22	23	3	600	Auto
35	21	22	4	600	Auto
36	22	15	4	800	Auto
37	21	20	3	600	Auto
38	20	19	5	600	Auto
39	19	17	3	1200	Auto
40	17	16	3	600	Auto
41	17	10	5	400	Auto
42	20	22	3	400	Auto
43	15	19	3	600	Auto
44	7	8	3	400	Auto
45	19	15	3	600	Auto
46	11	10	5	600	Auto
47	14	11	4	400	Auto
48	18	16	3	600	Auto
49	18	7	4	600	Auto
50	20	21	3	600	Auto
51	20	18	7	600	Auto
52	24	13	3	600	Auto
53	9	10	3	1500	Train
54	9	8	3	1500	Train
55	10	9	3	1500	Train
56	10	16	3	1500	Train

TABLE 9: Continued.

ID	T	H	Fft	Capa	Type
57	8	9	3	1500	Train
58	8	16	3	1500	Train
59	16	10	3	1500	Train
60	16	8	3	1500	Train
61	20	30	7	1000	Train
62	30	16	6	1000	Train
63	13	24	2	1000	Train
64	24	23	4	1000	Train
65	23	28	4	1000	Train
66	28	27	4	1000	Train
67	27	10	4	1000	Train
68	2	29	10	1000	Train
69	29	8	2	1000	Train
70	1	26	3	1000	Train
71	26	25	9	1000	Train
72	25	9	7	1000	Train
73	3	26	4	1000	RPR
74	6	29	4	1000	RPR
75	14	28	3	1000	RPR
76	19	30	3	1000	RPR

*T is tail node; H is head node; Fft denotes free flow time; Capa denotes capacity.

TABLE 10

ID	Origin	Destination	Demand
1	1	8	220
2	1	9	180
3	1	10	200
4	1	16	220
5	2	8	240
6	2	9	190
7	2	10	180
8	2	16	180
9	13	8	160
10	13	9	130
11	13	10	170
12	13	16	130
13	24	8	140
14	24	9	150
15	24	10	140
16	24	16	100
17	23	8	130
18	23	9	150
19	23	10	150
20	23	16	170
21	20	8	180
22	20	9	160
23	20	10	140
24	20	16	170

D. Attributes of Study Network

See Table 9.

E. OD Travel Demand of Study Network

See Table 10.

Conflicts of Interest

The authors declare that they have no conflicts of interest.

References

[1] X. Xu, A. Chen, and L. Cheng, "Reformulating environmentally constrained traffic equilibrium via a smooth gap function," *International Journal of Sustainable Transportation*, vol. 9, no. 6, pp. 419–430, 2015.

[2] S. Kitthamkesorn, A. Chen, X. Xu, and S. Ryu, "Modeling mode and route similarities in network equilibrium problem with go-green modes," *Networks and Spatial Economics*, vol. 16, no. 1, pp. 33–60, 2016.

[3] W. Y. Szeto, X. Jaber, and S. C. Wong, "Road network equilibrium approaches to environmental sustainability," *Transport Reviews*, vol. 32, no. 4, pp. 491–518, 2012.

[4] X. Chen, Z. Liu, I. Kim, and J. Leng, "Optimal Sitting and Sizing of the Remote Park-and-Ride Scheme in the Multimodal Transport Network," in *Proceedings of the Transportation Research Board 97th Annual Meeting*, Washington DC, USA, 2018.

[5] Z. Liu and Q. Meng, "Bus-based park-and-ride system: a stochastic model on multimodal network with congestion pricing schemes," *International Journal of Systems Science*, vol. 45, no. 5, pp. 994–1006, 2014.

[6] C. E. Wallace, K. Courage, D. Reaves, Schoene. G., and G. Euler, TRANSYT-7F user's manual.In.

[7] L. R. Rilett and C. M. Benedek, "Traffic assignment under environmental and equity objectives," *Transportation Research Record*, no. 1443, pp. 92–99, 1994.

[8] C. M. Benedek and L. R. Rilett, "Equitable traffic assignment with environmental cost functions," *Journal of Transportation Engineering*, vol. 124, no. 1, pp. 16–22, 1998.

[9] S. Sugawara and D. Niemeier, "How much can vehicle emissions be reduced? Exploratory analysis of an upper boundary using an emissions-optimized trip assignment," *Transportation Research Record Journal of the Transportation Research Board*, no. 1815, pp. 29–37, 2002.

[10] A. Chen, Z. Zhou, and S. Ryu, "Modeling physical and environmental side constraints in traffic equilibrium problem," *International Journal of Sustainable Transportation*, vol. 5, no. 3, pp. 172–197, 2011.

[11] Z.-C. Li, W. H. K. Lam, S. C. Wong, and A. Sumalee, "Environmentally sustainable toll design for congested road networks with uncertain demand," *International Journal of Sustainable Transportation*, vol. 6, no. 3, pp. 127–155, 2012.

[12] A. Nagurney, P. Ramanujam, and K. K. Dhanda, "A multimodal traffic network equilibrium model with emission pollution permits: complicance vs noncompliance," *Transportation Research Part D: Transport and Environment*, vol. 3, no. 5, pp. 349–374, 1998.

[13] E. Fernandez, J. de Cea, M. Florian, and E. Cabrera, "Network equilibrium models with combined modes," *Transportation Science*, vol. 28, no. 3, pp. 182–192, 1994.

[14] R. García and A. Marín, "Network equilibrium with combined modes: models and solution algorithms," *Transportation Research Part B: Methodological*, vol. 39, no. 3, pp. 223–254, 2005.

[15] Z.-C. Li, W. H. K. Lam, S. C. Wong, D.-L. Zhu, and H.-J. Huang, "Modeling park-and-ride services in a multimodal transport network with elastic demand," *Transportation Research Record: Journal of the Transportation Research Board*, no. 1994, pp. 101–109, 2007.

[16] W. H. Lam, Z. Li, S. C. Wong, and D. Zhu, "Modeling an elastic-demand bimodal transport network with park-and-ride trips," *Tsinghua Science and Technology*, vol. 12, no. 2, pp. 158–166, 2007.

[17] C. Pineda, C. E. Cortés, P. Jara-Moroni, and E. Moreno, "Integrated traffic-transit stochastic equilibrium model with park-and-ride facilities," *Transportation Research Part C: Emerging Technologies*, vol. 71, pp. 86–107, 2016.

[18] Z. Song, Y. He, and L. Zhang, "Integrated planning of park-and-ride facilities and transit service," *Transportation Research Part C: Emerging Technologies*, vol. 74, pp. 182–195, 2017.

[19] S. Kitthamkesorn and A. Chen, "Alternate weibit-based model for assessing green transport systems with combined mode and route travel choices," *Transportation Research Part B: Methodological*, vol. 103, pp. 291–310, 2017.

[20] Z. Liu, X. Chen, Q. Meng, and I. Kim, "Remote park-and-ride network equilibrium model and its applications," *Transportation Research Part B: Methodological*, vol. 117, pp. 37–62, 2018.

[21] S. Wang, Q. Meng, and H. Yang, "Global optimization methods for the discrete network design problem," *Transportation Research Part B: Methodological*, vol. 50, pp. 42–60, 2013.

[22] Z. Liu, S. Wang, and Q. Meng, "Optimal joint distance and time toll for cordon-based congestion pricing," *Transportation Research Part B: Methodological*, vol. 69, pp. 81–97, 2014.

[23] T. Liu, H. Huang, H. Yang, and X. Zhang, "Continuum modeling of park-and-ride services in a linear monocentric city with deterministic mode choice," *Transportation Research Part B: Methodological*, vol. 43, no. 6, pp. 692–707, 2009.

[24] D. Z. W. Wang and B. Du, "Reliability-based modeling of park-and-ride service on linear travel corridor," *Transportation Research Record: Journal of the Transportation Research Board*, vol. 2333, pp. 16–26, 2013.

[25] B. Du and D. Z. W. Wang, "Continuum modeling of park-and-ride services considering travel time reliability and heterogeneous commuters – A linear complementarity system approach," *Transportation Research Part E: Logistics and Transportation Review*, vol. 71, no. 0, pp. 58–81, 2014.

[26] Z. Liu, Q. Meng, and S. Wang, "Variational inequality model for cordon-based congestion pricing under side constrained stochastic user equilibrium conditions," *Transportmetrica A: Transport Science*, vol. 10, no. 8, pp. 693–704, 2014.

[27] T. Larsson and M. Patriksson, "Side constrained traffic equilibrium models— analysis, computation and applications," *Transportation Research Part B: Methodological*, vol. 33, no. 4, pp. 233–264, 1999.

[28] Q. Meng, Z. Liu, and S. Wang, "Asymmetric stochastic user equilibrium problem with elastic demand and link capacity constraints," *Transportmetrica A: Transport Science*, vol. 10, no. 4, pp. 304–326, 2014.

[29] C. F. Daganzo, "Stochastic network equilibrium with multiple vehicle types and asymmetric, indefinite link cost Jacobians," *Transportation Science*, vol. 17, no. 3, pp. 282–300, 1983.

[30] G. E. Cantarella, "A general fixed-point approach to multimode multi-user equilibrium assignment with elastic demand," *Transportation Science*, vol. 31, no. 2, pp. 107–128, 1997.

[31] A. Chen, X. Xu, S. Ryu, and Z. Zhou, "A self-adaptive Armijo stepsize strategy with application to traffic assignment models

and algorithms," *Transportmetrica A: Transport Science*, vol. 9, no. 8, pp. 695–712, 2011.

[32] A. Chen, Z. Zhou, and X. Xu, "A self-adaptive gradient projection algorithm for the nonadditive traffic equilibrium problem," *Computers & Operations Research*, vol. 39, no. 2, pp. 127–138, 2012.

[33] Z. Zhou, A. Chen, and S. C. Wong, "Alternative formulations of a combined trip generation, trip distribution, modal split, and trip assignment model," *European Journal of Operational Research*, vol. 198, no. 1, pp. 129–138, 2009.

[34] S. Ryu, A. Chen, and K. Choi, "Solving the combined modal split and traffic assignment problem with two types of transit impedance function," *European Journal of Operational Research*, vol. 257, no. 3, pp. 870–880, 2017.

[35] B. S. He, H. Yang, Q. Meng, and D. R. Han, "Modified Goldstein-Levitin-Polyak projection method for asymmetric strongly monotone variational inequalities," *Journal of Optimization Theory and Applications*, vol. 112, no. 1, pp. 129–143, 2002.

[36] D. Han and W. Sun, "A new modified Goldstein-Levitin-Polyakprojection method for variational inequality problems," *Computers Mathematics with Applications*, vol. 47, no. 12, pp. 1817–1825, 2004.

[37] Y. Yin and S. Lawphongpanich, "Internalizing emission externality on road networks," *Transportation Research Part D: Transport and Environment*, vol. 11, no. 4, pp. 292–301, 2006.

[38] A. Nagurney, Q. Qiang, and L. S. Nagurney, "Environmental impact assessment of transportation networks with degradable links in an era of climate change," *International Journal of Sustainable Transportation*, vol. 4, no. 3, pp. 154–171, 2010.

[39] X. Yang, X. J. Ban, and R. Ma, "Mixed equilibria with common constraints on transportation networks," *Networks and Spatial Economics*, vol. 17, no. 2, pp. 547–579, 2017.

[40] S. Ryu, A. Chen, and K. Choi, "Solving the stochastic multiclass traffic assignment problem with asymmetric interactions, route overlapping, and vehicle restrictions," *Journal of Advanced Transportation*, vol. 50, no. 2, pp. 255–270, 2016.

[41] J. M. Ortega and W. C. Rheinboldt, *Iterative Solution of Nonlinear Equations in Several Variables*, Academic Press, Thailand, 1970.

7

Integrated Optimization on Train Control and Timetable to Minimize Net Energy Consumption of Metro Lines

Yuhe Zhou,[1] **Yun Bai**,[1] **Jiajie Li**,[1] **Baohua Mao**,[1] **and Tang Li**[2]

[1]*MOE Key Laboratory for Urban Transportation Complex Systems Theory and Technology, Beijing Jiaotong University, Beijing 100044, China*
[2]*Department of Civil and Environmental Engineering, Imperial College London, London SW7 2AZ, UK*

Correspondence should be addressed to Yun Bai; yunbai@bjtu.edu.cn

Academic Editor: Andrea D'Ariano

Energy-efficient metro operation has received increasing attention because of the energy cost and environmental concerns. This paper developed an integrated optimization model on train control and timetable to minimize the net energy consumption. The extents of train motoring and braking as well as timetable configurations such as train headway and interstation runtime are optimized to minimize the net energy consumption with consideration of utilizing regenerative energy. An improved model on train control is proposed to reduce traction energy by allowing coasting on downhill slopes as much as possible. Variations of train mass due to the change of onboard passengers are taken into account. The brute force algorithm is applied to attain energy-efficient speed profiles and an NS-GSA algorithm is designed to attain the optimal extents of motoring/braking and timetable configurations. Case studies on Beijing Metro Line 5 illustrate that the improved train control approach can save traction energy consumption by 20% in the sections with steep downhill slopes, in comparison with the commonly adopted train control sequence in timetable optimization. Moreover, the integrated model is able to significantly prolong the overlapping time between motoring and braking trains, and the net energy consumption is accordingly reduced by 4.97%.

1. Introduction

Metro systems develop rapidly across the world because of their high reliability and large capacity. For example, the total length of metro lines in mainland China was only about 1500 km in 2010 and more than 3700 km in 2016 and is expected to reach 6000 km by 2020. Although the metro is one of the most energy-efficient transport means, the long operating mileage and high traffic volume make the total energy consumption of metro systems become significant. Metro operators are facing more pressure to save energy due to increasing environmental concerns and operational costs. Energy-efficient train control and service timetable optimization are preferred to reduce metro train energy consumption in existing urban rail systems, as significant energy saving could be achieved with relatively low capital investment and minor modifications [1].

Energy-efficient train control (EETC) has been widely applied to real-world train operation since the 1980s. It aims to find the optimal sequence of train control modes and the switching points among the modes. Ichikawa [2] pointed out that energy-efficient train control on level tracks includes maximum acceleration (MA), cruising (CR), coasting (CO), and maximum braking (MB). Strobel et al. [3] found that it is also applicable to trains running in downhill and uphill sections with small slopes, where partial motoring and braking might be implemented to maintain cruising. Milroy [4] concluded that cruising is not necessarily required in short interstation runs. Subsequently, Benjamin et al. [5] and Howlett et al. [6] demonstrated that coasting presents opportunities for energy saving especially in long interstation runs. Khmelnitsky [7] researched the optimal operation of a train on a variable grade profile subject to arbitrary speed restrictions and presented a numerical algorithm to find the optimal velocity profile. Liu and Golovitcher [8] gave the analytical solution of optimal control and equations to find the control change points; speed limit and steep gradients are taken into account. Dongen and Schuit [9] and Scheepmaker and Goverde [10] proposed two models to find the optimal cruising speed and the proper coasting location to minimize energy consumption for train interstation runs.

In recent years, Howlett et al. [11] proposed energy-efficient control approaches for freight trains on steep tracks. Their survey indicated that coasting before a steep downhill section is very important for energy saving. Bai et al. [12] found that up to 9% of the energy consumption for train interstation runs could be saved by applying coasting as much as possible before braking. Domínguez et al. [13] and Carvajal-Carreño et al. [14] presented approaches to attain the optimal schemes of the ATO speed profile to obtain the Pareto front between train runtime and energy consumption. Sicre et al. [15] considered a specific model for energy-efficient manual driving in high speed lines by means of fuzzy parameters; uncertainty in manual control and punctuality requirement are taken into account. Keskin and Karamancioglu [16] attached more importance to train mass and compared the effectiveness of different evolutionary algorithms in attaining solutions for energy-efficient train control. Shangguan et al. [17] developed a multiobjective optimization approach for single-train speed trajectory.

Energy-efficient train timetabling (EETT) aims to reduce the net energy consumption of metro lines via optimizing the timetable configurations, such as the distribution of runtime supplements among different interstations, dwell time at stations, and train headways. Albrecht and Oettich [18] applied dynamic programming to find the optimal timetable considering both train energy consumption and passenger waiting time. Sicre et al. [19] attained the Pareto curves of journey time and energy consumption for high speed train interstation itinerary and then optimized the distribution of runtime supplements among different interstations. Chevrier et al. [20] proposed a biobjective evolutionary algorithm to attain speed profiles for train interstation runs in which runtime and energy consumption were optimized concurrently.

EETT has become more popular in recent years with the application of regenerative braking [21], which is able to convert kinetic energy into electricity during train braking. Regenerative electricity could provide up to 40% of the energy supplied to trains [22]. The regenerative energy is firstly used to support onboard auxiliary devices of the braking train and then fed back into the overhead contact line to assist motoring and auxiliary equipment of other trains in the same power supply interval (PSI). If the feedback energy cannot be used in time, it will be consumed by protective resistor, as most metro systems have not installed energy storage devices. The motoring and braking trains could be synchronized for better utilization of regenerative energy by optimizing timetable configurations, which determine train motoring and braking timing when the train departs from and approaches stations. Albrecht [23] explored the optimal allocation of runtime supplement for train interstation runs, to minimize the net energy consumption and reduce the maximal load of power systems. Yang et al. [24] built a model to maximize the overlapping time between the motoring and braking of successive trains by regulating train headway and dwell time. Further, Yang et al. [25] and Le et al. [26] proposed models to maximize the utilization of regenerative braking energy (RBE) as well. Peña-Alcaraz et al. [27] designed a mathematical programming model to synchronize the braking of trains arriving at station with the acceleration of trains exiting from the same or another station.

More energy saving could be achieved by integrated optimization on train timetable and control scheme. Ding et al. [28] proposed a two-level optimization model to find the energy-efficient train trajectory as well as the distribution of runtime supplements to minimize the traction energy without changing the predefined journey time. Su et al. [29, 30] developed an integrated optimization model, consisting of both train control and timetable formulation, to minimize the total energy consumption of multiple trains. Yang et al. [31] developed a stochastic programming model for the integrated optimization on train timetable and speed profile, where train mass in each interstation was set as a stochastic variable with a given probability distribution. Li and Lo [32] developed an integrated model on train control and timetable formulation to minimize the net energy consumption, where regenerative braking was taken into account. Zhao et al. [33] presented an integrated optimization model on train trajectory and timetable to reduce the net energy consumption and peak power of substations; the effectiveness of the proposed model was verified by a multiple-train movement simulator.

Many previous researches have explored the integrated optimization of train operation for energy saving, including both train control and timetabling, especially for the metro lines where RBE is available. In these studies, the maximum traction and braking force are commonly applied in train control. However, partial traction and braking when trains depart from and approach stations are able to improve the utilization of regenerative energy via prolonging train motoring and braking time, at the expense of a slight increment of traction energy consumption. Moreover, the EETC applied in EETT is usually based on the control sequence of MA-CR-CO-MB. Howlett et al. [11] proved that CR-CO-CR saves more energy than the control scheme of only CR when the train runs on steep downhill slopes. Liu and Golovitcher [8] also presented an equation to derive the conditions where different control modes should be applied with consideration of steep downhill slopes. Nevertheless, these improved train control strategies have not been employed in energy-efficient timetable optimization problems.

In this paper, an integrated optimization model on train control and timetable is developed to minimize the system net energy consumption, considering both traction energy reduction and utilization of regenerative braking energy. Synchronization of motoring and braking trains to better use the regenerative braking energy is realized by optimizing the extents of train motoring and braking as well as timetable configurations including headway and scheduled interstation runtime. An improved train control model allowing coasting on downhill slopes as much as possible is proposed to further reduce traction energy consumption compared with the simple control sequence of MA-CR-CO-MB, which is commonly adopted in previous energy-efficient timetabling studies. The proposed model takes into account the practical operation conditions such as varied train mass in different interstations, power peak of traction power supply system,

Table 1: Literature on energy-efficient timetabling with RBE.

Publication	Objective(s)	Variables	Speed profile	Mechanical braking	Power peak	Auxiliary devices	Varied mass
Albrecht (2004)	Power peak + energy consumption	Trip time	MA-CO-MB		√		
Peña-Alcaraz et al. (2011)	Overlapping time	Arrival & departure time			√	√	
Li and Lo Hong (2014)	Net energy consumption	Speed profile, headway, arrival & departure time	MA-CO-MB				
Yang et al. (2015)	Net energy consumption	Dwell time	MA-CO-MB			√	
Yang et al. (2016)	Traction energy	Speed profile, arrival & departure time	MA-CO-MB				√
Su et al. (2016)	Traction energy	Varied train characteristic	MA-CO-MB	√	√	√	√
Zhao et al. (2017)	Substation energy consumption	Speed profile, interstation journey time, service intervals	MA-CO-MB		√		
This paper	Net energy consumption	Speed profile, headway, interstation runtime	MA-[CO-CR]$_n$-MB	√	√	√	√

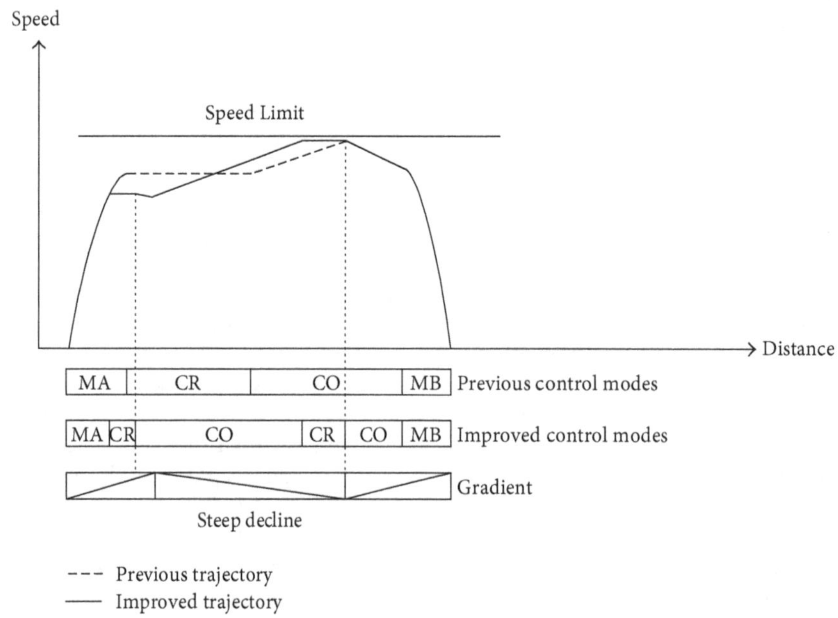

Figure 1: The improvements of the energy-efficient train control strategy.

and application of mechanical braking when the train adopts brakes at a very high speed. Table 1 gives a comparison between this study and the existing literatures.

The rest of this paper is organized as follows. In Section 2, an improved model on energy-efficient train control is presented. In Section 3, the integrated optimization model taking into account energy-efficient train control is developed to minimize the net energy consumption of a whole metro line. In Section 4, the brute force algorithm and an NS-GSA algorithm are presented to attain the optimal train trajectory and timetable, respectively. In Section 5, case studies on Beijing Metro Line 5 are conducted to verify the effectiveness of the proposed approach. Conclusions are provided in Section 6.

2. Energy-Efficient Train Control

There is usually more than one control scheme to drive a train from one station to the next even with the same runtime. The energy-efficient train control is to find the optimal control scheme leading to the minimal traction energy consumption while the scheduled runtime and speed limits are respected. Most previous studies proposed that the energy-efficient train control consists of the following four successive modes (i.e., MA, CR, CO, and MB) in the interstation with constant speed limit and no steep slopes, as illustrated by the dashed line in Figure 1.

With such a control, the train adopts maximum motoring followed by cruising at a constant speed and then coasting

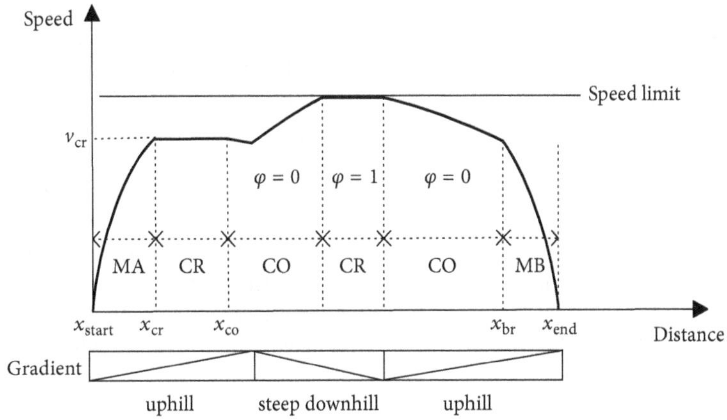

FIGURE 2: Improved energy-efficient train control scheme.

till braking for station stops. However, trains must brake to keep the speed constant in steep downhills, which is not preferred to minimize traction energy. The train could start cruising earlier and then adopt coasting on downhill slopes, followed by employing cruising and coasting again before stop braking, as illustrated by the solid line in Figure 1. As such, the energy consumption of the train interstation run could be saved, as the time for MA in the new trajectory is shortened while the interstation runtime remains the same. To this end, an improved model on train control is proposed to reduce the traction energy consumption by taking the full advantage of potential energy in downhill tracks, with the assumption that there is no change of speed limit in the whole interstation run.

2.1. Objective Function. The train adopts MA and MB in accelerating from and approaching stations in the proposed train control model, while the pairs of CR and CO are implemented in the interstation runs, as shown in Figure 2. It should be noted that Figure 2 is a sketch and switching points among different control modes might not be precise. With the proposed model, train control scheme is based on MA-[CR-CO]$_n$-MB and the positive integer n represents the number of CR and CO in one interstation journey. n is equal to 1 in most cases and the control scheme is MA-CR-CO-MB, which is the same as that attained by the previous model. n might be greater than 1 when the train is running on steep downhill tracks. Therefore, the proposed model is a supplement to previous work and it could contribute toward energy saving in the interstations with steep downhill profiles since more coasting is allowed.

The decision variables in the improved model are two switching points during one interstation run. The first point is the location where the first CR is applied, that is, x_{cr}. The second point is the location where the first CO is applied, that is, x_{co}. The other switching locations could be obtained once the decision variables are known. For example, the location to start the second CR is where the train reaches the speed limit. The train continues CR till it leaves the steep downhill slopes and CO is implemented again when the train does not accelerate with no traction power. The location to start applying MB, which is denoted by x_{br}, depends on the intersection point of coasting and braking profiles before the station stop. The objective function of the proposed model is to minimize the traction energy consumption for one interstation run, which is described as

$$\min \quad E_{trac}(x_{cr}, x_{co}) = \frac{\sum_{x=0}^{X} \max(\mathbf{F}(v,x), 0)}{3600 \cdot \eta_1}, \tag{1}$$

where E_{trac} is traction energy consumption with the given x_{cr} and x_{co} which are decision variables in EETT; η_1 is the energy conversion efficiency factor from electrical to mechanical energy; X is the length of one interstation; the units for distance, force, power, energy, and time are defined as m, kN, kW, kWh, and s in this paper; $\mathbf{F}(v, x)$ stands for the output force acting upon the train given by ATO. A positive value of $\mathbf{F}(v, x)$ denotes that the train is motoring. Zero indicates that the train is coasting and a negative value represents braking.

Traction force and braking force are expressed as a vector force that can be described as

$$\mathbf{F}(v,x) = \begin{cases} F_{tr}(v,x) & 0 \le x \le x_{cr} \\ F_{cr}(v,x) & x_{cr} < x \le x_{co} \\ \varphi \cdot F_{cr}(v,x) & x_{co} < x \le x_{br} \\ -F_{br}(v,x) & x_{br} < x \le X, \end{cases} \tag{2}$$

where $\mathbf{F}(v, x)$ is a vector force which is greater than 0 in traction mode or less than 0 in braking mode; $F_{tr}(v, x)$ is the maximum available traction force according to motor characteristic; $F_{cr}(v, x)$ is the force to keep cruising at position x; $F_{br}(v, x)$ is the maximum available braking force; φ is a binary variable with no unit.

$\varphi = 0$ indicates that the train control mode is CO, while $\varphi = 1$ stands for the case in which the control mode is CR. For example, coasting is implemented when the train arrives at x_{co} and φ is thus set as 0. When train speed increased to the limit, CR will be implemented again and φ becomes 1. Then resistance force F_R is used to decide the following

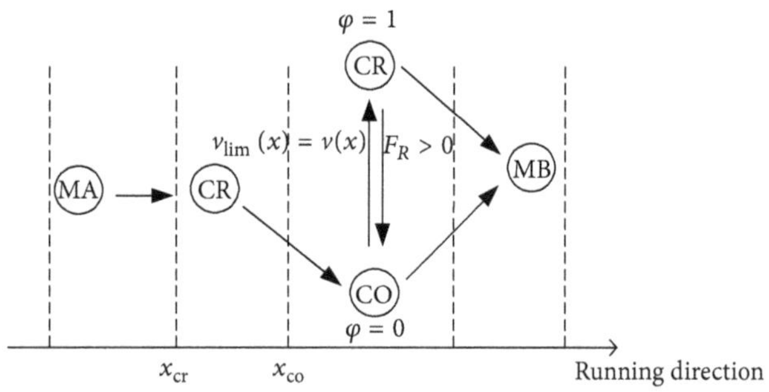

Figure 3: Transitions of train control modes with the improved model.

control regimes. φ can be therefore attained by the following equation:

$$\varphi(x) = \begin{cases} 0 & (v_{\lim}(x) - v(x)) + \varphi(x - \Delta x) \cdot F_R(v, x) > 0 \\ 1 & (v_{\lim}(x) - v(x)) + \varphi(x - \Delta x) \cdot F_R(v, x) \leq 0, \end{cases} \quad (3)$$

where $v_{\lim}(x)$ and $v(x)$ are speed limit and train speed at position x, respectively. $\varphi(x-\Delta x)$ represents the train control mode in the previous distance step, where Δx represents the length of distance step. $F_R(v,x)$ is resultant resistance, consisting of friction resistance, air drag, and additional resistance caused by grades and curves.

Figure 3 demonstrates the possible transitions among different control modes. At the beginning stage, MA should be applied till the location where the train starts cruising, which is the first decision variable x_{cr}. In the mode of CR, the effort acting upon the train by ATO is equal to the resistance to keep train speed constant. Train continues its travel at a constant speed till it reaches the location to start CO, which is the second decision variable x_{co}. The following driving mode depends on train speed and running resistance according to (3). If train speed $v(x)$ does not exceed the speed limit $v_{\lim}(x)$, coasting phase will be continued. If train speed exceeds the limit, cruising phase will be adopted again. The coasting will be applied again only when the resultant resistance is positive, which indicates train speed decreases when coasting is applied. Finally, MB is applied before the station stop and the braking curve is attained by backward calculation from the target stop point using the maximum available braking effort.

According to the characteristics of train motor, the traction force varies with train speed. When MA is applied, there are three traction force curves to be chosen automatically in response to three train weights, as shown in Figure 4(a). When the train is empty, the traction force curve of AW0, which is the lowest one among all the three curves, is chosen because the train does not need too much traction force for acceleration. In case the number of passengers on the train is equal to the nominal capacity, the traction force curve of AW1 is adopted to allow a bigger acceleration rate than that of AW0. The traction force curve of AW2, which is the highest one, is selected when the train carries the most passengers.

For the circumstances in which the actual train weight is not equal to one of the three weights above, the traction force is calculated as follows.

Firstly, we need to calculate train mass to obtain the available traction force.

$$m(x) = M_{\text{train}} + \mu \cdot \tau(n), \quad (4)$$

where $m(x)$ is the actual train mass; M_{train} is the rolling stock mass; $\tau(n)$ is the number of passengers on the train in the interstation of n; μ is the average mass of a person.

Secondly, linear interpolation is used to attain the practical traction force, as shown in

$$F_{tr}(v, x) = \begin{cases} F_0(v) + (F_1(v) - F_0(v)) \cdot \dfrac{m(x) - M_0}{M_1 - M_0} & M_0 \leq m(x) < M \\ F_1(v) + (F_2(v) - F_1(v)) \cdot \dfrac{m(x) - M_1}{M_2 - M_1} & M_1 \leq m(x) \leq M_2, \end{cases} \quad (5)$$

where M_0, M_1, and M_2 represent train masses when the train is empty, nominally loaded, and maximum loaded, respectively; $F_0(v)$, $F_1(v)$, and $F_2(v)$ indicate the traction force in the above three circumstances.

Train mass also has impacts on train braking force. There are usually three different braking force curves corresponding to three different load factors, that is, AW0-AW1-AW2. The braking force under different load factors can be calculated similarly. However, the braking force usually keeps constant when train speed varies [34–36]. It should be noted that the practical braking force (F_{br}) is the combination of both mechanical braking and regenerative braking. Figure 4(b) shows the components of braking force, taking the case of AW1 as an example. Regenerative braking force (F_{eb}) declines gradually when train speed becomes very large or very small, and mechanical braking (F_{mech}) is applied to compensate the shortage of braking force.

As the traction force $F_{tr}(v,x)$ and braking force $F_{br}(v,x)$ are obtained, the traction energy consumption can be calculated by (1) and (2) once we obtained the value of $F_{cr}(v,x)$. The resultant force $F(v,x)$ acting upon the train can be calculated as

$$\begin{aligned} F(v, x) &= \mathbf{F}(v, x) - F_R(v, x) \\ F_R(v, x) &= R_{\text{basic}}(v) + R_{\text{grad}}(x) + R_c(x), \end{aligned} \quad (6)$$

where $R_{\text{basic}}(v) = a + bv + cv^2$ and the coefficients of a, b, and c are usually given by rolling stock manufacturers; $R_{\text{grad}}(x)$ is the resistance caused by the gradient; $R_c(x)$ is the resistance caused by the curve.

When cruising is applied, the resultant force $F(v,x)$ should be equal to zero; then, $F_{cr}(v,x)$ can be obtained based on Newton's law of motion:

$$\begin{aligned} F_{cr}(v, x) &= F_R(v, x) \\ &= m(x)\, g \cdot \left[\left(a + bv + cv^2\right) + \sin\theta_x + \frac{600}{r_x} \right], \end{aligned} \quad (7)$$

where g is gravitational acceleration; θ_x is the angle of slope (negative means downhill); r_x is the radius of the curve at location x.

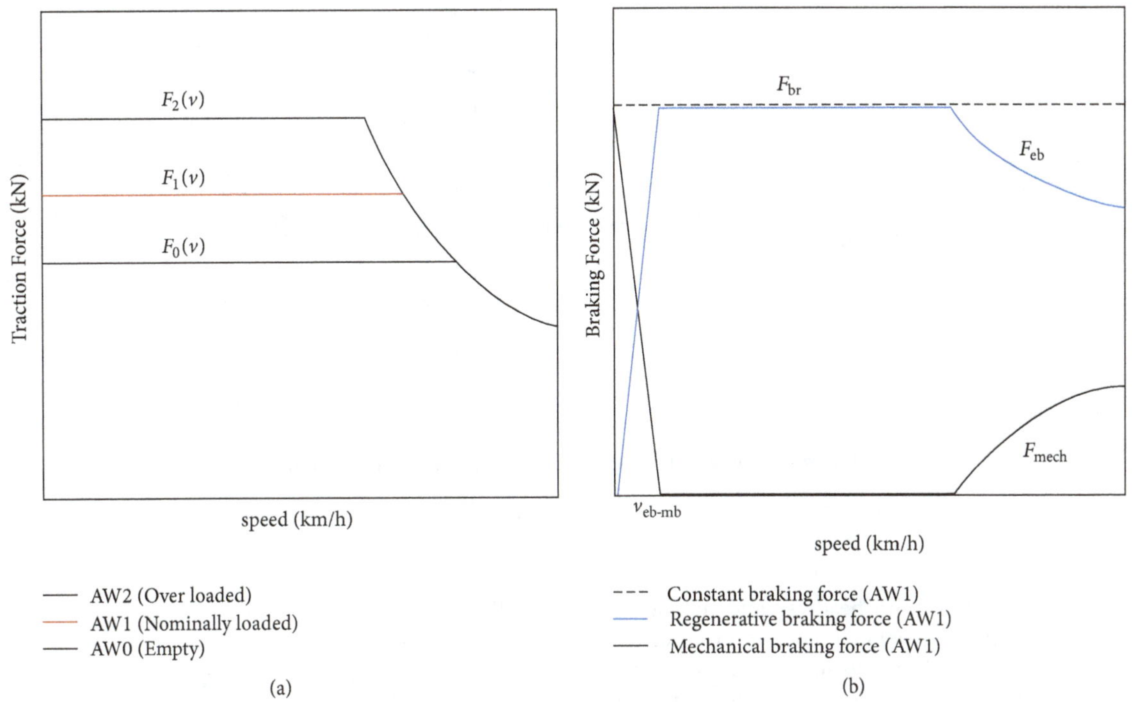

FIGURE 4: Maximum force against train speed. (a) Maximum traction force with different load factors. (b) Available braking forces in AW1.

2.2. Constraints. Train control schemes should be subject to the following constraints.

(1) For operational safety and passenger comfort, the acceleration should be limited within a proper range:

$$a_{\text{dec}} \leq \frac{dv(x)}{dt} \leq a_{\text{acc}}. \tag{8}$$

(2) The train must stop when it arrives at a station:

$$v(x_n) = 0, \quad \forall n \in [1, 2N]. \tag{9}$$

(3) Train velocity must not exceed the speed limit or be negative:

$$0 \leq v(x) \leq v_{\text{lim}}(x). \tag{10}$$

(4) The error between practical runtime and scheduled runtime should be less than a certain threshold. In this paper, we set this threshold as $\delta = 0.5$:

$$\left| \text{RT}_n - \sum_{x=x_n}^{x_{n+1}} \frac{\Delta x}{v(x)} \right| \leq \delta, \tag{11}$$

where RT_n is the scheduled runtime in interstation n; x_n refers to the location of station n.

(5) This constraint aims to calculate regenerated braking energy for timetable optimization in Section 3. As a matter of fact, regenerative braking force varies with train speed as well as load factor, and it will be entirely replaced by mechanical braking when train speed is lower than $v_{\text{eb-mb}}$.

$$F_{\text{eb}}(v, x) = \begin{cases} \min\left(F_{\text{eb}}(v, x), F_{\text{br}}\right), & v \geq v_{\text{eb-mb}} \\ 0, & v < v_{\text{eb-mb}}. \end{cases} \tag{12}$$

3. Integrated Optimization Model on Train Control and Timetable

With the application of regenerative braking, synchronization of motoring and braking trains in the same PSI for better utilization of regenerative energy becomes very important in minimizing the net energy consumption of metro lines, since the regenerative energy can provide a significant proportion of the total energy fed into rolling stocks. Train control also has a great influence on timetable optimization, as different control schemes lead to different energy consumption even when the interstation runtime remains the same.

An integrated model on train control and timetable optimization is proposed in this section to minimize the net energy consumption, taking into account the improved train control model described in Section 2. Train headway and interstation runtimes are optimized to improve the utilization of RBE as well as reduce the traction energy consumption by allocating the runtime supplements properly. In addition, power control system could be modified to constrain the output of traction and braking force when trains depart from and approach stations, which is able to enlarge the space for synchronization of motoring and braking trains via prolonging train motoring and braking time, although traction energy consumption increases slightly. Different from the previous studies, the new integrated model allows the train to adopt partial motoring and braking in accelerating from and approaching stations.

Figure 5 illustrates the influences on energy consumption by train headway, interstation runtime, and percentages of the full traction and braking force applied in train control. The motoring of train B and the braking of train A could be

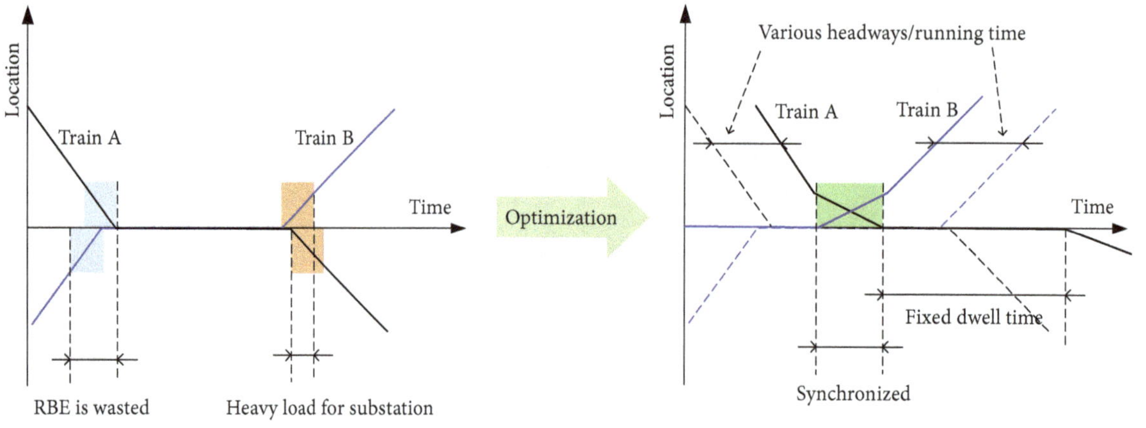

FIGURE 5: Energy-saving methods in integrated optimization.

synchronized by regulating train headway and interstation runtime; therefore, the regenerated energy produced by braking train A can be transferred to and utilized by traction train B. As a result, net energy consumption is reduced because the energy used in the accelerating phase of train B is provided by the regenerative braking of train A. The overlapping time, which means the time slot in which a train adopts regenerative braking while there is at least one train motoring in the same PSI, is further extended using partial traction and braking force.

Net energy consumption could be saved due to the much higher utilization of RBE, although the traction energy consumption may increase slightly. It should be pointed out that dwell time is not allowed to be changed in the integrated optimization model because it usually depends on the number of passengers getting on and off the train.

3.1. Objective Function. In this study, the whole problem of energy-efficient train operation is decomposed into train trajectory formulation and timetable optimization, which are successively processed. Firstly, train control scheme in each interstation run is optimized with the given runtime (RT), k_t, and k_b. Secondly, a database, including the optimal train control scheme and the corresponding energy consumption with all possible groups of RT, k_t, and k_b, is then built. Finally, the optimal HD, RT, k_t, and k_b are obtained by an evolutionary algorithm on the basis of the database. It should be noted that k_t and k_b are assumed as two constants for the sake of calculating efficiency and convenience of train control implementation.

The objective of the integrated model is to find the optimal utilization coefficients of traction and braking force (k_t, k_b), interstation runtime (RT), and headway (HD) to minimize the net energy consumption for all trains $j \in \{1, 2, 3, \ldots, J\}$ operating on the whole line, which is the difference between the required traction energy and the utilized regenerative energy. Additionally, energy consumed for auxiliary devices is also taken into account. The objective function is expressed as

$$\begin{aligned} \min \quad & E\left(k_t, k_b, \mathrm{RT}, \mathrm{HD}\right) \\ & = \sum_{j=1}^{J} \sum_{n=1}^{2N} E_{\mathrm{trac}}^{n,j}\left(k_t, k_b, \mathrm{RT}_n\right) + E_{\mathrm{aux}} \\ & \quad - \sum_{t=1}^{T_{\mathrm{total}}} \sum_{n_p=1}^{N_p} \sum_{j \in \kappa} E_{\mathrm{reg\_u}}\left(t, n_p, j, \mathrm{RT}, \mathrm{HD}\right), \end{aligned} \tag{13}$$

where k_t and k_b represent traction and braking force utilization coefficients; N_p is the number of PSIs; n_p represents a PSI on the metro line and $n_p \in [1, N_p]$; $E_{\mathrm{trac}}^{n,j}(k_t, k_b, \mathrm{RT}_n)$ is total traction energy consumption in interstation n for train j; E_{aux} is energy consumed by auxiliary devices; $E_{\mathrm{reg\_u}}(t, n_p, \mathrm{RT}, \mathrm{HD})$ is the amount of utilized RBE in n_p at time step t; T_{total} is the length of operation period; κ denotes the set of trains located in the same power supply interval n_p.

Partial traction and partial braking are applied to extend the motoring and braking phases, which result in a modification on the EETC model. It is known that the efficiency of traction systems usually increases with the extent of motoring [37] and a constant efficiency factor could lead to inaccurate results [38]. Therefore, energy conversion factor η_1 in (1) is defined as a variable calculated as follows [39]:

$$\eta_1 = \begin{cases} 1 & 0.95 \le k_t < 1 \\ 0.6620 + 0.3558 \cdot k_t & 0.5 \le k_t < 0.95 \\ 0.5714 + 0.537 \cdot k_t & k_t < 0.5. \end{cases} \tag{14}$$

Due to the force utilization coefficients, (2) in Section 2.1 is rewritten as

$$\mathbf{F}\left(k_t, k_b, v, x\right) = \begin{cases} k_t \cdot F_{\mathrm{tr}}\left(v, x\right) & 0 \le x \le x_{\mathrm{cr}} \\ F_{\mathrm{cr}}\left(v, x\right) & x_{\mathrm{cr}} < x \le x_{\mathrm{co}} \\ \varphi \cdot F_{\mathrm{cr}}\left(v, x\right) & x_{\mathrm{co}} < x \le x_{\mathrm{br}} \\ -k_b \cdot F_{\mathrm{br}}\left(v, x\right) & x_{\mathrm{br}} < x \le X. \end{cases} \tag{15}$$

Then, traction energy consumption could be calculated by attaining the optimal (x_{cr}, x_{co}) in energy-efficient train control model.

The traction energy comprises not only propulsion of the train, but also consumption of auxiliary systems aboard the train such as lighting, air conditioning, and the signal system. Usually, the power of onboard auxiliary devices P_{aux} remains constant during train movements. Hence, the total energy consumption for onboard auxiliary devices is only related to the cycle time, that is, the period required for one train to complete one cycle, which is expressed as C.

$$E_{aux} = \sum_{j=1}^{J} \frac{P_{aux}^{j} C}{3600}. \quad (16)$$

The calculation of utilized RBE in each PSI during the operation period is more complex and the framework on utilization of RBE is given in Figure 6. The RBE is sequentially used to support auxiliary devices on the braking trains and traction trains in the same PSI and auxiliary devices on the other trains except braking trains. It should be noted that the regenerative energy from braking trains can only be transmitted among trains which are operating in the same PSI.

The utilization of RBE consists of two parts. The first part is used to support onboard auxiliary devices in braking trains, and the second part is used by traction trains and onboard auxiliary devices on other trains except braking trains, which is distinguished by different lines in Figure 6. Therefore, utilized RBE for all trains j in the same PSI at time step t can be calculated as

$$E_{reg\_u}(\xi) = E_{reg\_u}^{self}(\xi) + \eta_2 \cdot E_{reg\_u}^{other}(\xi), \quad (17)$$

where $\xi = \{t, n_P, j, \mathrm{RT}, \mathrm{HD} \mid 1 \le t \le T_{total}, 1 \le n_P \le N_P, j \in \kappa\}$; η_2 is the transmission loss factor of the regenerative energy.

The first part of utilized RBE which is used by braking trains is calculated by (18). If train j is braking in power supply interval n_p at time t, the vector force $\mathbf{F}$ will be negative. This means the train is providing regenerative energy. A portion of RBE is used for onboard auxiliary devices on the braking train j firstly.

$$E_{reg\_u}^{self}(\xi) = \sum_{j\in\kappa} \min\left(\eta_3 \frac{\left|\min\left(\mathbf{F}_j(v,x)\, v\Delta t, 0\right)\right|}{3600}, \frac{P_{aux}^{j}\Delta t}{3600}\right), \quad (18)$$

where η_3 denotes the conversion factor from mechanical energy to regenerative energy.

The rest of the RBE, which can be attained by (19), is available to accelerate the motoring train followed by providing power for the auxiliary devices of other trains except braking trains in the same PSI. The energy required by motoring trains is calculated by (20).

$$E_{reg}^{avai}(\xi) = \sum_{j\in\kappa} \max\left(\eta_3 \frac{\left|\min\left(\mathbf{F}_j(v,x), 0\right)\right| v\Delta t}{3600} - \frac{P_{aux}^{j}\Delta t}{3600}, 0\right) \quad (19)$$

$$Ee_{trac}(\xi) = \sum_{j\in\kappa} \frac{\max\left(\mathbf{F}_j(v,x), 0\right) v\Delta t}{3600 \cdot \eta_1}. \quad (20)$$

If the available RBE is more than the total traction energy required by motoring train in the same PSI, the surplus RBE is then used to support auxiliary devices for other trains except the braking trains in the same PSI. The total energy required by all the trains in the same PSI except the braking trains can be calculated by (21). Finally, the total utilized regenerated energy by traction trains and other trains except the braking trains is calculated in (22).

$$Ee_{aux}(\xi) = \sum_{j\in\kappa} \frac{P_{aux}^{j}\Delta t}{3600}, \quad \forall \mathbf{F}_j(v,x) \ge 0 \quad (21)$$

$$E_{reg\_u}^{others}(\xi) = \begin{cases} Ee_{trac}(\xi) + \min\left(E_{reg}^{avai}(\xi) - Ee_{trac}(\xi), Ee_{aux}(\xi)\right) & Ee_{trac}(\xi) \le E_{reg}^{avai}(\xi) \\ E_{reg}^{avai}(\xi) & Ee_{trac}(\xi) > E_{reg}^{avai}(\xi), \end{cases} \quad (22)$$

where $Ee_{aux}(\xi)$ is the total energy consumption of onboard devices of trains in the same PSI, except the braking trains.

3.2. Constraints. In this paper, we consider the operation period from the time when the first train is put into operation in the up direction to the time when the last train returns to the depot from the down direction. During the operation period, all trains have the same cycle time, traction and braking force utilization coefficients, interstation runtime, and dwell time. The constraints for energy-efficient timetable are as follows:

(1) Runtime constraints: the interstation runtime should be limited within a certain range for the consideration of service quality and the maximum traction force.

$$\underline{T_{cn}} \le T_n \le \overline{T_{cn}}, \quad \forall n \in [1, N-1] \cup [N+1, 2N-1]. \quad (23)$$

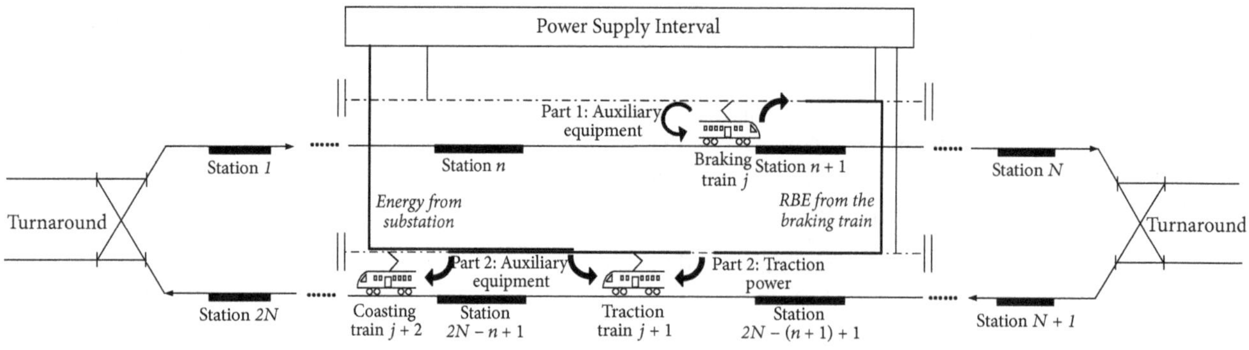

FIGURE 6: The utilization of regenerative braking energy.

(2) Turnaround time constraints: the turnaround time should be larger than the minimum time due to the track length of turnaround and limited travel speed, and less than a certain upper bound for the consideration of service quality and rolling stock operation.

$$\underline{T_{\text{turn}}} \le T_n \le \overline{T_{\text{turn}}}, \quad n \in \{N, 2N\}. \tag{24}$$

(3) Cycle time constraints: the total travel time for one cycle in the optimal timetable should be equal to the one in the original timetable, to keep the service frequency unchanged. In this study, we assume turnaround time as a special interstation runtime.

$$C = \sum_{n=1}^{2N} T_n + \sum_{n=1}^{2N} T_n^{\text{dwell}}. \tag{25}$$

(4) Headway constraints: the headway between any two successive trains must be less than the maximal headway required by service quality and larger than the minimal one for safety concerns.

$$\underline{H} \le H_j \le \overline{H}. \tag{26}$$

(5) Power peak: the total traction power of all trains located in the same PSI at each time moment should be no more than the maximal load of the power supply systems $P_{\text{sub}}^{\max}$.

$$\sum_j P_j(t) \le P_{\text{sub}}^{\max}. \tag{27}$$

(6) The percentages of full traction and braking force applied in motoring and braking should be limited in a reasonable range, considering the service quality requirement.

$$\begin{aligned} k_b &\in [0.5, 1] \\ k_f &\in [0.5, 1]. \end{aligned} \tag{28}$$

4. Solution Algorithm

Evolutionary algorithms, particularly the genetic algorithm (GA), have been widely adopted to solve such kind of problems especially the scheduling optimization [40–42]. However, it is easy to fall into a local solution rather than the global optimum if a standard GA is employed to solve such a complex integrated optimization problem. To improve the computing efficiency, the whole problem of energy-efficient train operation is decomposed into train trajectory formulation and timetable optimization, which are successively processed in this study. Figure 7 shows the whole structure of the optimization process.

In the single-train trajectory optimization, the brute force algorithm is applied to find the exact solution for train control with the given traction and braking force utilization coefficients as well as runtime in each interstation, and the results are saved in a database. Then, a genetic annealing algorithm with neighborhood search strategy is implemented in the integrated optimization to calculate the proper traction and braking force utilization coefficients, runtime, and headway. Train trajectories and traction energy consumption are directly taken from the database.

4.1. Brute Force Algorithms to Attain Train Trajectories. Brute force is an exact algorithm by searching all the possible solutions and evaluating their fitness, which has been successfully used in railway operation optimization [43, 44]. The main steps of brute force algorithm are as follows.

(1) Enumerating: the energy consumption and interstation runtime of all the possible solutions (x_{co}, x_{cr}) in the interstation run will be calculated, with the given force utilization coefficients (k_t, k_b) and running direction (d). The results will be stored in the database.

(2) Selection: the database may include a number of solutions with the same interstation runtime (RT), but the traction energy consumption is different. For each runtime, only the solution with the least traction energy consumption will be retained and other solutions with the same runtime will be removed from the database.

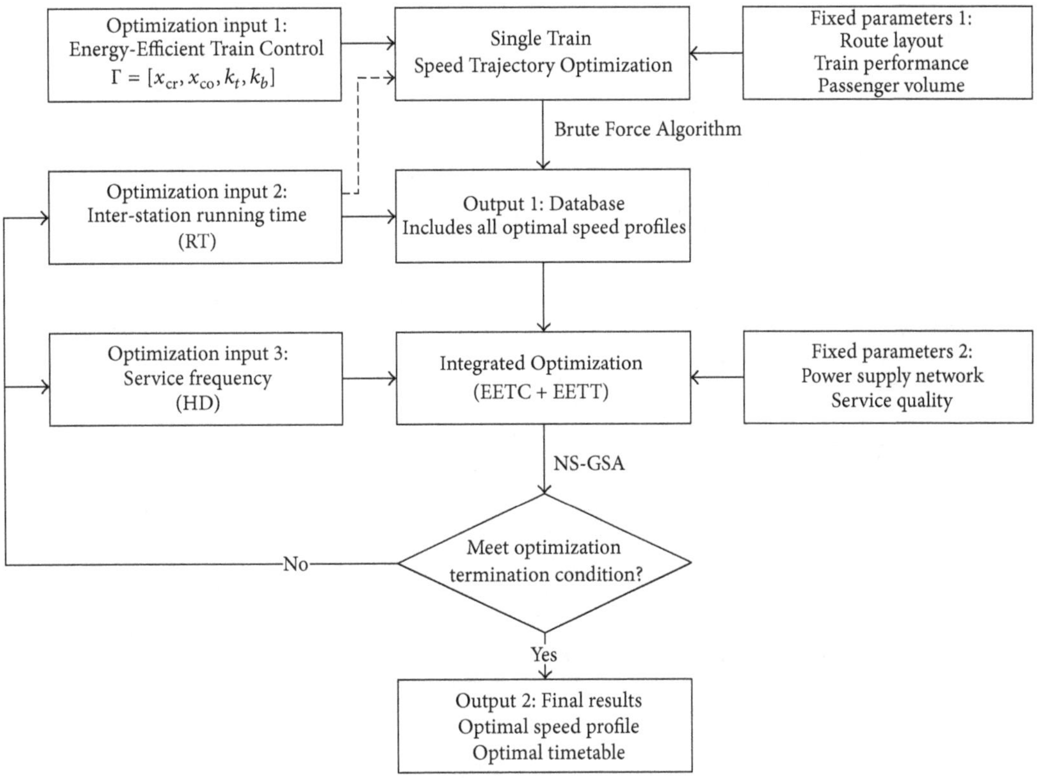

FIGURE 7: Flowchart of the integrated optimization.

(3) Updating: the force utilization coefficients (k_t, k_b) and running direction (d) will be updated, while the value should satisfy the constraint in (28). Then, the algorithm returns back to Step (1). If all the possible combinations of k_t, k_b, and d in this interstation are calculated, go to Step (4).

(4) Termination: the optimal train control schemes in the interstation with different runtimes, running directions, traction force utilization coefficients k_t, and braking force utilization coefficients k_b are obtained and stored in the database.

The algorithm takes about 1.1 hours to generate a database for one interstation, which gives the optimal solutions and the corresponding energy performance under 2160 different operation conditions, involving 6 traction coefficients, 6 braking coefficients, 30 runtimes, and 2 run directions.

4.2. NS-GSA Algorithm in Timetable Optimization. An improved genetic annealing algorithm with neighborhood search strategy (NS-GSA) is used to optimize the timetable. To overcome the premature convergence in the standard genetic algorithm, a simulated annealing (SA) algorithm is introduced to escape from the local optimum and approach the global optimum. Metropolis criterion of simulated annealing (SA) allows a decreasing probability of accepting worse solutions, which provides a diverse population for GA without compromising solution convergence. Additionally, the solution space is quite large in this integrated optimization problem, and the variables are continuous. Therefore, it is necessary to enhance the local search ability of genetic simulated annealing (GSA) algorithm, and neighborhood search strategy (or local search) is used which has been proved to be an effective search technique and is widely implemented in railway operation and management [45–48]. The procedure of NS-GSA is shown in Figure 8.

The main steps of the developed algorithm are as follows:

(1) Initialization: a random initial population of the solutions is produced to form the first generation. A number of individuals are included in the population and each individual represents a set of traction force utilization coefficient (k_t), braking force utilization coefficient (k_b), interstation runtime (RT), and headway (HD), which are the decision variables in the integrated optimization model.

(2) Neighborhood search: each individual explores its neighborhood to enhance the local search ability of the algorithm. The details on neighborhood search are presented below.

 (a) Initialize the set of neighborhood structures $\mathrm{SOL}(p,q)$, $p \in \{1,\ldots,P\}$, $q \in \{1,\ldots,Q\}$. P is the number of individuals and Q is the maximum number of local searches for each individual. Set $p = 1$, $q = 1$.

 (b) Until $q = Q + 1$, repeat the following operations. Through randomly generating a set of RT within a reasonable range, a set of solutions $\mathrm{SOL}(p,q)$ in the neighborhood of $\mathrm{SOL}(p)$ is

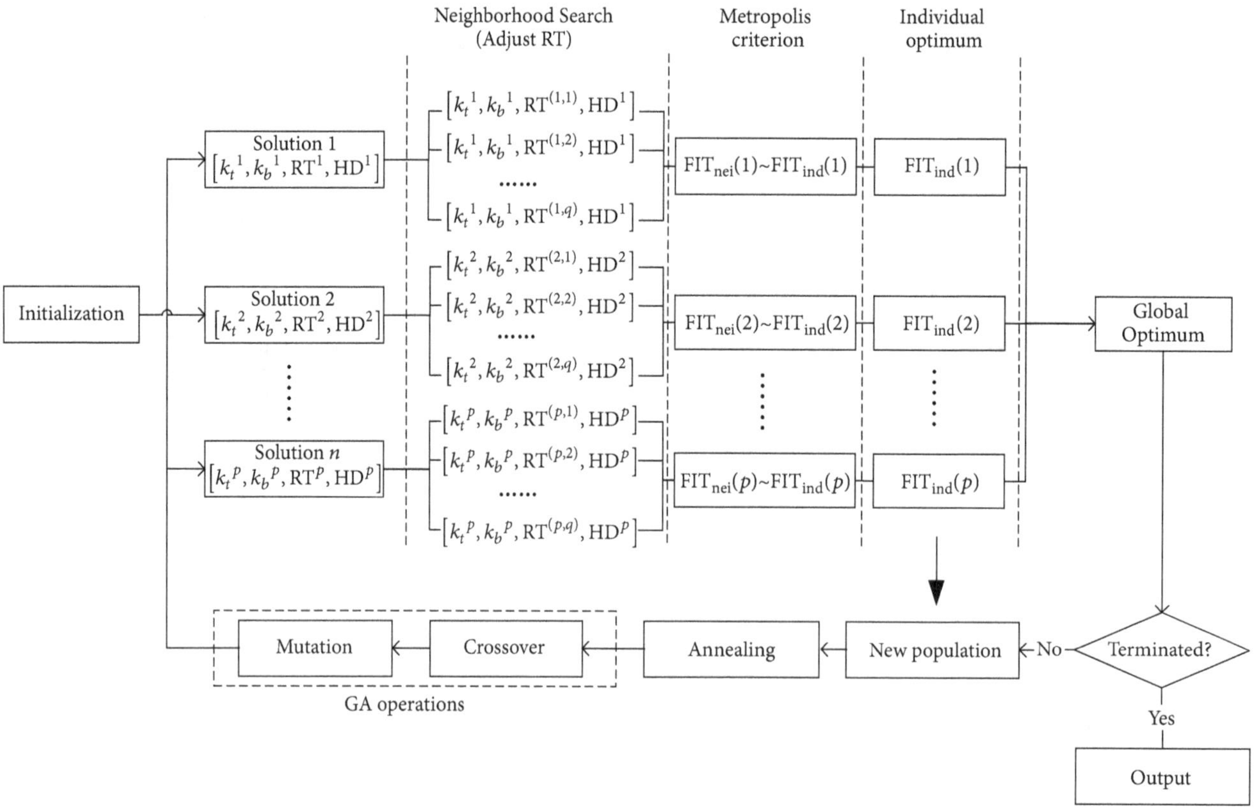

FIGURE 8: Flowchart of NS-GSA algorithm.

produced by replacing the RT in SOL(p) with the newly generated RT. It should be noted that the journey time should not be changed in each local searching. Set $q \leftarrow q + 1$.

(c) Set $p \leftarrow p+1$. If $p > P$, finish the neighborhood search. Otherwise, set $q = 1$ and go to Step (b).

(3) Evaluation: each solution in the population will be evaluated. Equation (13) is used to calculate the fitness FIT(p, q) of solution SOL(p, q).

(4) Acceptance: if the best neighborhood solution of the individual p with the fitness $\text{FIT}_{\text{nei}}(p)$ is better than the individual fitness $\text{FIT}_{\text{ind}}(p)$, the individual will be replaced by the neighborhood solution. Otherwise, the individual will be replaced by the best neighborhood solution according to the Metropolis criterion in simulated annealing algorithm. The acceptance criterion is listed as follows:

$$\begin{aligned} &\forall p \in P, \text{if } \text{FIT}_{\text{nei}}(p) < \text{FIT}_{\text{ind}}(p) \\ &\text{FIT}_{\text{ind}}(p) = \text{FIT}_{\text{nei}}(p) \\ &\text{elseif } e^{(\text{FIT}_{\text{ind}}(p) - \text{FIT}_{\text{nei}}(p))/\text{Temper}} < \text{Rand}(0,1) \\ &\text{FIT}_{\text{ind}}(p) = \text{FIT}_{\text{nei}}(p). \end{aligned} \quad (29)$$

(5) New generation: if the termination conditions are not satisfied, the optimal individuals after neighborhood searching will be used to create a new generation. The parameter Temper in Metropolis criterion decreases at a certain rate at the same time.

(6) Crossover: set the random number k_c ($0 \le k_c \le 1$) and the probability of crossover P_c. If $k_c \le P_c$, then crossover operation will randomly select three positions of genes in three parts of the chromosome, corresponding to force utilization coefficients, inter-station runtime, and headway. Then, two chromosomes in the population will exchange these genes. As shown in Figure 9, genes which have been exchanged in crossover are marked in green.

(7) Mutation: set random number k_m ($0 \le k_m \le 1$) and the probability of crossover P_m. If $k_m \le P_m$, the mutation operation will randomly select one gene from one chromosome and replace the gene with a random number within the reasonable range. It should be noted that constraints in Sections 2.2 and 3.2 should be satisfied.

(8) Termination: the algorithm returns to Step (2) and repeats until the maximum generation is reached.

The required CPU time of NS-GSA algorithm is about 20 hours to attain the energy-efficient timetable as well as the optimal control schemes using a computer with 3.2 GHz processor speed and 4 GB memory, when the number of individuals in GA is 20, the number of neighborhood solutions for each individual is 2, and the maximum generation is 40.

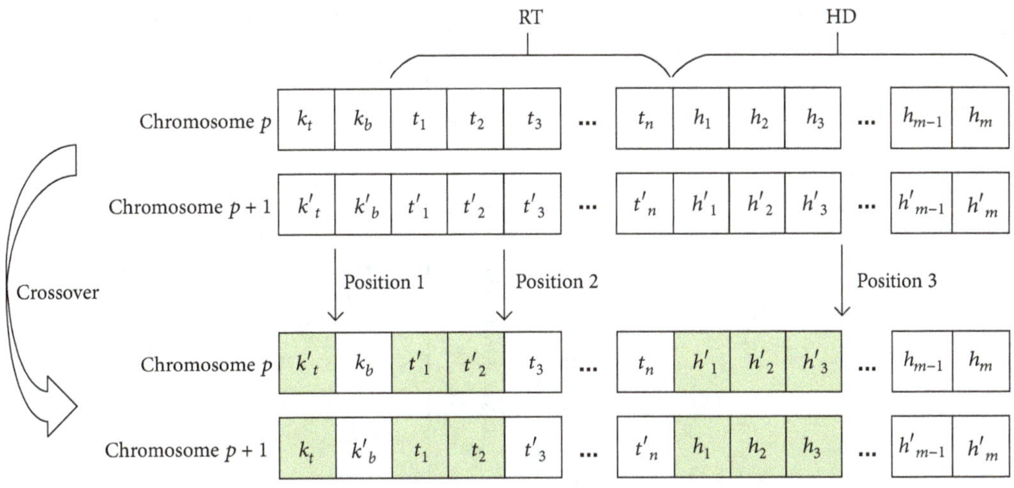

FIGURE 9: The structure of chromosome and crossover operation.

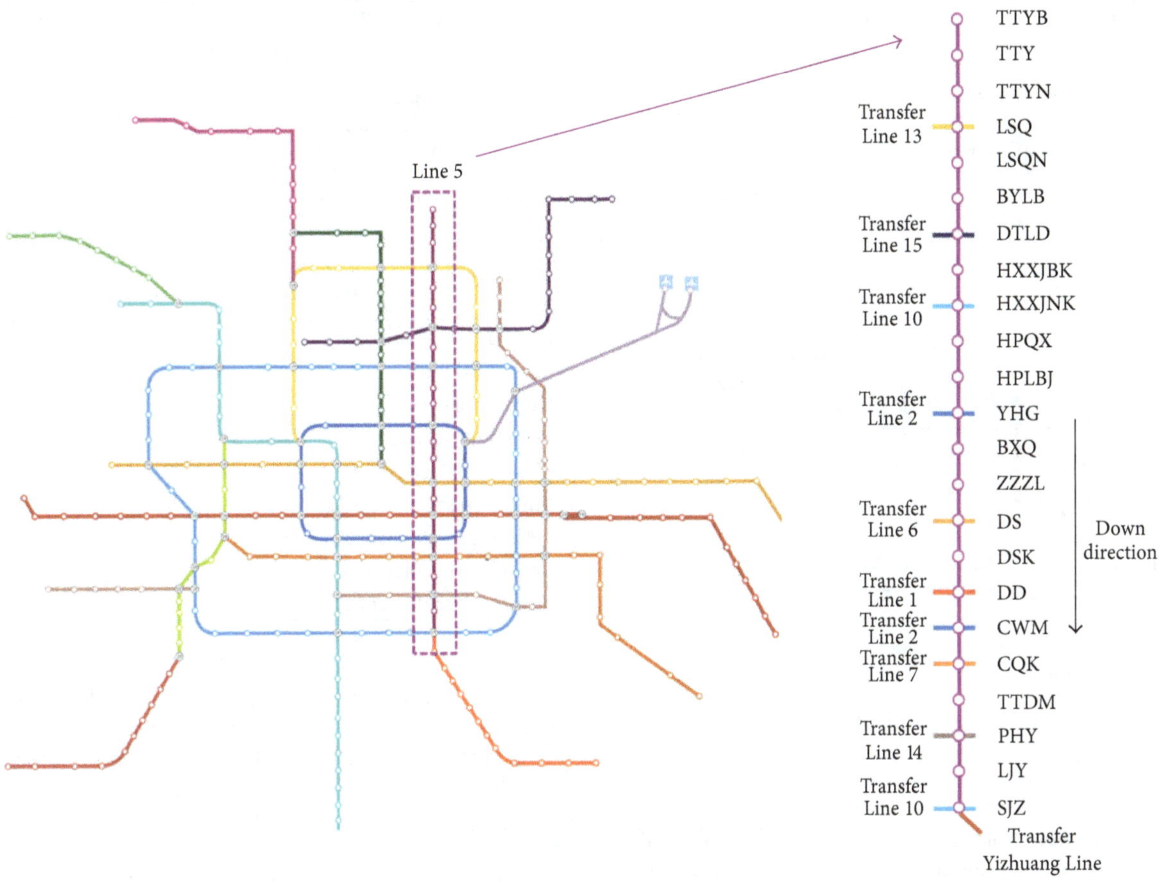

FIGURE 10: Beijing Metro network and Line 5.

5. Case Study

In this section, case studies on Beijing Metro Line 5 in peak hours are employed to illustrate the effectiveness of the improved train control and integrated optimization model.

5.1. Basic Data. The Beijing Metro Line 5 is 27.6 km long with 23 stations. It covers several commercial centers and links the residential districts in north and south of Beijing (see Figure 10). Table 2 gives the basic information of trains servicing on Beijing Metro Line 5 and Figure 11 shows the traction and regenerative braking characteristics of the train.

The number of passengers on the train in different interstation runs is shown in Figure 12, taking peak hour as an example. It is used to attain train mass, assuming the mass of each person is 60 kg. There are 11 PSIs on Beijing Metro

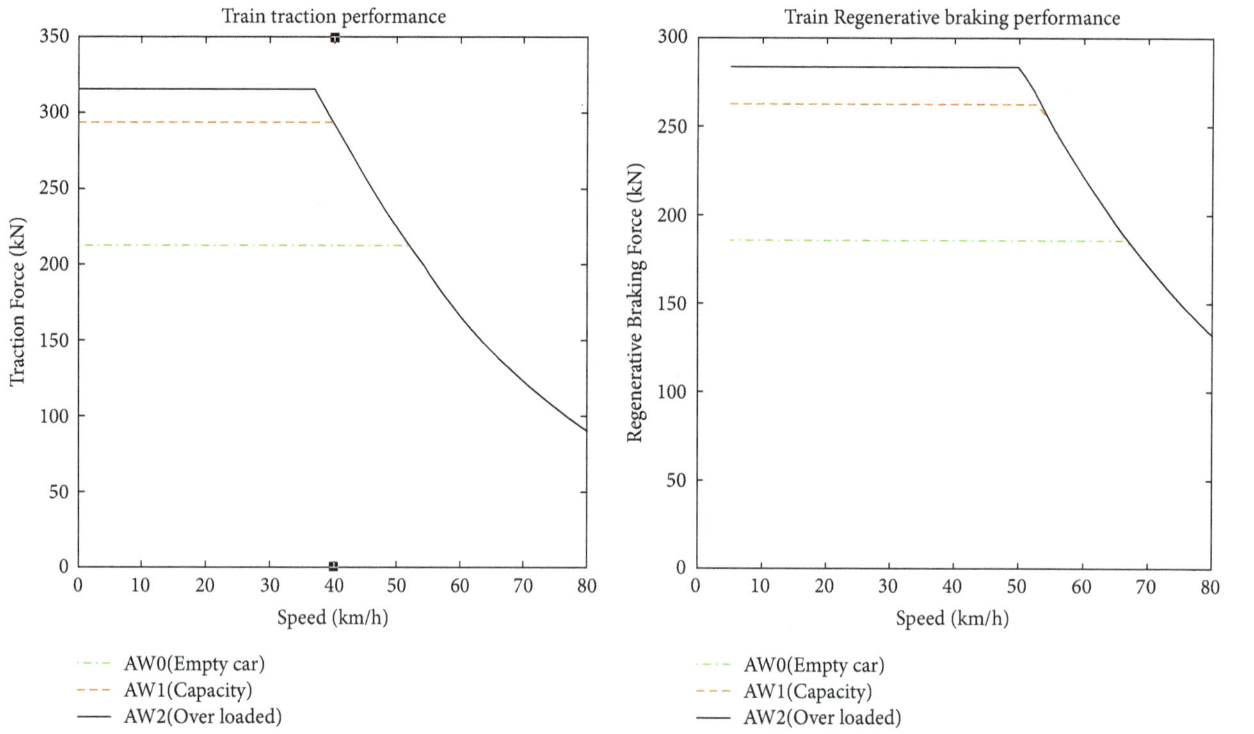

FIGURE 11: The power performance under different load factors.

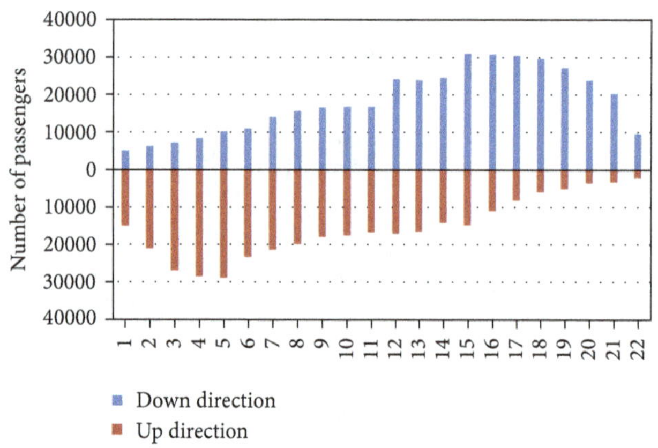

FIGURE 12: Number of passengers on the train in different interstations.

TABLE 2: Train characteristics.

Rolling stock mass	203,000 kg
Max speed limit	80 km/h
Number of cars	6 pcs
Maximum acceleration	1.0 m/s^2
Maximum deceleration	−1.0 m/s^2
Minimum speed for RBE	5 km/h
Resistance coefficient a	1.2414
Resistance coefficient b	0.0144
Resistance coefficient c	0.000221

Line 5, and power peak in every PSI is 8500 kW. For safety and quality of service, the maximal headway is 160 s and the minimal headway is 140 s. In the NS-GSA solution algorithm, we set population $P = 100$, $Q = 10$, $P_c = 0.8$, $P_m = 0.1$, the maximum generation as 50, and the annealing rate as 0.95 with an initial temperature of 100°C.

5.2. Energy Performance of Improved Energy-Efficient Train Control. Table 3 gives the energy performance of the improved model (MA-[CR-CO]$_n$-MB) and previous model (MA-CR-CO-MB) with the same runtime, from DTLD to HXXJNK in down direction. The improved train control strategy is able to reduce traction energy consumption by 1.07 kWh in the first interstation with steep downhill slopes and the potential annual traction energy saving in this interstation could be up to 50772 kWh. On the other hand, two driving strategies consumed the same traction energy in the second interstation where there is no steep downhill slope.

Figure 13 demonstrates the train trajectories of two driving strategies to further explain the reason of traction energy reduction by improved control model in a more intuitive way. In the first interstation, the trajectory obtained by the improved control consists of MA-[CR-CO]$_2$-MB; both CR and CO are implemented twice but the second CO is omitted because MB has to be implemented for station stop. In contrast, the trajectory attained by the previous model is sequentially composed of MA, CR, CO, and MB. The improved trajectory extends the coasting distance in steep downhill tracks in order to take full advantage of potential energy to increase the kinetic energy. As such, MA can be

TABLE 3: Energy performance of the improved train control model.

Interstation	DTLD-HXXJBK (with steep downhills)		HXXJBK-HXXJNK (without a steep downhill)	
Driving strategy	Improved	MA-CR-CO-MB	Improved	MA-CR-CO-MB
Runtime (s)	129	129	94	94
Traction energy consumption (kWh)	6.42	7.49	6.09	6.09

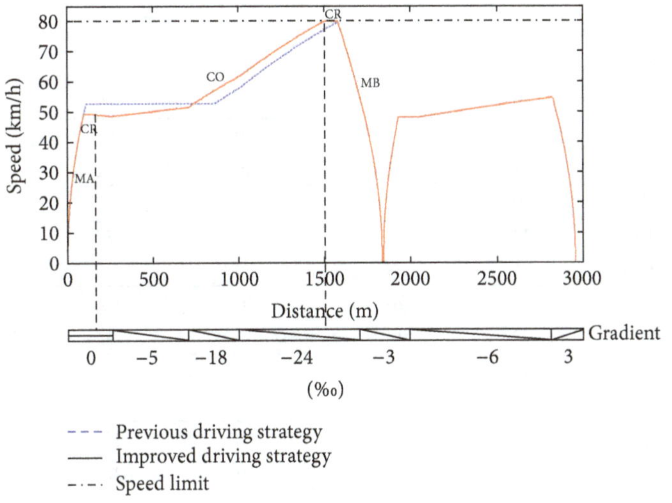

FIGURE 13: The comparison between two driving strategies.

shortened and the traction energy consumption is saved, while the interstation runtime remains the same. In the second interstation, two control strategies obtained the same trajectory as well as the traction energy consumption, as there is no steep downhill track in this interstation.

It can be concluded that the traction energy performance of the improved control strategy is better than or equal to the MA-CR-CO-MB strategy. Additionally, runtime supplements in each interstation run have significant influences on train trajectory optimization. We studied the energy-saving rates of the improved train control in comparison with the previous model under different runtimes. Figure 14 shows traction energy consumption of two control strategies in interstation DTLD-HXXJNK with different runtimes.

When the scheduled runtime is equal to the minimum interstation runtime, the train has to accelerate to speed limit using maximum traction force and keep the speed constant to satisfy the runtime constraint until the train brakes for station stop. Under this condition, two control strategies attained the same trajectory and consumed the same energy because there is no room to apply CO. Along with the increment of runtime supplements, the improved train control can realize increased energy saving because more CR is allowed in downhill sections, which reduces the traction energy consumption by taking advantage of potential energy. Once the interstation runtime reaches a certain level, the energy-saving ratio decreases because the motoring distances in both control strategies decrease. In case the scheduled runtime is large enough, the trajectories as well as the energy consumption of the two models are the same.

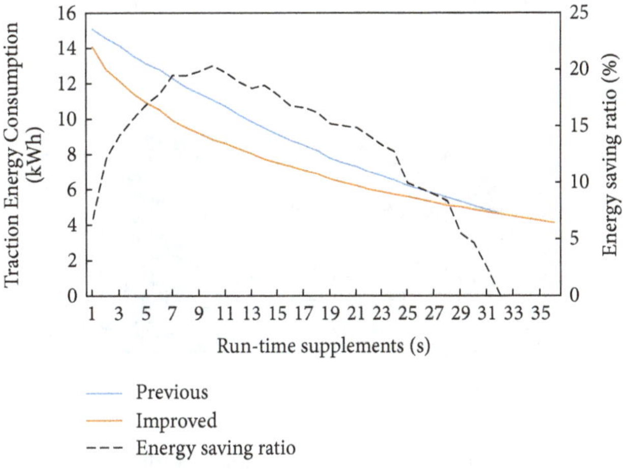

FIGURE 14: Energy-saving effect of the improved driving strategy at different runtimes.

In addition to interstation runtime and track profile, train mass also has a great impact on traction energy consumption because a large mass results in heavier resistance during the interstation runs and also helps the train accelerate in downhill slopes. We used the proposed control strategy and the one applied in previous studies to obtain the optimal trajectories in the interstation DTLD-HXXJBK, with different train masses and different interstation runtimes. The traction energy reduction by the improved control in comparison with the previous model is shown in Figure 15. It is found that

TABLE 4: The original timetable and the optimized timetable.

ID	Interstation	Length (m)	Up direction Runtime (s) Original	Up direction Runtime (s) Optimized	Up direction Dwell time (s)	Down direction Runtime (s) Original	Down direction Runtime (s) Optimized	Down direction Dwell time (s)	PSI
1	SJZ-LJY	1671	126	122	45	130	123		1
2	LJY-PHY	905	81	92	30	82	88	41	
3	PHY-TTDM	1900	133	140	30	133	127	30	2
4	TTDM-CQK	1183	98	96	45	97	92	30	3
5	CQK-CWM	877	80	80	50	79	84	45	
6	CWM-DD	822	79	92	60	79	75	50	
7	DD-DSK	945	83	93	30	82	88	60	4
8	DSK-DS	848	81	82	45	81	88	30	
9	DS-ZZZL	1017	89	86	30	87	84	45	5
10	ZZZL-BXQ	791	78	82	30	74	75	30	
11	BXQ-YHG	866	81	88	50	85	86	30	
12	YHG-HPLBJ	1151	98	96	30	103	91	50	6
13	HPLBJ-HPXQ	1059	93	90	30	93	91	30	
14	HPXQ-HXXJNK	1025	89	87	55	91	87	30	7
15	HXXKNK-HXXJBK	1122	99	92	30	94	97	55	
16	HXXJBK-DTLD	1838	133	137	40	129	127	30	8
17	DTLD-BYLB	3000	180	180	30	181	181	40	9
18	BYLB-LSQN	1286	110	110	30	108	100	30	10
19	LSQN-LSQ	1306	111	109	50	112	112	30	
20	LSQ-TTYN	1544	120	109	30	119	111	50	11
21	TTYN-TTY	966	86	93	30	86	84	30	
22	TTY-TTYB	939	85	94		86	85	40	
Total		27061	2213	2250	800	2211	2176	806	

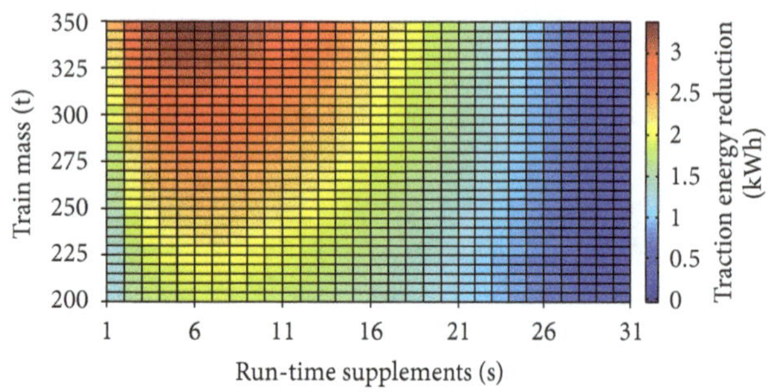

FIGURE 15: Traction energy reduction under various conditions using improved driving strategies.

the improved model achieves better energy saving when the train mass increases. The reason is that heavier trains acquire greater acceleration when coasting in steep downhills, and they can gain higher kinetic energy in steep downhill tracks and achieve better energy saving.

To sum up, the improved model is a supplement to the MA-CR-CO-MB control strategy in interstations with constant speed limit and it is able to save traction energy in downhill sections by taking advantage of potential energy. Energy saving is influenced by the scheduled runtime and train mass. The energy saving firstly increases with scheduled runtime and then decreases, finally reaching zero when the runtime is sufficient. Heavier trains can achieve a better traction energy reduction because the acceleration is larger when the train is coasting in downhills where the potential energy could help the train more in accelerating.

5.3. Energy Performance of Integrated Optimization. Separate train control to minimize the traction energy consumption of each train does not necessarily lead to the minimum net energy consumption of whole metro lines. The energy consumption of the whole metro lines can be further reduced by optimizing train control and timetable concurrently. Table 4 shows the original timetable and train control configurations (RT, HD, k_t, and k_b) provided by operators and the optimized one obtained by the proposed integrated optimization model. The original journey time is 2213 s in the up direction and 2211 s in the down direction. Compared with the original timetable, the optimized timetable increases by 37 s in the up direction and decreases by 35 s in the down direction. Therefore, the cycle time is extended by 2 s, which is acceptable. Additionally, train headway is regulated to improve the utilization of RBE. The headway is set as a fixed value (i.e., 150 s) in the original timetable. In the optimized one, the headway between every two adjacent trains varies within a reasonable range from 140 s to 159 s, without changing the number of service provisions and dwell time at each station.

TABLE 5: Comparison of energy consumption between the original timetable and optimized timetable.

	Original timetable	Optimized timetable	
$[k_t, k_b]$	[1, 1]	[0.9, 0.9]	
Net energy consumption (kWh)	15059	14311	(−4.97%)
Auxiliary devices (kWh)	530	530	(+0.00%)
Traction energy (kWh)	16567	17554	(+5.96%)
Total RBE (kWh)	13016	14360	(+10.33%)
Utilized RBE (kWh)			
Auxiliary (self)	52	57	(+9.62%)
Traction	1840	3570	(+94.02%)
Auxiliary (other)	146	147	(+0.68%)
Total	2038	3773	(+85.13%)
Overlapping time (s)	4194	4543	(+8.32%)
Utilization percentage of RBE (%)	15.66	26.27	(+67.81%)

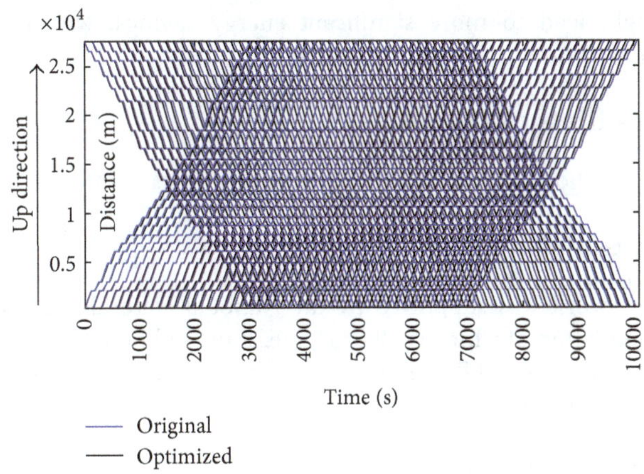

FIGURE 16: Comparison of the original and optimal timetable.

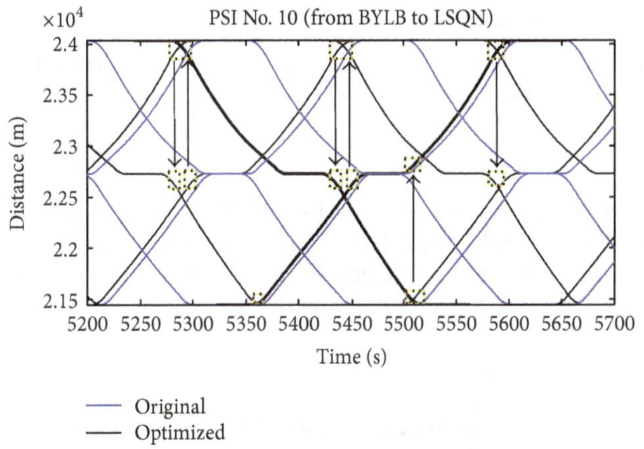

FIGURE 17: The utilization of the regenerative braking energy in the same power supply interval.

Table 5 shows the comparison of energy consumption between the original timetable with previous train control model and the optimized timetable with the improved train control model. Although traction energy consumption with the optimized timetable increases by 5.96%, the net energy consumption of the whole line decreases by 4.97% because of the significant improvements on utilization of RBE. This verifies the argument that minimum traction energy consumption does not necessarily lead to the minimum net energy consumption. Although partial traction and braking result in a slight increment in traction energy consumption, the utilization of RBE with the optimized timetable increases by 67.81% as the overlapping time increases by 8.32%. It is concluded that the integrated optimization model can notably improve the utilization of RBE, thus reducing the net energy consumption.

Figure 16 shows the difference between the original timetable and the optimized timetable. The total journey time and the operating time for service provision remain the same in the optimized timetable, although the interstation runtime (RT) and train headway (HD) are regulated slightly.

An enlarged view of part of Figure 16, taking a section from the station of BYLB to the station of LSQ in the same PSI as an example, is shown in Figure 17 to explicate how the utilization of RBE is optimized by the integrated model. The arrows in the figure illustrate the flow of regenerative braking energy; that is, the RBE generated by the braking train can be transmitted to the traction train. It is found that motoring and braking of trains in the same PSI could be synchronized by adjusting headway and interstation runtime in timetable formulation. Moreover, overlapping time could be further extended using partial traction/braking to prolong the motoring and braking time. As a result, the net energy consumption decreases, although the traction energy consumption is not the minimal one.

Figure 18 gives the trajectories of two successive trains moving in the same PSI. Under the original timetable with full motoring and braking control, the RBE is not used timely because motoring train and braking train are not synchronized well. By optimizing headway and interstation runtime, the departure of train 2 from DD station just located in the same time window when train 1 arrives at the TTDM station. Furthermore, the synchronized time between traction and braking train is prolonged by applying partial motoring and braking, which enable a larger overlapping

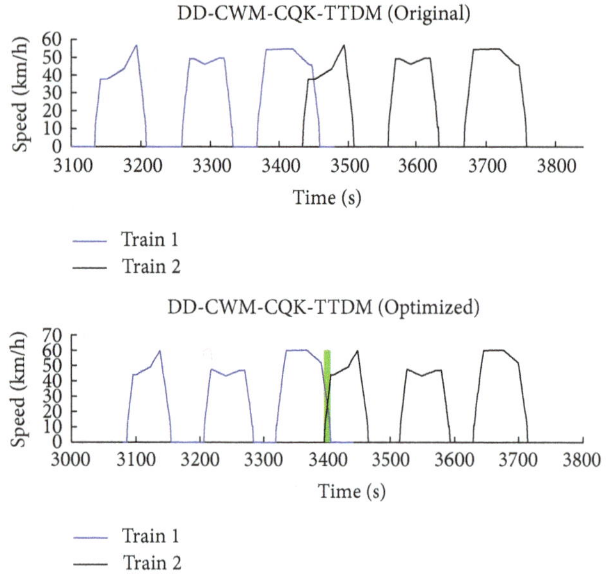

Figure 18: Synchronized traction and braking train in one power supply interval.

time as well as a better utilization of RBE. Thus, most of the RBE produced by train 1 is timely used by train 2, which helps to reduce the energy feeding from substations and in turn save the net energy consumption of the whole metro line.

6. Conclusion

An integrated optimization model on train control and timetable formulation is proposed in this study to minimize the net energy consumption of all trains servicing on the metro line. The proposed train control is still based on finding the optimal switching points among the control modes of MA, CR, CO, and MB to minimize the traction energy consumption, while cruising and coasting regimes might be adopted for more than one time according to track profiles. For better utilization of RBE, timetable configurations such as interstation runtime and train headway as well as the extents of motoring and braking in train control are optimized concurrently, taking into account the synchronization of motoring and braking trains in the same PSI. Practical operation condition and constraints, such as varied train mass in different interstations, the limitation on maximal loading of power system, are taken into consideration in the proposed model. The brute force algorithm is employed to attain the energy-efficient trajectory of train interstation runs and an NS-GSA algorithm is developed to attain the optimal extents of motoring/braking and timetable configurations.

Case studies on Beijing Metro Line 5 demonstrated the energy performance of the proposed integrated model. It is found that the improved train control can reduce the traction energy by 20% in the interstations with steep downhill slopes. The energy-saving rate of the improved train control in comparison with the previous model depends on track profiles and the scheduled runtime. The integrated optimization on train control and timetable formulation is able to save net energy consumption by 4.97% through regulating interstation runtime, train headway, and the extents of motoring and braking in train control. Although the traction energy slightly increases by 5.96% because of the application of partial motoring and braking in train control, the utilization of RBE is significantly improved by prolonging the overlapping time of motoring and braking trains in the same PSI.

This paper explored the energy-efficient train control with the constant speed limit, while the speed limit may vary during interstation runs due to curves or temporary maintenance. In such case, the proposed control strategy might not be the optimal one. One of our future researches is to take the varied speed limits into consideration in train control and apply a more general energy-efficient train control model in timetable optimization. Additionally, this study assumes that all trains have the same extents of motoring and braking in different interstation runs. Different extents of motoring and braking for different trains in different interstation runs might lead to more significant energy savings, which is another direction of future research.

Conflicts of Interest

The authors declare that they have no conflicts of interest.

Acknowledgments

This work was supported by the National Natural Science Foundation of China (71390332, 71621001, and 71571016) and the Fundamental Research Funds for the Central Universities (2016JBM031).

References

[1] A. González-Gil, R. Palacin, P. Batty, and J. P. Powell, "A systems approach to reduce urban rail energy consumption," *Energy Conversion and Management*, vol. 80, pp. 509–524, 2014.

[2] K. Ichikawa, "Application of optimization theory for bounded state variable problems to the operation of train," *Journal of the Japan Society of Mechanical Engineers*, vol. 11, no. 47, pp. 857–865, 1968.

[3] H. Strobel, P. Horn, and M. Kosemund, "A contribution to optimum computer-aided control of train operation," in *Proceedings of the 2nd FAC/IFIP/IFORS Symposium on Traffic Control and Transportation Systems*, pp. 377–387, 1974.

[4] I. P. Milroy, *Aspect of automatics train control [Ph.D. dissertation]*, Loughborough University of Technology, Sydney, Australia, 1980.

[5] B. R. Benjamin, A. M. Long, I. P. Milroy, R. L. Payne, and P. J. Pudney, "Control of railway vehicles for energy conservation and improved timekeeping," in *Proceedings of the Conference on Railway Engineering*, pp. 41–47, Perth, Australia, 1987.

[6] P. G. Howlett, I. P. Milroy, and P. J. Pudney, "Energy-efficient train control," *Control Engineering Practice*, vol. 2, no. 2, pp. 193–200, 1994.

[7] E. Khmelnitsky, "On an optimal control problem of train operation," *IEEE Transactions on Automatic Control*, vol. 45, no. 7, pp. 1257–1266, 2000.

[8] R. Liu and I. M. Golovitcher, "Energy-efficient operation of rail vehicles," *Transportation Research Part A: Policy and Practice*, vol. 37, no. 10, pp. 917–932, 2003.

[9] L. A. M. V. Dongen and J. H. Schuit, "Energy-efficient driving patterns in electric railway traction," in *Proceedings of the International Conference on Main Line Railway Electrification (IET '02)*, pp. 154–158, 2002.

[10] G. M. Scheepmaker and R. M. P. Goverde, "The interplay between energy-efficient train control and scheduled running time supplements," *Journal of Rail Transport Planning and Management*, vol. 5, no. 4, pp. 225–239, 2015.

[11] P. G. Howlett, P. J. Pudney, and X. Vu, "Local energy minimization in optimal train control," *Automatica*, vol. 45, no. 11, pp. 2692–2698, 2009.

[12] Y. Bai, B. H. Mao, F. M. Zhou, Y. Ding, and C. B. Dong, "Energy-efficient driving strategy for freight trains based on power consumption analysis," *Journal of Transportation Systems Engineering & Information Technology*, vol. 9, no. 3, pp. 43–50, 2009.

[13] M. Domínguez, A. Fernández-Cardador, A. P. Cucala, T. Gonsalves, and A. Fernández, "Multi objective particle swarm optimization algorithm for the design of efficient ATO speed profiles in metro lines," *Engineering Applications of Artificial Intelligence*, vol. 29, pp. 43–53, 2014.

[14] W. Carvajal-Carreño, A. P. Cucala, and A. Fernández-Cardador, "Optimal design of energy-efficient ATO CBTC driving for metro lines based on NSGA-II with fuzzy parameters," *Engineering Applications of Artificial Intelligence*, vol. 36, pp. 164–177, 2014.

[15] C. Sicre, A. P. Cucala, and A. Fernández-Cardador, "Real time regulation of efficient driving of high speed trains based on a genetic algorithm and a fuzzy model of manual driving," *Engineering Applications of Artificial Intelligence*, vol. 29, no. 29, pp. 79–92, 2014.

[16] K. Keskin and A. Karamancioglu, "Energy-efficient train operation using nature-inspired algorithms," *Journal of Advanced Transportation*, vol. 2017, Article ID 6173795, 12 pages, 2017.

[17] W. Shangguan, X.-H. Yan, B.-G. Cai, and J. Wang, "Multiobjective optimization for train speed trajectory in CTCS high-speed railway with hybrid evolutionary algorithm," *IEEE Transactions on Intelligent Transportation Systems*, vol. 16, no. 4, pp. 2215–2225, 2015.

[18] T. Albrecht and S. Oettich, *A New Integrated Approach to Dynamic Schedule Synchronization and Energy-Saving Train Control*, WIT Press, Southampton, UK, 2002.

[19] C. Sicre, P. Cucala, A. Fernández, J. A. Jiménez, I. Ribera, and A. Serrano, "A method to optimise train energy consumption combining manual energy efficient driving and scheduling," in *Proceedings of the 12th International Conference on Computer System Design and Operation in the Railways and other Transit Systems (COMPRAIL '10)*, pp. 549–560, September 2010.

[20] R. Chevrier, P. Pellegrini, and J. Rodriguez, "Energy saving in railway timetabling: a bi-objective evolutionary approach for computing alternative running times," *Transportation Research Part C: Emerging Technologies*, vol. 37, pp. 20–41, 2013.

[21] G. M. Scheepmaker, R. M. Goverde, and L. Kroon, "Review of energy-efficient train control and timetabling," *European Journal of Operational Research*, vol. 257, no. 2, pp. 355–376, 2017.

[22] R. Barrero, J. van Mierlo, and X. Tackoen, "Energy savings in public transport," *IEEE Vehicular Technology Magazine*, vol. 3, no. 3, pp. 26–36, 2008.

[23] T. Albrecht, "Reducing power peaks and energy consumption in rail transit systems by simultaneous train running time control," *Computers in Railways*, pp. 885–894, 2004.

[24] X. Yang, X. Li, Z. Gao, H. Wang, and T. Tang, "A cooperative scheduling model for timetable optimization in subway systems," *IEEE Transactions on Intelligent Transportation Systems*, vol. 14, no. 1, pp. 438–447, 2013.

[25] X. Yang, B. Ning, X. Li, and T. Tang, "A two-objective timetable optimization model in subway systems," *IEEE Transactions on Intelligent Transportation Systems*, vol. 15, no. 5, pp. 1913–1921, 2014.

[26] Z. Le, K. Li, J. Ye J, and X. Xu, "Optimizing the train timetable for a subway system," *Proceedings of the Institution of Mechanical Engineers Part F: Journal of Rail & Rapid Transit*, vol. 229, no. 8, pp. 2532–2542, 2015.

[27] M. Peña-Alcaraz, A. Fernández, A. P. Cucala, A. Ramos, and R. R. Pecharromán, "Optimal underground timetable design based on power flow for maximizing the use of regenerative-braking energy," *Proceedings of the Institution of Mechanical Engineers, Part F: Journal of Rail and Rapid Transit*, vol. 226, no. 4, pp. 397–408, 2011.

[28] Y. Ding, H. D. Liu, Y. Bai, and F. M. Zhou, "A two-level optimization model and algorithm for energy-efficient urban train operation," *Journal of Transportation Systems Engineering & Information Technology*, vol. 11, no. 1, pp. 96–101, 2011.

[29] S. Su, X. Li, T. Tang, and Z. Gao, "A subway train timetable optimization approach based on energy-efficient operation strategy," *IEEE Transactions on Intelligent Transportation Systems*, vol. 14, no. 2, pp. 883–893, 2013.

[30] S. Su, T. Tang, X. Li, and Z. Gao, "Optimization of multitrain operations in a subway system," *IEEE Transactions on Intelligent Transportation Systems*, vol. 15, no. 2, pp. 673–684, 2014.

[31] X. Yang, A. Chen, B. Ning, and T. Tang, "A stochastic model for the integrated optimization on metro timetable and speed profile with uncertain train mass," *Transportation Research Part B: Methodological*, vol. 91, pp. 424–445, 2016.

[32] X. Li and H. K. Lo, "An energy-efficient scheduling and speed control approach for metro rail operations," *Transportation Research Part B: Methodological*, vol. 64, pp. 73–89, 2014.

[33] N. Zhao, C. Roberts, S. Hillmansen, Z. Tian, P. Weston, and L. Chen, "An integrated metro operation optimization to minimize energy consumption," *Transportation Research Part C: Emerging Technologies*, vol. 75, pp. 168–182, 2017.

[34] S. Su, T. Tang, and Y. H. Wang, "Evaluation of strategies to reducing traction energy consumption of metro systems using an optimal train control simulation model," *Energies*, vol. 9, no. 2, pp. 105–123, 2016.

[35] S. Watanabe and T. Koseki, "Energy-saving train scheduling diagram for automatically operated electric railway," *Journal of Rail Transport Planning and Management*, vol. 5, no. 3, pp. 183–193, 2015.

[36] M. Miyatake and H. Ko, "Optimization of train speed profile for minimum energy consumption," *IEEJ Transactions on Electrical and Electronic Engineering*, vol. 5, no. 3, pp. 263–269, 2010.

[37] A. Albrecht, P. Howlett, P. Pudney, X. Vu, P. Zhou, and D. Chen, "Using maximum power to save energy," in *Proceedings of the 17th IEEE International Conference on Intelligent Transportation Systems*, pp. 1205–1208, Qingdao, China, November 2014.

[38] Á. J. López-López, R. R. Pecharromán, A. Fernández-Cardador, and A. P. Cucala, "Assessment of energy-saving techniques in direct-current-electrified mass transit systems," *Transportation Research Part C: Emerging Technologies*, vol. 38, pp. 85–100, 2014.

[39] M. Domínguez, A. Fernández, A. P. Cucala, and P. Lukaszewicz, "Optimal design of metro automatic train operation speed

profiles for reducing energy consumption," *Proceedings of the Institution of Mechanical Engineers, Part F: Journal of Rail and Rapid Transit*, vol. 225, no. 5, pp. 463–474, 2011.

[40] A. P. Cucala, A. Fernández, C. Sicre, and M. Domínguez, "Fuzzy optimal schedule of high speed train operation to minimize energy consumption with uncertain delays and drivers behavioral response," *Engineering Applications of Artificial Intelligence*, vol. 25, no. 8, pp. 1548–1557, 2012.

[41] X. Yang, X. Li, B. Ning, and T. Tang, "An optimisation method for train scheduling with minimum energy consumption and travel time in metro rail systems," *Transportmetrica B: Transport Dynamics*, vol. 3, no. 2, pp. 79–98, 2015.

[42] P. Shrivastava, S. L. Dhingra, and P. J. Gundaliya, "Application of genetic algorithm for scheduling and schedule coordination problems," *Journal of Advanced Transportation*, vol. 36, no. 1, pp. 23–41, 2002.

[43] B. Fan, C. Roberts, and P. Weston, "A comparison of algorithms for minimising delay costs in disturbed railway traffic scenarios," *Journal of Rail Transport Planning and Management*, vol. 2, no. 1-2, pp. 23–33, 2012.

[44] N. Zhao, C. Roberts, S. Hillmansen, and G. Nicholson, "A multiple train trajectory optimization to minimize energy consumption and delay," *IEEE Transactions on Intelligent Transportation Systems*, vol. 16, no. 5, pp. 2363–2372, 2015.

[45] S. Yang, K. Yang, Z. Gao, L. Yang, and J. Shi, "Last-train timetabling under transfer demand uncertainty: mean-variance model and heuristic solution," *Journal of Advanced Transportation*, vol. 2017, Article ID 5095021, 13 pages, 2017.

[46] M. Samà, A. D'Ariano, F. Corman, and D. Pacciarelli, "A variable neighbourhood search for fast train scheduling and routing during disturbed railway traffic situations," *Computers & Operations Research*, vol. 78, pp. 480–499, 2017.

[47] A. D'Ariano, F. Corman, D. Pacciarelli, and M. Pranzo, "Reordering and local rerouting strategies to manage train traffic in real time," *Transportation Science*, vol. 42, no. 4, pp. 405–419, 2008.

[48] P. Pellegrini and M. Birattari, "Implementation effort and performance," in *Engineering Stochastic Local Search Algorithms. Designing, Implementing and Analyzing Effective Heuristics, International Workshop, SLS 2007, Brussels, Belgium, September 2007*, Springer, 2007.

8

Robust Train Scheduling Problem with Optimized Maintenance Planning on High-Speed Railway Corridors

Chuntian Zhang,[1] Yuan Gao,[1] Wenjie Li,[2] Lixing Yang,[1] and Ziyou Gao[1]

[1] *State Key Laboratory of Rail Traffic Control and Safety, Beijing Jiaotong University, Beijing 100044, China*
[2] *Sid and Reva Department of Civil, Environmental and Infrastructure Engineering, George Mason University, Fairfax, VA 22030, USA*

Correspondence should be addressed to Yuan Gao; gaoyuan@bjtu.edu.cn

Academic Editor: Emanuele Crisostomi

Simultaneously considering train scheduling problem and maintenance planning problem with uncertain travel time, we propose a two-stage integrated optimization model for the sunset-departure and sunrise-arrival trains (SDSA-trains). Specifically, in the first stage, we obtain an optimal solution of the SDSA-trains under each scenario, which leads to the minimum total travel time. In the second stage, a robust SDSA-train schedule is generated based on the optimal solutions of the first stage. The key is that we consider two operation modes to solve the conflict between the SDSA-trains and the maintenances. Some state variables are used to deal with train operation mode selection. Furthermore, some linearization techniques are used to formulate a mixed-integer linear programming (MILP) model. Finally, numerical experiments are implemented to prove the effectiveness of the proposed model and optimization method.

1. Introduction

Desirable train schedules are important to ensure the efficient operation of the high-speed railway system, but some uncertain events (e.g., bad weather conditions, equipment failure) often occur due to the complexity of the real-world conditions and affect the nominal train schedules. Therefore, finding a robust train schedule is very important for the high-speed railway system. Its robustness is not only related to the trains' anti-interference ability during its operation but also the efficiency of train operation. In general, the uncertain events will directly affect the travel time of each train, which makes the travel time uncertain on each section. Thus, an effective method to obtain a robust train schedule with the minimum deviation from the expected travel time to reduce the delay propagation is very crucial for train operatives and managers.

In addition to finding a robust train schedule, the railway infrastructure also should be in good condition to ensure that tracks are in appropriate state for trains. In order to ensure this condition, regular maintenance works (i.e., maintenance windows) on high-speed railway are executed at night in China. As is known, when regular maintenance is executed, the whole railway system does not work. Especially, high-speed railway in China is the longest and busiest high-speed railway system in the world. The regular maintenance works from 0 : 00 to 6 : 00 am for high-speed railway system become a hindrance of efficient operation. Moreover, at this situation, some long-distance passengers have to spend most of daytime in high-speed trains, which reduces their attractiveness to long-distance passengers.

In recent years, sunset-departure and sunrise-arrival trains (SDSA-trains) on high-speed railway have been introduced to attract more long-distance passengers. A SDSA-train departs from the origin station in the evening and arrives at the destination station in the next morning. The travel distance of the SDSA-train is usually between 1500 km and 2500 km. It should be mentioned that when traveling at night, a SDSA-train is not only a means of transport but also a hotel for passengers, which is more in accord with passenger travel behavior.

Obviously, there is a conflict between the SDSA-train schedules and maintenance window plans on high-speed railway. Therefore, maintenance window must be considered when generating a robust train schedule. By combining these aforementioned aspects, we intend to solve the problem in the following three aspects. First, in order to generate a robust train schedule, we propose a two-stage integrated optimization model. In the first stage, we generate an integrated optimization model of train scheduling and maintenance planning on high-speed railway corridors under each scenario. In the second stage, we formulate a biobjective optimization model to minimize total travel time of all SDSA-trains in the premise of meeting the robustness requirement. Note that the second stage also should satisfy the constraints established in the first stage. Second, we consider two operation modes to solve the conflict between the SDSA-train scheduling and maintenance planning, namely, (1) dwelling and waiting at high-speed railway station (Mode 1) and (2) switching from high-speed railway to *normal-speed* railway (Mode 2). Note that paralleling along each main high-speed railway corridor, there is an old railway corridor, which is referred to as *normal-speed* railway. In order to solve SDSA-train operation modes selection problems, we introduce some state variables and analyze the relationship among these variables. Third, in order to make the model be handled easily, some linearization techniques are used and nonlinear constraints and objective function are equivalently replaced by linear constraints and objective function. To verify the proposed integrated optimization model, we carry out a series of numerical examples. The computational result shows that the proposed model can be solved to optimality.

The rest of this paper is organized as follows. Section 2 reviews relevant literature on robust train scheduling and maintenance planning problems. Section 3 describes in detail the SDSA-trains' operation modes, when there is a conflict with maintenance window. In Section 4, a two-stage integrated optimization model is formulated to generate robust train scheduling. In Section 5, the proposed model is applied to the SDSA-trains on Beijing-Guangzhou high-speed railway corridor. The numerical experiments are implemented to verify the effectiveness and efficiency of the model. Finally, some conclusions and further works are discussed in Section 6.

2. Literature Review

2.1. Nominal and Robust Train Scheduling. Due to the importance of train schedule in the operation of railway systems, lots of researches have been conducted on train scheduling problem in the past decades. In general, there are mainly two study versions dealing with the train timetabling problem, *i.e.*, *nominal* and *robust* versions. Roughly speaking, the nominal version aims to optimize an objective function (*e.g.*, minimizing the passenger travel time, maximizing the passenger satisfaction) in the premise of satisfying the track capacity constraints, whereas the robust version mainly aims to reduce the delay propagation when there is a disturbance or disruption on the railway network.

In the past years, there have been many works which have treated the nominal version of the problem. Cacchiani et al. [1] pointed out that the nominal timetable problem can be studied mainly from two aspects of cyclic (periodic) timetable and noncyclic (nonperiodic) timetable. Nachtigall and Voget [2] generated a timetable with minimum passenger waiting times for periodically served railway networks. They proposed a method based on a genetic algorithm combined with a greedy heuristic and an optimization procedure to solve the problem. In order to achieve more potential for optimization, Liebchen and Möhring [3] integrated the decisions of network planning, line planning and vehicle scheduling into the task of periodic timetabling and showed how to extend the periodic event scheduling problem (PESP) model to consider this integration. Caimi et al. [4] extended the well-known periodic event scheduling problem (PESP) to the flexible periodic event scheduling problem (FPESP), allowing the departure and arrival times to be flexible time slots rather than the exact time. The computational results on real-world instance proved that the model was effective. Besides, Kroon and Peeters [5], Lindner [6] and Peeters and Kroon [7] formulated an integer linear programming (ILP) model to solve the periodic timetable problem. Some researchers, however, studied noncyclic timetabling problems. Generally speaking, noncyclic version had a greater advantage in a competitive market, which aimed to generate the optimal timetables by maximizing the overall profit. Brännlund et al. [8] proposed an ILP model to obtain a profit maximization timetable for a set of trains. They used a Lagrangian relaxation solution approach to solve the problem, where the track capacity constraints were relaxed. Caprara et al. [9] proposed a graph theoretical formulation to formulate an ILP model and aimed to maximize the sum of the profits of the scheduled trains. This formulation was relaxed in a Lagrangian way. The relaxed constraints were associated only with nodes of the aforementioned graph. Cacchiani et al. [10] considered the customary formulation of noncyclic train timetabling and aimed to maximize profits for compatible paths in a suitable graph. They analyzed the existing ILP models and proposed some new models to solve the problem. Cacchiani et al. [11] solved the nonperiodic version and the train platforming problem (TPP) from a tactical perspective. Cacchiani et al. [11] showed some standard but successful solution approaches based on integer programming. Through considering the dynamic characteristics of passenger flow, Shi et al. [12] proposed an effective method to simultaneously optimize the train timetable and accurate passenger flow control strategies on an oversaturated metro line.

Furthermore, finding robust version of the problem is a major practical issue that has received a lot of attention in recent years. Delays occurring at an operational level make nominal timetable infeasible, but the train schedules with better robustness can reduce the deviation degree between the train adjustment plan and the basic plan. There are many aspects of robust train timetabling problems to be considered. Kroon et al. [13] stressed that real-time railway operations were subject to stochastic disturbances. They proposed a stochastic optimization model to improve the robustness of a given cyclic railway timetable by allocating time supplements and buffer times in a given timetable. The experimental results showed that the average delays of trains can be significantly reduced by making relatively minor modifications to a given timetable. Khan and Zhou [14] considered various stochastic disturbances and developed a two-stage stochastic optimization model. They aimed to minimize the total planned trip time and reduce the expected schedule delay. Furthermore, they tested the effectiveness of the model on the real data of Beijing-Shanghai high-speed rail corridor in China. Cacchiani et al. [15] pointed out that Lagrangian heuristic was a very effective way to solve robust schedule issues. Besides, they showed how to modify the existing Lagrange heuristic to generate robust solutions. Shafia et al. [16] proposed a robust train timetabling model, which could deal with the disturbances existed among traveling times. They used a branch-and-bound (B&B) algorithm and a new heuristic beam search (BS) algorithm to solve the large-scale problems in a rational amount of time. Besides, Yang et al. [17, 18] studied robust train timetabling problems by considering the uncertainty of passenger. Yang et al. [17] studied the timetabling problem with fuzzy passenger demand. They considered the two objectives of fuzzy total passengers' time and total delay time. Yang et al. [18] regarded the number of passengers boarding/leaving each train at each station as a random variable. They formulated a 0-1 mixed-integer programming model and designed a branch-and-bound algorithm to solve the model.

2.2. Consideration of Train Scheduling and Maintenance Planning at the Same Time. In addition to the study of the railway timetable problem, railway infrastructure maintenance problem has also attracted a lot of attentions from researchers in the past twenty years. In the early research of railway maintenance plan, Higgins [19] aimed at determining the optimal maintenance activities and crews allocation on a single track line so as to minimize the round-trip scheduled trains and reduce completion time. The proposed nonlinear model was solved by a tabu search heuristic. Budai et al. [20, 21] also studied the railway infrastructure maintenance problem. Budai et al. [20] presented an optimization model to improve railway maintenance decisions. Some greedy heuristics were proposed to solve the model aiming to minimize possession costs. Later, genetic algorithms, memetic algorithms and the two-phase opportunity-based heuristic were proposed by Budai [21] to solve the preventive maintenance scheduling problem. Lidén and Joborn [22] established maintenance cost model, freight traffic cost model and passenger traffic cost model to evaluate the effects of maintenance windows on the maintenance costs and the train traffic and transportation costs. In addition, they also explored the effects of different maintenance window widths on maintenance costs and freight traffic costs. Zhou et al. [23] proposed a train-set circulation optimization model to minimize the total connection time and maintenance costs, and this model was solved by an efficient multiple-population genetic algorithm (MPGA). A realistic high-speed railway case was given to show the effectiveness of the proposed model and algorithm.

In recent five years, some researchers have begun to consider the train scheduling problem and maintenance planning problem at the same time. Aken et al. [24] studied the train timetable adjustment problem (TTAP) under the condition of maintenances. They proposed a mixed-integer linear programming model and used a row generation approach to generate an alternative timetable. The alternative timetable has the minimum deviation from the original timetable. Lidén and Joborn [25] presented a mixed-integer programming model for long term tactical plan, which simultaneously considered train services and railway network maintenances. The experimental results showed that the model could handle large-scale instances. Luan et al. [26] solved the problem of integrated optimization of train scheduling and preventive maintenance planning from a micro perspective. They proposed a novel integrated mixed-integer linear programming model by regarding a maintenance task as a virtual train, and the model was solved by a heuristic based on Lagrangian relaxation. Arenas et al. [27] presented a mixed-integer linear programming formulation that rearranged a timetable to cope with the capacity consumption produced by maintenance activities. They also considered a microscopic approach to solve the problem.

Note that Vansteenwegen et al. [28] took the influences of the planned infrastructure works into account and obtained a more robust timetable. They aimed to adjust the published timetable as small as possible in case of a planned infrastructure unavailable to keep the level of passenger service as high as possible, which was implemented by an iterative approach. The developed algorithm allowed small modifications to the routing and the published timetable; besides they improved the robustness of solutions. In this paper, we also consider the impact of maintenance planning on train scheduling and generate a robust train schedule by establishing a two-stage integrated optimization model.

3. Problem Description and Assumptions

3.1. Problem Description

3.1.1. Railway Corridors for SDSA-Trains. When solving the problem of SDSA-trains timetable optimization, the impact of high-speed railway maintenance window must be considered. The SDSA-train cannot pass directly when there is a maintenance window, so it has to wait at the high-speed railway station or switch to the normal-speed railway, as shown in Figure 1. *Case*1 represents SDSA-trains select wait at the high-speed railway station to avoid the conflict with the maintenance window, and *Case*2 represents SDSA-trains switch from high-speed railway to normal-speed railway when the next segment high-speed railway undergoes maintenance window. Note that the normal-speed railway is parallel to the corresponding high-speed railway. There is a physical connection between them at major stations. Besides, the maintenance window is not considered on normal-speed railway.

Figure 2 clearly describes the connecting relationship between high-speed railway and its parallel normal-speed railway. There are 3 major stations (*S*1, *S*2, and *S*3) belonging to both high-speed railway and its parallel normal-speed railway. It provide the possibility for SDSA-trains to switch from high-speed railway to normal-speed railway. Besides, only major stations can serve as origin stations and destination stations. Figure 2 shows that the 3 major stations divide the

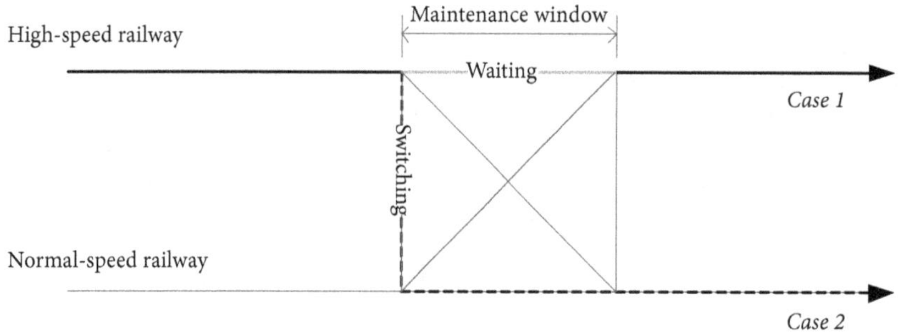

FIGURE 1: Waiting and switching of SDSA-trains to avoid the conflict with the maintenance window.

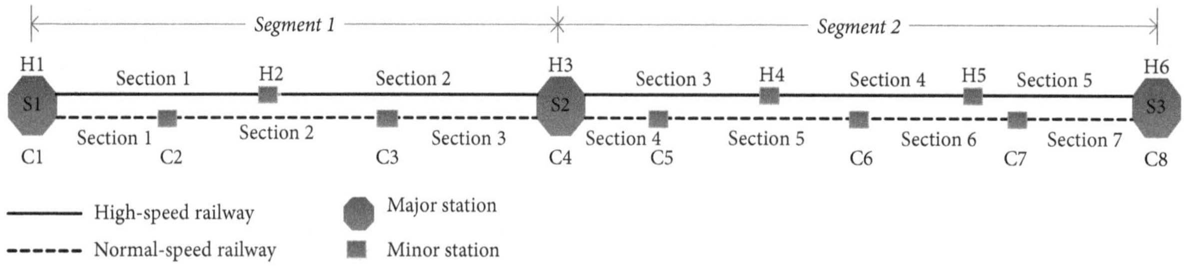

FIGURE 2: A high-speed railway corridor and its parallel normal-speed railway corridor.

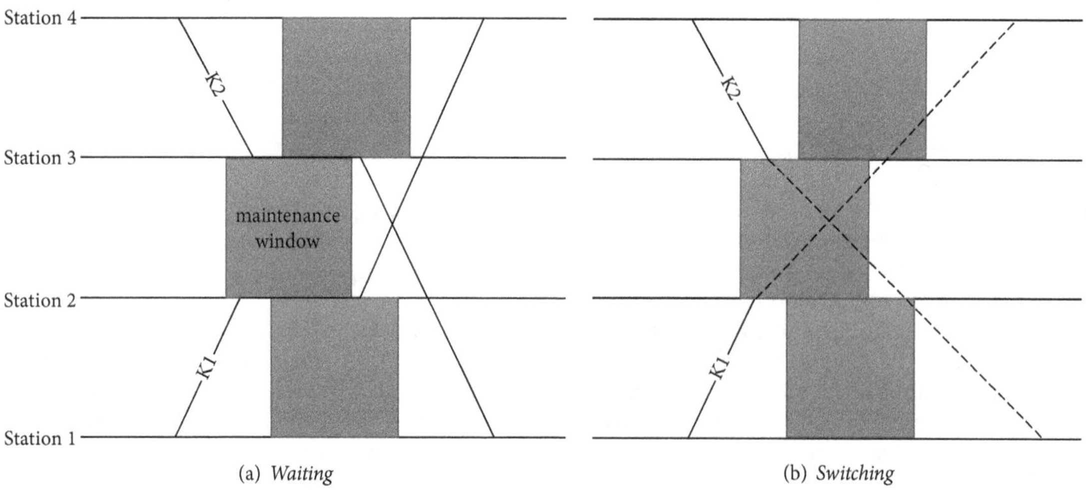

FIGURE 3: Illustration of SDSA-trains operation modes.

high-speed railway corridor into 2 segments, while all high-speed railway stations divide the high-speed railway corridor into 5 sections. Similarly, there are 2 segments and 7 sections on the normal-speed railway.

In this paper, we consider that the maintenance window on the whole line is asynchronous. It is synchronous on each segment. For example, Section 1 and Section 2 have the same maintenance window on high-speed railway, because they all belong to Segment 1. However, the maintenance windows on Segment 1 and Segment 2 are asynchronous.

3.1.2. Scheduling SDSA-Trains by Considering Conflicts with Maintenance Windows. Based on above analysis, there are two operation modes to be selected for SDSA-train when there is a conflict with maintenance window. Figure 3 shows these two operation modes (only major stations are considered) are simply named Mode 1 and Mode 2. Figure 3(a) shows Mode 1, which corresponds to *Case*1 in Figure 1. It means that the SDSA-train operates only on the high-speed railway; namely, when the next segment high-speed railway undergoes maintenance, the SDSA-train just waits at a major station until the maintenance is finished. For example, in Figure 3(a), train $K1$ waits for the second segment maintenance window at station 2 until the maintenance is finished. Figure 3(b) shows Mode 2, which corresponds to *Case*2 in Figure 1. A SDSA-train switches from high-speed railway to normal-speed railway at a proper major station if the next segment maintenance is executed. Then the SDSA-train runs on the normal-speed railway until it reaches its destination, seeing train $K1$ in Figure 3(b). The same analysis

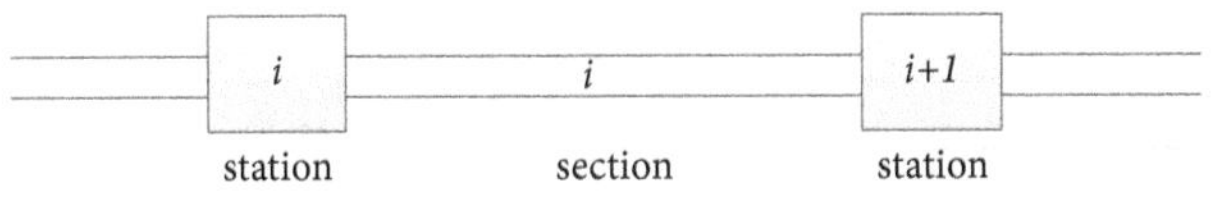

FIGURE 4: The relationship between section indexes and station indexes.

is also reasonable for the opposite train $K2$ in Figures 3(a) and 3(b).

3.2. Assumptions. To better illustrate the model, the following basic assumptions are made in this study:

Assumption 1. In order to ensure safety, maintenance on a segment cannot be separated. The widths of maintenance window are the same and predetermined on all segments.

Assumption 2. We do not consider the extra acceleration/deceleration times of SDSA-trains, because the consideration of acceleration/deceleration times does not make the problem be solved difficultly, but it makes the formula tedious.

Assumption 3. The detailed number of passengers boarding/alighting of each SDSA-train at each station are ignored, while the maximum capacity of each SDSA-train and the number of passenger demands at each major station are considered.

Assumption 4. The time cost of switching from high-speed railway to normal-speed at major stations is tiny compared to the total travel time. For simplicity, we assume that the time of switching is 0.

Assumption 5. From a safety perspective, when a train runs from origin station to destination station, the operation of switching from one railway line to another is allowed to occur at most once.

4. Mathematical Formulations

4.1. Notations and Parameters. For convenience, the notations used in this paper are listed in Table 1.

Remark 6. In Figure 2, we have $N_H = \{H1, H2, \cdots, H6\}$ and $N_C = \{C1, C2, \cdots, C8\}$. Since there are 3 major stations ($S1$, $S2$, and $S3$), the set of major stations is denoted by $N_m = \{S1, S2, S3\}$. Accordingly, mappings $H(i)$ and $C(i)$, $i \in N_m$, are defined as follows:

$$\begin{aligned} H(S1) &= H1, \\ H(S2) &= H3, \\ \text{and } H(S3) &= H6, \\ C(S1) &= C1, \\ C(S2) &= C4, \\ \text{and } C(S3) &= C8. \end{aligned} \tag{1}$$

Remark 7. Figure 4 shows the relationship between section indexes and station indexes.

4.2. Decision Variables. Note that this paper aims to formulate a two-stage integrated optimization model to schedule SDSA-trains and plan maintenance windows. Clearly, the set of decision variables include the start and end times of the maintenance windows and the arrival and departure times of the SDSA-trains, all of which are continuous variables. Moreover, some binary variables are used to describe stop pattern, departure orders of trains, and the operation modes selection of SDSA-trains. Table 2 lists all the decision variables and the corresponding definitions.

4.3. First Stage: Generation of Train Scheduling and Maintenance Planning under Each Scenario

4.3.1. Objective Function. Typically, the railway company aims to provide passengers with more convenient and quick train. Thus, in the first stage, we use objective function (2) to minimize the total travel time of SDSA-trains in order to achieve greater social benefits under different scenarios.

$$\min F_1 = \sum_{k=1}^{T_n} \left(a^H_{s,k,D(k)} - d^H_{s,k,O(k)} \right). \tag{2}$$

4.3.2. Constraints

(a) Basic Constraints of SDSA-Train

$$\sum_{k=1}^{T_n} x^H_{s,k,i} \cdot Cap_k \geq Q_i, \quad i \in N_m. \tag{3}$$

$$d^-_{k,O(k)} \leq d^H_{s,k,O(k)} \leq d^+_{k,O(k)}, \quad k \in T_n. \tag{4}$$

$$a^-_{k,D(k)} \leq a^H_{s,k,D(k)} \leq a^+_{k,D(k)}, \quad k \in T_n. \tag{5}$$

$$a^H_{s,k,H(i)} = a^C_{s,k,C(i)}, \quad k \in T_n, \ i \in N_m, \ H(i) \in N_H, \ C(i) \in N_C. \tag{6}$$

$$d^H_{s,k,H(i)} = d^C_{s,k,C(i)}, \quad k \in T_n, \ i \in N_m, \ H(i) \in N_H, \ C(i) \in N_C. \tag{7}$$

TABLE 1: Notations.

Notations	Definition
N	set of all stations.
N_H	set of high-speed railway stations.
N_C	set of normal-speed railway stations.
N_m	set of major stations.
i, j	index of stations.
$H(i)$	mapping from set N_m to set N_H, *i.e.*, $i \in N_m$, $H(i) \in N_H$.
$C(i)$	mapping from set N_m to set N_C, *i.e.*, $i \in N_m$, $C(i) \in N_C$.
T	set of all trains.
T_n	set of SDSA-trains.
T_h	set of existing high-speed trains.
T_c	set of existing normal-speed trains.
k, l	index of trains.
S	set of all scenarios.
s	index of scenarios.
$O(k)/D(k)$	origin/destination station of SDSA-train k.
$r^H_{i,min}/r^C_{i,min}$	minimum running time on section i on high-speed railway/normal-speed railway.
$r^H_{i,max}/r^C_{i,max}$	maximum running time on section i on high-speed railway/normal-speed railway.
$t^H_{s,k,i}/t^C_{s,k,i}$	running time of train k on section i on high-speed railway/normal-speed railway under scenario s.
h^{aH}_i	minimal headway between consecutive arrival at station i on high-speed railway.
h^{dH}_i	minimal headway between consecutive departure from station i on high-speed railway.
h^{aC}_i	minimal headway between consecutive arrival at station i on normal-speed railway.
h^{dC}_i	minimal headway between consecutive departure from station i on normal-speed railway.
τ^H_i	the minimum dwelling time of each train at station i on high-speed railway.
τ^C_i	the minimum dwelling time of each train at station i on normal-speed railway.
$[d^-_{k,O(k)}, d^+_{k,O(k)}]$	time window of SDSA-train k departure from origin station.
$[a^-_{k,D(k)}, a^+_{k,D(k)}]$	time window of SDSA-train k arrival at destination station.
$[w^-_i, w^+_i]$	time window of maintenance at high-speed railway station i.
Q_i	the total passenger demand at major station i on high-speed railway.
Cap_k	the maximum capacity of SDSA-train k.
D	the width of maintenance window on high-speed railway.
M	a large positive number.

TABLE 2: Decision variables.

Variables	Definition
$s^w_{s,i}/e^w_{s,i}$	actual start/end time of maintenance window at major station i on high-speed railway under scenario s, $s \in S, i \in N_m$.
$d^H_{s,k,i}/d^C_{s,k,i}$	departure time of train k on high-speed railway/normal-speed railway from station i under scenario s, $s \in S$.
$a^H_{s,k,i}/a^C_{s,k,i}$	arrival time of train k on high-speed railway/normal-speed railway at station i under scenario s, $s \in S$.
$x^H_{s,k,i}$	= 1 if train k stops at station i under scenario s; =0, otherwise. $s \in S, k \in T_n, i \in N_H$.
$x^C_{s,k,i}$	= 1 if train k stops at station i under scenario s; =0, otherwise. $s \in S, k \in T_n, i \in N_C$.
$O_{s,k,l,i}$	= 1 if train k departs from station i earlier than train l under scenario s; =0, otherwise. $s \in S$.
$\rho_{s,k,i}$	= 1 if train k runs on high-speed railway from major station i to major station $i + 1$ under scenario s; =0, otherwise. $s \in S, k \in T_n, i \in N_m$.
$Z^{(0)}_{s,k,i}$	= 1 if train k waits at major station i when there is a conflict with maintenance under scenario s; =0, otherwise. $s \in S, k \in T_n, i \in N_m$.
$Z^{(1)}_{s,k,i}$	= 1 if train k switches from high-speed railway to normal-speed railway at major station i under scenario s; =0, otherwise. $s \in S, k \in T_n, i \in N_m$.

$$d^{H}_{s,k,i} - a^{H}_{s,k,i} \geq \tau^{H}_{i} \cdot x^{H}_{s,k,i}, \quad k \in T_n, \ i \in N_H. \tag{8}$$

$$x^{H}_{s,k,i} \leq \rho_{s,k,j}, \quad k \in T_n, \ i \in N_H, \ j \in N_m, \ i \in \{H(j), H(j)+1, \ldots, H(j+1)-1\}. \tag{9}$$

$$d^{C}_{s,k,i} - a^{C}_{s,k,i} \geq \tau^{C}_{i} \cdot x^{C}_{s,k,i}, \quad k \in T_n, \ i \in N_C. \tag{10}$$

$$x^{C}_{s,k,i} \leq 1 - \rho_{s,k,j}, \quad k \in T_n, \ i \in N_C, \ j \in N_m, \ i \in \{C(j), C(j)+1, \ldots, C(j+1)-1\}. \tag{11}$$

$$a^{H}_{s,k,i+1} - d^{H}_{s,k,i} = t^{H}_{s,k,i} \cdot \rho_{s,k,j} + \left(a^{C}_{s,k,i+1} - d^{C}_{s,k,i}\right) \cdot \left(1 - \rho_{s,k,j}\right), \quad i \in N_H, \ j \in N_m. \tag{12}$$

$$a^{C}_{s,k,i+1} - d^{C}_{s,k,i} = t^{C}_{s,k,i} \cdot \left(1 - \rho_{s,k,j}\right) + \left(a^{H}_{s,k,i+1} - d^{H}_{s,k,i}\right) \cdot \rho_{s,k,j}, \quad i \in N_C, \ j \in N_m. \tag{13}$$

$$d^{H}_{s,l,i} - d^{H}_{s,k,i} \geq h^{dH}_{i} - M \cdot \left(3 - O_{s,k,l,i} - \rho_{s,l,j} - \rho_{s,k,j}\right), \quad k,l \in T_n \cup T_h, \ i \in N_H, \ j \in N_m. \tag{14}$$

$$d^{H}_{s,k,i} - d^{H}_{s,l,i} \geq h^{dH}_{i} - M \cdot \left(3 - O_{s,l,k,i} - \rho_{s,l,j} - \rho_{s,k,j}\right), \quad k,l \in T_n \cup T_h, \ i \in N_H, \ j \in N_m. \tag{15}$$

$$a^{H}_{s,l,i} - a^{H}_{s,k,i} \geq h^{aH}_{i} - M \cdot \left(3 - O_{s,k,l,i-1} - \rho_{s,l,j} - \rho_{s,k,j}\right), \quad k,l \in T_n \cup T_h, \ i \in N_H, \ j \in N_m. \tag{16}$$

$$a^{H}_{s,k,i} - a^{H}_{s,l,i} \geq h^{aH}_{i} - M \cdot \left(3 - O_{s,l,k,i-1} - \rho_{s,l,j} - \rho_{s,k,j}\right), \quad k,l \in T_n \cup T_h, \ i \in N_H, \ j \in N_m. \tag{17}$$

$$d^{C}_{s,l,i} - d^{C}_{s,k,i} \geq h^{dC}_{i} - M \cdot \left(1 - O_{s,k,l,i} + \rho_{s,l,j} + \rho_{s,k,j}\right), \quad k,l \in T_n \cup T_c, \ i \in N_C, \ j \in N_m. \tag{18}$$

$$d^{C}_{s,k,i} - d^{C}_{s,l,i} \geq h^{dC}_{i} - M \cdot \left(1 - O_{s,l,k,i} + \rho_{s,l,j} + \rho_{s,k,j}\right), \quad k,l \in T_n \cup T_c, \ i \in N_C, \ j \in N_m. \tag{19}$$

$$a^{aC}_{s,l,i} - a^{C}_{s,k,i} \geq h^{aC}_{i} - M \cdot \left(1 - O_{s,k,l,i-1} + \rho_{s,l,j} + \rho_{s,k,j}\right), \quad k,l \in T_n \cup T_c, \ i \in N_C, \ j \in N_m. \tag{20}$$

$$a^{C}_{s,k,i} - a^{C}_{s,l,i} \geq h^{aC}_{i} - M \cdot \left(1 - O_{s,l,k,i-1} + \rho_{s,l,j} + \rho_{s,k,j}\right), \quad k,l \in T_n \cup T_c, \ i \in N_C, \ j \in N_m. \tag{21}$$

Constraint (3) is established to satisfy the passenger demands, which is the most important constraint for SDSA-trains stop plan. Constraint (4) is used to guarantee the departure time from origin station in a suitable time range to make SDSA-trains more attractive. The same applies to constraint (5). Constraints (6) and (7) are high-speed railway and normal-speed railway connection constraints at major stations. According to Assumption 4 and mappings $H(i)$ and $C(i)$, $i \in N_m$, it is equivalent to say that SDSA-train k "arrives at major station i," "arrives at high-speed railway station $H(i)$," or "arrives at normal-speed railway station $C(i)$." In other words, at major station i, we have $a^{H}_{s,k,H(i)} = a^{C}_{s,k,C(i)}$; that is, constraint (6) holds. Similarly, constraint (7) holds.

Constraints (8)-(11) are established to ensure that the dwelling time is long enough to complete passengers' boarding and alighting if the SDSA-train k stops at the station i. Constraints (8) and (10) are the basic dwelling time constraints. Constraints (9) and (11) are established for the operation of SDSA-trains switch from high-speed railway to normal-speed railway at major station. Obviously, for constraint (9), if the SDSA-train k runs on normal-speed railway (i.e., $\rho_{s,k,j} = 0$), then the SDSA-train k cannot dwell at high-speed railway station (i.e., $x^{H}_{s,k,i} = 0$); if the SDSA-train k runs on high-speed railway (i.e., $\rho_{s,k,j} = 1$), then the SDSA-train k may dwell at high-speed railway station (i.e., $x^{H}_{s,k,i} \leq 1$). Similarly, constraint (11) holds.

Constraint (12) expresses the running time between two consecutive high-speed railway stations. If $\rho_{k,j} = 1$, the running time of SDSA-train k from station i to station $i+1$ on high-speed railway under scenario s is equivalent to $t^{H}_{s,k,i}$; if $\rho_{k,j} = 0$, $a^{H}_{s,k,i+1} - d^{H}_{s,k,i} = a^{C}_{s,k,i+1} - d^{C}_{s,k,i}$ always holds. The same applies to constraint (13). Besides, constraints (14)-(21) are the necessary headway constraints (including the arrival headway constraints and departure headway constraints) to guarantee the safe operations on the high-speed railway corridor and normal-speed railway corridor.

(b) Additional Constraints of SDSA-Train When There Is a Conflict with Maintenance

$$w^{-}_{i} \leq s^{w}_{s,i} \leq e^{w}_{s,i} \leq w^{+}_{i}, \quad i \in N_m. \tag{22}$$

$$e^{w}_{s,i} = s^{w}_{s,i} + D, \quad i \in N_m. \tag{23}$$

$$Z^{(0)}_{s,k,i} + Z^{(1)}_{s,k,i} \geq 1, \quad \text{if } \rho_{s,k,i-1} = 1 \text{ and } s^{w}_{s,i} \leq a^{H}_{s,k,H(i)} < e^{w}_{s,i}, \ k \in T_n, \ i \in N_m. \tag{24}$$

$$Z^{(0)}_{s,k,i} + Z^{(1)}_{s,k,i} \geq 1, \quad \text{if } \rho_{s,k,i-1} = 1, \ a^{H}_{s,k,H(i)} < s^{w}_{s,i} \text{ and } a^{H}_{s,k,H(i+1)} > s^{w}_{s,i}, \ k \in T_n, \ i \in N_m. \tag{25}$$

$$d^{H}_{s,k,H(i)} \geq e^{w}_{s,i} \text{ and } \rho_{s,k,i} = 1, \quad \text{if } Z^{(0)}_{s,k,i} = 1, \ k \in T_n, \ i \in N_m. \tag{26}$$

$$\rho_{s,k,i} \leq \rho_{s,k,i-1}, \quad k \in T_n, \ i \in N_m. \tag{27}$$

$$Z^{(0)}_{s,k,i} + Z^{(1)}_{s,k,i} \leq 1, \quad k \in T_n, \ i \in N_m. \tag{28}$$

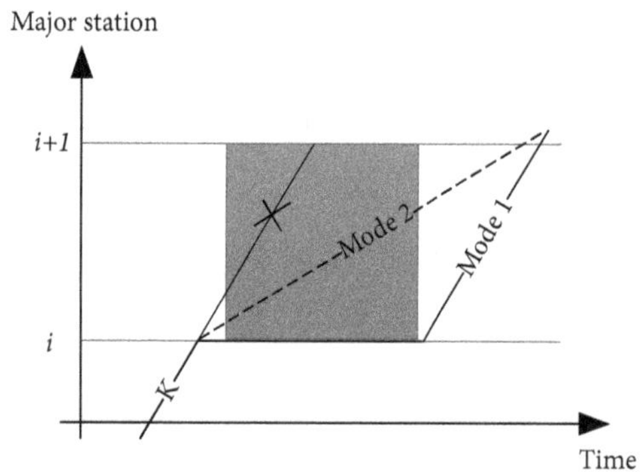

Figure 5: An illustration of constraint (25).

$$Z^{(0)}_{s,k,i} \le \rho_{s,k,i-1} \text{ and } Z^{(0)}_{s,k,i} \le \rho_{s,k,i}, \quad k \in T_n,\ i \in N_m \tag{29}$$

$$Z^{(1)}_{s,k,i} \ge \rho_{s,k,i-1} - \rho_{s,k,i}, \quad k \in T_n,\ i \in N_m. \tag{30}$$

$$\rho_{s,k,i} \le 1 - Z^{(1)}_{s,k,i}, \quad k \in T_n,\ i \in N_m. \tag{31}$$

Constraint (22) ensures that maintenance starting time $s^w_{s,i}$ and ending time $e^w_{s,i}$ are in a reasonable time interval $[w^-_i, w^+_i]$. Constraint (23) means that the widths of maintenance window (*i.e.*, D) are the same and predetermined on each segment. Constraints (24)-(26) are the operation modes selection constraints. Constraint (24) indicates the fact that when SDSA-train k arrives at major station i, if the maintenance on segment from major station i to major station $i+1$ is being executed, then SDSA-train k must select an operation mode (i.e., Mode 1 or Mode 2) as shown in Figure 3. Constraint (25) indicates the fact that when SDSA-train k arrives at major station i, there is no conflict with the maintenance window, but it will conflict with the maintenance window at major station $i+1$ if SDSA-train k continues to run, which is illustrated in Figure 5. Then SDSA-train k must select Mode 1 or Mode 2, namely, waiting at major station i or switching to normal-speed railway corridor. Furthermore, if Mode 1 is selected for SDSA-train k to avoid the conflict with the maintenance window at major station i, then its waiting end time should satisfy constraint (26).

Constraints (27)-(31) are state-mode constraints. Obviously, when a SDSA-train conflicts with maintenance window at a major station, the railway line which it will run on may change. Binary variables $\rho_{s,k,i}$, $Z^{(0)}_{s,k,i}$, and $Z^{(1)}_{s,k,i}$ record the selection of SDSA-trains together. From a security perspective, we do not allow the SDSA-trains to switch back to high-speed railway when Mode 2 is selected, so constraint (27) is given. Constraint (28) indicates that at most one operation mode can be selected for each SDSA-train at major stations. Constraint (29) ensures that the SDSA-trains can only run on high-speed railway if Mode 1 is selected. Constraints (30)-(31) ensure that Mode 2 can be successfully selected. Obviously, only when $\rho_{s,k,i-1} = 1$ and $\rho_{s,k,i} = 0$ are satisfied at the same time will the operation of switching from high-speed railway to normal-speed railway happen. Moreover, when $\rho_{s,k,i-1}$ and $\rho_{s,k,i}$ take other values, constraint (30) always holds. Similarly, constraint (31) always holds.

4.4. Second Stage: Generation of Robust Train Scheduling and Maintenance Planning. With the train scheduling and maintenance planning generated in the first stage under different scenarios, the second stage aims to generate a robust train scheduling and maintenance planning. We aim to achieve the following two objectives in this stage, namely, (1) the total travel time of all SDSA-trains is minimized and (2) the generated train scheduling is more robust (i.e., the new train scheduling has the minimum deviation from the train scheduling under different scenarios). Thus, the objective function of the second stage is formulated as follows:

$$\begin{aligned} \min F_2 = {} & \omega_1 \cdot \sum_{k=1}^{T_n} \left(a^H_{k,D(k)} - d^H_{k,O(k)}\right) + \omega_2 \\ & \cdot \left(\sum_{s=1}^{S} \sum_{k=1}^{T_n} \left| \left(a^H_{k,D(k)} - d^H_{k,O(k)}\right) - \left(a^H_{s,k,D(k)} - d^H_{s,k,O(k)}\right) \right| \right). \end{aligned} \tag{32}$$

$d^H_{k,O(k)}$ is the departure time of train k on high-speed railway from origin station $O(k)$. $a^H_{k,D(k)}$ is the arrival time of train k on high-speed railway at destination station $D(k)$. These two variables are applicable to the optimization of the second stage.

The new train scheduling and maintenance planning also should satisfy the same constraints (3)-(31) set in the first stage to ensure safe operations of trains and meet the travel demands of passengers. Note that we consider the travel time as a decision variable in the second stage to formulate a biobjective optimization model, which is different from the first stage. The corresponding running time constraints of the second stage are shown as follows:

$$\begin{aligned} & r^H_{i,min} - M \cdot \left(1 - \rho_{s,k,j}\right) \le a^H_{s,k,i+1} - d^H_{s,k,i} \\ & \quad \le r^H_{i,max} + M \cdot \left(1 - \rho_{s,k,j}\right), \quad i \in N_H,\ j \in N_m. \end{aligned} \tag{33}$$

$$\begin{aligned} & r^C_{i,min} - M \cdot \rho_{s,k,j} \le a^C_{s,k,i+1} - d^C_{s,k,i} \le r^C_{i,max} + M \cdot \rho_{s,k,j}, \\ & \qquad i \in N_C,\ j \in N_m. \end{aligned} \tag{34}$$

Based on the above analysis, the second-stage optimization model can be formulated as follows:

$$\begin{aligned} \min \quad & F_2 \\ \text{s.t.} \quad & \text{Constraints (3)} - \text{(11), constraints (14)} \\ & - \text{(31), and constraints (33)} - \text{(34)}. \end{aligned} \tag{35}$$

4.5. Linearization. Note that constraints (12)-(13) and (24)-(26) and the second-stage objective function F_2 are nonlinear, so we will linearize them with some linearization techniques.

4.5.1. Linearize the Nonlinear Constraints. We find that constraints (12) and (13) are not easy to linearize, but we can replace them with the following two identical linearization formulas:

$$
\begin{aligned}
& t^H_{s,k,i} - M \cdot \left(1 - \rho_{s,k,j}\right) \le a^H_{s,k,i+1} - d^H_{s,k,i} \\
& \quad \le t^H_{s,k,i} + M \cdot \left(1 - \rho_{s,k,j}\right), \quad i \in N_H, \ j \in N_m. \\
& t^C_{s,k,i} - M \cdot \rho_{s,k,j} \le a^C_{s,k,i+1} - d^C_{s,k,i} \le t^C_{s,k,i} + M \cdot \rho_{s,k,j}, \\
& \qquad i \in N_C, \ j \in N_m.
\end{aligned} \tag{36}
$$

Some extra binary variables (i.e., α_1, α_2, β_1, and β_2) are used to linearize constraints (24)-(26) in the integrated optimization model. The following lemmas are easy to verify, so we do not give a detailed proof in this paper.

Lemma 8. *For $i \in N_m$, $H(i) \in N_H$, $i \ge 2$, and $k \in T_n$, constraint (24) can be equivalently replaced by the following linear constraint:*

$$
Z^{(0)}_{s,k,i} + Z^{(1)}_{s,k,i} \ge -2 + \alpha_1 + \alpha_2 + \rho_{s,k,i-1}, \tag{37}
$$

where α_1 is a binary variable, indicating the order relation between $a^H_{s,k,H(i)}$ and $s^w_{s,i}$. Similarly, α_2 is a binary variable, indicating the order relation between $a^H_{s,k,H(i)}$ and $e^w_{s,i}$, i.e.,

$$
\begin{aligned}
\alpha_1 &= \begin{cases} 1, & if \ a^H_{s,k,H(i)} \ge s^w_{s,i} \\ 0, & if \ a^H_{s,k,H(i)} < s^w_{s,i} \end{cases} \\
\alpha_2 &= \begin{cases} 1, & if \ a^H_{s,k,H(i)} < e^w_{s,i} \\ 0, & if \ a^H_{s,k,H(i)} \ge e^w_{s,i} \end{cases}
\end{aligned} \tag{38}
$$

It is easy to verify that definitions of α_1 and α_2 can be equivalently transformed to the following linear and binary constraints:

$$
\begin{aligned}
\alpha_1 &> \frac{a^H_{s,k,H(i)} - s^w_{s,i}}{M}, \\
\alpha_1 &\le \frac{a^H_{s,k,H(i)} - s^w_{s,i}}{M} + 1; \\
\alpha_2 &\ge \frac{e^w_{s,i} - a^H_{s,k,H(i)}}{M}, \\
\alpha_2 &< \frac{e^w_{s,i} - a^H_{s,k,H(i)}}{M} + 1; \\
& \qquad \alpha_1, \alpha_2 \in \{0, 1\},
\end{aligned} \tag{39}
$$

where M is a sufficiently large positive value.

Similarly, constraint (25) can be linearized according to the following lemma.

Lemma 9. *For $i \in N_m$, $H(i) \in N_H$, $i \ge 2$, and $k \in T_n$, constraint (25) can be equivalently replaced by the following linear constraint:*

$$
Z^{(0)}_{s,k,i} + Z^{(1)}_{s,k,i} \ge -2 + \beta_1 + \beta_2 + \rho_{s,k,i-1}, \tag{40}
$$

where β_1 is a binary variable, indicating the order relation between $a^H_{s,k,H(i)}$ and $s^w_{s,i}$. Similarly, β_2 is a binary variable, indicating the order relation between $a^H_{s,k,H(i+1)}$ and $s^w_{s,i}$.

Further, constraint (26) can be linearized directly without introducing extra additional variables, and it can be linearized according to the following lemma.

Lemma 10. *For $i \in N_m$, $H(i) \in N_H$, and $k \in T_n$, constraint (26) can be equivalently replaced by the following linear constraints:*

$$
d^H_{s,k,H(i)} \ge e^w_{s,i} - M \cdot \left(1 - Z^{(0)}_{s,k,i}\right) - M \cdot \left(1 - \rho_{s,k,i}\right). \tag{41}
$$

4.5.2. Linearize the Nonlinear Objective Function. Different from the nonlinear constraints, the nonlinear objective function F_2 is caused by the absolute value. In order to linearize it, we use the following lemma.

Lemma 11. *Assume that $\mathbf{X}$ is the feasible set and $v_0(x)$ and $v_1(x)$ are linear functions of $x \in \mathbf{X}$.*
Model

$$
\begin{aligned}
\max_{x} \quad & v_0(x) + |v_1(x)| \\
\text{s.t.} \quad & \mathbf{x} \in \mathbf{X},
\end{aligned} \tag{42}
$$

is equivalent to the following model with linear objective function:

$$
\begin{aligned}
\max_{x, C_0, C_1} \quad & C_0 + C_1 \\
\text{s.t.} \quad & C_0 = v_0(x), \\
& C_1 \ge v_1(x), \\
& C_1 \ge -v_1(x), \\
& \mathbf{x} \in \mathbf{X}.
\end{aligned} \tag{43}
$$

The proof of Lemma 11 is simple. According to Lemma 11, the second-stage objective function F_2 of model (32) is reformulated as

$$
\min \widetilde{F_2} = \sum_{k=1}^{T_n} \left(\omega_1 \cdot C^{tra}_k + \omega_2 \cdot \sum_{s=1}^{S} C^{tra}_{s,k} \right). \tag{44}
$$

where C^{tra}_k and $C^{tra}_{s,k}$ satisfy the following constraints:

$$
C^{tra}_k = a^H_{k,D(k)} - d^H_{k,O(k)}, \quad k \in T_n \tag{45}
$$

$$
C^{tra}_{s,k} \ge \left(a^H_{k,D(k)} - d^H_{k,O(k)}\right) - \left(a^H_{s,k,D(k)} - d^H_{s,k,O(k)}\right), \quad k \in T_n, \ s \in S \tag{46}
$$

$$
C^{tra}_{s,k} \ge -\left(\left(a^H_{k,D(k)} - d^H_{k,O(k)}\right) - \left(a^H_{s,k,D(k)} - d^H_{s,k,O(k)}\right)\right), \quad k \in T_n, \ s \in S. \tag{47}
$$

Obviously, the second-stage objective function F_2 can be equivalently replaced by $\widetilde{F_2}$. This stage model also should satisfy constraints (45)-(47) in addition to the constraints contained in model (35).

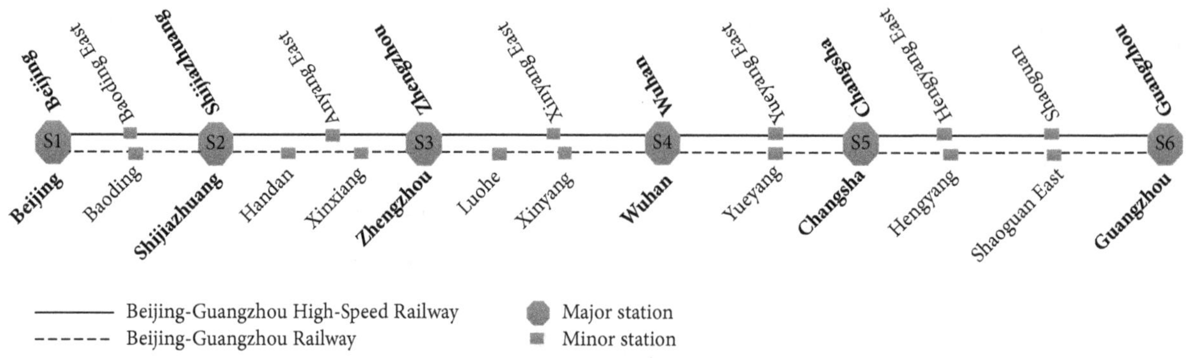

FIGURE 6: Topological relations of Beijing-Guangzhou High-Speed Railway and Beijing-Guangzhou Railway.

5. Numerical Experiments

In this section, we carry out computational experiment based on the Beijing-Guangzhou High-Speed Railway and the Beijing-Guangzhou Railway in China to show the effectiveness and efficiency of our proposed model. All the experiments are carried out on a computer with Intel(R) Core(TM) i5-3337U@1.80GHz CPU and 8.00GB RAM using Microsoft Windows 7(64bit) OS. The programming is coded in MATLAB 2014b. All MILP models are solved by CPLEX 12.5 to optimality with a gap of 0.01%.

5.1. Setup. In this experiment, we plan to insert 10 SDSA-trains (5 downstream and 5 upstream) into the existing timetable under each scenario. Besides, we will consider the double track Beijing-Guangzhou High-Speed Railway and the Beijing-Guangzhou Railway. Beijing-Guangzhou High-Speed Railway is the longest high-speed railway line in the world. It consists of 36 stations with a total length 2, 298 km. However, some of the stations are very small. In order to make the calculation simpler and without loss of generality, we only select 12 typical stations to carry out numerical experiments. Similarly, we also select 14 typical stations on Beijing-Guangzhou Railway to carry out numerical experiments. The topological relations between the two railway lines are shown in Figure 6.

Obviously, there are 6 major stations in the selected corridor, which divide Beijing-Guangzhou Railway corridor into 5 segments. According to Section 3, we know that there are 5 different maintenance works on the high-speed railway corridor. Besides, the Beijing-Guangzhou High-Speed Railway is divided into 11 sections and the Beijing-Guangzhou Railway is divided into 13 sections, depending on all the selected stations. For more clarity, Tables 3 and 4 are given to show the distances and running times of each section. Note that the minimum and maximum running times of each section are listed. For example, in the first row of Table 3, (37, 49) means that the minimum running time of section 1 is 37 minutes and the maximum running time of section 1 is 49 minutes.

In order to keep a safe distance of consecutive trains, the minimum arrival headway and departure headway should be set in advance. On the high-speed railway, at minor stations, the minimum arrival headway is set to 2 minutes and the minimum departure headway is set to 3 minutes, i.e., h_i^{aH}=2 and h_i^{dH}=3; at major stations, h_i^{aH}=3 and h_i^{dH}=5. Similarly, on the normal-speed railway, at minor stations, we set h_i^{aC}=3 and h_i^{dC}=4; at major stations, we set h_i^{aC}=4 and h_i^{dC}=5. The minimum dwelling time should be set, and we have $\tau_i^H = 2$ and $\tau_i^C = 3$. Besides, we are interested in giving the passenger demand at each major station, which determines the SDSA-train stop plan to some extent. The downstream passenger demands at major stations are 1000, 550, 600, 600, 650, and 800; the upstream passenger demands at major stations are 800, 650, 600, 600, 550, and 1000. Though, this may have a little difference from the actual situation, and the capacity of the SDSA-train at each station is still set to 630, *i.e.*,Cap_k = 630.

The original timetable we use is the actual timetable on April 15st, 2017. Note that we only consider the existing trains that may have conflict with SDSA-trains. After processing, 73 downstream existing trains and 58 upstream existing trains on Beijing-Guangzhou High-Speed Railway remained; 32 downstream existing trains and 23 upstream existing trains on Beijing-Guangzhou Railway remained.

The numerical experiments are divided into two stages. In the first stage, we generate 10 different train schedules and maintenance plans under different scenarios. In the second stage, a robust train scheduling is designed based on the first stage.

5.2. The Numerical Experiments of the First Stage. In the first stage, we generate 10 different scenarios. Table 5 clearly shows some related parameters, "Dep." represents departure time window, and "Arr." represents arrival time window. Because a SDSA-train departs from its origin station on the first night and arrives at its destination station on the next morning, we simply set the start of the second day as 1440. In addition, we assume that the width of maintenance window on each segment is 240 minutes, *i.e.*, D = 240; the time window of maintenance at each station is from 0: 00 to 6: 00 am, i.e., $[w_i^-, w_i^+] = [1440, 1800]$. Note that the above parameters are the same under each scenario.

Further, we carry out numerical experiments based on the different disturbance times obtained in 10 different scenarios.

Table 3: Information of distance and running time on the high-speed railway.

Downstream (Beijing ⟶ Guangzhou)			Upstream (Guangzhou ⟶ Beijing)		
Section	Distance (km)	Running time (min)	Section	Distance (km)	Running time (min)
1	139	(35, 49)	1	227	(38, 50)
2	142	(33, 41)	2	303	(61, 69)
3	235	(50, 68)	3	177	(40, 46)
4	177	(39, 50)	4	147	(31, 39)
5	337	(70, 86)	5	215	(46, 56)
6	199	(39, 50)	6	199	(39, 46)
7	215	(45, 60)	7	337	(66, 86)
8	147	(30, 42)	8	177	(34, 48)
9	177	(34, 42)	9	235	(45, 63)
10	303	(56, 67)	10	142	(28, 39)
11	227	(49, 63)	11	139	(28, 46)

Table 4: Information of distance and running time on the normal-speed railway.

Downstream (Beijing ⟶ Guangzhou)			Upstream (Guangzhou ⟶ Beijing)		
Section	Distance (km)	Running time (min)	Section	Distance (km)	Running time (min)
1	150	(72, 97)	1	221	(133, 143)
2	131	(64, 78)	2	300	(186, 204)
3	165	(79, 98)	3	186	(114, 137)
4	157	(77, 122)	4	147	(76, 84)
5	90	(43, 52)	5	215	(117, 136)
6	140	(64, 85)	6	234	(119, 154)
7	162	(78, 97)	7	162	(80, 101)
8	234	(116, 139)	8	140	(69, 115)
9	215	(115, 142)	9	90	(42, 56)
10	147	(78, 88)	10	157	(75, 98)
11	186	(113, 128)	11	165	(79, 95)
12	300	(186, 207)	12	131	(69, 76)
13	221	(138, 180)	13	150	(82, 105)

Each scenario is an assumption of a set of running time of the trains on the sections. In order to consider all possible disturbances more comprehensively, the value of the running time of different scenarios should include all cases between the minimum running time and the maximum running time as much as possible. We first get the minimum running time and the maximum running time of the section by processing the obtained actual timetable as shown in Tables 3 and 4. Secondly, the running time of each section is divided into 10 statistical intervals according to the increasing order of running time. Finally, different scenarios are generated based on the 10 statistical intervals. To more clearly illustrate the method of generating the scenarios, we give an example of how to make the parameter value of different scenarios in the downstream of the high-speed railway, as shown in Table 6.

The value of the parameter R_s is the same under each scenario s, as shown in Table 6. That is to say, the scenarios parameters are generated randomly only once for each section of the railway under each scenario. Note that the generated scenarios parameters not only apply to the high-speed railway downstream but also the high-speed railway upstream, the normal-speed railway downstream, and the normal-speed railway upstream.

Table 7 shows the objectives and computation times of the experiments of each scenario. Each scenario involves 12636 decision variables and 52392 constraints.

To further analyze the influence of scenarios parameters on the objectives, Figure 7 is given.

Figure 7 presents the change of the objectives under different scenarios parameters. It is reasonable to see that the objective function value increases with the increase of the scenarios parameter values. For example, the parameter value generated in scenario 1 is the smallest in all 10 scenarios, corresponding to the shortest running time in all 10 scenarios (i.e., the maximum running speed). Generally, the total travel time increases with the increases of running time of section.

5.3. The Numerical Experiments of the Second Stage. Obviously, there are two objectives in the second stage. The weights of the two objectives are represented by ω_1 and ω_2, respectively. In this paper, we consider that the two objectives

TABLE 5: The parameters of SDSA-trains.

No.	Downstream (Beijing ⟶ Guangzhou)			Upstream (Guangzhou ⟶ Beijing)		
	Trip	Dep.	Arr.	Trip	Dep.	Arr.
1	Beijing⟶Guangzhou	[1020, 1140]	[1740, 2040]	Guangzhou⟶Beijing	[1020, 1140]	[1740, 2040]
2	Beijing⟶Guangzhou	[1160, 1200]	[1740, 2040]	Guangzhou⟶Beijing	[1160, 1200]	[1740, 2040]
3	Beijing⟶Changsha	[1220, 1260]	[1740, 2040]	Guangzhou⟶Shijiazhuang	[1220, 1260]	[1740, 2040]
4	Beijing⟶Guangzhou	[1280, 1320]	[1740, 2040]	Guangzhou⟶Beijing	[1280, 1320]	[1740, 2040]
5	Shijiazhuang⟶Guangzhou	[1260, 1320]	[1740, 2040]	Changsha⟶Beijing	[1260, 1320]	[1740, 2040]

TABLE 6: Scenarios setting example: downstream of high-speed railway.

Section	Running time(min)	Scenario 1	Scenario 2	......	Scenario 9	Scenario 10
1	(35, 49)	$35+R_1\cdot(49-35)$	$35+R_2\cdot(49-35)$	......	$35+R_9\cdot(49-35)$	$35+R_{10}\cdot(49-35)$
2	(33, 41)	$33+R_1\cdot(41-33)$	$33+R_2\cdot(41-33)$	......	$33+R_9\cdot(41-33)$	$33+R_{10}\cdot(41-33)$
3	(50, 68)	$50+R_1\cdot(68-50)$	$50+R_2\cdot(68-50)$	......	$50+R_9\cdot(68-50)$	$50+R_{10}\cdot(68-50)$
4	(39, 50)	$39+R_1\cdot(50-39)$	$39+R_2\cdot(50-39)$	......	$39+R_9\cdot(50-39)$	$39+R_{10}\cdot(50-39)$
5	(70, 86)	$70+R_1\cdot(86-70)$	$70+R_2\cdot(86-70)$	......	$70+R_9\cdot(86-70)$	$70+R_{10}\cdot(86-70)$
6	(39, 50)	$39+R_1\cdot(50-39)$	$39+R_2\cdot(50-39)$	......	$39+R_9\cdot(50-39)$	$39+R_{10}\cdot(50-39)$
7	(45, 60)	$45+R_1\cdot(60-45)$	$45+R_2\cdot(60-45)$	......	$45+R_9\cdot(60-45)$	$45+R_{10}\cdot(60-45)$
8	(30, 42)	$30+R_1\cdot(42-30)$	$30+R_2\cdot(42-30)$	......	$30+R_9\cdot(42-30)$	$30+R_{10}\cdot(42-30)$
9	(34, 42)	$34+R_1\cdot(42-34)$	$34+R_2\cdot(42-34)$	......	$34+R_9\cdot(42-34)$	$34+R_{10}\cdot(42-34)$
10	(56, 67)	$56+R_1\cdot(67-56)$	$56+R_2\cdot(67-56)$	......	$56+R_9\cdot(67-56)$	$56+R_{10}\cdot(67-56)$
11	(49, 63)	$49+R_1\cdot(63-49)$	$49+R_2\cdot(63-49)$	......	$49+R_9\cdot(63-49)$	$49+R_{10}\cdot(63-49)$

R_s: scenarios parameters.
$R_1 \in [0, 0.1]$; $R_2 \in [0.1, 0.2]$; $R_3 \in [0.2, 0.3]$; $R_6 \in [0.3, 0.4]$; $R_5 \in [0.4, 0.5]$;
$R_6 \in [0.5, 0.6]$; $R_7 \in [0.6, 0.7]$; $R_8 \in [0.7, 0.8]$; $R_9 \in [0.8, 0.9]$; $R_{10} \in [0.9, 1]$.

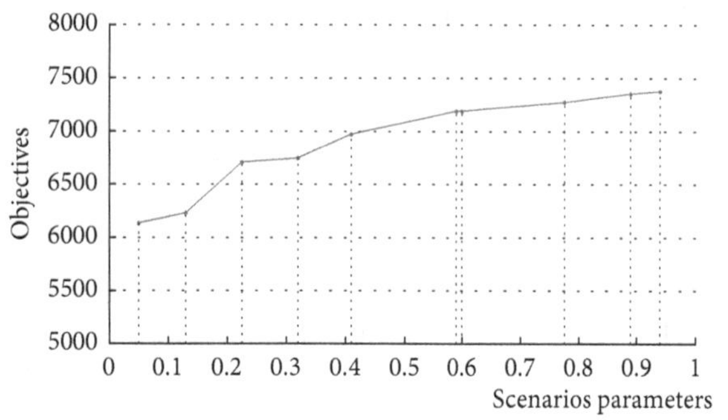

FIGURE 7: The change of the objectives under different scenarios parameters.

have the same importance; namely, the weight of the two objectives in the second stage are all set as 0.5, i.e., $\omega_1 = \omega_2 = 0.5$. This is the robust optimization model proposed in this paper, which is abbreviated with RO-II. In order to show the difference between the solutions of robust optimization and other cases, we have done the following two additional numerical experiments. One is to optimize the model without considering the above 10 scenarios. That means only the first item in the objective function (32) of the second stage is considered. This case can be called nonrobust optimization, which is abbreviated with NRO-II. The other is to consider the average running time of the 10 scenarios in the first stage. It is solved by the first stage model, which is abbreviated with AS-I. Table 8 shows the objectives, computation times, and total deviation from 10 scenarios of these experiments.

In order to show the deviation between these contrast experiments and each scenario, Figure 8 is given.

In Figure 8, the deviation from each curve is the absolute value of the difference between the contrast experiment and each scenario. Obviously, nonrobust optimization has the least total travel time, but the biggest total deviation. Trains tend to choose the maximum operating speed in this case. Therefore, the solution of nonrobust optimization has a small deviation from scenario 1 and scenario 2 but a large deviation from other scenarios. The timetable generated by nonrobust optimization may require a large adjustment when encountering some uncertain events. However, the robust optimization timetable generated in this paper can better adapt to different running times. It needs less adjustment when running in a specific scenario. In addition, the deviation of robust optimization is very close to the average running time, and the total travel time of robust optimization is shorter than the average running time. Furthermore, we also analyzed the deviations of some trains, as shown in Figure 9.

Obviously, in Figure 9, for each SDSA-train, the solution of robust optimization can also better adapt to different running times.

Figures 10, 11(a), and 11(b) depict the time-space diagrams of SDSA-trains and existing trains in the case of robust optimization with $\omega_1 = \omega_2 = 0.5$. Figure 10 mainly depicts the

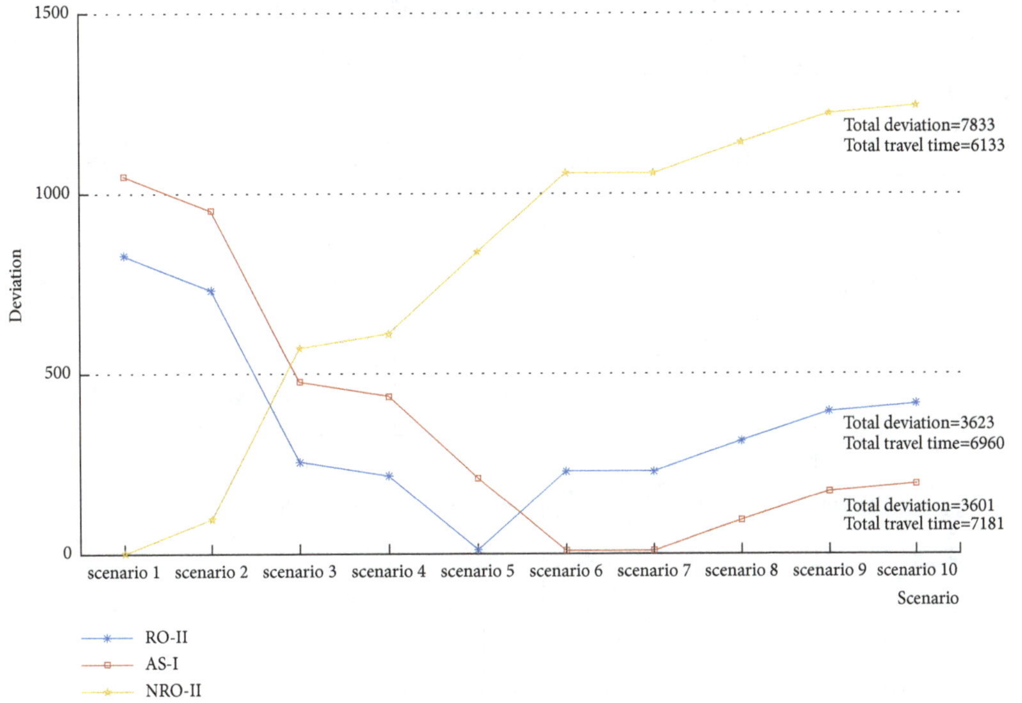

FIGURE 8: The deviation between the three sets of contrast experiments and each scenario.

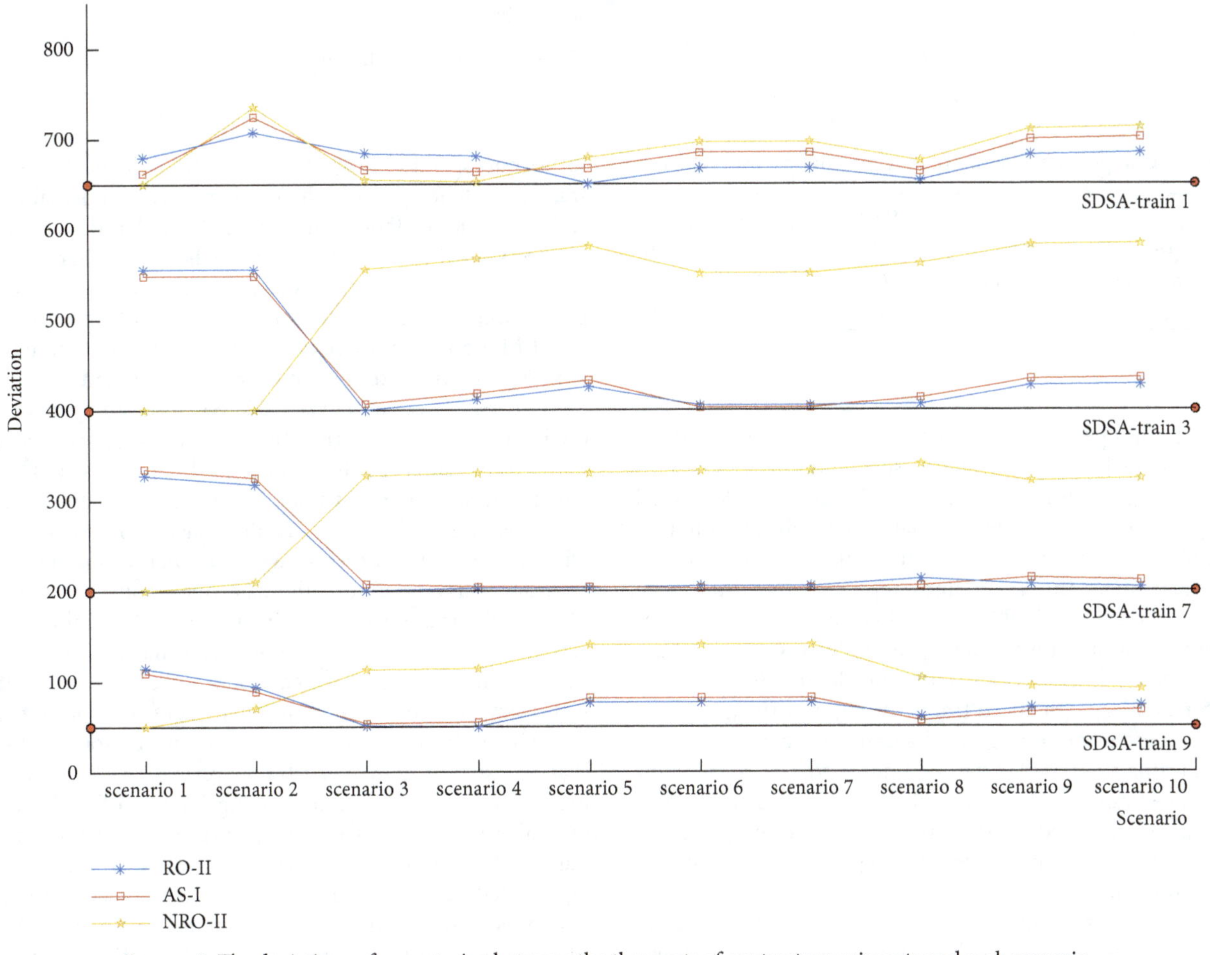

FIGURE 9: The deviations of some trains between the three sets of contrast experiments and each scenario.

Table 7: Computational results under different scenarios.

	Scenario 1	Scenario 2	Scenario 3	Scenario 4	Scenario 5
Objectives (unit: min)	6133	6229	6704	6744	6971
Computation time (unit: s)	22	25	17	20	23
	Scenario 6	Scenario 7	Scenario 8	Scenario 9	Scenario 10
Objectives (unit: min)	7189	7189	7274	7354	7376
Computation time (unit: s)	9	13	16	5	4

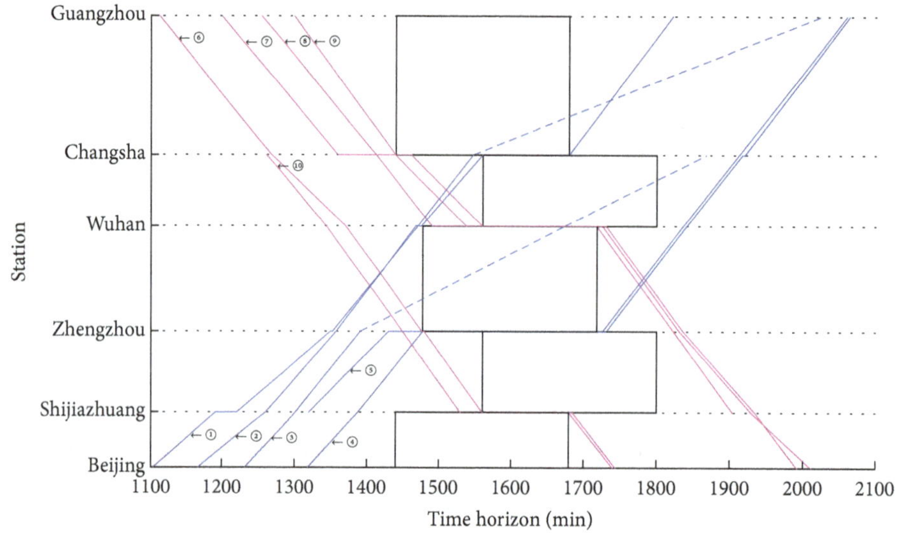

Figure 10: Time-space diagram of SDSA-trains at major stations.

Table 8: Computational results of different contrast experiments.

	RO-II	NRO-II	AS-I
Total travel time (unit: min)	6960	6133	7181
Total deviation from 10 scenarios	3623	7833	3601
Computation time (unit: s)	6	41	17

SDSA-trains operation at major stations and the maintenance windows on high-speed railway segments, in which solid lines represent running on high-speed railway, while dashed lines represent running on normal-speed railway. Note that maintenance windows are asynchronous on each segment. In Figure 11(a), the green lines represent the time-space diagrams of existing high-speed trains. The blue dashed lines represent running on normal-speed railway. Obviously, the high-speed railway corridor is unavailable from 0: 00 to 6: 00 am for existing high-speed trains. In Figure 11(b), the green lines represent the time-space diagrams of existing normal-speed trains and the blue dashed lines represent running on high-speed railway. Note that only SDSA-train 2 and SDSA-train 3 are involved in the operation of running on the normal-speed railway, so we only depict these two SDSA-trains in Figure 11(b).

We can see that only SDSA-train 2 and SDSA-train 3 select Mode 2 when there is a conflict with maintenance windows. Specifically, SDSA-train 2 switches from high-speed railway to normal-speed railway at major station Changsha; SDSA-train 3 switches from high-speed railway to normal-speed railway at major station Zhengzhou. Note that if SDSA-train 2 selects Mode 1 at major station Changsha, there will be a shorter travel time. However, it is not conducive to the generation of robust train scheduling. Other SDSA-trains all select Mode 1 when there is a conflict with the maintenance windows on the next high-speed railway segment.

Furthermore, we analyze the SDSA-train stop plans based on Figures 11(a) and 11(b). There are three cases that may make trains stop at station. Specifically, the first case is to ensure the safe operations of all trains. The second case is to meet passengers' demands. The third case is to avoid conflicts with the maintenance windows. Note that, in the first case, it is necessary to ensure the safe operations between the SDSA-train and the SDSA-train and the SDSA-train and the existing train. It should be emphasized that we assume that the existing train schedules cannot be affected, so the SDSA-trains have to stop at stations when there is a conflict with the existing train schedules. It is clear and reasonable to see that an existing high-speed train overtakes SDSA-train 1 at high-speed railway station Baoding East and SDSA-train 2 overtakes SDSA-train 1 at high-speed railway station Xinyang East. In the second case, the reason that SDSA-train 6 stops at high-speed railway station Zhengzhou is to meet passengers' demands. In the third case, the reason that SDSA-trains select Mode 1 (i.e., waiting at high-speed railway station) is to avoid conflicts with the maintenance windows.

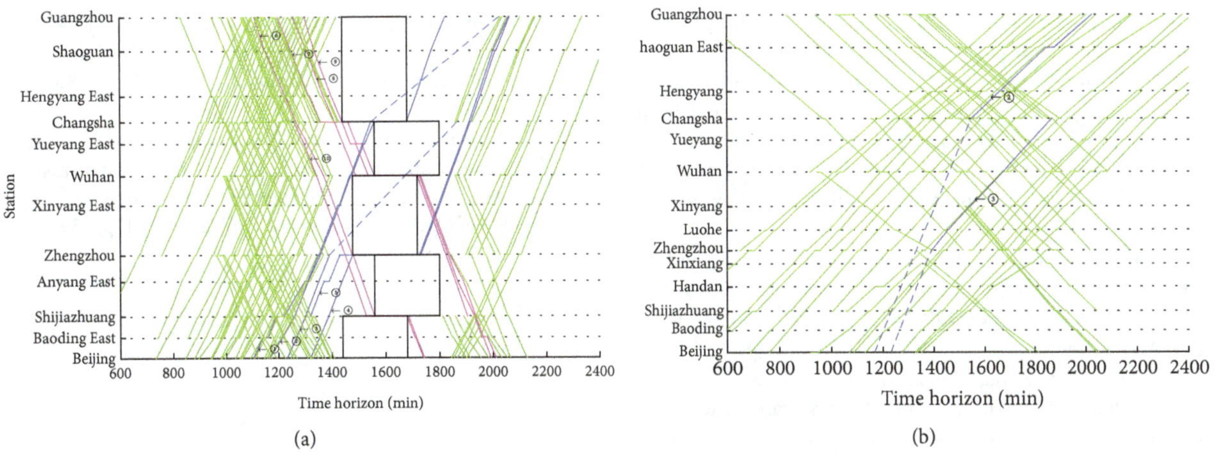

FIGURE 11: Time-space diagrams of SDSA-trains and existing trains on the high-speed railway (a) and the normal-speed railway (b).

6. Conclusion and Future Work

Taking into account the disturbances in the actual operations, we propose a two-stage method to generate a robust train scheduling and maintenance planning. In the first stage, we generate an integrated optimization model of train scheduling and maintenance planning on high-speed railway corridors under each scenario. Note that we consider two operation modes of SDSA-trains under each scenario. The selection of operation modes is realized by introducing binary state variables. In the second stage, a robust SDSA-train schedule is generated based on the optimal solutions of the considered scenarios. Besides, in order to make the model be handled easily, some linearization techniques are used. The nonlinear constraints and objective function are equivalently replaced by linear constraints and objective function. Finally, numerical experiments show the effectiveness of the proposed model and optimization method.

The following aspects should also be considered in future research. First, we do not consider the uncertainty of passengers' demands and the number of passengers boarding/alighting of each train at each station. These are two very important issues for future research. Second, line planning can be integrated into the issue. Due to the line planning, the origin/destination stations and stop plans can be determined in advance. Third, we can use other modeling methods and propose some efficient algorithms to solve this problem. For example, we can build a space-time network model and use Lagrangian-relaxation-based algorithms to solve it. These aspects can be studied in depth in the future.

Conflicts of Interest

The authors declare that they have no conflicts of interest regarding the publication of this paper.

Acknowledgments

The research was supported by the National Natural Science Foundation of China (nos. 71401008, 71771016, and 71621001) and State Key Laboratory of Rail Traffic Control and Safety, Beijing Jiaotong University (nos. RCS2017ZT002 and RCS2017ZZ01).

References

[1] V. Cacchiani and P. Toth, "Nominal and robust train timetabling problems," *European Journal of Operational Research*, vol. 219, no. 3, pp. 727–737, 2012.

[2] K. Nachtigall and S. Voget, "A genetic algorithm approach to periodic railway synchronization," *Computers & Operations Research*, vol. 23, no. 5, pp. 453–463, 1996.

[3] C. Liebchen and R. H. Möhring, "The modeling power of the periodic event scheduling problem: Railway timetables - and beyond," *Lecture Notes in Economics and Mathematical Systems*, vol. 600, pp. 117–150, 2008.

[4] G. Caimi, M. Fuchsberger, M. Laumanns, and K. Schüpbach, "Periodic railway timetabling with event flexibility," *Networks*, vol. 57, no. 1, pp. 3–18, 2011.

[5] L. G. Kroon and L. W. P. Peeters, "A variable trip time model for cyclic railway timetabling," *Transportation Science*, vol. 37, no. 2, pp. 198–212, 2003.

[6] T. Lindner, *Train schedule optimization in public rail transport [PhD thesis]*, University of Technology, Braunschweig, Germany, 2000.

[7] L. Peeters and L. Kroon, "A cycle based optimization model for the cyclic railway timetabling problem," *Computer-Aided Scheduling of Public Transport*, vol. 505, pp. 275–296, 2001.

[8] U. Brännlund, P. O. Lindberg, A. Nõu, and J.-E. Nilsson, "Railway timetabling using Lagrangian relaxation," *Transportation Science*, vol. 32, no. 4, pp. 358–369, 1998.

[9] A. Caprara, M. Fischetti, and P. Toth, "Modeling and solving the train timetabling problem," *Operations Research*, vol. 50, no. 5, pp. 851–861, 2002.

[10] V. Cacchiani, A. Caprara, and P. Toth, "Non-cyclic train timetabling and comparability graphs," *Operations Research Letters*, vol. 38, no. 3, pp. 179–184, 2010.

[11] V. Cacchiani, L. Galli, and P. Toth, "A tutorial on non-periodic train timetabling and platforming problems," *Euro Journal on Transportation Logistics*, vol. 4, no. 3, pp. 285–320, 2015.

[12] J. Shi, L. Yang, J. Yang, and Z. Gao, "Service-oriented train timetabling with collaborative passenger flow control on an oversaturated metro line: An integer linear optimization approach," *Transportation Research Part B: Methodological*, vol. 110, pp. 26–59, 2018.

[13] L. Kroon, G. Maróti, M. R. Helmrich, M. Vromans, and R. Dekker, "Stochastic improvement of cyclic railway timetables," *Transportation Research Part B: Methodological*, vol. 42, no. 6, pp. 553–570, 2008.

[14] M. B. Khan and X. Zhou, "Stochastic optimization model and solution algorithm for robust double-track train-timetabling problem," *IEEE Transactions on Intelligent Transportation Systems*, vol. 11, no. 1, pp. 81–89, 2010.

[15] V. Cacchiani, A. Caprara, and M. Fischetti, "A lagrangian heuristic for robustness, with an application to train timetabling," *Transportation Science*, vol. 46, no. 1, pp. 124–133, 2012.

[16] M. A. Shafia, M. Pourseyed Aghaee, S. J. Sadjadi, and A. Jamili, "Robust train timetabling problem: Mathematical model and branch and bound algorithm," *IEEE Transactions on Intelligent Transportation Systems*, vol. 13, no. 1, pp. 307–317, 2012.

[17] L. X. Yang, K. P. Li, and Z. Y. Gao, "Train timetable problem on a single-line railway with fuzzy passenger demand," *IEEE Transactions on Fuzzy Systems*, vol. 17, no. 3, pp. 617–629, 2009.

[18] L. Yang, Z. Gao, and K. Li, "Passenger train scheduling on a single-track or partially double-track railway with stochastic information," *Engineering Optimization*, vol. 42, no. 11, pp. 1003–1022, 2010.

[19] A. Higgins, "Scheduling of railway track maintenance activities and crews," *Journal of the Operational Research Society*, vol. 49, no. 10, pp. 1026–1033, 1998.

[20] G. Budai, D. Huisman, and R. Dekker, "Scheduling preventive railway maintenance activities," *Journal of the Operational Research Society*, vol. 57, no. 9, pp. 1035–1044, 2006.

[21] G. Budai, "Operations research models for scheduling railway infrastructure maintenance," *Public Performance Management Review*, 2009.

[22] T. Lidén and M. Joborn, "Dimensioning windows for railway infrastructure maintenance: Cost efficiency versus traffic impact," *Journal of Rail Transport Planning and Management*, vol. 6, no. 1, pp. 32–47, 2016.

[23] Y. Zhou, L. Zhou, Y. Wang, Z. Yang, and J. Wu, "Application of multiple-population genetic algorithm in optimizing the train-set circulation plan problem," *Complexity*, vol. 2017, no. 7, Article ID 3717654, 14 pages, 2017.

[24] S. Van Aken, N. Bešinović, and R. M. P. Goverde, "Designing alternative railway timetables under infrastructure maintenance possessions," *Transportation Research Part B: Methodological*, vol. 98, pp. 224–238, 2017.

[25] T. Lidén and M. Joborn, "An optimization model for integrated planning of railway traffic and network maintenance," *Transportation Research Part C: Emerging Technologies*, vol. 74, pp. 327–347, 2017.

[26] X. Luan, J. Miao, L. Meng, F. Corman, and G. Lodewijks, "Integrated optimization on train scheduling and preventive maintenance time slots planning," *Transportation Research Part C: Emerging Technologies*, vol. 80, pp. 329–359, 2017.

[27] D. Arenas, P. Pellegrini, S. d. Hanafi, and J. Rodriguez, "Timetable rearrangement to cope with railway maintenance activities," *Computers & Operations Research*, vol. 95, pp. 123–138, 2018.

[28] P. Vansteenwegen, T. Dewilde, S. Burggraeve, and D. Cattrysse, "An iterative approach for reducing the impact of infrastructure maintenance on the performance of railway systems," *European Journal of Operational Research*, vol. 252, no. 1, pp. 39–53, 2016.

9

Testing the Generality of a Passenger Disregarded Train Dwell Time Estimation Model at Short Stops: Both Comparison and Theoretical Approaches

Dewei Li,[1] **Yonghao Yin**,[1] **and Hong He**[2]

[1]*State Key Lab of Rail Traffic Control and Safety, Beijing Jiaotong University, Beijing 100044, China*
[2]*Guangzhou Railway Polytechnic, Guangzhou, Guangdong 510430, China*

Correspondence should be addressed to Dewei Li; dwli.bjtu@163.com

Academic Editor: Zhi-Chun Li

Train dwell time estimation is a critical issue in both scheduling and rescheduling phases. In a previous paper, the authors proposed a novel dwell time estimation model at short stops which did not require the passenger data. This model shows promising results when applied to Dutch railway stations. This paper focuses on testing and improving the generality of the model by two steps: first, the model is tested by applying more independent datasets from another city and comparing the estimation accuracy with the previous Dutch case; second, the model's generality is tested by a theoretical approach through the analysis of individual model parameters, variables, model scenarios, and model structure as well as work conditions. The validation results during peak hours show that the MAPE of the model is 11.4%, which is slightly better than the results for the Dutch railway stations. A more generalized predictor called "dwell time at the associated station" is used to replace the square root term in the original model. The improved model can estimate train dwell time in all the investigated stations during both peak and off-peak periods. We conclude that the proposed train dwell time estimation model is generic in the given condition.

1. Introduction

Train dwell time estimation is a critical issue for both the scheduling and rescheduling phases. In the scheduling phase, accurate estimation of the train dwell time provides necessary inputs to the timetable and allows the timetable to match the passenger demand. In the rescheduling phase, the estimation results of the train dwell time can be used to predict potential conflicts between train lines.

To estimate dwell time accurately, the development of a dwell time estimation model is necessary. One of the most critical issues for broad applications of the train dwell time estimation model is its generality. The generality of a model can be defined as a model that is generic rather than a more specific or detailed one [1]. In other words, a generalized dwell time estimation model can be used to estimate dwell times in different instances with acceptable accuracy.

To test the generality of a train dwell time estimation model, there are two possible approaches. One intuitive methodology is to apply the model to more or a wider range of independent cases; if the accuracies under these different cases are acceptable, then the model can be classified as a relatively general model. This approach is called a comparison approach. However, comparison approaches may face problems like how many cases they should cover and what the wider range means. Another methodology is to analyze the parameters and variables of the model theoretically to prove its generality [1], which is called a measure theoretical approach. The generality of a model is the measure assigned to the set of possibilities to which it applies. Such a method can provide an account of the minimum formal machinery required to make precise, quantitative assessments of generality.

The main methodology of this paper is as follows: first, a train dwell time estimation model is selected; second, the generality of the selected model is analyzed by using both comparison and theoretical approaches. In the comparison

Table 1: Main features of existing dwell time estimation models.

Source	Passenger traffic dependency	Model methodology	Validation: Stations and lines	Validation: Country	Railway mode	Station type
Lam, 1988	Pr	Lr	3S on 3L	HKG	Subway	Ich
Lin, 1992	Pr	NLr	2S on 1L	USA	Light rail	—
Puong, 2000	Pr	NLr	2S on 1L	USA	Subway	Ich and Ime
Buchmueller et al., 2008 [16]	Pr	Di	—	SWE	Hsr	—
Zhang et al., 2008 [3]	Pr	Mi	3S on 3L	CHN	Subway	Ich and Ime
Hansen, 2010	Pd	Lr	2S on 1L	NED	Hsr	Ime
Jong, 2011	Pr	Lr	1S on 1L	TPE	Hsr	Ich
Sourd, 2011	Pr	Mi	1S on 1L	FRA	Hsr	—
Yamamura, 2013	Pr	Mi	—	JPN	Subway	—
Kecman, 2015	Pd	Lr	—	NED	Hsr	Ich and Ime
Seriani, 2015	Pr	Mi	—	—	Subway	—
Hänseler et al., 2015 [17]	Pr	Mi	1S on 1L	SWE	Hsr	Ich
Li, 2016	Pd	NLr	1S on 1L	NED	Cr	Ime
This paper	Pd	NLr	1S on 1L and 7S on 1L	NED and CHN	Cr and subway	Ich and Ime

Note. Pr: passenger regarded; Pd: passenger disregarded; Lr: linear regression; NLr: nonlinear regression; Mi: microscopic simulation; Di: distribution; 3S on 3L: 3 stations on 3 lines (other abbreviations et cetera); HKG/USA/CHN and others: the acronym of countries; Hsr: high-speed railway; Cr: commuter railway; Ich: interchange station; Ime: intermediate station.

approach, an independent dataset is applied, and the estimation result is compared with the previous one. In the theoretical approach, the model is analyzed from four aspects: variable analysis, parameter analysis, scenario analysis, and structure and condition analysis. Finally, based on the testing of generality, an improvement is made to the model.

The remainder of this paper is organized as follows. In Section 2, a literature review is presented. Section 3 introduces the selection of the train dwell time estimation model. In Section 4, the comparison approach is applied to the selected model. A new dataset from the Beijing urban rail transit is applied and then compared to the original result, which had been validated by data from Dutch railway stations [2], and the estimation accuracy is analyzed by comparing the results of the models. In Section 5, the train dwell time estimation model is analyzed by the theoretical approach. Section 6 improves the generality of the proposed model. In Section 7, three controversial issues are discussed to explain the condition of the model. Section 8 presents several conclusions.

2. Literature Review

In the literature, many types of train dwell time estimation models have been proposed. The existing models can be classified according to different criteria such as passenger traffic dependency, modeling methodology, validation, railway mode, and type of station, all of which are listed in Table 1.

From Table 1, there are two distinct models that are interrelated and can be distinguished from the research direction. One is a passenger regarded model and the other is a passenger disregarded model.

In the passenger regarded model, the dwell time is estimated based on the alighting and boarding time of the passengers. The numbers of boarding and alighting passengers are the key inputs of this kind of model. The approaches of the passenger regarded train dwell time estimation mainly include two types: a microscopic simulation model and a regression model.

Among the microscopic simulation models, these studies focused on the microlevel movement during alighting and boarding processes by simulation modeling. Most of the microscopic simulation studies are based on field data collected in test stations and lines. They are used to explain the effect of the uncertainties of passenger behaviors on dwell times. Zhang et al. [3] presented a cellular automata-based alighting and boarding microsimulation model for passengers to estimate the dwell time, which considers individual desired speed, pressure from passengers behind, personal activity, and behavioral tendencies (dependent on gender, age, etc.). Seriani and Fernandez [4] determined the effect of passenger traffic management in the boarding and alighting time at metro stations by simulations and experiments. Yamamura et al. [5] developed a simulator to estimate the dwell times of trains when various improvements for reducing dwell time were taken. The simulator was based on a multiagent model and simulated each passenger's behavior both in a train and on a platform. Sourd et al. [6] presented a behavioral based simulation model dedicated to the evaluation of the dwell time for trains, which separate the environmental model and can be computed easier and faster. The results of these studies showed that passenger behaviors vary from different countries or districts because of diversity characteristics and service levels and other factors that may influence pedestrian movement. These models consumed a lot of time during the simulation phase and their inputs were varied; thus they can be hardly used in real-time applications.

Among the passenger regarded regression models, the train dwell times have usually been estimated using a linear or nonlinear function of the number of alighting and boarding passengers. Lam et al. [7] and Wirasinghe and Szplett [8] both presented linear regression models to estimate the dwell time, with the difference being that the former assume that the boarding passengers on the platform take on a uniform distribution and the latter consider a nonuniform distribution of the boarding passengers. Lin and Wilson [9], Parkinson and Fisher [10], and Puong [11] developed a nonlinear estimation model, and they considered the number of standing passengers in the vehicle and their interactions with boarding and alighting passengers. These models were usually validated for specific rolling stocks [12, 13], stations [14], and the number of alighting and boarding passengers [15]. Specifically, Harris [12] found that the interior layout of the train typically improves the dwell time estimation model accuracy. Jong and Chang [13] estimated the alighting and boarding times at a specific station as a function of the numbers of alighting and boarding passengers and different train services that implied the influence of rolling stock. Xenia and Nick [14] discussed the influence of step height between the train and platform and the door width to the dwell time estimation. Daamen et al. [15] found that the number of boarding and alighting passengers was the main determinant of the dwell times. The problems with these models were that many background variables were not included, such as the composition of the passenger population configuration of the rolling stock and the type of station (for details, refer to [2]), which had an irrefutable impact on the dwell time. However, none of the existing models fully took all of the influencing factors into account.

Among the passenger disregarded models, the dwell time was estimated using many variables that were not related to passenger demand. To the best of our knowledge, so far, regression is the only approach that has been used in passenger disregarded models. Hansen et al. [18] and Kecman and Goverde [19] estimated the train dwell time as a function of its arrival delay, which was derived from the track occupancy data of the Dutch railway. Li et al. [2] proposed a train dwell time estimation model that improved the accuracy of these models. This model predicted the train dwell time using several substitute variables, including the actual dwell time of consecutive trains at the target station and the dwell time of the target train at consecutive stations, which could indirectly reflect the passenger demand [2]. However, the generality of the models was not validated for cases from other countries. Meanwhile, the result of the research was not good enough for an estimate model. There is still space to improve the accuracy of the dwell time estimation regression model.

With regard to the validation of the existing models, most of the former studies included only a few stations belonging to one railway (as shown in Table 1). Instead of validating the model in one specific railway in most of the former works, this paper focuses on two different independent cases: a commuter railway case in Netherlands and a subway case in China. However, despite the lack of a theoretical approach for all the existing models, this paper bridges the gap by providing a general framework for testing the generality of a dwell time model through theoretical analysis.

3. Model Description

To select a possible generic train dwell time estimation model, both the input data availability and model characteristics should be considered. We noticed that all the passenger regarded models could not be selected because the data, such as the number of boarding and alighting passengers, were usually not available, especially in real time. However, the generality of the passenger disregarded model, especially the model proposed by Li et al., is worthy of discussion because the variables considered in Li et al.'s model do not depend on specific types of rolling stocks and traffic conditions [2]. Therefore, this paper analyzes the extent to which the model can be generalized by both comparison and theoretical approaches.

Li et al.'s model studied the dwell time estimation problem particularly for short stops without passenger demand by means of a statistical analysis of track occupation data from Netherlands. In their paper, the factors of influence on dwell time are classified into five categories: passenger, rolling stock, station, operation, and external factors. The five categories include the majority of influencing factors. Then, with the analysis of these factors, 10 potential predictors, including time variation, length of target train, length of preceding train, departure delay at previous station, dwell time of target train at previous station, second previous station, dwell time of preceding train at target and previous station, and dwell time of the same train during the last week, are selected. Various combinations (including linear items and nonlinear items) are tried to establish 10 different parametric regression models. With the correlation analysis and experimental testing, the model shown in formula (1) was finally developed; this model was found to perform the best among their 10 regression models. For details, refer to Li et al., 2016.

$$\begin{aligned} \mathrm{DT}_k^s &= c + \beta_1 L_k^s + \beta_2 L_{k-1}^s + \beta_3 \mathrm{DT}_{k-1}^s \\ &\quad + \beta_4 \sqrt{\mathrm{DT}_k^{s-1} * \mathrm{DT}_k^{s-2}}, \end{aligned} \tag{1}$$

where DT_k^s indicates the dwell time of train k at station s, L_k^s indicates the length of train k at station s, L_k^{s-1} indicates the length of train k at station $s-1$, DT_{k-1}^s indicates the dwell time of the previous train at the target station, and DT_k^{s-1} and DT_k^{s-2} indicate the dwell time of the target train at the previous station and at the second previous station, respectively. c, β_1, β_2, β_3, and β_4 are the parameters of the variables.

The reason why this model can work well in practice lies in the selected predictors. Unlike other models, Li et al.'s model does not consider the station layout, rolling stock configuration, or passenger behavior directly because these factors are involved in independent variables in the model and act on the dependent variable indirectly. For example, the predictors of the model are mainly the dwell time of the preceding train DT_{k-1}^{s} and the dwell time at the previous station DT_{k}^{s-1}. The common characteristic of the preceding train and the target train is that both trains stopped at the same station, such that parameter β_2 can reflect the effect of the station. The common characteristic of the dwell time at the previous station and the dwell time at the target station is that both times are for the same train, such that β_3 can reflect the effect of the rolling stock. Passenger demand appears in two ways: one is simply separating the peak and off-peak hour models, and the other is to reflect the demand difference via the different lengths of the trains. It should be noted that the variance of passenger demand between different stations is inevitable and is mainly indicated by parameter c. Other remaining influencing factors are also implied in this parameter. Accordingly, the model avoids the direct input of all the influencing factors, which would otherwise be difficult to obtain. However, to what extent the relationship among these variables holds is not clear. Therefore, the generality of such relationships is worthy of discussion.

To test the generality of the model, in the following two sections, two different approaches are used. First, a Beijing dataset is applied to the model that was established based on the Dutch railway station, and the estimation accuracy is analyzed by comparing the two results. Because the two datasets are independent in both time and space, it can be inferred that the validated result reflects the generality of the model to some extent. Second, the quantitative generality of the model is analyzed by a theoretical approach.

4. Generality Analysis by Comparison Approach

The original dataset for Li et al.'s model development is from Dutch railway stations. To test whether it is still applicable for a totally independent case, a new dataset, which is dependent on the original validated case, is selected from the Beijing urban railway. First, an empirical study is conducted by analyzing the difference between the scheduled and actual dwell time in the Beijing urban rail transit, while the distributions of the actual dwell time are also analyzed. Then, these results are compared with the results from Netherlands. Second, the dataset of the Beijing urban rail stations is used to regress the original model that was previously validated for a station in the Dutch railway system, and the regression parameters and the accuracy of the results are compared with the proposed model. In this way, the generality of the proposed model is analyzed.

4.1. Dataset Resources

4.1.1. Dataset Collection. To study the generality of the model by a comparison approach, datasets from China are carefully selected. To obtain precise datasets from the Beijing railway stations, a line that has been operating for a relatively long amount of time is selected so that the passenger demand is stable. Thus, line four of the Beijing urban rail transit is selected for this empirical study; the stations of this line are shown in Figure 1.

Table 2: The selected stations and their classifications.

Number	Target station	Classification
1	Renmin University	Intermediate station
2	Weigongcun	Intermediate station
3	National Library	Interchange station
4	Beijing Zoo	Intermediate station
5	Xizhimen	Interchange station
6	Xinjiekou	Intermediate station
7	Ping'anli	Interchange station

This line opened in 2009, has a length of 50 km, and comprises 35 stations; all trains stop at every station, and no train overtaking occurs at any station. In this railway, Anheqiao North is the first station, and Tiangongyuan is the last station, where the trains are able to turn around. The interchange stations are Haidianhuangzhuang, National Library, Xizhimen, Ping'anli, Xidan, Xuanwumen, Caishikou, Beijing South Railway Station, and Jiaomen West, where passengers can transfer to another line; however, passenger connections are not considered in the timetable, so we neglect the "adhere to schedule effect" which may exist in some transfer stations in other cases. The remaining stations are intermediate stations.

In this paper, four intermediate stations and three interchange stations are selected, and all of the selected stations are consecutive. The selected stations are listed in Table 2.

There are two different ways to obtain the dwell time for the two cases. In the Dutch case, almost three months of data were used to analyze the distribution of the actual dwell time in different periods. The dwell times at the selected stops and trains were estimated based on the track occupation data. In Netherlands, track occupation data were collected using a train describer system (TROTS), which provided the exact time of occupation and clearance of track sections [20]. By using a dwell time estimation algorithm [21], a total of 17,306 trains running from 1 September 2012 to 30 November 2012 were processed and analyzed. The sample was separated from the off-peak period ([9:00, 16:00]∪[18:30, 20:00]) and the peak period ([6:30, 9:00]∪[16:00, 18:30]). The peak period was further classified as a morning peak period ([6:30, 9:00]) and an afternoon peak period ([16:00, 18:30]). However, it is difficult to automatically obtain the actual arrival and departure time of the trains in the Beijing urban rail transit system, such that it was necessary to investigate manually the dwell times of selected stations. The dwell times were investigated for stations and trains. At first, the investigators were distributed among consecutive stations from Anheqiao to Xuanwumen station, on 25 September 2015, which is a workday, from 7:28 am to 11:11 am to include data from both the peak hours [7:28, 9:00] and off-peak hours [9:00, 11:11]. In

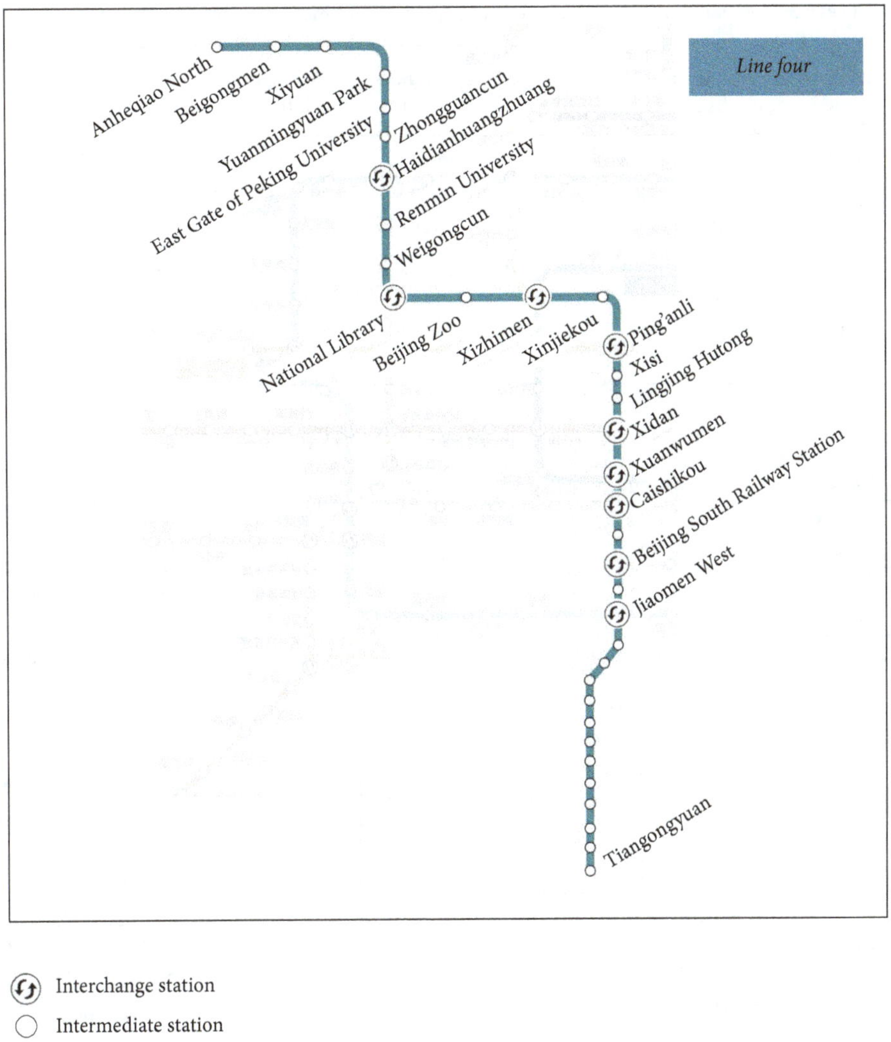

FIGURE 1: Stations of line four of the Beijing urban rail transit system.

total, 71 train arrival and departure times were recorded by using a stop watch. Second, the investigators were requested to board the same number of trains successively departing from the start station of the line, and each investigator recorded the dwell time of the train at every station along the line in one direction by a stop watch. The dwell time is defined as the time from the moment the train stops at the station to the moment that the train begins to leave. The scheduled timetable for line four was also obtained from the operating company. Using the original recorded dataset, we are now able to calculate the dwell time of every train at every station.

4.1.2. Dataset Comparison between the Actual and Scheduled Dwell Times. The original recorded datasets of the investigated stations in Beijing are processed and analyzed, and we find that the actual and scheduled dwell times are significantly different, whether by more or less time. The analysis results are shown in Table 3. The maximum dwell time is longer than the scheduled dwell time by more than 20 s at every station. The minimum dwell time is shorter than the scheduled dwell time by no more than 10 s at most of stations. In addition, the average dwell time is longer than the scheduled dwell time at every station. We find that the actual dwell time fluctuates significantly, and the difference between the maximum and scheduled dwell times is larger than the difference between the minimum and scheduled dwell times. Next, we compare the datasets between the two cases.

4.1.3. Dataset Comparison between the Two Cases. In this section, the datasets from two countries are compared. At first, the two datasets are statistically described; and then the dwell time distributions and the differences between the actual and scheduled dwell times of the two cases are analyzed.

(1) Data Statistical Descriptions. The average, minimum value, maximum value, standard deviation, and variance of the two datasets are compared. Moreover, the datasets in the two cases are compared by applying the two-sample Kolmogorov-Smirnov test (K-S test) to test whether the two datasets are significantly different. The result is listed in Table 4. This table shows that the two datasets are significantly different. Thus, the two datasets can test the generality of the model.

Table 3: Comparison of the actual and scheduled dwell times of the investigated stations.

Stations	Max dwell time	Min dwell time	Average actual dwell time	Scheduled dwell time	Average absolute error	Average relative error	Root mean square error	Max absolute error
Renmin University	69 s	25 s	34.76 s	30/35 s	5.51 s	17.69%	9.31 s	39 s
Weigongcun	67 s	25 s	35.76 s	30/35 s	6.2 s	19.94%	10.42 s	37 s
National Library	69 s	29 s	46.55 s	35 s	12.59 s	35.98%	15.5 s	34 s
Beijing Zoo	65 s	24 s	43.37 s	40 s	8.55 s	21.37%	10.65 s	25 s
Xizhimen	67 s	33 s	46.62 s	45 s	6.38 s	14.18%	8.36 s	22 s
Xinjiekou	70 s	27 s	38.68 s	35 s	6.1 s	17.42%	11.62 s	35 s
Ping'anli	95 s	24 s	45.45 s	35 s	11.77 s	33.64%	16.72 s	60 s

Table 4: Comparison between Dutch and Beijing datasets.

Dataset	Average	Minimum	Maximum	SD	Variance	K-S test
D_peak	52.67 s	40 s	71 s	10.28 s	105.75	0.375
B_peak	47.68 s	29 s	65 s	8.99 s	80.84	
D_off-peak	41.67 s	17 s	77 s	16.46 s	271.06	0.505
B_off-peak	38.68 s	24 s	65 s	9.38 s	88.04	

Note. "D" denotes Dutch case; "B" denotes Beijing case; "SD" denotes standard deviation; "K-S test" denotes Kolmogorov-Smirnov test.

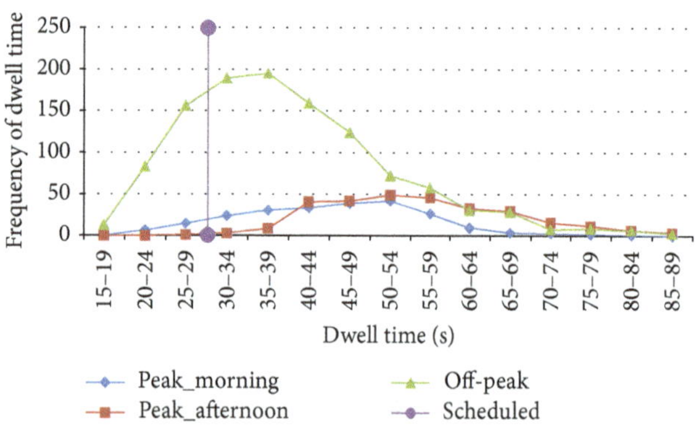

Figure 2: Distribution of the actual and scheduled dwell times in the Dutch railway stations.

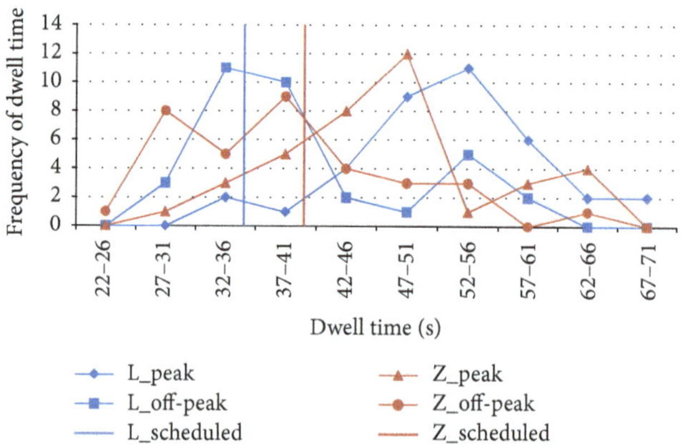

Figure 3: Distribution of the actual and scheduled dwell times in the Beijing urban railway stations.

(2) Dwell Time Distributions. Because the distribution of the actual dwell time varies significantly during different periods [21], previewing the data and studying the generality of the model through a comparison require datasets from the two countries to be compared. The dwell time distributions of the two cases during peak and off-peak hours are compared, as shown in Figures 2 and 3, respectively.

In the Dutch case, as shown in Figure 2, the maximum frequency value of the actual dwell time (broken line) during both the peak and off-peak hours was higher than the scheduled dwell time (vertical line) because the actual passenger demand could not be satisfied. Therefore, to match the actual passenger demand, the scheduled dwell times were extended.

In the Beijing case, the dwell time is collected manually by investigators. The distribution of the dwell time is analyzed for two typical types of stations during both peak and off-peak hours. The National Library interchange station and the Beijing Zoo intermediate station are selected. The result is shown in Figure 3. The vertical line indicates the scheduled dwell time, and the broken line denotes the actual dwell time frequency distribution. The maximum frequency value of the actual dwell time during the peak period is higher than that during the off-peak period at both the Beijing Zoo station (Z_off-peak) and National Library station (L_off-peak). The dwell time at the National Library (L_peak) interchange station is larger than that of the Beijing Zoo (Z_peak) intermediate station during peak hours. Conversely, during off-peak hours, the dwell time at the interchange station is less than that of the intermediate station. The primary reason for the greater dwell time is that the passenger demand at the Beijing Zoo station is higher than National Library station; however, the number of transfer passengers is high during the peak period, which also explains why the scheduled dwell time for the Beijing Zoo station is larger than that of the National Library station. Furthermore, the maximum frequency value of the actual dwell time during the peak hours is higher than the maximum value during off-peak hours in both the Beijing and Dutch railway stations. That is, compared with the dwell time in the off-peak hours, the dwell time is likely to increase during peak hours.

In summary, the distributions of the actual dwell time in the Dutch railway stations and in the Beijing urban rail transit have some common characteristics: regarding the maximum frequency value, the actual dwell time during the peak period is longer than that during the off-peak period. Meanwhile, the maximum frequency values in the two cases are similar. For example, during the peak period, the maximum value of the actual dwell time is 47–51 s at the Beijing Zoo station and 52–56 s at the National Library station, which is similar to the dwell time of 50–54 s during the morning peak period in the Dutch railway stations.

(3) Difference between the Actual and Scheduled Dwell Times. In general, the train drivers considered in this study strive to drive in accordance with the timetable. However, differences between the scheduled and actual dwell times do occur, making it necessary to analyze the difference between the scheduled and actual dwell times in both the Dutch railway stations and the Beijing railway stations.

The time period is separated into the peak period and the off-peak period: the peak period lasts from 7:30 to 9:00 am, and the off-peak period lasts from 9:00 to 10:30 am. Two typical stations, the National Library interchange station and the Beijing Zoo intermediate station, are selected to analyze the differences between the scheduled and actual dwell times during the peak and off-peak hours, respectively.

Figure 4 compares the actual and scheduled dwell times in the Dutch railway stations during different periods of a workday on Dutch railway stations. The different line types in the figure represent the dwell time during different periods. Most of the actual dwell time is greater than the scheduled dwell time because the passenger demand is so high that the scheduled time cannot meet the actual demand.

Figure 5 compares the actual and scheduled dwell times of the two stations during peak and off-peak hours in the Beijing railway stations. The different line types in the figure represent the different stations and the dwell time during different periods. During the peak hours, most of the actual dwell times are clearly larger than the scheduled dwell times. In contrast, during the off-peak hours, the actual and scheduled dwell times are closer, and the actual dwell time may even be shorter than the scheduled dwell time because the passenger demand during the off-peak hours is less than that during the peak hours. Furthermore, the dwell time of the National Library interchange station is longer than that of the Beijing Zoo intermediate station during peak hours; conversely, during off-peak hours, the dwell time of the interchange station is less than that of the intermediate station. The primary reason for this result is that the number of transfer passengers is high during the peak period; however, the passenger demand at the Beijing Zoo station is higher than that of the National Library station during the off-peak period, which also explains why the

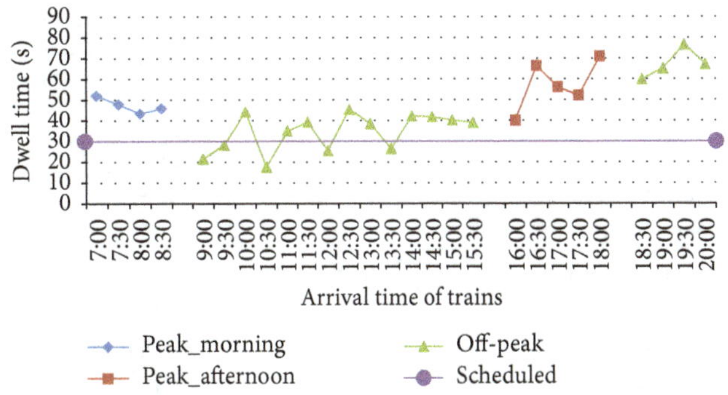

FIGURE 4: Comparison of the actual and scheduled dwell times in the Dutch railway stations.

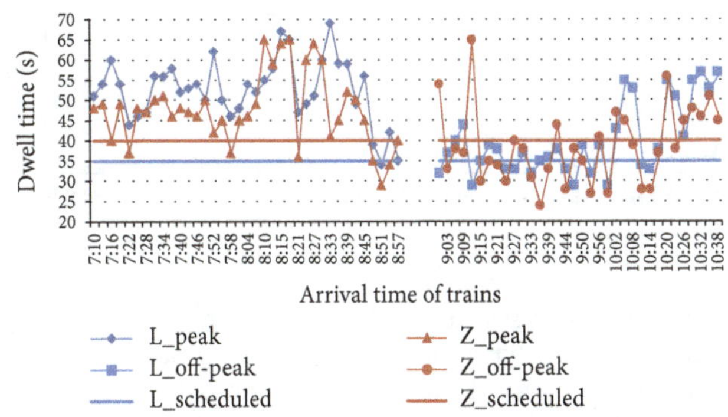

FIGURE 5: Comparison of the actual and scheduled dwell times in the Beijing railway stations.

scheduled dwell time for the Beijing Zoo station is greater than that of the National Library station.

In summary, the differences between the actual and scheduled dwell times in the Beijing urban rail transit system and in the Dutch railway stations have common characteristics: most of the actual dwell times are longer than the scheduled dwell times during the peak period. The actual and the scheduled dwell times are closer during the off-peak period, and the actual dwell time is likely to increase during the peak period.

4.2. Generalization Analysis Approach Based on Comparison. In this section, the generality of the selected train dwell time estimation model is analyzed by the comparison approach. The collected data are separated into two parts, the first of which is used to calibrate the model. The other part is used to validate and measure the error of the model. Ten records are selected randomly from the data as the validation set, and the remainder are used as the learning sample in the calibration part. In other words, the model is validated by applying the model to the Beijing urban railway dataset; the results are compared with the original values. Accordingly, this method verifies whether the model can cover wider scenarios.

4.2.1. Model Preparation. First, we consider the predictors that are used in the original model [2]; some small adjustments are made to adapt these parameters to the Beijing dataset. The lengths of the trains on line four of the Beijing

TABLE 5: The correlations between the input variables and the dwell time.

Stations	Periods					
	Peak			Off-peak		
	DT^s_{k-1}	DT^{s-1}_k	DT^{s-2}_k	DT^s_{k-1}	DT^{s-1}_k	DT^{s-2}_k
Renmin University	—	—	—	—	—	0.36
Weigongcun	0.41	0.84	—	—	0.53	—
National Library	0.54	0.54	0.35	0.54	0.36	—
Beijing Zoo	0.46	0.55	—	0.46	0.77	0.56
Xizhimen	0.47	0.49	0.34	—	0.51	—
Xinjiekou	0.73	0.68	0.60	—	0.43	0.48
Ping'anli	0.35	0.59	0.44	—	0.36	—

TABLE 6: Performance measures.

Indicators	Peak period	Off-peak period
Adj-R^2	0.594	0.162
RMSE	7.21 s	6.21 s
MAPE	11.4%	13.1%

Note. "Adj-R^2" denotes the adjusted coefficient of determination; "RMSE" denotes the root mean square error; "MAPE" denotes the mean absolute percentage error.

urban rail transit system are the same during the operating period, and the types of all of the rolling stock are the same. However, the Dutch railway stations have different train lengths at the same railway stations. To accommodate this fact, the trains' length predictors are neglected and transformed into a constant in the regression model. Thus, in this paper, the predictors are DT^s_{k-1}, DT^{s-1}_k, and DT^{s-2}_k.

Second, a simple correlation analysis is performed between the selected predictors and the dwell times at different stations. The result in Table 5 shows that most of the predictors are significantly different from zero ($\alpha = 0.01$) at different stations during different periods. A relatively strong linear relationship exists for Weigongcun during the peak hours and Beijing Zoo during the off-peak hours. Most of the predictors have correlation coefficients larger than 0.5, especially during the peak periods. In the Dutch cases, correlations between these three variables for the merged data of all stations are 0.376, 0.456, and 0.381, respectively. Therefore, the relationship between the variables is weak but still holds even if the instances are different.

Third, the dataset of the Beijing railway stations is used to calibrate the parameters in formula (1). The trains' length is neglected and treated as a constant. Therefore, to maintain the stability of formula (1), we set $\beta_1 = 0$ and $\beta_2 = 1.6$ for the six-car train based on the Dutch dataset (because the trains' length on line four in the Beijing urban rail transit is six cars). The parametric regression results for the peak and off-peak periods are presented in formulas (2) and (3), respectively. The models are regressed using the calibration dataset, which includes the predictors for the 7 selected stations in the Beijing railway stations. Formula (2) and formula (3) can predict the dwell time of all 7 stations during the peak and off-peak hours, respectively.

$$DT^s_k = -0.83 + 1.6 \times L^s_{k-1} + 0.76 \times DT^s_{k-1} + 0.04 \times \sqrt{DT^{s-1}_k * DT^{s-2}_k}, \quad (2)$$

$$DT^s_k = 7.02 + 1.6 \times L^s_{k-1} + 0.30 \times DT^s_{k-1} + 0.24 \times \sqrt{DT^{s-1}_k * DT^{s-2}_k}. \quad (3)$$

Three indicators are introduced to evaluate the estimation accuracy of the results: the adjusted coefficient of determination (adj-R^2), the root mean square error (RMSE), and the mean absolute percentage error (MAPE). Adj-R^2 reflects the level to which the regression model is interpreted by the variables: the larger this indicator is, the better the datasets are explained. The other two indicators are taken from the literature [2]; the RMSE reflects the actual error, and the MAPE is used to evaluate the estimation accuracy. The performance measures of the results based on the Beijing dataset are shown in Table 6.

Table 6 shows that the model accuracy for the Beijing case is better and that the accuracy of the adapted model based on the data for the investigated stations of the urban railway line 4 in Beijing is satisfactory, in particular for the peak period.

4.2.2. Parameter Comparison. The parameters of the model that which based on the two datasets are shown in Table 7, which indicates that the parameters in both cases have similar tendencies, such as the constants during the peak period being negative numbers. The different parameter values indicate the contributions of the predictors to the dwell time, as stated in the literature [2]: β_3 reflects the effects of the stations, and β_4 reflects the effects of the rolling stock. When neglecting train length, in the Dutch dataset, the value of β_4 is greater than that of β_3 during the peak period; in other words, the contribution of dwell time at the previous station is positive, whereas the dwell time of previous trains has almost no effect. The variables of the same trains at different stations have a strong relationship, indicating that the rolling stock factor plays an important role in the dwell time of the Dutch railway which is reflected in reality. In our investigation,

Table 7: Model parameters in the different datasets.

Parameters	Dutch dataset	Beijing dataset			
	Peak period	Peak period	SD	Off-peak period	SD
c	−13.07 (0.01)	−0.83 (0.00)	0.23	7.02 (0.00)	0.37
β_1	—	—	—	—	—
β_2	1.60 (0.00)	1.60 (—)	—	1.60 (—)	—
β_3	0.03 (—)	0.76 (0.00)	0.04	0.30 (0.00)	0.06
β_4	1.11 (0.00)	0.04 (0.00)	0.05	0.24 (0.00)	0.07

Note. p values are shown in brackets; "SD" denotes standard deviation.

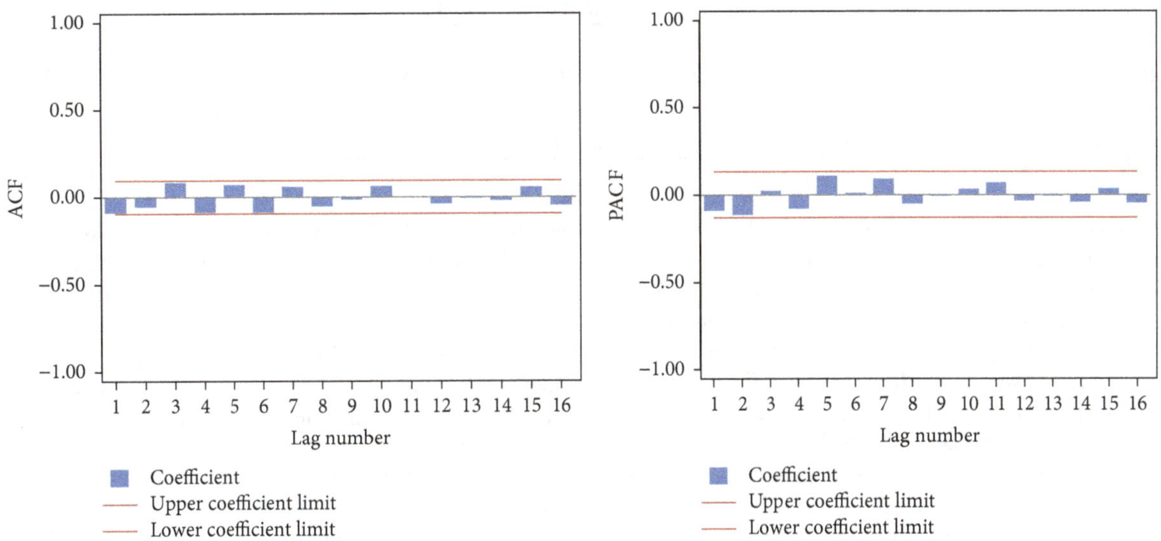

(a) ACF

(b) PACF

Figure 6: Residuals analysis by ACF and PACF.

the yard layout and the sizes of the selected stations in the Dutch railway stations are similar, but the trains' lengths differ significantly; thus, the trains' lengths play a key role in the Dutch model. Conversely, the situation in the Beijing urban rail transit during peak hours differs from that in the Dutch railway stations. The trains' lengths on line four of the Beijing urban rail transit are all the same, but the platform layouts, for example, the positions of the stairs and escalators, and the sizes of the selected stations vary; thus, the station factors play key roles in the model based on the Beijing dataset.

The p values and standard deviation of the predictors are listed in Table 7. It is shown that the p values for all parameters are smaller than 0.05, which indicates that the values of the parameters are significantly different from zero. Meanwhile, we also analyze the residuals based on model diagnostics for the time series. The autocorrelation function (ACF) and partial autocorrelation function (PACF) are shown in Figure 6. The ACF and PACF of the residuals show that the residuals do not correlate statistically and accord with a white noise sequence. In conclusion, the regression model can be used in the Beijing cases.

In addition, the model fails during the off-peak hours for the Dutch case but is still applicable during the off-peak hours for the Beijing case. One possible reason for this is that the scheduled headway is large during the off-peak hours in the Dutch case, which tends to decorrelate the dwell times. Another reason is that the passenger demand in the Dutch case during the off-peak period is so small that the random factors play a key role in the dwell time. However, the headway (Figure 9) and the passenger demand of the Beijing case during the off-peak period are similar to those of the Dutch case during the peak period, meaning that the influencing factors of the dwell time may be similar. The results show that the values of β_3 and β_4 in the Beijing dataset are small during the off-peak period. Because the variation in the passenger demand is tolerable, the influencing factors have a weak effect on the dwell time; thus, the dwell time at the target station tends to be stable. In summary, the parameters can be justified by analyzing the actual situation.

4.2.3. Performance Comparison. The accuracy of the parametric model based on the Beijing dataset is compared with those of existing models reported in literature (see Table 8). The datasets published in various reports differ from this article. The comparison could have some bias, yet it could reflect the effectiveness and potential of the selected model.

The results from the comparison of the author's model in the different cases are discussed in further detail. First, adj-R^2 is high in the model based on the Beijing dataset during peak hours, which shows that the model can explain most

Table 8: Comparison of the regression results for different datasets.

Indicators	Puong [11]	Hansen et al. [18]	Kecman (2014)	Dutch dataset		Beijing dataset	
				Peak	Off-peak	Peak	Off-peak
Adj-R^2	—	—	—	0.577	—	0.594	0.162
RMSE	4.04 s	16.6 s	—	6.2 s–8.8 s	8.49 s	7.21 s	6.21 s
MAPE	14.55%	—	15%	11.5%–14.2%	19.9%	11.4%	13.1%

Note. "Adj-R^2" denotes the adjusted coefficient of determination; "RMSE" denotes the root mean square error; "MAPE" denotes the mean absolute percentage error.

Table 9: The distribution of the worldwide subway dwell time.

City	Station	Date	Dwell time	Flow/dwell	Passenger	Average headway
San Francisco	Montgomery	pm peak	30 s–60 s	38%	586	153 s
New York	Grand Central	am peak	35 s–120 s	64%	1143	160 s
Toronto	Commission King	am peak	25 s–45 s	31%	428	168 s
British Columbia	SkyTrain Burrard	am peak	20 s–30 s	40%	562	151 s

of the samples in the dataset. In other words, the model can estimate dwell time very well in the Beijing urban rail transit system during peak hours. Second, the RMSE and MAPE are tolerable in both the Dutch dataset and the Beijing dataset during the investigated periods. Therefore, the errors between the predicted and observed dwell times are sufficiently small that the model can be used to estimate the dwell time for selected intermediate ordinary stations in the Dutch railway network and urban railway line 4 in Beijing. Third, adj-R^2 in the model based on the Beijing dataset during the off-peak period is low in the model based on the Beijing dataset, where the cases in Netherlands reveal that the characteristics of the off-peak passenger flow in the Beijing dataset cannot be reflected well by the model.

5. Generality Analysis Using a Theoretical Approach

The comparison approach in the previous section performed well with regard to the generality of the model. However, this approach may face problems such as how many cases should be covered, although it is usually impossible to cover all cases for a model in practice. To solve this problem, the following section shows that the generality of the model is tested using a theoretical approach. The theoretical approach includes four aspects. These aspects include variable analysis, parameter analysis, scenario analysis, and model structure and condition analysis.

5.1. Variable Analysis. First of all, the generality of the model is discussed based on the variables.

Assumption 1. Under the premise that the parameters are fixed, the range of the values of the variables (definition domain) should be within a certain trusted value domain. In other words, it has a lower bound and an upper bound.

Based on Assumption 1, we can discuss how the estimation result will change and in which range can the result be trusted. When the variables change within the domain of the definition, we can determine the range of the model's variables, establishing the applicability of the model. For the model used in this research, the variables include two types, the length of the train L_k^s and L_{k-1}^s and the train dwell time DT_{k-1}^s, DT_k^{s-1}, and DT_k^{s-2}.

For the length of the train, literature [10] shows that the range of the length is 2–12 cars in the majority of the urban rail transit. For the dwell time, previous research [10] presented the distribution of the worldwide rail transit dwell time, as shown in Table 9. In this table, the authors introduced the four major rail transit systems and the range of the dwell time was recorded from 20 s to 120 s. So, we considered dwell times within this range to be acceptable.

Next, we discuss the result of the model when the variables change within the domain. The domain variables are introduced into the model which is regressed by the Beijing datasets to obtain the results. Figure 7 shows the results and it reflects that the dwell time of the target train in the target station falls within the limited range (20–120 s). The results for the variables from the domain are acceptable. Therefore, the model is universally relevant to the Beijing datasets when the variables are in the domain. The results obtained from the model regressed by the Dutch datasets are shown in Figure 8; most of the results are acceptable when the variables are from the domain. Only a few parts extend beyond the range, but the model still has a certain general relevance.

5.2. Parameter Analysis. In this section, the value ranges of the parameters are discussed. The values of the parameters can be inferred from both the structure of the model and the domain of the definition of the variables.

At first, there are two contributors, and the results are for the dwell time in the model. If the trains' length does not change in the modeling scenario, the value of the dwell time at the target station can be calculated approximately as a weighted average of the dwell times of the preceding train, at the previous station and the second previous station. Thus, we consider that the values of β_3 and β_4 which are the parameters of the contributors should lie in $[0,1]$, which also ensures that the dwell time ranges of the different stations in a given railway are consistent.

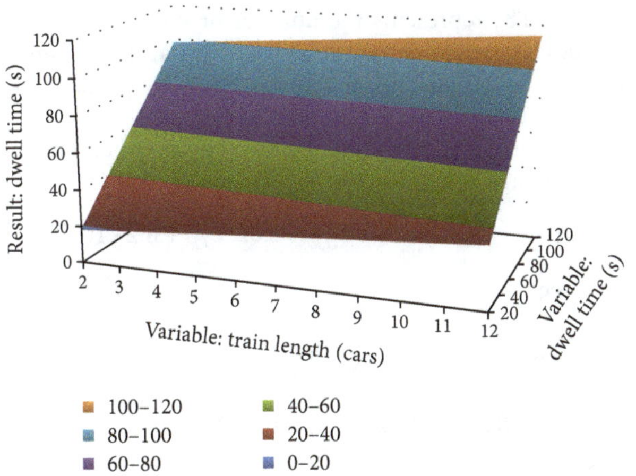

FIGURE 7: The results from changing the variables in the Beijing railway stations.

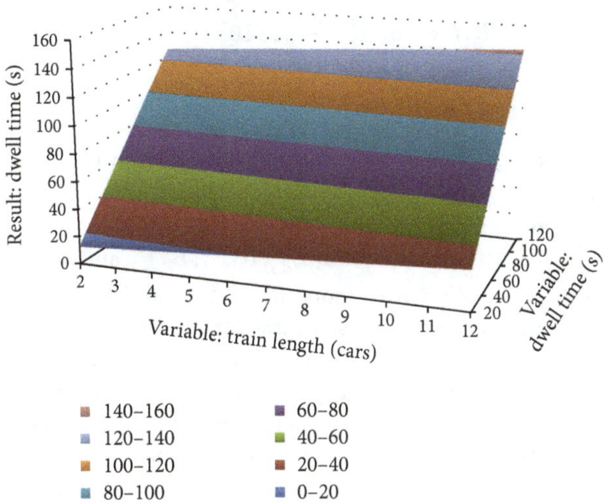

FIGURE 8: The results from changing the variables in the Dutch railway stations.

It can be deduced from the meaning of the parameter that the value of c is mainly related to the variance of passenger demand between the different stations and trains. Because this variation can be significantly different from station to station, the value can be deduced only by model regression. Formula (2) can be manipulated to obtain a new formula for parameter c, as shown in formula (4). In this formula, the dwell times of different trains at different stations are considered the same. In addition, the range of the dwell time is $[20, 120]$ and the range of the train lengths is $[2, 12]$; thus the range of parameter c is obtained as $[-220, 120]$. When the train lengths and the dwell time of other trains at other stations have a weak influence on the target, the dwell time is similar to parameter c. As a result, we limit the range of c in $[-20, 100]$.

$$c = \mathrm{DT}_k^s - \beta_2 L_{k-1}^s - \beta_3 \mathrm{DT}_{k-1}^s - \beta_4 \sqrt{\mathrm{DT}_k^{s-1} * \mathrm{DT}_k^{s-2}}. \quad (4)$$

Finally, the parameters concerning the train lengths are discussed. Although the ranges of the other parameters and variables are already limited, nonetheless we can still obtain the ranges of β_1 and β_2. The ranges of these parameters are too large, and the relationships between train length and dwell time are not clear. The train length parameters are difficult to determine.

In summary, when the parameters are in their established ranges, the results can be trusted, and the ranges of the parameters are within the model scope of the application.

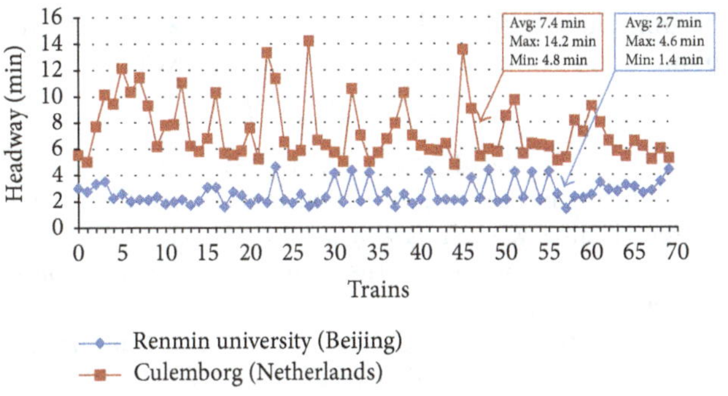

FIGURE 9: The headway of two cases.

5.3. Scenario Analysis. The scenario is usually defined by several indicators, such as headway, type of station, and train speed. The scenario is one of the most important factors of influence when we discuss the generality of a model.

Assumption 2. The model is general if it can apply to more scenarios.

To study whether the model can be applied to different scenarios, it is important to discuss the effect of these scenario indicators on the model.

First, the effect of headway is twofold. On one hand, the headway influences the relationship between the dwell time of the target train at the target station DT_k^s and the dwell time of the previous train at the target station DT_{k-1}^s. The larger the headway, the weaker the relationship. The results of the model when regressed by the Beijing and Dutch datasets separately reflect this conclusion, and the headways of the two cases are shown in Figure 9. The red line denotes the headways of departing trains at the Culemborg station in the Dutch railway, and the blue line denotes the headways of departing trains at the Renmin University station in Beijing. The average headway of the Dutch dataset is 7.4 minutes, and DT_k^s and DT_{k-1}^s have a weak relationship; thus $\beta = 0.03$. The average headway of the Beijing dataset is 2.7 minutes, and the relationship between DT_k^s and DT_{k-1}^s is strong; thus $\beta = 0.76$ and the set DT_{k-1}^s has a strong impact on DT_k^s. This result can be explained by the fact that the relationship between passenger demand and consecutive trains is stronger when the headway is small. On the other hand, the minimum headway is determined based on the dwell time, operating margin, and safety separation time, as shown in formula (5). $h_{\min}$ denotes the minimum headway, DT_k^s means the dwell time, T_{om} stands for the operating margin time, and T_{ss} identifies the safety separation time. The dwell time and the minimum headway have a positive correlation, so the minimum headway has an influence on the dwell time.

$$h_{\min} = \mathrm{DT}_k^s + T_{\mathrm{om}} + T_{\mathrm{ss}}. \tag{5}$$

However, the headways in Figure 9 from either Netherlands (during peak hours) or China are less than 15 min. For a larger headway, such as a headway greater than 15 min, formula (5) does not hold. The evidence can also be found in Li et al. (2016).

Second, the effect of different types of stations is also an indicator for determining which scenario is relevant. At intermediate stations, where a train cannot overtake another, the dwell time is mainly determined by the dwell time of the previous train, the minimum headway, and the passengers. At interchange stations, the previous train's dwell time is one of the most significant influencing factors. Moreover, whenever the train timetable is required for synchronization between two lines, the connections for transfer passengers should be considered. That is, the dwell time of the target train is related to the departure and arrival times of other trains in the interchange station. At the terminal station, the dwell time is mainly determined by the time when the train turns back. The turn back time is related to the train's length and turn back technique. The shorter the train and the higher its efficiency in turning back, the shorter the dwell time.

The average headway, minimum headways, train length, and dwell time (the target and previous train at the target and previous station) can all be used to describe different scenarios. The model is applicable to those scenarios, if the indicators that determine the scenarios are considered in the model. Based on our analysis, the model is applicable to more scenarios because the correlation between the headway and dwell time is high, and we can also determine from formula (5) that the dwell time is a component of the minimum headway. As a result, the headway is not considered in the regression model.

5.4. Structure and Condition Analysis. In the selected model, passenger demand is disregarded because passenger demand can be reflected by the substitute variables such as the dwell time of the preceding train and the dwell time at the previous station, so its assumption is not proven to be true. To demonstrate the generality of the model, it is necessary to know to what extent this assumption holds. To answer this question, the dwell time variables of the model can be replaced by an equation for the other variables that have already been proven to be relevant from formulations in previous research. Then, we can establish a new model with new variables and discuss the applicability of the new model. If the new model is plausible, then the assumption is proven.

In previous research, Puong [11] developed a nonlinear passenger regarded model to estimate train dwell time. The variables of the model included the number of standing passengers in the vehicle and their interactions with the boarding and alighting passengers, as shown in formula (6).

$$\mathrm{DT} = 12.22 + 2.27B_d + 1.82A_d + 6.2 \cdot 10^{-4} \cdot \mathrm{TS}_d^3 B_d, \tag{6}$$

where A_d represents the number of alighting passengers per door, B_d represents the number of boarding passengers per door, and TS_d represents the number of standees per door. The model is used to estimate the dwell time at stations on the MBTA Red Line. The result shows that R^2 is approximately 0.9; the accuracy of the estimation is high.

We enter formula (5) into formula (2) and then obtain a new formula, as shown in formula (7).

$$\begin{aligned}
\text{Left part} &= 12.22 + 2.27B_{d,k}^s + 1.82A_{d,k}^s + 6.2 \cdot 10^{-4} \\
&\quad \cdot {\mathrm{TS}_{d,k}^s}^3 B_{d,k}^s, \\
\text{Right part} &= -0.83 + 1.6 \times L_{k-1}^s + 0.76 \times \big(12.22 \\
&\quad + 2.27B_{d,k-1}^s + 1.82A_{d,k-1}^s + 6.2 \cdot 10^{-4} \\
&\quad \cdot {\mathrm{TS}_{d,k-1}^s}^3 B_{d,k-1}^s\big) + 0.04 \times \big[12.22 + 2.27B_{d,k}^{s-1} \\
&\quad + 1.82A_{d,k}^{s-1} + 6.2 \cdot 10^{-4} \cdot \left(\mathrm{TS}_{d,k}^{s-1}\right)^3 B_{d,k}^{s-1} * 12.22 \\
&\quad + 2.27B_{d,k}^{s-2} + 1.82A_{d,k}^{s-2} + 6.2 \cdot 10^{-4} \\
&\quad \cdot \left(\mathrm{TS}_{d,k}^{s-2}\right)^3 B_{d,k}^{s-2}\big]^{0.5},
\end{aligned} \tag{7}$$

where $B_{d,s}^k$, $A_{d,k}^s$, and $\mathrm{TS}_{d,k}^s$ indicate the number of boarding passengers, alighting passengers, and standees per door of train k at station s.

In this formula, apparently $B_{d,s}^k$, $B_{d,s-1}^k$, $B_{d,s-2}^k$, and $B_{d,s}^{k-1}$ are different. The difference among $B_{d,s}^k$, $B_{d,s-1}^k$, and $B_{d,s-2}^k$ is the boarding passengers of the same train at different stations and is influenced by the passenger OD demand. The difference between $B_{d,s}^k$ and $B_{d,s}^{k-1}$ is the boarding passengers of different trains at the same station and is influenced by the period of time. We enter the value ranges of A_d, B_d, and TS_d in Puong [11] into formula (7), and the result is shown in Figure 10. It can be seen that the value of the left part and the value of the right part are close in half of the cases, whereas the value of the left part of the formula is not equal to the value of the right in the remaining cases. By applying a two-sample t-test, it can be shown that $H = 0$ means that the two samples are independent random samples from normal distributions with equal means at a 5% significance level, demonstrating that the model is reliable.

6. Generality Improvement

From the generality testing, it can be seen that the relationships between passenger demand at different stations and trains significantly impact the estimation result. However, such a relationship is not necessarily based on consecutive stations or consecutive trains. Based on this inference, we introduce a more generalized predictor to improve the generality of the proposed model, and the results of the improved model are compared with the results of the proposed model in this section.

6.1. A More Generalized Predictor. The generality of our model can still be improved. Past research analyzed the relationships among the dwell times of three consecutive stations

TABLE 10: Correlation coefficients of the dwell time in peak periods.

Stations	Pervious stations										
	1	2	3	4	5	6	7	8	9	10	11
Renmin University	0.22	**0.47**	0.44	0.16	0.12	—	—	—	—	—	—
Weigongcun	0.22	0.42	0.29	0.27	0.41	**0.77**	—	—	—	—	—
National Library	0.06	0.36	0.05	**0.61**	−0.12	0.33	0.54	—	—	—	—
Beijing Zoo	0.02	−0.06	−0.21	0.31	−0.13	0.04	0.16	**0.75**	—	—	—
Xizhimen	−0.07	−0.03	−0.15	0.16	−0.15	−0.13	0.12	0.44	**0.49**	—	—
Xinjiekou	−0.03	−0.13	−0.23	0.14	−0.04	0.00	0.08	0.28	0.59	**0.71**	—
Ping'anli	−0.15	−0.14	−0.15	−0.08	0.34	−0.09	0.02	0.07	0.24	0.33	**0.53**

Notes. Numbers 1–11 denote stations on line 4 from Xiyuan to Xinjiekou.

TABLE 11: Correlation coefficients of the dwell time in off-peak periods.

Stations	Pervious stations										
	1	2	3	4	5	6	7	8	9	10	11
Renmin University	0.33	0.12	**0.38**	0.06	0.12	—	—	—	—	—	—
Weigongcun	**0.31**	0.06	0.14	0.12	0.06	0.23	—	—	—	—	—
National Library	−0.11	0.21	0.1	**0.41**	0.15	0.22	0.12	—	—	—	—
Beijing Zoo	−0.12	0.03	−0.14	0.08	−0.15	−0.05	**0.64**	0.03	—	—	—
Xizhimen	**0.32**	−0.03	−0.1	0.27	0.13	0.09	−0.24	0.1	0.2	—	—
Xinjiekou	−0.06	0.18	0.22	0.39	0.18	0.17	0.07	0.24	0.28	**0.46**	—
Ping'anli	0.08	0.1	0.13	0.15	0.1	0.16	0.09	0.12	0.08	0.1	**0.17**

Notes. Numbers 1–11 denote stations on line 4 from Xiyuan to Xinjiekou.

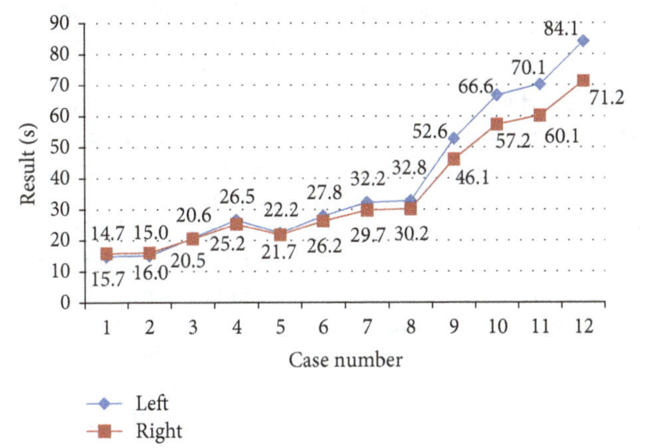

FIGURE 10: The comparison of the values of the left part and right part in formula (7).

in the Dutch railway lines. Because of the data consistency problem, some data were missing due to incorrect track identifiers resulting from track changes after maintenance. However, the Beijing dataset includes the dwell times of all the stations on the selected line, which inspires our refinement of the model.

We can analyze the relationships between the dwell times of the target station and previous stations. The correlation of dwell time between the target station and previous stations in peak and off-peak periods is analyzed and the result is shown in Tables 10 and 11, respectively. Through correlation analysis, we can find the most correlated previous station, which is called the associated station (the boldface number in the table). The dwell time of the associated station (denoted as DT_k^{s*}) impacts the target station more than the square root of the product of the dwell time at the previous station and at the second previous station on ($\sqrt{\mathrm{DT}_k^{s-1} * \mathrm{DT}_k^{s-2}}$), which is used in the original model.

To test the above idea, the correlation coefficients between the dwell time at the target station and the dwell time at the associated stations are obtained and are compared with the square root term used in the original model. The results from the Beijing dataset are shown in Table 12.

Table 12 shows that the variable DT_k^{s*} has a stronger relationship with the dwell times of the target train at the target station than the original square root term. Thus, the predictor DT_k^{s*} is a better substitute for the predictor $\sqrt{\mathrm{DT}_k^{s-1} * \mathrm{DT}_k^{s-2}}$. The boldface indicates that there was a significant correlation (>0.3) between the predictors and the dwell time.

Based on this improvement, a more generalized model is obtained, as shown in the following formula:

$$\mathrm{DT}_k^s = c + \beta_1 L_k^s + \beta_2 L_{k-1}^s + \beta_3 \mathrm{DT}_{k-1}^s + \beta_4 \mathrm{DT}_k^{s*}. \quad (8)$$

The Beijing dataset is used to calibrate formula (8), and the results from the peak and off-peak periods are shown in formula (9) and formula (10), respectively.

$$\mathrm{DT}_k^s = -8.04 + 1.6 \times L_{k-1}^s + 0.61 \times \mathrm{DT}_{k-1}^s + 0.34 \times \mathrm{DT}_k^{s*}, \quad (9)$$

$$\mathrm{DT}_k^s = 0.17 + 1.6 \times L_{k-1}^s + 0.47 \times \mathrm{DT}_{k-1}^s + 0.26 \times \mathrm{DT}_k^{s*}. \quad (10)$$

TABLE 12: Correlation coefficients of the predictors.

Stations	Peak periods			Off-peak periods		
	DT^{s}_{k-1}	$\sqrt{DT^{s-1}_{k} * DT^{s-2}_{k}}$	DT^{s*}_{k}	DT^{s}_{k-1}	$\sqrt{DT^{s-1}_{k} * DT^{s-2}_{k}}$	DT^{s*}_{k}
Renmin University	0.20	0.14	**0.47**	0.12	0.08	**0.38**
Weigongcun	**0.41**	**0.56**	**0.77**	0.15	0.12	**0.31**
National Library	**0.54**	**0.42**	**0.61**	**0.58**	0.16	**0.41**
Beijing Zoo	**0.46**	**0.35**	**0.75**	0.07	0.15	**0.64**
Xizhimen	**0.47**	**0.47**	**0.49**	0.23	0.14	**0.32**
Xinjiekou	**0.73**	**0.65**	**0.71**	0.02	**0.36**	**0.46**
Ping'anli	0.28	**0.42**	**0.53**	0.07	0.13	0.17
All stations	**0.87**	**0.42**	**0.74**	**0.37**	0.27	**0.46**

TABLE 13: Comparison with the improved parametric model.

Dataset	Dutch		Beijing			
Model	PM	NPM	PM		IPM	
Period	Peak	Off-peak	Peak	Off-peak	Peak	Off-peak
Indicators						
Adj-R^2	0.577	—	0.594	0.162	0.860	0.455
RMSE	6.2 s–8.8 s	8.49 s	7.21 s	6.21 s	4.19 s	4.14 s
MAPE	11.5%–14.2%	19.9%	11.4%	13.1%	7.8%	9.9%

6.2. Estimation Performance. To test the performance of the improved model and compare the estimation accuracy of the improved model with that of the original model, adj-R^2, RMSE, and MAPE indicators are used. The accuracy of the improved parametric model (IPM) is compared with that of the parametric model (PM) based on the Beijing dataset for the peak and off-peak periods (see Table 13).

In the IPM, adj-R^2 is higher than that in the PM during both the peak and off-peak hours. Thus, the IPM can reflect more characteristics of the Beijing dataset. However, adj-R^2 is still low in the IPM during the off-peak periods, and improving the accuracy of the off-peak period by adding new influencing factors is difficult. Other prediction methods for the off-peak hours may require testing. Moreover, the RMSE and MAPE of the IPM are smaller than those of the original model, indicating that there are smaller differences between the estimated and actual dwell times. Therefore, the results of the IPM are better than the results of the PM. In other words, using the IPM improves the accuracy of the results.

Due to the data consistency problem in the Dutch railway, the associated stations are not analyzed further. The square root term can be a special case of the associated station; therefore, it can be deduced that using the associated station item would not reduce the model's accuracy.

7. Discussion

7.1. Scenario of the Model. Based on the test datasets in this paper, it is worth noting that the generality of the estimation model indicates that the model with the same parameters and predictors can fit for all railway lines, regardless of whether the line is a subway or commuter rail or other railway types. Certain conditions must be satisfied to use the model.

The original model is specifically designed for short stops. Short stops are stops on the open track where sidings are usually not available and where trains dwell only for alighting and boarding, after which they immediately continue their journey. In both cases, all stations are short stops without sidings, and whether the model fits stations with sidings and large stations with passenger connections remains unclear.

The estimation model with the same predictors can regress the dwell time of different stations on different lines even in different countries, and the result is valid. However, the values can be different in different cases. In other words, the model followed the specific cases, and the parameters vary from case to case. The main reason is that the influence level of the predictors may differ in different cases. Therefore, when applying the model to a new case, the parameters should still be regressed. Accordingly, it is better to use the model at rescheduling phase where the train system is running and necessary datasets such as previous dwell times and train length are available.

In summary, the approach in this paper is convincing. However, it is difficult to establish a model that can fit all scenarios.

7.2. Dataset Requirement of the Model. The dataset should follow several constraints when studying specific cases. To improve the model's performance, the following principles should be considered.

The first issue concerns the data investigation site:

(1) The original model focused on short stops where sidings are not usually available and trains dwell for only alighting and boarding. That is, no overtaking occurs in the station. Therefore, the selected stations should be small stations with the same characteristics.

(2) In the original case study, the passengers check in, wait, and board the train without barriers. Therefore, the new dataset should have the same system.

(3) Three consecutive stations were selected in the original model to examine the relationship between the dwell times of consecutive stations; therefore, the new dataset should contain the dwell times from at least three stations.

(4) In the original case, the rolling stock is sprinters with wider doors and small gap differences between the rolling stock and the station to ensure fast alighting and boarding processes. These trains stop at every station without skipping any stops; therefore, the new case of the rolling stock should also be used for the rapid boarding and alighting of passengers.

Another issue is measurement. The measurements in the two cases are different, which could cause bias. The dwell time is estimated from track occupation data in the Dutch case; in that case, the train dwell time is estimated by an algorithm according to the stop sign of a train and the time-dependent occupation states of a track [21]. The exact position of a train is not clear. In fact, it is not possible for all drivers to stop exactly in front of the stop sign. From this point of view, the automatic measurements may underestimate the dwell time. In the Chinese case, the dwell times are measured manually via stop watches. The time is recorded as soon as a train standstill or departure occurs. Despite the error from the investigators, the accuracy could be higher than in the Dutch case. Based on the analysis, the error distributions of the two cases would be significantly influenced by the measurement. The theoretical issues associated with using automated versus manual dwell time measurements should be considered in future research.

7.3. Restrictions of the Model. Station dwell times are the major component of headways at close train frequencies. The existing literature suggests that the best achievable headways under these circumstances are in the range of 110 to 125 seconds [10]. Thus, it is reasonable that, as the major part of the headway, the dwell time also lies within a loosely identified interval. We discuss the ranges of train lengths, dwell times, and headways under variable analysis in Section 5.1, according to a widely selected dataset from the references [10]. In addition, we found that the model is applicable within this range. However, it should be noted that this inference does not necessarily hold at low railway frequencies, especially for long distance railways. Because the headways in the two datasets are less 15 min, we can only conclude that the model is applicable for close frequencies. Furthermore, we can also obtain the evidence from former research [2] to prove that formula (5) may not hold in larger headway scenarios (larger than 15 minutes).

8. Conclusions

The main contribution of this paper is to propose a systematic approach for testing the generality of a train dwell time estimation model and improving the model performance by introducing a more generalized predictor.

The generality of the model is tested by using two approaches, namely, a comparison approach and a theoretical approach. In the comparison approach, a dwell time estimation is selected and applied to a completely new independent test bed. The test results from two datasets from two different countries show relatively high accuracy and equivalent effectiveness. The regression parameters are compared using the dataset of line four in the Beijing urban rail transit system. The parameters are regressed using the datasets of these two cases. In the theoretical approach, the train dwell time estimation model is analyzed by four steps: variable analysis, parameter analysis, scenario analysis, and model structure and condition analysis. We conclude that the test model is general in the given condition, namely, short stops with a headway less than 15 min.

Furthermore, a more generalized predictor, dwell times at the associated station, is introduced to improve the performance of the model. The performance indicators show that the result of the improved model is better than the result of the original model. Thus, the associated station would be relevant for estimating train delays during operation, and using the associated station item would improve the generality of the results.

Usually, generality and accuracy are two contrary indicators for a model. If the accuracy of the model is higher, the range of the model's applicability is narrower, meaning that the model can be applicable only in certain scenarios. In contrast, if the model is applicable to many different scenarios, in other words, it is more general, it is difficult to maintain high accuracy at the same time as generality. The estimation error of the model we test is acceptable. Meanwhile, it is proven that the model is relatively general; therefore, the model can balance the tradeoff between generality and accuracy well.

The focus of future research should be on two directions. On one hand, more datasets from different cases can be tested to verify the model. On the other hand, the accuracy and the generality of the model should be improved, particularly for long headways. Randomness and the weak relationship between the dwell time and passenger demand under this condition make dwell time estimation more difficult. To improve the accuracy of the model, other nonparameter methods, such as kernel regression and the nearest neighbors method, which do not require a significant amount of data, can be tested. More influencing factors, such as the transfer rule, weather, and accident, could also be considered if such data can be collected by using sensors. These factors will be extensively studied in the near future.

Conflicts of Interest

The authors declare that they have no conflicts of interest.

Acknowledgments

This research is supported by the National Natural Science Foundation of China (U1434207), the Fundamental Research Funds for the Central Universities of China (2016JBM030), the National Key Research and Development Plan (2016YFE0201700), and the Beijing Chaoyang District Science and Technology Commission (CYXC1607). The authors wish to acknowledge these agencies.

References

[1] C. T. Lewis and C. Belanger, "The generality of scientific models: a measure theoretic approach," *Synthese. An International Journal for Epistemology, Methodology and Philosophy of Science*, vol. 192, no. 1, pp. 269–285, 2015.

[2] D. Li, W. Daamen, and R. M. P. Goverde, "Estimation of train dwell time at short stops based on track occupation event data: A study at a Dutch railway station," *Journal of Advanced Transportation*, vol. 50, no. 5, pp. 877–896, 2016.

[3] Q. Zhang, B. Han, and D. Li, "Modeling and simulation of passenger alighting and boarding movement in Beijing metro stations," *Transportation Research Part C: Emerging Technologies*, vol. 16, no. 5, pp. 635–649, 2008.

[4] S. Seriani and R. Fernandez, "Pedestrian traffic management of boarding and alighting in metro stations," *Transportation Research Part C: Emerging Technologies*, vol. 53, pp. 76–92, 2015.

[5] A. Yamamura, M. Koresawa, T. Inagi, and N. Tomii, "Dwell time analysis in railway lines using multi agent simulation," in *Proceedings of the 13th World Conference on Transportation Research (WCTR)*, Rio de Janeiro, Brazil, 2013.

[6] F. Sourd, C. Talotte, C. Ambroise et al., Modelling of pedestrian flows during dwelling: development of a simulator to evaluate rolling stock and platform flow performance[C]//WCRR2011-World Congress on Railway Research, Lille France. 2011.

[7] W. H. K. Lam, C. Y. Cheung, and Y. F. Poon, "A study of train dwelling time at the hong kong mass transit railway system," *Journal of Advanced Transportation*, vol. 32, no. 3, pp. 285–296, 1998.

[8] S. C. Wirasinghe and D. Szplett, "An investigation of Passenger interchange and train standing time at LRT stations: (II) estimation of standing time," *Journal of Advanced Transportation*, vol. 18, no. 1, pp. 13–24, 1984.

[9] T. M. Lin and N. H. M. Wilson, Dwell time relationships for light rail systems, 1992.

[10] T. Parkinson and I. Fisher, "Rail transit capacity," TCRP Report, 13, Transportation Research Board, Washington, DC.

[11] A. Puong, *Dwell Time Model and Analysis for the mbta Red Line*, Massachusetts Institute of Technology Research Memo, Cambridge, UK, 2000.

[12] N. G. Harris, "Train boarding and alighting rates at high passenger loads," *Journal of Advanced Transportation*, vol. 40, no. 3, pp. 249–263, 2006.

[13] J. C. Jong and E. F. Chang, "Investigation and estimation of train dwell time for timetable planning," *Office of Scientific &Technical Information Technical Reports*, vol. 4141, pp. 178–185, 2011.

[14] X. Karekla and N. Tyler, "Reduced dwell times resulting from train-platform improvements: The costs and benefits of improving passenger accessibility to metro trains," *Transportation Planning and Technology*, vol. 35, no. 5, pp. 525–543, 2012.

[15] W. Daamen, Y.-C. Lee, and P. Wiggenraad, "Boarding and alighting experiments: Overview of setup and performance and some preliminary results," *Transportation Research Record*, no. 2042, pp. 71–81, 2008.

[16] S. Buchmueller, U. Weidmann, and A. Nash, "Development of a dwell time calculation model for timetable planning," in *Proceedings of the 11th International Conference on Computer System Design and Operation in the Railway and Other Transit Systems, COMPRAIL 2008*, pp. 525–534, Spain, September 2008.

[17] F. S. Hänseler, M. Bierlaire, and R. Scarinci, "Assessing the usage and level-of-service of pedestrian facilities in train stations: a swiss case study," *Transportation Research Part A: Policy and Practice*, vol. 89, pp. 106–123, 2016.

[18] I. A. Hansen, R. M. P. Goverde, and D. J. Van Der Meer, "Online train delay recognition and running time prediction," in *Proceedings of the 13th International IEEE Conference on Intelligent Transportation Systems, ITSC 2010*, pp. 1783–1788, Portugal, September 2010.

[19] P. Kecman and R. M. P. Goverde, "Predictive modelling of running and dwell times in railway traffic," *Public Transport*, vol. 7, no. 3, pp. 295–319, 2015.

[20] P. Kecman and R. M. P. Goverde, "Process mining of train describer event data and automatic conflict identification," in *Computers in Railways XIII, WIT Transactions on the Built Environment*, C. A. Brebbia, N. Tomii, P. Tzieropoulos, and J. M. Mera, Eds., vol. 127, pp. 227–238, 2012.

[21] D. Li, R. M. P. Goverde, W. Daamen, and H. He, "Train dwell time distributions at short stop stations," in *Proceedings of the 2014 17th IEEE International Conference on Intelligent Transportation Systems, ITSC 2014*, pp. 2410–2415, China, October 2014.

10

Forecasting of Short-Term Metro Ridership with Support Vector Machine Online Model

Xuemei Wang,[1] Ning Zhang,[2] Yunlong Zhang,[3] and Zhuangbin Shi[2]

[1] *Department of Transportation Engineering, Tongji Zhejiang College, Jiaxing 314000, China*
[2] *Intelligent Transportation System Institute of Ministry of Education, Southeast University, Nanjing 211189, China*
[3] *Zachry Department of Civil Engineering, Texas A&M University, 3136 TAMU, College Station, USA*

Correspondence should be addressed to Ning Zhang; ningzhang1972@gmail.com

Academic Editor: David F. Llorca

Forecasting for short-term ridership is the foundation of metro operation and management. A prediction model is necessary to seize the weekly periodicity and nonlinearity characteristics of short-term ridership in real-time. First, this research captures the inherent periodicity of ridership via seasonal autoregressive integrated moving average model (SARIMA) and proposes a support vector machine overall online model (SVMOOL) which insets the weekly periodic characteristics and trains the updated data day by day. Then, this research captures the nonlinear characteristics of the ridership via successive ridership value inputs and proposes a support vector machine partial online model (SVMPOL) which insets the nonlinear characteristics and trains the updated data of the predicted day by time interval (such as 5-min). Afterwards, to avoid the drawbacks and to take advantages of the strengths of the two individual online models, this research takes the average predicted values of two models as the final predicted values, which are called support vector machine combined online model (SVMCOL). Finally, this research uses the 5-min ridership at Zhujianglu and Sanshanjie Stations of Nanjing Metro to compare the SVMCOL model with three well-known prediction models including SARIMA, back-propagation neural network (BPNN), and SVM models. The resultant performance comparisons suggest that SARIMA is superior for the stable weekday ridership to other models. Yet the SVMCOL model is the best performer for the unstable weekend ridership and holiday ridership. It shows that for metro operation manager that gear toward timely response to real-world unstable and abnormal situations, the SVMCOL may be a better tool than the three well-known models.

1. Introduction

Short-term ridership forecasting is a vital component of metro operation and management. Accurate predictions can reflect real-time changes in ridership. The prediction results can become important inputs for decision-making in evaluating rail transit service level and system operating status and provide an important basis for station passenger crowd regulation and emergency response. In addition, short-term ridership forecasting is the key to the success of revenue management for railway operators [1].

In the last two decades, traditional metro ridership forecasting is based on travel demand forecasting models including the steps of trip generation, trip distribution, mode choice, and assignment [2, 3]. This type of long-term forecasting has been applied in the planning and construction of metro, but it cannot be adapted to the needs of the operations management.

Though the spatial-temporal characteristics of metro ridership are not completely the same as those for vehicle traffic flow [4], short-term forecasting methods can also be divided into two categories: the theory driven method and the data driven method. Theory driven method is based on traffic flow mechanism to investigate traffic dynamics [5, 6]. The data driven method on the other hand is based on the data of traffic flow series itself to construct models and make predictions. The data driven model is the main method of short-term prediction and can be divided into linear, nonlinear, and hybrid forecasting methods. The linear forecasting method mainly includes time series model [7–9] and Kalman filtering model [10–12]. The nonlinear forecasting method includes nonparametric regression [13, 14], neural network algorithm [15–17], support vector machine [18–20], and Gaussian maximum likelihood model [21]. The hybrid forecasting method combines at least two methods for prediction to achieve better performance in accuracy and reliability. Hybrid models mainly include wavelet decomposition hybrid model [22, 23], Bayesian decomposition hybrid model [24, 25], empirical mode decomposition hybrid model [26], neural network hybrid model [27, 28], and support vector machine hybrid model [29–35].

Whether it is traffic flow or passenger, time series model has become one of the classic models of short-term flow prediction [36]. Of all the time series models, seasonal autoregressive integrated moving average (SARIMA) model considers the periodicity feature of the time series, so it can capture the inherent periodicity of traffic flow data. Williams et al. [9–11] used the SARIMA model for short-term traffic flow prediction and verified its good performance. But time series model is a linear model, and its prediction performance may worsen significantly if the time series are nonstationary and nonlinear. Nevertheless SARIMA model is widely used to be the benchmark to evaluate the forecasting performance of a novel model.

Neural networks are among the most widely used nonlinear models. A neural network trains neurons based on historical data, maps the complicated nonlinear relation between input and output data, and uses the relationship for predictions for given inputs. Neural network algorithms have the adaptive and learning advantages and are flexible without the need to construct detailed and explicit models like other methods. Vlahogianni et al. [37] optimized neural network structure to forecast urban traffic flow parameters. But the neural network algorithm cannot make expected risk minimization because of the empirical risk minimization principle that may also lead to two major drawbacks: local minima and overfitting [38]. The local minima are associated with the training process of neural network, which is to minimize the difference between the predicted outputs and the observed outputs by optimizing the network weights. Overfitting leads to poor generalization ability and may produce inaccurate predictions with some particular testing data.

Compared with neural network algorithm, support vector machine (SVM) model can strike a compromise between prediction accuracy and generalization ability based on the structural risk minimization principle. With the help of intelligent use of kernel function, SVM can solve the problems of small sample, nonlinearity and the curse of dimensionality, overfitting, and local minima. Zhang and Xie [19] proposed a υ-support vector machine model for short-term traffic volume prediction and showed that it outperformed the multilayer feed-forward neural network (MLFNN) model. Zhang et al. [30] proposed a novel hybrid model that identified the SVM input dimensions via SARIMA model to forecast short-term traffic volume, taking advantage of the individual strengths of the two models. Hong [33] presented a traffic flow forecasting model to forecast interurban traffic flow, which combines the seasonal support vector regression model with chaotic immune algorithm (SSVRCIA), and yielded more accurate forecasting results than the SARIMA, BPNN, and seasonal Holt–Winters models. Wang and Shi [34] constructed a new kernel function using a wavelet function to capture the nonstationary characteristics of the short-term traffic speed data, proposed a short-term traffic speed forecasting hybrid model (Chaos-Wavelet Analysis-Support Vector Machine model, C-WSVM), and achieved the encouraging results. Chen et al. [35] proposed an approach which hybridizes SVR model with adaptive genetic algorithm (AGA) and the seasonal index adjustment, namely, AGA-SSVR, to forecast holiday daily tourist flow.

The research of short-term metro ridership forecasting is a rather new undertaking. Tsai et al. [1] proposed two novel neural network structures based on temporal feature extraction and successfully applied them in railway short-term passenger demand forecasting in Taiwan. Wei and Chen [26] used empirical mode decomposition to extract neural network input variables to forecast the short-term ridership of Taipei Rapid Transit Muzha Line. Sun et al. [29] proposed a novel hybrid model Wavelet-SVM, and the experimental results showed that the approach has appeared to be the promising and robust. These studies indicated that metro ridership has significant characteristics of periodicity and nonlinearity reflecting a variety of factors; however, how these characteristics are embedded into the model without affecting the computational complexity of the model is worth discussing. And, for neural network or support vector model, the previous literature also did not discuss the training time to see if it meets the demand of practical operation. If the training time is too long and leads to serious forecasting delay, the prediction model cannot meet the demand of practical operation even if it has good prediction performance. In addition, most existing research on short-term metro ridership forecasting focused mainly on normal situations; it is not clear how the applicability and the prediction accuracy of the model is when it comes to holidays, inclement weather, large sports events, or emergencies. Sun et al. [29] selected the data including a Valentine's Day (not a major holiday) as training data, not as a predictor. Finally, the short-term prediction interval is long (i.e., 15-min) in these literatures, and, for the actual operation of the metro, it cannot meet the requirement of the operator because the departure intervals are short.

The reliability and the operability of the models play a crucial role in the accuracy and real-time implementation of the prediction, so the choice of the model is very important in a practical application. Since the characteristics of metro ridership are quite different from those in other transportation systems, most of forecasting models provide unsatisfactory prediction effectiveness. After comparing time series model, neural network model, and SVM model, this paper selects SVM model as the base short-term prediction model, considering capturing in real-time the periodicity and nonlinearity characteristics of short-term ridership as mentioned previously. With this base model, this paper proposes a support vector machine overall online (SVMOOL) model, which extracts input features via SARIMA model, trains the updated data by day, and optimizes the parameters by a particle swarm optimization (PSO) algorithm, to capture the periodicity of ridership in real-time. This paper also proposes a support vector machine partial online (SVMPOL) model, which extracts input features based on the temporal continuity of ridership model, trains the updated data by time intervals (such as 5-min), and also optimizes the parameters by a PSO algorithm to capture the nonlinearity of ridership. Afterwards, the support vector machine combined online (SVMCOL) model is proposed by combining the SVMOOL model and the SVMPOL model.

The main contributions of this paper are as follows.

(1) This paper proposes a novel hybrid model combining the SVMOOL model and the SVMPOL model for short-term ridership forecasting that better captures the periodicity and nonlinearity characteristics by the updated data set. The SVMCOL model takes advantages of the individual strengths of the two models. The actual results of 5-min short-term ridership forecasting show the feasibility and effectiveness of the proposed combined model in real-time implementation.

(2) While the SARIMA model is superior for the stable weekday ridership to other models, experiments results indicate that the SVMOOL model is superior to SARIMA model, BPNN model, or SVM model in terms of MAE and RMSE for the weekend and holiday ridership test. It should be noted that the prediction of ridership under abnormal situations (such as holiday) is evidently more challenging than doing so under normal conditions (such as weekday ridership) and, hence, is much desired by the operator. Therefore, the proposed SVMCOL model is found to be suitable and useful in real-world operations.

(3) The experiments using LibSVM package on desktop computers indicate that the SVMOOL model needs about one hour for three weeks' data (4284 observations) to construct the prediction function and the forecasting time takes less than 1 second for a one-step prediction using SVM. In the process of the implementation experiments, the SVMPOL model needs less than 1 s to construct due to the small data sample and the forecasting time needs less than 1 s for a one-step prediction. Therefore, the training time and the forecasting time can meet the real-time demand for the one-step prediction in implementation as well.

(4) In general, short-term forecasting represents prediction for a specific time interval, such as 5 min, 10 min, and 15 min. For metro ridership, 5-min interval will be more useful for metro operation and management because the departure interval of the metro vehicle is really short. In addition, it is obvious that ridership during workdays is different from that on weekends or holidays. As discussed by Chen et al., some prediction models that work well for workdays data may yield unsatisfactory results for weekends or holidays data. In order to discuss the applicability of the proposed model, three samples were selected. The first sample contains weekdays, weekends, and no holidays, and the second and third samples contain weekdays, weekends, and holidays.

This paper attempts to develop an online hybrid model to improve the forecasting performance of metro ridership. The rest of this paper is organized in the following manner. A brief theoretical background of the SVM model is presented first, followed by detailed description on SVMOOL model, SVMPOL model, and SVMCOL model. After that, a brief description of the data source and the implementation of the models are given. Finally, results analysis and conclusions are presented.

2. Methodology

To introduce the SVMOOL, SVMPOL, and SVMCOL models, SVM model is illustrated here first.

2.1. Support Vector Machine for Regression. A detailed description of SVM algorithm is given in Vapnik [38]. Assume that training input data and the corresponding training output data are (x_i, y_i) $(i = 1, 2, \ldots, m)$, where $x_i \in X \subseteq R^n$ and $y_i \in Y \subseteq R^n$, and m denotes the total number of data. The basic idea of SVM is to map the low-dimensional input space to the high-dimensional feature space using a function $\Phi(x_i)$. The linear regression function can be stated as

$$f(x) = w^T \Phi(x_i) + b, \quad (1)$$

where w and b are coefficients. For SVM, these coefficients can be obtained by solving the following optimization problems:

$$\begin{aligned} \min \quad & \frac{1}{2} w^T w + C \sum_{i=1}^{m} (\xi_i^+ + \xi_i^-) \\ \text{subject to} \quad & y_i - w^T \Phi(x_i) - b \le \varepsilon + \xi_i^+ \\ & w^T \Phi(x_i) + b - y_i \le \varepsilon + \xi_i^- \\ & \xi_i^+ \ge 0, \ \xi_i^- \ge 0, \ i = 1, 2, \ldots, n, \end{aligned} \quad (2)$$

where $\varepsilon(\ge 0)$ is the insensitive loss function, ξ_i^+ and ξ_i^- are slack variables, and C is a regularization parameter. The maximal dual function in (2) has the following form:

$$\begin{aligned} \max \quad & \sum_{i=1}^{n} y_i (\alpha_i^* - \alpha_i) - \varepsilon \sum_{i=1}^{n} (\alpha_i^* + \alpha_i) \\ & - \frac{1}{2} \sum_{i=1}^{n} \sum_{j=1}^{n} (\alpha_i^* - \alpha_i)(\alpha_j^* - \alpha_j) K(x_i, x_j) \\ \text{subject to} \quad & \sum_{i=1}^{n} (\alpha_i^* - \alpha_i) = 0 \\ & \alpha_i \ge 0, \ \alpha_i^* \le C, \ i = 1, 2, \ldots, n, \end{aligned} \quad (3)$$

where α_i^* and α_i are Lagrange multipliers.

Ultimately, the decision function given by (1) has the explicit form:

$$\begin{aligned} f(x) &= \sum_{i=1}^{n} (\alpha_i^* - \alpha_i)(\Phi(x_i) \bullet \Phi(x)) + b \\ &= \sum_{i=1}^{n} (\alpha_i^* - \alpha_i) K(x_i, x) + b, \end{aligned} \quad (4)$$

where $K(x_i, x)$ is the kernel function. There are several types of kernel functions, including polynomial, radial basis, and sigmoid. Generally, a Gaussian radial basis function (see (5)) is widely used because of better prediction performance:

$$K(x_i, x) = \exp\left(-\gamma |x - x_i|^2\right). \quad (5)$$

2.2. Input Features and Parameter Optimization. Identifying input features is crucial step in SVM modeling. Metro ridership has significant characteristics of periodicity and nonlinearity. Abe [39] discovered that excessive features caused not only long training time but also poor generalization ability. Some researchers documented in detail the identification of input features. For example, Zhang et al. [30] identified the SVM input dimensions via SARIMA. Wu et al. [40] extracted input features from successive actual values before the prediction time; that is to say, if the value y_t of future time t is regarded as output, then the real values $y_{t-1}, y_{t-2}, \ldots, y_{t-m}$ of past time $t-1, t-2, \ldots, t-m$ serve as inputs. Cao et al. [41] used principal component analysis, kernel principal component analysis, and independent component analysis for inputs extraction. Huang and Wang [42] and Lin et al. [43] used genetic algorithm (GA) and particle swarm optimization (PSO) algorithm to extract input features, respectively.

Parameter optimization is to obtain better forecasting accuracy of the SVM model. The parameters optimized are mainly the penalty coefficient, the insensitive loss coefficient, and the corresponding parameters of kernel function. The LibSVM package [44] uses the grid-searching algorithm combined by cross-validation to determine these parameters but the process takes lengthy computation time. Hong et al. [45], Lin et al. [43], and Hong et al. [45] successfully used GA, PSO, and the ant colony optimization (ACO) algorithm to find the most optimal parameters, respectively. The advantages of PSO lie in easier application, fewer parameters to adjust, and faster convergence to optimum. As a result, PSO is used to optimize the parameters in this study. PSO simulates social behavior, like birds flocking to a promising position, to achieve precise objectives in a multidimensional space [46]. PSO gains the optimal solution through collaboration between individuals.

2.3. Support Vector Machine Online Model

2.3.1. Support Vector Machine Overall Online Model. Support vector machine overall online (SVMOOL) model is based on the theory of SVM, to extract input features, to train the batched updated training data, to use intelligent algorithms, to find the optimal parameters, and to get time-varying prediction function to realize the short-term forecasting.

Due to apparent periodicity feature of the rail transit ridership, SARIMA model is used to extract input features because SARIMA model is able to capture the periodicity of time series. A time series $\{X_t, t = 1, 2, \ldots, n\}$ is generated by the SARIMA(p,d,q)(P,D,Q)$_s$ process of Box and Jenkins as described by Williams et al. [8, 10] and Zhang et al. [30] described the process how to extract the features via SARIMA model in detail.

Considering the computation time of the training data and the real-time demand of the one-step prediction, the SVMOOL model is constructed by updating the training data day by day. That is to say, the training data is updated by adding the ridership data of the most recent day, and the time-varying prediction function is then constructed. Stating in simpler words, assume that y_{ij} denotes the ridership value at time j of day i, $i \in \{1, 2, \ldots, i, i+1, i+2, \ldots\}$, $j \in \{1, 2, \ldots, j, j+1, j+2, \ldots, n\}$, where n denotes the number of the data points each day. All of the prediction values of ridership after day i are forecasted by the training data of $i \times n$ ridership values. According to SVMOOL model described above, the prediction function is obtained by using the SARIMA model to extract input features from the training data and using PSO algorithm to optimize parameters, then forecasting value of every time interval j, $j = 1, 2, \ldots, n$, until the real values of day $i+1$ are totally obtained. After that, the training data is updated by adding the actual ridership values of day $i+1$. New prediction function is then constructed to forecast every value of day $i+2$, by retraining data and updating the parameters, and the process repeats. This process of constructing SVMOOL model is shown in Figure 1.

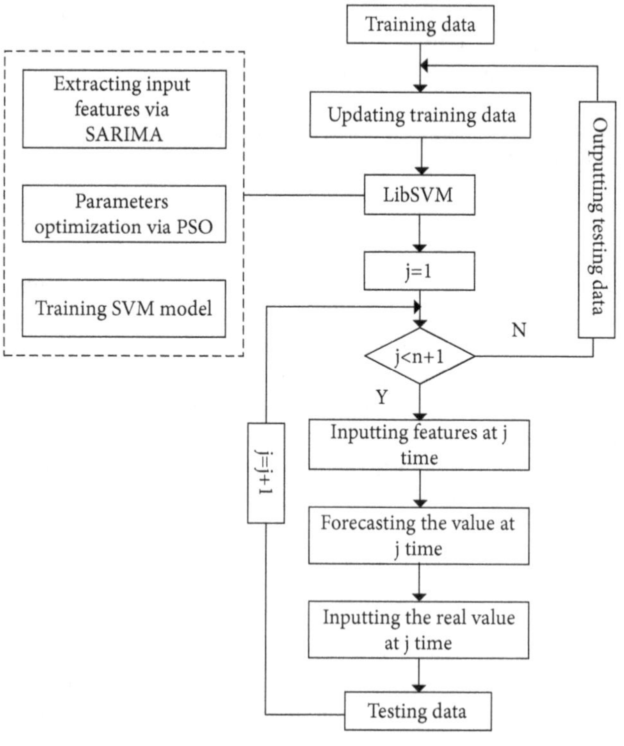

FIGURE 1: The process of constructing SVMOOL model.

2.3.2. Support Vector Machine Partial Online Model. Support vector machine partial online (SVMPOL) model is also based on the theory of SVM, to extract input features, to train the real-time updated testing data, to use intelligent algorithm, to find the optimal parameters, and to get real-time prediction function to realize the short-term forecasting.

According to the input feature extraction approaches mentioned previously and considering the temporal continuity of the real-time data, SVMPOL model extracts input features from successive actual values before the prediction time to capture nonlinear features of the ridership. In addition, parameters are also optimized by PSO.

The SVMPOL model makes full use of the temporal continuity of ridership data and takes advantage of SVM's capability of addressing small samples. The testing data is updated

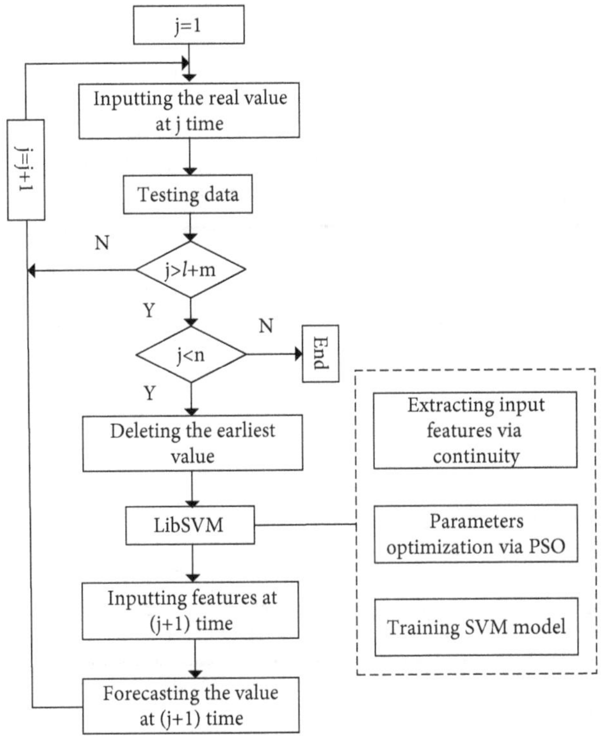

FIGURE 2: The process of constructing SVMPOL model.

by adding the ridership value of every time interval of the prediction day at same time deleting the earliest value. The real-time forecast function is obtained by training updated data and optimizing parameters in real-time to predict the value in the next time until the end of the prediction day. Stating in simpler words, assume that y_j denotes the ridership value at time j of the prediction day, $j \in \{1, 2, \ldots, j, j+1, j+2, \ldots, n\}$, where n denotes the number of the data points every day. The rest of the ridership after time j needs to be forecasted with the j passenger values. According to SVMPOL model as described above, the prediction function is obtained by extracting successive ridership values prior to the prediction time as the inputs and using PSO to optimize parameters, then the ridership corresponding output in time $j+1$ is achieved. After that, the testing data is updated by adding the actual value of time $j+1$ and deleting the earliest data. New prediction function is then constructed by retraining data and updating the parameters to forecast the ridership values in time $j+2$, and the process continues. This process of constructing SVMPOL model is shown in Figure 2, where l denotes the size of the moving window and m denotes the number of the input features via continuity.

2.3.3. Support Vector Machine Combined Online Model. As described previously, this paper proposes a SVMOOL model to address the periodicity of ridership and a SVMPOL model to address the nonlinearity of ridership. But the SVMOOL model updates the training data day by day and cannot capture the real-time local variations of ridership on the day being predicted. And considering the computation time of the testing data and the real-time demand of the one-step prediction, the testing data contains one-day data at most for constructing the SVMPOL model and the internal mechanism of metro ridership to study is insufficient. To avoid the drawbacks and to take advantages of the strengths of the two individual online models, the average predicted values of two models are the final results, which are called support vector combined online (SVMCOL) model.

3. Data Set and Evaluation Criteria

3.1. Data Set. At present, Automatic Fare Collection (AFC) System has been able to realize real-time data collection of metro passengers in and out station records [47] (though there is a slight delay in data transmission.). By simple statistics, the ridership data can be achieved for the required time interval. That is to say, the short-term ridership data of metro can be collected online, which puts forward higher requirements for short-time prediction. Operators expect faster and more accurate predictions, in order to plan ahead to accommodate the changes in passenger flow.

A ridership dataset of metro is collected to investigate the validity of the proposed SVMOOL, SVMPOL, and SVMCOL model for forecasting short-term ridership. The dataset is collected from the entrance transaction records of Nanjing Metro's Automatic Fare Collection (AFC) Systems. In general, short-term forecasting represents prediction for a specific time interval, such as 5 min, 10 min, and 15 min. For metro ridership, 5-min interval will be more useful for metro operation and management because the departure interval of the metro vehicle is really short. Taking the operation time of Nanjing Metro into consideration, the time period of data collection for each day is from 6:00 AM to 11:00 PM. There are 204 observations collected with a 5-min interval every day. The collected data is divided into two sets of training data plus testing data. In addition, it is obvious that ridership during workdays is different from that on weekends or holidays. As discussed by [48], some prediction models that work well for workdays data may yield unsatisfactory results for weekends or holidays data. In order to discuss the applicability of the proposed model, three samples were selected. The first sample contains no holidays, and the second and third samples contain holidays of Ching-Ming Festival and May Day. The specific sample information is as follows.

3.1.1. Sample 1. The dataset is collected from the entrance transaction records of the Sanshanjie station during the period from November 5 to December 2, 2012, so there are 5712 observations in total for these 28 days. The first training data set is data collected from November 5 to November 25, and the first testing data set contains the remaining seven days' ridership values, or 1428 observations, as shown in Figure 3. The weekend ridership pattern is different from weekday's obviously and the metro ridership shows the weekly periodic characteristics. This is because the weekday ridership is mainly composed of the commuted passenger flow, which is more stable. To the contrary, the weekend ridership mainly consists of the leisure and travel passenger flow, which is relatively fluctuant and has the obvious nonlinear characteristics.

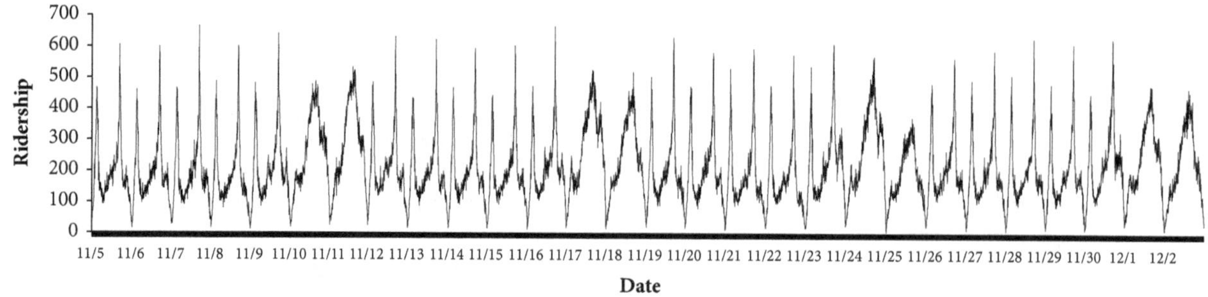

FIGURE 3: The origin entrance ridership time series at Sanshanjie Station of Nanjing Metro from November 5 to December 2, 2012.

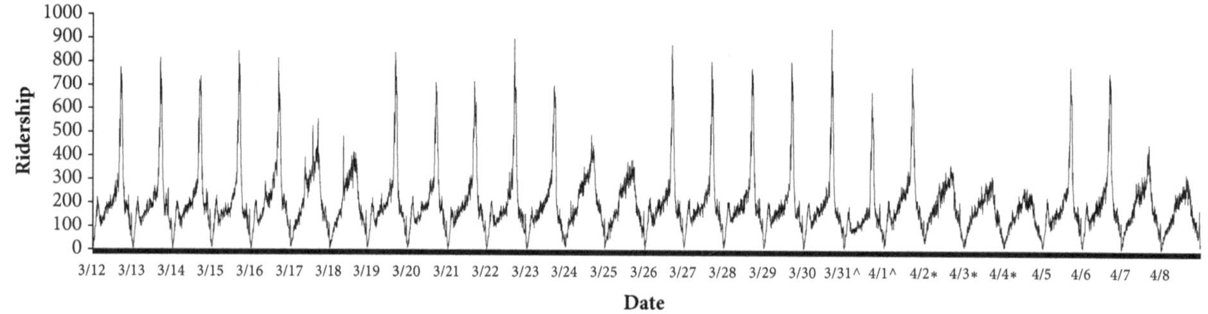

FIGURE 4: The origin entrance ridership time series at Zhujianglu Station of Nanjing Metro from March 12 to April 8, 2012. *Remark.* ∧ denotes that the weekend changes the weekday; ∗ denotes that of the holiday.

3.1.2. Samples 2 and 3. The dataset is collected from the entrance transaction records of the Zhujinglu station during the period from March 12 to May 6, 2012, so there are 11424 observations in total for these 56 days. The second training data set is data collected from March 12 to April 1, and the third training data set is data collected from April 9 to April 29. Both two training data sets contain three weeks' ridership values, or 4284 observations. Both two testing data sets contain the remaining seven days' ridership values, respectively, or 1428 observations, as shown in Figures 4 and 5. It must be noted that the Ching-Ming Festival is on April 2 (Monday) to April 4 (Wednesday); meanwhile March 30 (Saturday) and April 1 (Sunday) change weekday. The May Day is on April 29 (Sunday) to May 1 (Tuesday); meanwhile April 28 (Saturday) changes weekday. The holiday ridership pattern is similar to weekend's as a whole but is different from the local.

3.2. Data Normalization. Usually, normalizing raw input data can improve the convergence rate and performance of an SVM model. A common practice of data normalization was used to transform the raw data into a range $[-1, 1]$. In this study, each input data point is scaled according to

$$x_i^* = \frac{x_i - 0.5\left(x_{\max} + x_{\min}\right)}{0.5\left(x_{\max} - x_{\min}\right)}, \tag{6}$$

where x_i^* is the normalized value, x_i is any input vector of ridership data, $x_{\max}$ and $x_{\min}$ are, respectively, the maximum value and the minimum value of the training data in the period of training data.

3.3. Performance Indices. The mean absolute error (MAE), the mean absolute percent error (MAPE), and the root mean square error (RMSE) are commonly used criteria to evaluate the forecasting model. Generally, the smaller the MAE, MAPE and RMSE values, the better the prediction performance. The three performance criteria are, respectively, defined as

$$\text{MAE} = \frac{1}{N}\sum_{t=1}^{N}\left|y_t - \hat{y}_t\right|, \tag{7}$$

$$\text{MAPE} = \frac{1}{N}\sum_{t=1}^{N}\left|\frac{y_t - \hat{y}_t}{y_t}\right|, \tag{8}$$

$$\text{RMSE} = \sqrt{\frac{1}{N}\sum_{t=1}^{N}\left(y_t - \hat{y}_t\right)^2}, \tag{9}$$

where y_t is the actual observed value in time t and $\hat{y}_t$ is the forecasting value in time t, and N is the number of the observations every day.

4. Model Implementation

In this section, specific applications of the SVMOOL, the SVMPOL, and the SVMCOL models described previously are addressed.

In the methodology section, several methods of choosing the appropriate input features are introduced. The SVMOOL model's input features are extracted using the SARIMA

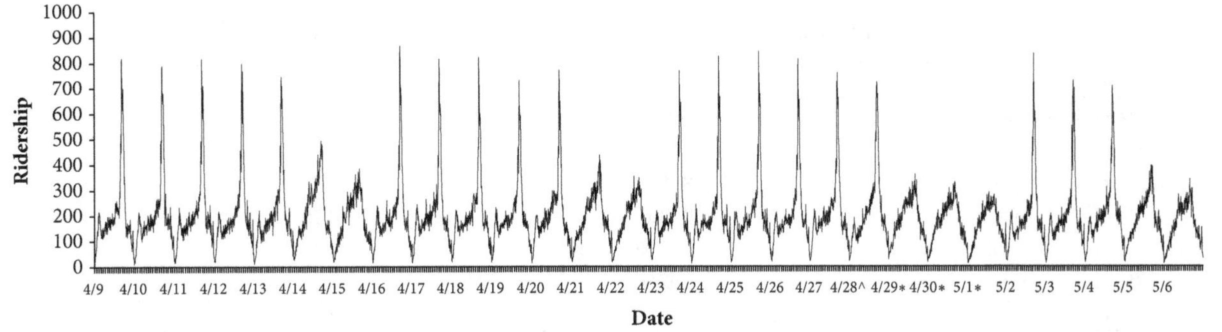

FIGURE 5: The origin entrance ridership time series at Zhujianglu Station of Nanjing Metro from April 9 to May 6, 2012.

model. The SARIMA model is formulated with statistical software SAS. The model forms generated from the three training data sets are all $\mathrm{SARIMA}(1,0,1)(0,1,1)_{1428}$. For example, the specific equation is shown as the following, which constructs by the second training data set at Zhujianglu station:

$$\begin{aligned} y_t = {} & y_{t-1428} + 0.93904_1 \left(y_{t-1} - y_{t-1429}\right) + \varepsilon_t \\ & - 0.52176\varepsilon_{t-1} - 0.49147\varepsilon_{t-1428} + 0.25643\varepsilon_{t-1429}, \end{aligned} \tag{10}$$

where y_t is the real value in time t, ε_t is error between the real value y_t, and the predicted value $\hat{y}_t$ in time t.

Therefore, for the prediction at time t, the real values for time $t-1429$, $t-1428$, $t-1$ serve as inputs. Afterwards, the ε-SVM model and the Gaussian radial basis function are implemented using the LibSVM software package developed by Wu et al. [40]. The Python codes were developed to integrate the LibSVM package with the PSO algorithm for parameters optimizing. The fivefold cross-validation technique and PSO are applied to obtain the optimal parameters (shown in Table 1) with the training data to construct the final ε-SVM model for future forecasting. The testing data are then used as input to the final ε-SVM model to produce predicted outputs.

The SVMOOL model updates training data set day by day. For example, using the second training data set from March 12 to April 1, predictions of ridership for every time interval on April 2 are made, then actual observed values of April 2 are added to the initial training data set to produce an updated training data set. Then the updated training data set from March 12 to April 2 is used to forecast the ridership of every interval on April 3 and the process repeats.

For the SVMPOL model, the testing data is updated by time interval (i.e., 5-min) for the day being predicted, and the number of input features extracted via continuity, or m value is determined to be 4 through several trails. In each of the 7 testing days, the first 10 data points (from 6:05 am to 6:50) were used as the testing data, with the 11th data point (at 6:55 am) being the target. Then 10-point window "walks", incorporating the 11th data point, which results on a new 10-point window (from 6:10 am to 6:55), having then the 12th data point (at 7:00 am) as the target. The process continues until the last observation (at 23:00 pm) becomes the target.

For the combined model, after the values from SVMOOL and SVMPOL models are calculated, the final prediction value is the average prediction of the previous two models.

5. Results Analysis

After the SVMOOL, SVMPOL, and SVMCOL models are implemented with the data sets, this research selects SARIMA, SVM, and BPNN models (i.e., back-propagation neural network) as the benchmark for one-step prediction are shown in Tables 2, 3, and 4.

5.1. Weekday Ridership Forecasting Results. In addition, the pattern of weekday's ridership is similar, so Table 1 shows the forecasting ridership results of three weekdays. As shown in Table 2, the SARIMA model is the best among them in terms of forecasting accuracy for weekday's ridership. It is demonstrated that the SARIMA model is good at predicting the ridership with periodic and stability characteristics as shown in Figure 6. The SVMOOL model is superior to the two models (BPNN and SVM models) because the updating data set, but the performance of the SVMCOL model, is not very satisfactory and is affected by the SVMPOL model, which is not applicable to the weekday ridership forecasting independently.

5.2. Weekend Ridership Forecasting Results. Table 3 shows that the SVMCOL model is the best among them in terms of forecasting accuracy, which the performance improves 40% compared with SARIMA model and improves 10% compared with the BPNN and SVM models for the value RMSE and MAE. It confirms that the combined model captures the weekly periodic and nonlinear characteristics of time series data for the estimation of short-term ridership (as shown in Figure 7). Though the SVMPOL model is not getting good results, it is much better than the SARIMA model. The SVM and BPNN models are both better than the SARIMA model for the weekend ridership forecasting, which also demonstrate that SVM and BPNN are suitable for nonlinear and fluctuant passenger flow.

5.3. Holiday Ridership Forecasting Results. As shown in Table 4 and Figure 8, it is not difficult to find that the SARIMA model is the worst performance for the holiday ridership and cannot meet the accuracy of short-time prediction. This is because the ridership data in samples 2 and 3 contains the unstable holiday passenger flow of the Ching-Ming Festival and May Day and demonstrates that the

TABLE 1: The optimal parameter sets of the second training data set at Zhujianglu station.

Training Data Set	C	γ	ε
Data From March 12 to April 1	597.69	0.42	0.57
Data From March 12 to April 2	870.02	0.77	0.76
Data From March 12 to April 3	996.26	0.32	0.57
Data From March 12 to April 4	1001.00	0.28	0.76
Data From March 12 to April 5	577.02	0.46	0.48
Data From March 12 to April 6	797.80	0.47	0.90
Data From March 12 to April 7	1001.00	0.07	1.00

TABLE 2: Weekday and weekend performance comparison for one-step prediction.

Model	RMSE (ridership/5-min)	MAE (ridership/5-min)	MAPE (%)
	November 27, Tuesday, 2012 at Sanshanjie Station		
SARIMA	*23.52*	*18.38*	*11.00*
BPNN	34.84	26.82	15.99
SVM	34.85	26.82	15.93
SVMOOL	34.75	26.73	15.89
SVMPOL	45.29	32.82	20.48
SVMCOL	36.50	26.31	16.52
	April 5, Thursday, 2012 at Zhujianglu Station		
SARIMA	*25.38*	*19.05*	*11.79*
BPNN	33.84	23.53	13.89
SVM	33.78	23.58	13.93
SVMOOL	33.68	23.41	13.81
SVMPOL	57.75	32.86	17.90
SVMCOL	41.31	25.76	14.58
	May 3, Thursday, 2012 at Zhujianglu Station		
SARIMA	*25.72*	*18.46*	*11.70*
BPNN	39.05	26.99	15.09
SVM	39.18	27.10	15.13
SVMOOL	38.88	26.80	14.62
SVMPOL	58.45	32.02	15.92
SVMCOL	40.06	27.41	15.26

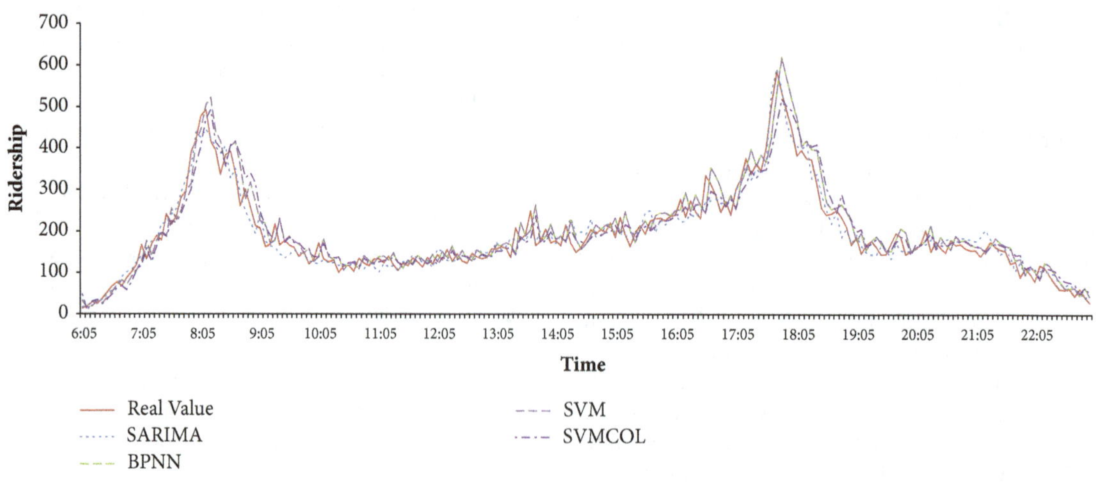

FIGURE 6: The comparison between the real value and the predicted values by 4 different models using origin ridership time series at Sanshanjie Station on November 27, 2012.

TABLE 3: weekend performance comparison for one-step prediction.

Model	RMSE (ridership/5-min)	MAE (ridership/5-min)	MAPE (%)
	September 1, Saturday, 2012 at Sanshanjie Station		
SARIMA	44.17	33.50	17.07
BPNN	30.99	24.35	12.60
SVM	30.81	24.20	12.53
SVMOOL	29.68	23.32	*12.03*
SVMPOL	32.33	26.20	15.20
SVMCOL	*27.11*	*22.52*	12.23
	April 7, Saturday, 2012 at Zhujianglu Station		
SARIMA	46.98	34.65	17.91
BPNN	30.67	23.41	16.84
SVM	30.66	23.43	16.85
SVMOOL	30.25	23.04	*16.51*
SVMPOL	32.04	25.60	21.75
SVMCOL	*28.51*	*21.61*	17.56
	May 5, Saturday, 2012 at Zhujianglu Station		
SARIMA	52.09	33.29	18.69
BPNN	30.02	22.78	15.48
SVM	31.48	23.29	15.63
SVMOOL	29.71	22.53	*15.32*
SVMPOL	31.88	23.83	18.87
SVMCOL	*27.89*	*21.25*	15.64

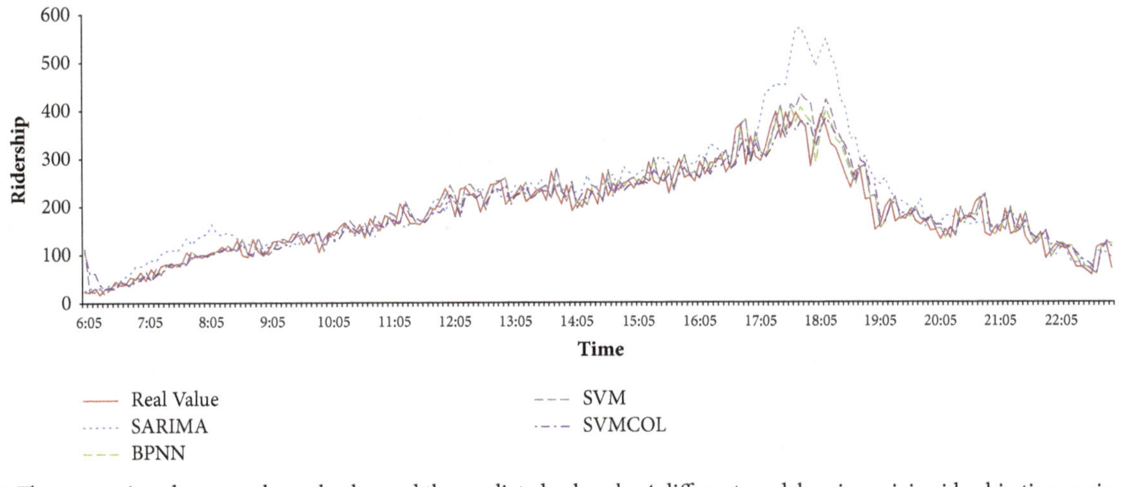

FIGURE 7: The comparison between the real value and the predicted values by 4 different models using origin ridership time series at Zhujianglu Station on May 5, 2012.

SARIMA model does not apply to nonstationary ridership. The SVMOOL and SVMCOL two models both outperform the three models (such as SARIMA, BPNN, and SVM models) and the SVMCOL model is the best among them in terms of forecasting accuracy, because the two models are constructed on the updated data set and more responsive to the change of passenger flow. The results of the SVMPOL model outperform the SARIMA and SVM models in the case of a small sample (with only 10 samples) as shown in Table 4. It is demonstrated that the SVMPOL model can capture the change of passenger flow in real-time and has special advantages for small sample prediction. In a word, compared with the offline models, the online models achieve better prediction performance. Of course, the prediction performance of BPNN is slightly better than the SVMPOL model, which is due to the small sample information. The results demonstrate the effectiveness of the proposed model. It is noted that the predicted performance of the SVMOOL

TABLE 4: Ching-Ming Festival and May Day performance comparison for one-step prediction.

Model	RMSE	MAE	MAPE (%)
	April 2, Monday, 2012 at Zhujianglu Station		
SARIMA	103.92	62.34	34.38
BPNN	29.01	22.98	14.39
SVM	30.48	23.98	14.68
SVMOOL	30.48	23.98	14.68
SVMPOL	31.37	25.06	14.89
SVMCOL	*27.87*	*22.01*	*14.09*
	April 3, Tuesday, 2012 at Zhujianglu Station		
SARIMA	102.98	54.73	32.40
BPNN	27.61	21.32	15.32
SVM	29.53	22.55	15.78
SVMOOL	27.42	21.21	15.26
SVMPOL	27.01	22.01	16.25
SVMCOL	*24.84*	*19.48*	*14.18*
	April 4, Saturday, 2012 at Zhujianglu Station		
SARIMA	121.96	69.05	51.69
BPNN	24.06	18.26	15.42
SVM	28.67	20.36	16.38
SVMOOL	23.88	18.08	*15.35*
SVMPOL	23.74	18.70	18.07
SVMCOL	*21.90*	*16.95*	15.64
	April 30, Monday, 2012 at Zhujianglu Station		
SARIMA	106.88	60.02	38.55
BPNN	28.68	22.17	15.42
SVM	32.25	24.51	16.22
SVMOOL	32.25	24.51	16.22
SVMPOL	28.46	22.58	*14.02*
SVMCOL	*26.92*	*21.22*	14.73
	May 1, Tuesday, 2012 at Zhujianglu Station		
SARIMA	117.53	65.01	53.22
BPNN	24.98	19.16	16.66
SVM	32.13	22.54	18.16
SVMOOL	24.28	18.80	16.34
SVMPOL	26.08	20.48	19.25
SVMCOL	*21.16*	*16.24*	*15.59*

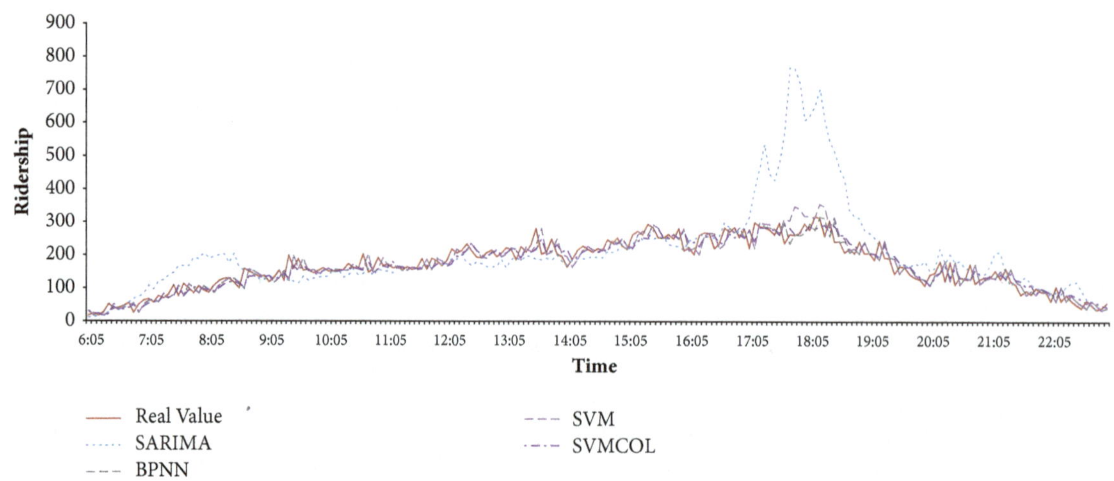

FIGURE 8: The comparison between the real value and the predicted values by 4 different models using origin ridership time series at Zhujianglu Station on April 3, 2012.

model on April 2 and 30, 2012, at Zhujianglu station is equal to the SVM model because the two models both own the same training sample.

6. Discussion

The training time and the forecasting time are the keys to real-time implementation. The experiments using LibSVM package on desktop computers indicate that the training time needs about one hour for three weeks' data (4284 observations) to construct the prediction function and the forecasting time needs less than 1 second for a one-step prediction using SVM. According to the SVMOOL model updating testing data set by day, the SVMOOL model uses the training data sample size from 21 days' observations to 22 days' or 27 days' observations, but the training time only increases 10 min and the forecasting time needs less than 1 s. Because the SVMOOL model is retrained once a day, the obtained forecasting function can be used for one-step predictions for the day, therefore real-time implementation is possible. The SVMPOL model is retained in real-time in 5-min interval. The obtained forecasting function can be used to one-step prediction for the next 5-min. In the process of the implementation experiments, the SVMPOL model needs less than 1 s to construct due to the small data sample and the forecasting time needs less than 1 s for one-step prediction. Therefore, the training time and the forecasting time can meet the real-time demand for the one-step prediction in the implementation as well.

7. Conclusions

The key to metro operation and management is based on the changes of the ridership to effectively deploy and use the system resources and to timely adjust operation strategy to ensure that metro is safe to complete the transportation service task. The results of short-term ridership forecasting can provide useful information to decision makers of metro system, and the prediction accuracy directly influence the legitimacy and effectiveness of any changes in operations, such as adjustments to headway, train dispatching, and the activation of station passenger crowd regulation plan or emergency response plan.

This paper proposes a novel hybrid model combining the SVMOOL model and the SVMPOL model for short-term ridership forecasting that better captures the periodicity and nonlinearity characteristics by the updated data set. The SVMCOL model takes advantages of the individual strengths of the two models. While the SARIMA model is superior for the stable weekday ridership to other models, experiments results indicate that the SVMOOL model is superior to SARIMA model, BPNN model, or SVM model in terms of MAE and RMSE for the weekend and holiday ridership test. The actual results of 5-min short-term ridership forecasting show the feasibility and effectiveness of the proposed combined model in real-time implementation.

It should be noted that the prediction of ridership under abnormal situations (such as holiday) is evidently more challenging than doing so under normal conditions (such as weekday ridership), and hence, much desired by the operator. Therefore, the proposed SVMCOL model is found to be suitable and useful in real-world operations, particularly in prediction under abnormal conditions. And, further studies need apply the proposed model to other abnormal situations (such as horrible weather, large sports events or emergencies, this study chooses the weekday, weekend, and holiday ridership as the demonstration). In addition, different characteristics (the impact of different meteorological conditions, the number of metro station entrances, etc.) can be considered as the input features in further studies. Jia et al. [49] indicate that, with the consideration of additional rainfall factor, the traffic flow prediction accuracy is improved.

Disclosure

An earlier version of this paper has been presented in the Transportation Research Board 92nd Annual Meeting (Washington DC, 2013).

Conflicts of Interest

The authors declare that they have no conflicts of interest.

Acknowledgments

This research has been supported by the Fundamental Research Funds for the Central Universities (no. KYLX16_0270). The authors thank the Nanjing Metro for providing the data used in this research. Thanks are due to Chih-Chung Chang and Chih-Jen Lin for permission to use the LibSVM package in this research.

References

[1] T.-H. Tsai, C.-K. Lee, and C.-H. Wei, "Neural network based temporal feature models for short-term railway passenger demand forecasting," *Expert Systems with Applications*, vol. 36, no. 2, pp. 3728–3736, 2009.

[2] H. Bar-Gera and D. Boyce, "Origin-based algorithms for combined travel forecasting models," *Transportation Research Part B: Methodological*, vol. 37, no. 5, pp. 405–422, 2003.

[3] G. Jovicic and C. O. Hansen, "A passenger travel demand model for Copenhagen," *Transportation Research Part A: Policy and Practice*, vol. 37, no. 4, pp. 333–349, 2003.

[4] M.-C. Chen and Y. Wei, "Exploring time variants for short-term passenger flow," *Journal of Transport Geography*, vol. 19, no. 4, pp. 488–498, 2011.

[5] L. Chang-Jen and S.-P. Miaou, "Real-time prediction of traffic flows using dynamic generalized linear models," *Transportation Research Record*, no. 1678, pp. 168–178, 1999.

[6] S. Sundaram, H. N. Koutsopoulos, M. Ben-Akiva, C. Antoniou, and R. Balakrishna, "Simulation-based dynamic traffic assignment for short-term planning applications," *Simulation Modelling Practice and Theory*, vol. 19, no. 1, pp. 450–462, 2011.

[7] B. M. Williams, P. K. Durvasula, and D. E. Brown, "Urban freeway traffic flow prediction: application of seasonal autoregressive integrated moving average and exponential smoothing models," *Transportation Research Record*, no. 1644, pp. 132–141, 1998.

[8] B. M. Williams, "Multivariate vehicular traffic flow prediction: evaluation of ARIMAX modeling," *Transportation Research Record*, no. 1776, pp. 194–200, 2001.

[9] B. M. Williams and L. A. Hoel, "Modeling and forecasting vehicular traffic flow as a seasonal ARIMA process: theoretical basis and empirical results," *Journal of Transportation Engineering*, vol. 129, no. 6, pp. 664–672, 2003.

[10] Z. R. Ye, Y. L. Zhang, and D. R. Middleton, "Unscented Kalman filter method for speed estimation using single loop detector data," *Transportation Research Record*, no. 1968, pp. 117–125, 2006.

[11] J. Xia, M. Chen, and Z. Qian, "Predicting freeway travel time under incident conditions," *Transportation Research Record*, no. 2178, pp. 58–66, 2010.

[12] Z. Ye and Y. Zhang, "Speed estimation from single loop data using an unscented particle filter," *Computer-Aided Civil and Infrastructure Engineering*, vol. 25, no. 7, pp. 494–503, 2010.

[13] B. L. Smith, B. M. Williams, and R. K. Oswald, "Comparison of parametric and nonparametric models for traffic flow forecasting," *Transportation Research Part C: Emerging Technologies*, vol. 10, no. 4, pp. 303–321, 2002.

[14] S. Clark, "Traffic prediction using multivariate nonparametric regression," *Journal of Transportation Engineering*, vol. 129, no. 2, pp. 161–168, 2003.

[15] T. Tsung-Hsien, L. Chi-Kang, and W. Chien-Hung, "Design of dynamic neural networks to forecast short-term railway passenger demand," *Journal of the Eastern Asia Society for Transportation Studies*, no. 6, pp. 1651–1666, 2005.

[16] Y. Mo and Y. Su, "Neural networks based real-time transit passenger volume prediction," in *Proceedings of the 2009 2nd Conference on Power Electronics and Intelligent Transportation System, PEITS 2009*, pp. 303–306, December 2009.

[17] Y. Huang and H. Pan, "Short-term prediction of railway passenger flow based on RBF neural network," in *Proceedings of the 4th International Joint Conference on Computational Sciences and Optimization, CSO 2011*, pp. 594–597, April 2011.

[18] Y. Zhang and Y. Xie, "Forecasting of short-term freeway volume with v-support vector machines," *Transportation Research Record*, vol. 2024, no. 1, pp. 92–99, 2007.

[19] Y. Zhang and Y. Xie, "Travel mode choice modeling with support vector machines," *Transportation Research Record*, no. 2076, pp. 141–150, 2008.

[20] Q. Chen, W. Li, and J. Zhao, "The use of LS-SVM for short-term passenger flow prediction," *Transport*, vol. 26, no. 1, pp. 5–10, 2011.

[21] W.-H. Lin, "A Gaussian maximum likelihood formulation for short-term forecasting of traffic flow," in *Proceedings of the 2001 IEEE Intelligent Transportation Systems*, pp. 150–155, Oakland, Calif, USA, 2001.

[22] Y. Xie, Y. Zhang, and Z. Ye, "Short-term traffic volume forecasting using Kalman filter with discrete wavelet decomposition," *Computer-Aided Civil and Infrastructure Engineering*, vol. 22, no. 5, pp. 326–334, 2007.

[23] D. Boto-Giralda, F. J. Díaz-Pernas, D. González-Ortega et al., "Wavelet-based denoising for traffic volume time series forecasting with self-organizing neural networks," *Computer-Aided Civil and Infrastructure Engineering*, vol. 25, no. 7, pp. 530–545, 2010.

[24] C. P. I. van Hinsbergen, J. W. C. van Lint, and H. J. van Zuylen, "Bayesian committee of neural networks to predict travel times with confidence intervals," *Transportation Research Part C: Emerging Technologies*, vol. 17, no. 5, pp. 498–509, 2009.

[25] X. Fei, C. C. Lu, and K. Liu, "A bayesian dynamic linear model approach for real-time short-term freeway travel time prediction," *Transportation Research Part C: Emerging Technologies*, vol. 19, no. 6, pp. 1306–1318, 2011.

[26] Y. Wei and M. Chen, "Forecasting the short-term metro passenger flow with empirical mode decomposition and neural networks," *Transportation Research Part C: Emerging Technologies*, vol. 21, no. 1, pp. 148–162, 2012.

[27] H. Xiao, H. Sun, and B. Ran, "Special factor adjustment model using fuzzy-neural network in traffic prediction," *Transportation Research Record*, no. 1879, pp. 17–23, 2004.

[28] Y. Zhang and Z. Ye, "Short-term traffic flow forecasting using fuzzy logic system methods," *Journal of Intelligent Transportation Systems: Technology, Planning, and Operations*, vol. 12, no. 3, pp. 102–112, 2008.

[29] Y. Sun, B. Leng, and W. Guan, "A novel wavelet-SVM short-time passenger flow prediction in Beijing subway system," *Neurocomputing*, vol. 166, pp. 109–121, 2015.

[30] N. Zhang, Y. Zhang, and H. Lu, "Seasonal autoregressive integrated moving average and support vector machine models: prediction of short-term traffic flow on freeways," *Transportation Research Record*, vol. 2215, pp. 85–92, 2011.

[31] W.-C. Hong, Y. Dong, F. Zheng, and S. Y. Wei, "Hybrid evolutionary algorithms in a SVR traffic flow forecasting model," *Applied Mathematics and Computation*, vol. 217, no. 15, pp. 6733–6747, 2011.

[32] W.-C. Hong, "Traffic flow forecasting by seasonal SVR with chaotic simulated annealing algorithm," *Neurocomputing*, vol. 74, no. 12-13, pp. 2096–2107, 2011.

[33] W.-C. Hong, "Application of seasonal SVR with chaotic immune algorithm in traffic flow forecasting," *Neural Computing and Applications*, vol. 21, no. 3, pp. 583–593, 2012.

[34] J. Wang and Q. Shi, "Short-term traffic speed forecasting hybrid model based on chaos–wavelet analysis-support vector machine theory," *Transportation Research Part C: Emerging Technologies*, vol. 27, no. 1, pp. 219–232, 2013.

[35] R. Chen, C.-Y. Liang, W.-C. Hong, and D.-X. Gu, "Forecasting holiday daily tourist flow based on seasonal support vector regression with adaptive genetic algorithm," *Applied Soft Computing*, vol. 26, pp. 435–443, 2015.

[36] S. Ishak and H. Al-Deek, "Performance evaluation of short-term time-series traffic prediction model," *Journal of Transportation Engineering*, vol. 128, no. 6, pp. 490–498, 2002.

[37] E. I. Vlahogianni, M. G. Karlaftis, and J. C. Golias, "Optimized and meta-optimized neural networks for short-term traffic flow prediction: a genetic approach," *Transportation Research Part C: Emerging Technologies*, vol. 13, no. 3, pp. 211–234, 2005.

[38] V. N. Vapnik, *The Nature of Statistical Learning Theory*, Springer, New York, NY, USA, 1995.

[39] S. Abe, "Analysis of support vector machine," in *IEEE Signal Processing Society Workshop*, pp. 88-89, IEEE Press, New York, NY, USA, 2002.

[40] C.-H. Wu, J.-M. Ho, and D. T. Lee, "Travel-time prediction with support vector regression," *IEEE Transactions on Intelligent Transportation Systems*, vol. 5, no. 4, pp. 276–281, 2004.

[41] L. J. Cao, K. S. Chua, W. K. Chong, H. P. Lee, and Q. M. Gu, "A comparison of PCA, KPCA and ICA for dimensionality reduction in support vector machine," *Neurocomputing*, vol. 55, no. 1-2, pp. 321–336, 2003.

[42] C. Huang and C. Wang, "A GA-based feature selection and parameters optimizationfor support vector machines," *Expert Systems with Applications*, vol. 31, no. 2, pp. 231–240, 2006.

[43] S.-W. Lin, K.-C. Ying, S.-C. Chen, and Z.-J. Lee, "Particle swarm optimization for parameter determination and feature selection of support vector machines," *Expert Systems with Applications*, vol. 35, no. 4, pp. 1817–1824, 2008.

[44] C. C. Chang and C. J. Lin, "LIBSVM: A Library for Support Vector Machines," Department of Computer Science and Information Engineering, National Taiwan University, Taipei, Taiwan, 2001.

[45] W.-C. Hong, Y. Dong, F. Zheng, and C.-Y. Lai, "Forecasting urban traffic flow by SVR with continuous ACO," *Applied Mathematical Modelling: Simulation and Computation for Engineering and Environmental Systems*, vol. 35, no. 3, pp. 1282–1291, 2011.

[46] J. Kennedy and R. Eberhart, "Particle swarm optimization," in *Proceedings of the IEEE International Conference on Neural Networks (ICNN' 95)*, vol. 4, pp. 1942–1948, Perth, Western Australia, November-December 1995.

[47] W. Zhu, W. Wang, and Z. Huang, "Estimating train choices of rail transit passengers with real timetable and automatic fare collection data," *Journal of Advanced Transportation*, vol. 2017, Article ID 5824051, 12 pages, 2017.

[48] C. Chen, Y. Wang, L. Li, J. Hu, and Z. Zhang, "The retrieval of intra-day trend and its influence on traffic prediction," *Transportation Research Part C: Emerging Technologies*, vol. 22, pp. 103–118, 2012.

[49] Y. Jia, J. Wu, and M. Xu, "Traffic flow prediction with rainfall impact using a deep learning method," *Journal of Advanced Transportation*, vol. 2017, Article ID 6575947, 10 pages, 2017.

11

Does Urban Rail Transit Discourage People from Owning and Using Cars? Evidence from Beijing, China

Shasha Liu,[1] **Enjian Yao**,[1] **and Toshiyuki Yamamoto**[2]

[1] *MOE Key Laboratory for Urban Transportation Complex Systems Theory and Technology, Beijing Jiaotong University, Beijing, China*
[2] *Institute of Materials and Systems for Sustainability, Nagoya University, Nagoya, Japan*

Correspondence should be addressed to Enjian Yao; enjyao@bjtu.edu.cn

Academic Editor: Gonçalo Homem de Almeida Correia

With the rapid urbanization and motorization, many cities are developing urban rail transit (URT) to reduce car dependence. This paper explores the URT effect on car ownership and use based on the home-based work tour data in Beijing, China. Considering the mediating effects of car ownership and travel distance simultaneously, we develop a structural equation model to examine the complex relationship among URT, car ownership, travel distance, and car use. The results indicate that URT plays an important role in reducing car dependence. Living within URT catchment areas by itself is not significantly associated with car ownership and use, but if the workplace is near a URT station, people are less likely to own and use cars. People who both live and work near URT station areas have lower probability of owning and using cars. Moreover, car ownership and travel distance mediate the relationship between URT and car use, and the mediating effect of car ownership is greater than travel distance. Our study verifies that URT does discourage people from owning and using cars, which may have important implications for developing cities to make response to the ongoing motorization.

1. Introduction

Increasing car travel demand has resulted in numerous problems, such as traffic congestion, environmental pollution, road accidents, and global warming [1, 2]. In an attempt to get people out of their cars and address the above problems, great importance has been attached to urban rail transit (URT) [3, 4]. Many cities have invested heavily on URT systems, especially in fast-growing developing countries [5]. In China, more than 30 cities have URT lines in operation, and some cities are planning or constructing URT systems by the end of 2017. In India, metro is operational in 10 cities, under construction in 5 cities, and planned in 17 cities [6]. For these cities that are experiencing fast-paced urbanization, if empirical studies verify the importance of URT in reducing car dependence, this would provide great potential to intervene the development of urbanization [7] and make effective response to the continuing motorization.

Although many studies have examined the influence of URT on car ownership and use, the significance of URT effect varies a lot in different empirical studies. Some studies verify the negative impact of URT proximity on car ownership and use after controlling for other factors [8–12], but other studies show few effects [13, 14]. Moreover, limited efforts have been made to explore whether living within URT catchment areas influences car ownership and use differently compared with working near a URT station. Besides, most existing studies explore the relationship between URT and different dimensions of travel behavior (e.g., car ownership, mode choice and travel distance) separately [15]. Some consider car ownership and travel distance as the endogenous variables explained by sociodemographics and built environment [1, 16], while other studies treat them as the exogenous variables to explain travel mode choice behavior [4, 17]. To our knowledge, few studies consider the intermediary nature of car ownership and travel distance simultaneously when modeling the URT effect on mode choice behavior.

This paper attempts to address the above limitations. Considering the mediating effects of car ownership and travel distance simultaneously, we utilize structural equation modeling (SEM) to explore the complex relationship among URT, car ownership, travel distance, and car use based on home-based work tour data in Beijing, China. We endeavor to answer the following questions. First, does URT play a role in reducing car dependence? Does living near a URT station affect car ownership and use differently than working near a URT station? Second, do car ownership and travel distance medicate the relationship between built environment and travel mode choice? How does URT influence car use through the medicating variables car ownership and travel distance?

The remaining part of this paper is structured as follows. The next section reviews previous work on the URT effect on travel behavior. The third section describes the data, the variables, and the modeling approach. The subsequent section focuses on the empirical results. The fifth section discusses the important findings. The conclusion is then given in the final section.

2. Literature Review

Many studies have explored the relationship between built environment and travel behavior [19–23], in an attempt to guide urban planning practice to reduce automobile dependence and related environment and social costs [24]. Built environment is usually measured as 6Ds [25], including density, diversity, design, destination accessibility (e.g., accessibility to jobs, distance to destination), distance to transit, and demand management. These studies shed light on the relationship between transit and travel behavior [13, 16, 26, 27]. With the rapid development of URT, the influence of URT on travel behavior started to attract wide attention [10–12, 14].

Although some studies have examined the URT effect on travel behavior, they produce mixed results. Some verify that URT plays an important role in reducing car ownership and use after controlling for other factors [10–12, 17]. Zhang et al. [10] and Huang et al. [11] found that households are less likely to own cars if the residential neighborhoods have URT service. Shen et al. [4] proved that people who live or work near a URT station prefer URT rather than car in Shanghai, China. Zegras [16] found that living near URT is associated with low car ownership and short VKT (vehicle kilometers travelled) in Santiago de Chile. However, other studies find few effect of URT on car ownership and use. Cao and Cao [14] found that LRT (light rail transit) does not have an independent effect on car ownership after controlling for neighborhood design and self-selection in the Minneapolis-St. Paul metropolitan area, USA. Shen et al. [4] detected no significant effect of residential URT proximity and workplace URT proximity on car ownership in Shanghai, China. Li and Zhao [1] concluded that residential distance to URT station is not significantly associated with work VKT for private cars. To our knowledge, there is no consensus on the URT effects on car ownership and use.

Apart from the mixed outcomes, previous studies have several limitations. First, most of these studies focus on how residential URT proximity influences travel behavior. Limited efforts have been made to examine differences in travel behavior between people who live near URT station areas versus people who work within URT catchment areas. Shiftan and Barlach [28] found that built environment at the workplace has significant effects on the commute mode choice. Further, some studies concluded that built environment at the workplace plays a more important role than that at the residence in the travel decisions [29–31]. Therefore, it is necessary to take workplace URT proximity as well as residential URT proximity into account when evaluating the URT effect on travel behavior.

Second, many studies use trip-based models [10–12, 14], while people typically decide which mode to use for the entire tour (including the outbound and return trips plus any necessary intermediate stops) before leaving home [32]. So tour-based models match the travel decision-making process more closely. Tour complexity is usually studied by two different approaches: one is a categorical classification and the other focuses on the number of activities or trip legs within a tour [33]. It is found that more stops within a tour are associated with higher probability of using car rather than public transit [30, 33, 34].

Third, most of these studies examine the relationship between URT and different dimensions of travel behavior (e.g., car ownership, mode choice, and travel distance) separately. Some studies treat car ownership and travel distance as the endogenous variables explained by sociodemographics and built environment [1, 16], while other studies consider them as the exogenous variables to explain mode choice behavior [4, 17]. However, Van Acker and Witlox [35] concluded that car ownership mediates the relationship between built environment and travel mode choice. Ignoring the mediating effects of car ownership may result in a misspecification of the built environment effects on car use. Van Acker and Witlox [34] and Ding et al. [15] found that travel distance also mediates the connection between built environment and mode choice decision. Nevertheless, there are very limited studies that consider the intermediary nature of car ownership and travel distance simultaneously [15, 34].

This study attempts to address these limitations and shed light on how residential and workplace URT proximity affect car ownership and use. Considering the intermediary nature of car ownership and travel distance simultaneously, we apply the SEM to explore the complex relationship among URT, car ownership, travel distance, and car use based on home-based work tour data in Beijing, China.

3. Methodology

3.1. Study Area. To explore the URT effects on car ownership and use, we undertake an empirical study of Beijing, the capital of China. It is about 16410 km^2 and is divided into 16 administrative districts, 6 of which are in the urban area, as presented in Figure 1. Beijing has experienced fast-paced urbanization in the past decades. The population grew from 10.860 million to 21.516 million, the GDP per capital increased from 4,611 RMB to 99,139 RMB [36], and the built area increased from 397 km^2 to 1386 km^2 during the decade of 1990-2014 [37]. The fast-paced urban expansion and rising income level result in the increasing car dependence. At the end of 2014, Beijing had 5.591 million vehicles, of which 78.20% were privately owned. And 45% households had cars [36]. Car has become an important travel mode in individual

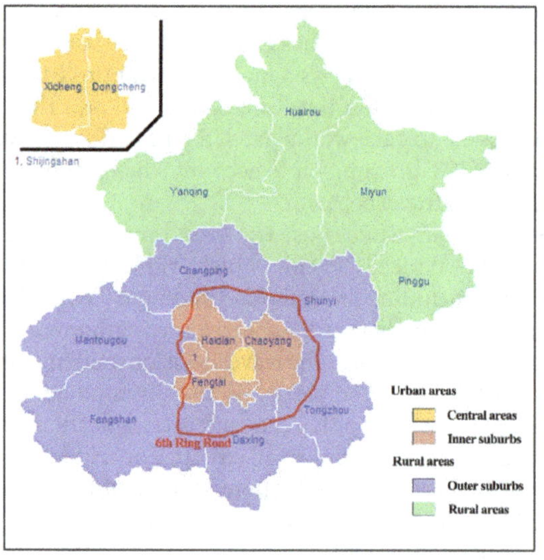

FIGURE 1: The study area in Beijing, China.

daily life. According to the fifth household travel survey conducted by Beijing Municipal Commission of Transport in 2014, car trips accounted for 31.96% of all trips in Beijing.

In order to reduce car dependence and address the problems such as traffic congestion and air pollution, great importance has been attached to urban rail transit (URT). By October 2014, Beijing URT system had shaped an extensive network with 17 lines, 231 stations, and 465.4 km in length, since the first URT line was opened in 1981. Despite the rapid development of URT, Beijing still faced heavy traffic congestion and serious air pollution, for which the increasing car dependence is responsible.

Considering most of the URT stations were located within 6th ring road in 2014, we focus on the area within 6th ring road (about 2267 km^2) in Beijing, as shown in Figures 1 and 2.

3.2. Data

3.2.1. Tour Data. A tour is defined as a sequence of trips that starts and ends at home [30, 34]. Tour data used in the study comes from Beijing Household Travel Survey (BHTS). This survey was conducted between July and October in 2014 by Beijing Municipal Commission of Transport. In order to develop a representative sample, 16 districts, 999 communities, and 40,003 households with 101,815 respondents were included in the survey. The surveyed households accounted for 0.60% of total households in Beijing in 2014. All respondents were interviewed face-to-face at their home and were asked about household sociodemographic information and their travel diary which includes trip purpose, origin, destination, travel mode, duration, frequency, etc.

In order to tackle the residential self-selection problem [10, 13, 38], we consider the characteristics of different housing types in China's unique diversified housing system and focus on a subsample with exogenous residential locations. *Danwei* housing is provided by *danwei* ("*Danwei* is a generic term denoting the Chinese socialist workplace and the specific range of practice that is embodies" [39]) in which households have virtually no freedom to choose residential locations [40]. Thus households living in *danwei* housings which accounted for 16.06% of all surveyed households were selected. Only fully responding households within 6th ring road were chosen for analysis, reducing the dataset to 5676 households.

Based on the individual travel diary of the selected households, 29139 unlinked person trips were included and chained into home-based tours. A total of 12721 tours were created. A single tour may involve more than one travel mode. A tour's main mode was assigned using the following priority order: URT, bus, car, cycling, walking, and others. This priority order is for two reasons: one is that higher priority modes have higher probability of taking up the longest part of the tour; the other is that the lower priority modes can be treated as feeder modes [33]. Then, we selected the sample using the following criteria: (1) there is at least one work activity within the tour; (2) tours which travel distance less than 5 km are excluded, since in short-distance tours, people prefer walking and cycling rather than public transit and car. Using these criteria, 3077 home-based work tours were selected for our study. The residential locations and workplaces are shown in Figure 2.

3.2.2. Built Environment Data. With the development of location based technology, point of interests (POIs) of the electronic map have almost covered every building in the urban area. In this study, we try to build a land use dataset based on the POI information. First, all POIs within 6th ring road of Beijing were crawled from Baidu.com. After cleaning the data, there are a total of 198,608 POIs, which are classified into 17 categories and 109 subcategories according to their land use characteristics. Then, based on POI names and addresses, gross floor area data of POIs were crawled from fang.com and baidu.com. As for the POIs that were not supplemented with floor area data in the above step, the mean floor areas of POIs in the same subcategory would be taken as their floor areas. After completing POI information, neighborhood land use characteristics can be estimated based on POI types and floor areas.

We also collected street network and bus network data from Beijing Municipal Commission of Transport and beijingcitylab.com, respectively. Coupled with URT network and land use data coded by authors, a built environment database was established in the ArcGIS 10.3.

Then, in order to append each individual with built environment data, we first matched residential and workplace addresses with coordinates by which each individual was geocoded into the ArcGIS platform. Then, we defined 1 km buffer zone as a neighborhood at a given address and joined built environment data onto buffer zones linked to each individual. The 1 km buffer radius is based on about a ten-minute walking distance, approximating what can be accessed on the road network [32].

3.3. Variables. Variables used in this study include residential and workplace built environment characteristics, household,

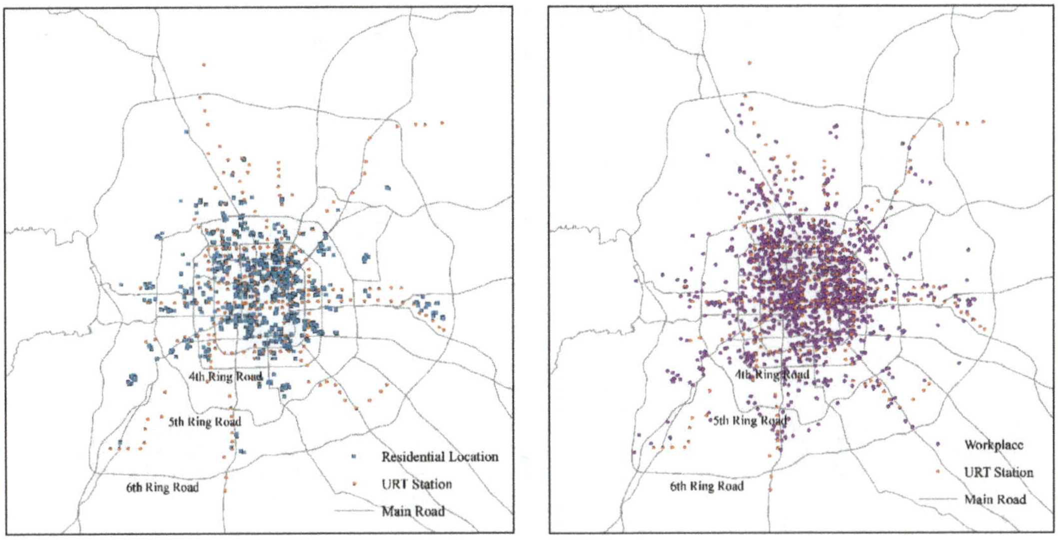

(a) Residential locations (b) Workplaces

FIGURE 2: Residential locations and workplaces in this study.

and individual characteristics and aspects of travel behavior. Table 1 presents the detail information on the variable names and definitions as well as the descriptive statistics. In order to explore the URT effect on car ownership and use, we used three dummy variables to evaluate residential and workplace proximity to URT. Since the service level of URT (e.g., transportation capacity, scheduling.) is highly related to the travel demand and the travel demand is associated with the surrounding land use, the effect of URT service level might be successfully captured by these explanatory variables (such as land use density and diversity) in our study. Therefore, we did not consider the URT service level further, which is in line with previous studies [4, 10, 41, 42].

As shown in Table 1, for the sampled households, the annual income is mainly between 50 and 150 thousand CNY, and 59.1% of them have one or more cars. There is an average of less than one child per household but more than one licensed drivers. For the sampled individuals, the average age is 39.448. 55.8% of them are male, 58.8% have a driving license, and 29.6% choose to travel by cars. Besides, 40.2 % of the sampled individuals can access the URT service both in their residential and workplace neighborhoods, and 43.9% either live or work within URT catchment areas. As for the tours, the average travel distance is 21.088 km, and an average of 1.169 stops is included in a tour.

3.4. Model Specification. We assume that car use behavior is directly affected by sociodemographics, built environment characteristics, car ownership, and travel characteristics (e.g., travel distance). Meanwhile, car ownership and travel distance themselves are also influenced by sociodemographics and built environment, which results in the indirect effects of built environment on car use through the mediating variables car ownership and travel distance, as shown in Figure 3. Car ownership and travel distance are the endogenous variables

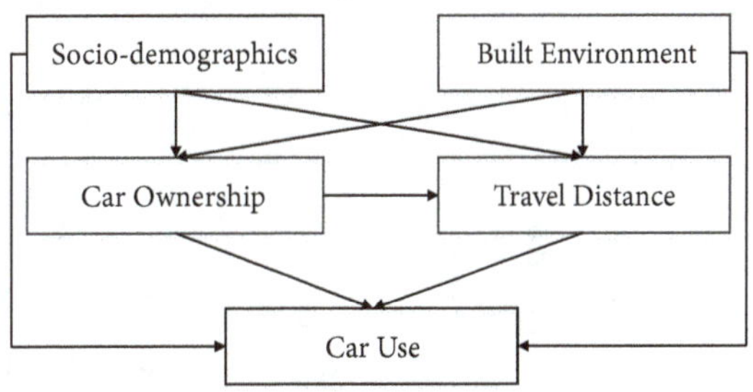

FIGURE 3: Potential relationship between built environment and travel behavior.

in some relationships and also the exogenous variables of car use behavior.

Considering the mediating effects of car ownership and travel distance simultaneously, structural equation modeling (SEM), which has been frequently used to explore the complex relationship between built environment and travel behavior [15, 34, 35], was applied in this study. Compared to discrete choice models and multivariate regression models, SEM is more capable of handling the complex relationship among endogenous and exogenous variables and identifying the direct, indirect, and total effects [35]. An SEM for observed endogenous and exogenous variables can be defined as follows [35]:

$$\eta = B\eta + \Gamma\xi + \zeta \tag{1}$$

with $\eta = L \times 1$ matrix of endogenous variables, $\xi = K \times 1$ matrix of exogenous variables, $B = L \times L$ matrix of coefficients of the endogenous variables, $\Gamma = L \times K$ matrix of coefficients of the exogenous variables, and $\zeta = L \times 1$ matrix of residuals of endogenous variables. In this study, the endogenous and exogenous variables are all observed.

Table 1: Definitions and descriptive statistics of variables used in the study.

Variable	Definition	Mean	Std.	Min	Max
URT proximity					
RW proximity to URT	Residential and workplace neighborhoods (1 km buffer zones) both have a URT station, 1 for yes and 0 otherwise	0.402	0.490	0	1
R proximity to URT	Only residential neighborhood has a URT station, 1 for yes and 0 otherwise.	0.236	0.425	0	1
W proximity to URT	Only workplace neighborhood has a URT station, 1 for yes and 0 otherwise.	0.203	0.402	0	1
Residential built environment					
Floor area ratio	The ratio of total floor area to total land area	1.139	0.497	0.066	2.466
Land use diversity	It is captured by a land use mix diversity index (Bhat and Gossen, 2004)[1]	0.193	0.081	0.018	0.484
Bus stop density	Number of bus stops within the residential neighborhood, in number per km	13.821	4.948	0	44
Density of crossings	Number of crossing facilities versus total street length, in number per km	3.166	0.891	0.287	6.494
Distance to business district	Distance from residence to the nearest central or sub-central business districts[2], in km	5.249	3.913	0.152	27.105
Workplace built environment					
Floor area ratio	The ratio of total floor area to total land area	1.130	0.495	0.025	2.503
Land use diversity	It is captured by a land use mix diversity index (Bhat and Gossen, 2004)[1]	0.241	0.094	0.004	0.633
Bus stop density	Number of bus stops within the workplace neighborhood, in number per km	8.211	3.662	0	36
Density of crossings	Number of crossing facilities versus total street length, in number per km	1.924	0.597	0.183	4.261
Within 4th ring road	1 for the workplace within 4th ring road, 0 otherwise	0.702	0.458	0	1
Commuting characteristics					
Travel distance	Total travel distance in a tour, in km	21.088	13.623	5.013	118.04
Joint household travel	1 for joint travel with other family members, 0 otherwise.	0.042	0.200	0	1
Number of stops	Number of stops in a tour	1.169	0.556	1	5
Household characteristics					
Annual Income	1:<50; 2:[50,100); 3:[100,150); 4:[150,200); 5:[200,250); 6:[250,300); 7:[300,500); 8:≥ 500 thousand CNY	2.579	1.144	1	8
Housing Area	The area of the family housing, in m^2	61.023	24.936	8	180
Children	Number of children (≤16)	0.275	0.459	0	2
Licensed Drivers	Number of licensed drivers	1.214	0.924	0	4
Bikes	Number of bicycles, e-bikes and motorcycles	1.444	1.134	0	8
Car ownership	1 for having one or more cars, 0 otherwise	0.591	0.492	0	1
Personal characteristics					
Gender	1 for male, 0 for female	0.558	0.497	0	1
Age	Individual age, in years	39.448	10.603	18	74
Education level	1 for having a college degree, 0 otherwise	0.734	0.442	0	1

TABLE 1: Continued.

Variable	Definition	Mean	Std.	Min	Max
Driving license	1 for having a driving license, 0 otherwise	0.588	0.492	0	1
Car use	1 for travelling by cars, 0 otherwise	0.296	0.457	0	1

Notes: 1. Following Bhat and Gossen [18], land use diversity is captured by a land use mix diversity index *DI*, which varies between 0 and 1, as follows. $DI = 1 - (|res/L - 1/5| + |com/L - 1/5| + |off/L - 1/5| + |pub/L - 1/5| + |rec/L - 1/5|)/(8/5)$, where $L = res + com + off + pub + rec$; *res, com, off, pub, and rec* are the floor area in residential use, commercial use, official use, public use and recreational use respectively, m^2.

2. The central business district CBD and three sub-central business districts including Zhongguancun, Jinrongjie, Yayuncun were considered in this study.

TABLE 2: Model fit indices for the car ownership and use model.

Model fit indices	Cut-off value	Model-based value
χ^2 (df), p	p > 0.05	68.853(18), p ≤ 0.001
RMSEA (root mean square error of approximation)	<0.05	0.030
WRMR (weighted root mean square residual)	<1.00	0.953
CFI (comparative fit index)	>0.90	0.986
TLI (Tucker-Lewis index)	>0.90	0.941

As for the SEM estimation, the maximum likelihood (ML) method is generally used. But a basic assumption of the ML estimator is the multivariate normal distribution of all continuous endogenous variables [43], which is not always met in reality. Besides, the variables car ownership and car use are both categorical. So the weighted least squares mean and variance adjusted (WLSMV) estimator, which accounts for nonnormally distributed data, was applied in this study. Then, the estimation of the SEM model was conducted using the software M-plus.

4. Results

The SEM model was estimated using WLSMV method. The links from household and personal characteristics with no statistical significance (p > 0.1) were removed, as well as the links from workplace built environment to car ownership, since the workplace built environment variables have no significantly direct effects on car ownership and no significantly indirect effects on travel distance and car use through car ownership. Then, the model was re-estimated. Table 2 presents the model fit indices and their corresponding cutoff values. The $\chi^2 - statistic$, which compares the observed and predicted variance-covariance matrices, is highly dependent on sample size. With large sample size, it is difficult to accept the null hypothesis. In other words, this index might be meaningless to large samples. So we can ignore this index in this study for the large sample size. Other indices suggest that the model fits the data well.

The standardized direct, indirect and total effects on car ownership, travel distance and car use are shown in Table 3. In this paper, we not only focus on whether URT plays a role in car ownership and use decision, but attempt to explore whether living near a URT station influences car ownership and use differently compared with working near a URT station as well. Meanwhile, we want to know how URT influences car use considering the mediating effects of car ownership and travel distance simultaneously.

4.1. Effects on Car Ownership. After controlling sociodemographics and other built environment variables, URT still has a significant impact on car ownership, which echoes other recent studies [10, 11, 16]. Living near URT by itself is not associated with car ownership, but if the workplace is near a URT station, people are less likely to own cars. People who live and work both within URT catchment areas have lower probability of owning cars.

With respect to other residential built environment variables, lower density of crossings and longer distance to business district are associated with higher probability of owning cars. Floor area ratio and land use diversity, however, show no significant effect on car ownership.

As for sociodemographic characteristics, they play a major role in determining car ownership level, which is consistent with previous studies [7, 44]. Higher annual income, larger housing area, more children, and licensed drivers are significantly associated with higher probability of owning cars, while number of bikes has significantly negative impacts on car ownership.

4.2. Effects on Travel Distance. After controlling sociodemographics and other built environment variables, URT significantly influences travel distance. Either living near a URT station, working near a URT station, or living and working close to URT all have significant and positive effects on travel distance, while the indirect effects through the mediating variable car ownership are not significant.

As for other residential built environment variables, land use diversity has a negative impact on travel distance, while longer distance to business district is related to longer distance travelled. Contrary to the negative effects of residential bus stop density, bus stop density in workplace neighborhoods is positively associated with travel distance. This might be because people probably include more activities in a tour after work if the bus has higher accessibility at the workplaces, which may lengthen the travel distance, whereas residential neighborhoods with high bus stop density usually have high land use density and diversity, which provides more employment opportunities and might shorten commuting distance. In addition, working within 4th ring road is significantly associated with short travel distance.

In terms of commuting characteristics, a tour which is a joint travel with other family members and includes more stops is more likely to have a longer travel distance. This is in line with our expectation.

Interestingly, the effects of household characteristics on travel distance mainly produce through car ownership. Namely, travel distance seems to be influenced only indirectly by annual income, number of children, and number of licensed drivers. Higher annual income, more children, and licensed drivers are indirectly associated with longer travel distance. Thus, ignoring car ownership as a mediating variable may result in a misspecification of sociodemographic effects on travel distance. In addition, housing area and number of bikes show significantly direct impacts on travel

Table 3: Standardized direct, indirect, and total effects on car ownership, travel distance, and car use.

Variable	Car ownership		Travel Distance			Car use	
	Direct	Total	Direct	Indirect	Total	Direct	Indirect
URT proximity							
RW proximity to URT	-0.082**	0.064**	0.069**	-0.005	-0.103***	-0.036	-0.068**
R proximity to URT	-0.046	0.105***	0.108***	-0.003	-0.005	0.027	-0.032
W proximity to URT	-0.056*	0.078***	0.081***	-0.003	-0.045*	-0.002	-0.043
Residential built environment							
Floor area ratio	0.037	-0.007	-0.009	0.002	0.008	-0.025	0.033
Land use diversity	0.026	-0.046**	-0.047**	0.001	-0.003	-0.022	0.019
Bus stop density	0.019	-0.057***	-0.058***	0.001	0.005	-0.006	0.011
Density of crossings	-0.085***	0.0002	0.005	-0.0048*	-0.061***	0.016	-0.077***
Distance to business district	0.051*	0.117***	0.114***	0.003	0.011	-0.046*	0.057**
Workplace built environment							
Floor area ratio	--	-0.032	-0.032	--	-0.093***	-0.090***	-0.003
Land use diversity	--	-0.020	-0.020	--	-0.038*	-0.036*	-0.002
Bus stop density	--	0.042***	0.042***	--	0.008	0.004	0.004**
Density of crossings	--	-0.019	-0.019	--	-0.008	-0.006	-0.002
Within 4th ring road	--	-0.221***	-0.221***	--	0.022	0.043*	-0.021***
Commuting characteristics							
Travel distance	--	--	--	--	0.094***	0.094***	--
Joint household travel	--	0.028*	0.028*	--	0.466***	0.463***	0.003
Number of stops	--	0.049***	0.049***	--	0.037*	0.032	0.005**
Household characteristics							
Annual Income	0.147***	0.008**	--	0.008**	0.133***	--	0.133***
Housing area	0.071***	0.03**	0.026**	0.004*	0.013	-0.054**	0.067***
Children	0.086***	0.005**	--	0.005**	0.029	-0.049**	0.078***
Licensed Drivers	0.520***	0.029**	--	0.029**	0.101***	-0.371***	0.472***
Bikes	-0.054**	-0.089***	-0.086***	-0.003*	-0.133***	-0.076***	-0.057***
Car ownership	--	0.056**	0.056**	--	0.907***	0.902***	0.005***
Personal characteristics							
Gender	--	--	--	--	0.135***	0.135***	--
Age	--	-0.040**	-0.040**	--	0.054**	0.058**	-0.004*
Education level	--	0.049**	0.049**	--	0.005**	--	0.005**
Driving license	--	--	--	--	0.235***	0.235***	--

Note. * significant at 10% level, ** significant at 5% level, and * * * significant at 1% level. "--" indicates the links that are not included in the model.

distance. Owning cars contributes to long travel distance. As for personal characteristics, age has a negative effect on travel distance, and higher education level is associated with longer travel distance.

4.3. Effects on Car Use. After controlling sociodemographics and other built environment variables, URT still plays an important role in car use behavior. Living within URT catchment areas by itself has no significant effect on car use, but if the workplace is near a URT station, people have lower probability of using cars. People who both live and work near URT are less likely to travel by car. This is in line with the finding of Kwoka et al. [45]. Moreover, the URT effects on car use seem to mainly produce through the mediating variable car ownership. The indirect effects of RW proximity to URT, R proximity to URT,and W proximity to URT through car ownership are -0.074, -0.042, and -0.050, while these indirect effects through travel distance are 0.006, 0.010, and 0.008, respectively. So the mediating impact of car ownership is greater than travel distance. If we only focus on the direct effects of URT, we may get a misspecification of URT effects on car use.

It is noteworthy that car use is mainly affected by residential built environment characteristics indirectly, while it is mainly influenced by workplace built environment factors directly. Density of crossing in residential neighborhood has a negative effect on car use. In addition, higher floor area ratio and higher land use diversity at the workplace are significantly associated with lower probability of using car, while the residential land use density and diversity show no significant impact on car use. Namely, land use characteristics at the workplace are more influential on mode choice than at the residence, which is consistent with the previous studies [2, 31].

For the commuting characteristics, longer travel distance increases the probability of using car. A tour which is a joint travel with other family members is more likely to travel by car, and number of stops in a tour is positively associated with car use. This is consistent with the previous studies [31, 33, 34].

With regard to household characteristic variables, the signs of some variables' direct and indirect effects are opposite. So the combination of the direct and indirect effects determines the total effects. We found that the indirect effects of household characteristics seem to play a more important role in determining the total effects. Annual income, number of licensed drivers, and owning cars have significantly positive effects on car use, while number of bikes is negatively related to car use. Interestingly, the annual income effect on car use seems to exist only through car ownership, which is consistent with the finding of Van Acker and Witlox [35]. As for personal characteristic variables, people who are male and older are more likely to use cars. Besides, having a driving license or having a higher level of education can increase the probability of travelling by cars.

5. Discussion

Previous studies mainly focus on the effects of residential proximity to URT on car ownership and use [1, 10, 11, 16], while we not only explore whether living within URT catchment areas affects car ownership and use differently from working near a URT station but also investigate how URT influences car use through the mediating variables car ownership and travel distance as well in this study.

After controlling for sociodemographics and other built environment factors, we find that URT plays a role in reducing car ownership level, which echoes other recent studies [10, 11, 16]. In these studies, URT proximity is all treated as a dummy variable. If there is a URT station within a certain distance (e.g., 500 m, 800 m, and 1 km), the value of the dummy variable is 1 and 0 otherwise. In contrast, Li and Zhao [1] and Shen et al. [4] used the distance to the closest URT station to evaluate URT proximity and found no significant effect on car ownership. So the quantification of URT proximity influences the results to a great degree. The URT effect on car ownership may occur within a simple threshold.

We also find that URT has a significant and negative impact on car use, which is in line with the previous studies [4, 16, 17]. Further, the URT effects on car use seem to mainly produce through the mediating variable car ownership, and the mediating effect of travel distance is relatively small. To our knowledge, very few studies have explored the URT influence on travel behavior considering the intermediary nature of car ownership and travel distance simultaneously. Our study fills this gap and identifies the mediating role of car ownership and travel distance.

Moreover, different from the previous studies that focus on the effect of residential proximity to URT [1, 10, 11, 16], we use three dummy variables to indicate residential and workplace proximity to URT and examine their effects on travel behavior. It is found that living within URT catchment areas by itself does not influence car ownership and use significantly, but if working near URT station areas, people have lower probability of owning and using cars. People with residences and workplaces both near URT are less likely to own and use cars. To our knowledge, only Kwoka et al. [45] explored whether working near a URT station affects travel behaviors of workers differently than living near a URT station. But they only demonstrated the travel behavior differences from the aggregate level using simple statistical analysis. By contrast, we control sociodemographics and other built environment factors and examine the relationship between residential and workplace proximity to URT and travel behavior based on disaggregate data, which better explains the differences in travel behavior.

Admittedly, we fail to explore the causal relationship between built environment and car ownership and use based on a cross-sectional sample. More advanced modeling techniques or longitudinal data will be considered in further study. In addition, due to data availability, travel-related attitudes or subjective perceptions of built environment attributes are not taken into account. Some studies used attitude and perception variables to address the residential self-selection problem, while we focus a subsample with exogenous residential locations, which can also tackle this problem.

6. Conclusion

This paper explored the URT effects on car ownership and use considering the intermediary nature of car ownership and travel distance simultaneously. We took advantage of the unique diversified housing system in China,and selected a subsample with exogenous residential locations to address the residential self-selection issue. Based on 3077 home-based work tour data, we developed a SEM model to investigate the complex relationship among URT, car ownership, travel distance, and car use. The results indicate that URT has important effects on car ownership and use. Living within URT catchment areas by itself is not significantly associated with car ownership and use, but if the workplace is near a URT station, people are less likely to own and use cars. People who both live and work near URT have lower probability of owning and using cars. Moreover, the URT effect on car use seems to mainly produce through the mediating variable car ownership, and the mediating impact of travel distance is relatively small. If we only focus on the direct effects of URT, we may get a misspecification of URT effects on car use.

As for the influence of other built environment variables, residential land use density (FAR) and diversity seem more influential on car ownership than those at the workplaces, while they play a less important role in mode choice behavior. Besides, lower density of crossings at the residences is associated with lower probability of owning cars and using cars. Bus stop density at the residential and workplace neighborhoods show no significant effect on car ownership and use. With respect to the commuting characteristics, a tour which is a joint travel with other family members and includes more stops has higher probability of using cars, and travel distance is positively associated with car use. For sociodemographics, the annual income effect on car use seems to exist only through car ownership.

Our findings verify that URT-supported urban growth can be an effective strategy to discourage people from owning and using cars, which have important implications for transportation and land use planning in response to the motorization and urbanization in Beijing and in other large developing cities.

Conflicts of Interest

The authors declare that they have no conflicts of interest.

Acknowledgments

This research is supported by the Fundamental Research Funds for the Central Universities (no. 2017YJS116) and Beijing Natural Science Foundation (no. 8171003).

References

[1] S. Li and P. Zhao, "Exploring car ownership and car use in neighborhoods near metro stations in Beijing: Does the neighborhood built environment matter?" *Transportation Research Part D: Transport and Environment*, vol. 56, pp. 1–17, 2017.

[2] A. Soltani, "Social and urban form determinants of vehicle ownership; evidence from a developing country," *Transportation Research Part A: Policy and Practice*, vol. 96, pp. 90–100, 2017.

[3] R. Cervero and J. Day, "Suburbanization and transit-oriented development in China," *Transport Policy*, vol. 15, no. 5, pp. 315–323, 2008.

[4] Q. Shen, P. Chen, and H. Pan, "Factors affecting car ownership and mode choice in rail transit-supported suburbs of a large Chinese city," *Transportation Research Part A: Policy and Practice*, vol. 94, pp. 31–44, 2016.

[5] X. Huang, X. J. Cao, X. Cao, and J. Yin, "How does the propensity of living near rail transit moderate the influence of rail transit on transit trip frequency in Xi'an?" *Journal of Transport Geography*, vol. 54, pp. 194–204, 2016.

[6] Urban rail transit in India, wikipedia, 2018.

[7] Y. Jiang, P. Gu, Y. Chen, D. He, and Q. Mao, "Influence of land use and street characteristics on car ownership and use: Evidence from Jinan, China," *Transportation Research Part D: Transport and Environment*, vol. 52, pp. 518–534, 2017.

[8] G. Giuliano and J. Dargay, "Car ownership, travel and land use: A comparison of the US and Great Britain," *Transportation Research Part A: Policy and Practice*, vol. 40, no. 2, pp. 106–124, 2006.

[9] P. Anastasopoulos, M. Karlaftis, J. Haddock, and F. Mannering, "Household automobile and motorcycle ownership analyzed with random parameters bivariate ordered probit model," *Transportation Research Record*, no. 2279, pp. 12–20, 2012.

[10] Y. Zhang, S. Zheng, C. Sun, and R. Wang, "Does subway proximity discourage automobility? Evidence from Beijing," *Transportation Research Part D: Transport and Environment*, vol. 52, pp. 506–517, 2017.

[11] X. Huang, X. Cao, J. Yin, and X. Cao, "Effects of metro transit on the ownership of mobility instruments in Xi'an, China," *Transportation Research Part D: Transport and Environment*, vol. 52, pp. 495–505, 2017.

[12] X. Huang, X. (Jason) Cao, J. Yin, and X. Cao, "Can metro transit reduce driving? Evidence from Xi'an, China," *Transport Policy*, 2018.

[13] C. R. Bhat and J. Y. Guo, "A comprehensive analysis of built environment characteristics on household residential choice and auto ownership levels," *Transportation Research Part B: Methodological*, vol. 41, no. 5, pp. 506–526, 2007.

[14] J. Cao and X. Cao, "The Impacts of LRT, Neighbourhood Characteristics, and Self-selection on Auto Ownership: Evidence from Minneapolis-St. Paul," *Urban Studies*, vol. 51, no. 10, pp. 2068–2087, 2014.

[15] C. Ding, D. Wang, C. Liu, Y. Zhang, and J. Yang, "Exploring the influence of built environment on travel mode choice considering the mediating effects of car ownership and travel distance," *Transportation Research Part A: Policy and Practice*, vol. 100, pp. 65–80, 2017.

[16] C. Zegras, "The built environment and motor vehicle ownership and use: Evidence from Santiago de Chile," *Urban Studies*, vol. 47, no. 8, pp. 1793–1817, 2010.

[17] H. Pan, Q. Shen, and T. Zhao, "Travel and car ownership of residents near new suburban metro stations in Shanghai, China," *Transportation Research Record*, no. 2394, pp. 63–69, 2013.

[18] C. R. Bhat and R. Gossen, "A mixed multinomial logit model analysis of weekend recreational episode type choice," *Transportation Research Part B: Methodological*, vol. 38, no. 9, pp. 767–787, 2004.

[19] A. Soltani and A. Shams, "Analyzing the influence of neighborhood development pattern on modal choice," *Journal of Advanced Transportation*, vol. 2017, 2017.

[20] M. Boarnet and R. Crane, "The influence of land use on travel behavior: Specification and estimation strategies," *Transportation Research Part A: Policy and Practice*, vol. 35, no. 9, pp. 823–845, 2001.

[21] S. Handy, X. Cao, and P. Mokhtarian, "Correlation or causality between the built environment and travel behavior? Evidence from Northern California," *Transportation Research Part D: Transport and Environment*, vol. 10, no. 6, pp. 427–444, 2005.

[22] M. Yazdanpanah and M. Hadji Hosseinlou, "The impact of perception toward the built environment in airport access mode choice using hybrid choice modeling," *Journal of Advanced Transportation*, vol. 2017, 13 pages, 2017.

[23] J. Hong, Q. Shen, and L. Zhang, "How do built-environment factors affect travel behavior? A spatial analysis at different geographic scales," *Transportation*, vol. 41, no. 3, pp. 419–440, 2014.

[24] S. Handy, "Methodologies for exploring the link between urban form and travel behavior," *Transportation Research Part D: Transport and Environment*, vol. 1, no. 2, pp. 151–165, 1996.

[25] R. Ewing and R. Cervero, "Travel and the built environment: a meta-analysis," *Journal of the American Planning Association*, vol. 76, no. 3, pp. 265–294, 2010.

[26] N. Limtanakool, M. Dijst, and T. Schwanen, "The influence of socioeconomic characteristics, land use and travel time considerations on mode choice for medium- and longer-distance trips," *Journal of Transport Geography*, vol. 14, no. 5, pp. 327–341, 2006.

[27] P. Zhao, "The Impact of the Built Environment on Individual Workers' Commuting Behavior in Beijing," *International Journal of Sustainable Transportation*, vol. 7, no. 5, pp. 389–415, 2013.

[28] Y. Shiftan and Y. Barlach, "Effect of employment site characteristics on commute mode choice," *Transportation Research Record*, no. 1781, pp. 19–25, 2002.

[29] R. Cervero, "Built environments and mode choice. Toward a normative framework," *Transportation Research Part D: Transport and Environment*, vol. 7, no. 4, pp. 265–284, 2002.

[30] C. Chen, H. Gong, and R. Paaswell, "Role of the built environment on mode choice decisions: additional evidence on the impact of density," *Transportation*, vol. 35, no. 3, pp. 285–299, 2008.

[31] C. Ding, Y. Lin, and C. Liu, "Exploring the influence of built environment on tour-based commuter mode choice: a cross-classified multilevel modeling approach," *Transportation Research Part D: Transport and Environment*, vol. 32, pp. 230–238, 2014.

[32] L. Frank, M. Bradley, S. Kavage, J. Chapman, and T. K. Lawton, "Urban form, travel time, and cost relationships with tour complexity and mode choice," *Transportation*, vol. 35, no. 1, pp. 37–54, 2008.

[33] C. Q. Ho and C. Mulley, "Multiple purposes at single destination: a key to a better understanding of the relationship between tour complexity and mode choice," *Transportation Research Part A: Policy and Practice*, vol. 49, pp. 206–219, 2013.

[34] V. van Acker and F. Witlox, "Commuting trips within tours: How is commuting related to land use?" *Transportation*, vol. 38, no. 3, pp. 465–486, 2011.

[35] V. Van Acker and F. Witlox, "Car ownership as a mediating variable in car travel behaviour research using a structural equation modelling approach to identify its dual relationship," *Journal of Transport Geography*, vol. 18, no. 1, pp. 65–74, 2010.

[36] Q. Jia, *Beijing Statistical Yearbook, China*, China Statistical Press, Beijing, China, 2015.

[37] X. Chen, *China City Statistical Yearbook*, China Statistical Press, Beijing, 2015.

[38] P. L. Mokhtarian and X. Cao, "Examining the impacts of residential self-selection on travel behavior: a focus on methodologies," *Transportation Research Part B: Methodological*, vol. 42, no. 3, pp. 204–228, 2008.

[39] D. Bray, *Social space and governance in urban China: The danwei system from origins to reform*, Stanford University Press, 2005.

[40] D. Wang and T. Lin, "Residential self-selection, built environment, and travel behavior in the chinese context," *Journal of Transport and Land Use*, vol. 7, no. 3, pp. 5–14, 2014.

[41] R. Cervero and M. Duncan, "Residential self-selection and rail commuting: a nested logit analysis," Working paper, University of California, Berkeley, 2002.

[42] D. Houston, M. G. Boarnet, G. Ferguson, and S. Spears, "Can compact rail transit corridors transform the automobile city? Planning for more sustainable travel in Los Angeles," *Urban Studies*, vol. 52, no. 5, pp. 938–959, 2015.

[43] R. B. Kline, *Principles and Practice of Structural Equation Modeling*, Guilford Press, New York, NY, USA, 2005.

[44] C. R. Bhat and V. Pulugurta, "A comparison of two alternative behavioral choice mechanisms for household auto ownership decisions," *Transportation Research Part B: Methodological*, vol. 32, no. 1, pp. 61–75, 1998.

[45] G. J. Kwoka, E. E. Boschmann, and A. R. Goetz, "The impact of transit station areas on the travel behaviors of workers in Denver, Colorado," *Transportation Research Part A: Policy and Practice*, vol. 80, pp. 277–287, 2015.

12

Impact of a New Metro Line: Analysis of Metro Passenger Flow and Travel Time Based on Smart Card Data

Xiao Fu [1] **and Yu Gu** [2]

[1] *School of Transportation, Southeast University, Nanjing 210096, China*
[2] *Jiangsu Key Laboratory of Urban ITS, Jiangsu Province Collaborative Innovation Centre of Modern Urban Traffic Technologies, School of Transportation, Southeast University, Nanjing 210096, China*

Correspondence should be addressed to Xiao Fu; fuxiao@seu.edu.cn

Academic Editor: David Z. W. Wang

Over the past few decades, massive volumes of smart card data from metro systems have been used to investigate passengers' mobility patterns and assess the performance of metro network. With the rapid development of urban rail transit in densely populated areas, new metro lines are constantly designed and operated in recent years. The appearance of new metro lines may significantly affect passenger flow and travel time in the metro network. In this study, smart card data of metro system from Nanjing, China, are used to study the changes of metro passenger flow and travel time due to the operation of a new metro line (i.e., Line 4, opened on 18 January 2017). The impact of the new metro line on passenger flow distribution and travel time in the metro network is first analysed. As commuting is one of the major purposes of metro trips, the impact of the new metro line on commuters' trips is then explicitly investigated. The results show that the new metro line influences passenger flow, travel time, and travel time reliability in the metro network and has different impacts on different categories of commuters.

1. Introduction

Multimodal transit networks provide convenient and efficient ways for individuals' daily travel in congested urban areas. Metro system, as a major component of multimodal transit system, has been significantly developed in recent years in many cities with large populations. Taking Nanjing, one of the most densely populated cities in China, for instance, 6 new metro lines have been constructed and operated in Nanjing in the past decade and 11 new metro lines are planned to be developed in the next five years.

To study the passenger flow characteristics and evaluate the performance of urban transit network, a substantial body of literature has been developed on analysing the distribution of passenger flow and travel time based on smart card data. However, limited effort has been devoted to investigating the impact of new metro line on metro network and passengers. The development of new metro lines changes the topology of metro network and influences the accessibility of city area. The operation of new metro lines may affect passenger flows of existing metro stations. Passengers' travel times and travel time reliability may also change as a result of the expansion of metro network. As commuters are the majority of metro users in peak hours and are expected to benefit from the opening of new lines, the impact of new metro line on commuters' trips should be explicitly studied. In the literature, little attention has been given to the changes of commuters' spatial/temporal distributions and travel times after the opening of new metro lines.

In this study, a comprehensive analysis of metro passenger flow and travel time is presented on the basis of smart card data from Nanjing, China. The impact of new metro line on metro passenger flow and travel time is first analysed on the level of whole metro network. Spatial and temporal distributions of metro passenger flow in peak and nonpeak hours are investigated before and after the opening of the new line. The changes of passenger travel time and travel time reliability are also studied. To investigate the impact of the new metro line on commuters' trips, this study identifies and classifies the passengers commuting by metro and investigates the impact of the new metro line on flow distribution, travel time, and travel time reliability of different categories of metro commuters.

The outline of this paper is as follows. Section 2 presents a review of studies using smart card data and studies discussing the impact of new rail transit lines. Data used in this study are described in Section 3. Impacts of new metro line on passenger flow and passenger travel time are investigated in Sections 4 and 5. Conclusions and recommendations for future research are given in the last section.

2. Literature Review

2.1. Applications of Smart Card Data. Smart card data, which are passively collected by automated fare collection systems with the use of smart cards for transit services, are important data source for transport researchers. Because of the large quantity and high precision of smart card data, many researchers believe that such innovative data source can complement or even replace the traditional collected transport data such as travel survey data [1, 2]. Existing transport studies based on smart card data mainly focus on three areas: analysis of passenger flow patterns, investigation of travel behaviours, and assessment of transit network performance.

Many studies are devoted to investigating the characteristics of metro passenger flow patterns based on smart card data. Smart card data record the entrance/exit stations and entrance/exit times of metro passengers, so the data are widely used to investigate and visualize the spatial and temporal distributions of metro passenger flow [3–5]. Based on the historical entrance and exit passenger flow records, the future passenger flow can also be predicted with smart card data [6, 7].

Smart card data can also be used to analyse travel behaviours and extract travel patterns of transit users. Travel behaviours in transit networks can be investigated such as transit ridership and origin-destination (OD) choices [8, 9], train and route choices [10, 11], and transfer behaviours [12]. Based on the investigation of travel behaviours, many studies are conducted to extract individuals' spatial/temporal travel patterns [13–15] and mobility patterns [16–18]. The dynamics of travel behaviours and the variation of travel pattern were also studied [19–22]. Some other studies focused on the regularity of individual travel pattern, using smart card data to classify transit passengers [23, 24].

In addition, smart card data can also be applied to investigate transit network performance such as the passenger travel time and the service reliability of transit systems. For instance, some studies investigated the spatial and temporal distribution of trip travel times [25], travel time differences between different transit modes [12], and splitting and estimation of metro travel time [26]. The service reliability of transit systems can also be derived based on smart card data. Comparing passengers' entrance/exit times to the transit schedules, researchers can select schedule deviation of vehicles or at-stop waiting time of passengers to measure the service reliability [27]. The concept of travel time reliability is widely employed to evaluate service reliability of transit systems from the perspective of individual passengers. The travel time reliability metrics include the schedule-based metrics such as excess journey time and reliability buffer time [28, 29], as well as statistical metrics such as standard deviation (SD) and coefficient of variation (CV) of travel time [4, 30–33].

2.2. Impact of New Rail Transit Lines. Construction of a new rail transit line may result in changes of individuals' travel choice behaviour. The change in mode choice attracts much attention in the literature. Senior [34] analysed the modal shift from bus and private vehicles to the light rail transit based on household survey and census data. Cao and Schoner [35] explored the impact of new light rail transit line on transit use. Harvey et al. [36] investigated people's willingness to travel by the newly built high speed rail line. Gadziński and Radzimski [37] combined household survey data with property sales transactions data to investigate the effect of a new tram on households' mode choice behaviour. In addition to studying the mode choice change, Saidi, Wirasinghe, and Kattan [38] assessed the changes in passengers' cost after introducing a new rail line based on job distribution data and origin-destination data. Hong, Boarnet, and Houston [39] adopted sociodemographic data, travel survey data, and activity data to investigate the impact of a new light rail transit line on individuals' travel behaviour.

Besides the studies focusing on the change of travel behaviour, a few studies reveal the impact of new metro lines on passenger flow. Li [40] analysed changes in passenger flow pattern caused by different types of new metro lines. Based on smart card data, Liu, Yao, and Li [41] analysed the relationship between metro network expansion, land use of station neighbourhood, and the growth of station passenger flow.

However, existing studies mainly focus on the change of passenger flow distribution on the network level, while little attention has been paid to the impact of new lines on different individuals. Transportation analyses which are not conducted from individuals' perspective may be biased as they fail to depict individuals' real travel experience [42]. Some studies have been made to investigate the performance of urban transportation network from the perspective of individual travellers using different kinds of data sources [33, 43]. In densely populated urban areas, as a large proportion of passengers' daily travels are made by metro, the opening of a new metro line may significantly influence some individuals' daily travel patterns while other individuals may not be affected. For giving a comprehensive analysis of the impact of new metro line on metro passenger flow and travel time, the analysis in this study is first conducted on the network level to present an overview of the metro system and investigate the changes of passenger flow and travel time after the new line opening. Further analysis focuses on metro commuters. Effort has been devoted to categorizing commuters based on their OD choices and evaluating the impact of the new metro line on different categories of commuters.

3. Study Area and Data Collection

Nanjing is one of the largest cities in China with a population of about 8 million. In 2005, the first metro line

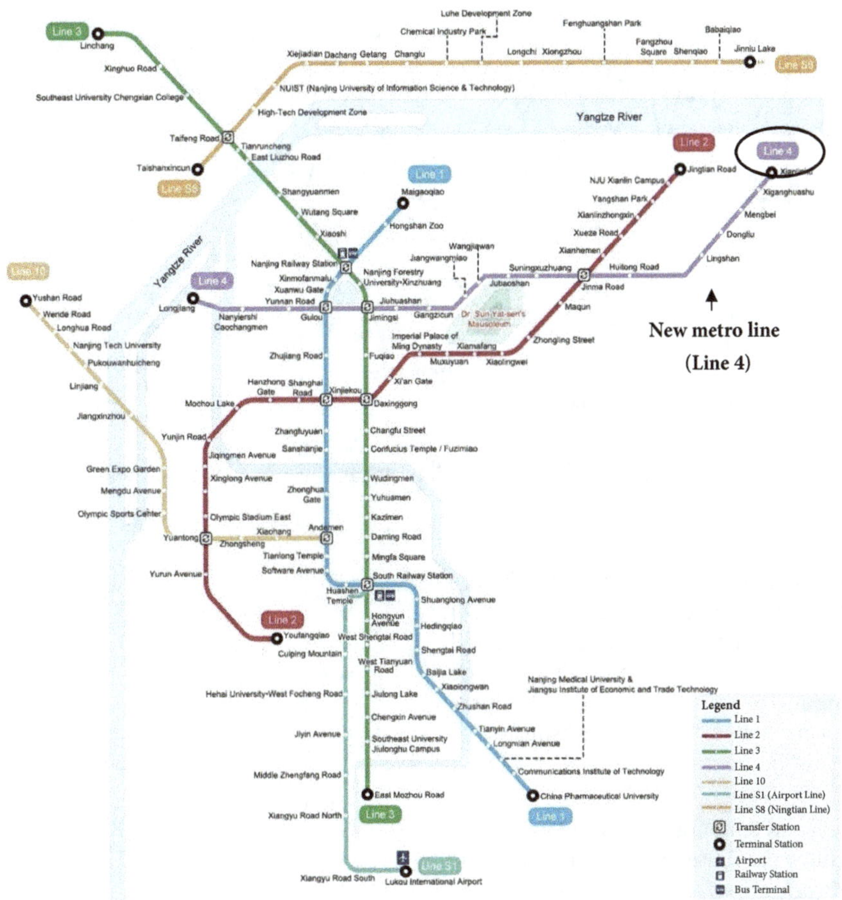

FIGURE 1: Nanjing metro network.

TABLE 1: Examples of smart card data used in this study.

Time of Entry	Station of Entrance	Time of Exit	Station of Exit	Card ID
2017/1/1 10:10:21	14	2017/1/1 10:30:21	7	976072959331
2017/1/1 20:27:50	14	2017/1/1 20:34:52	16	976072959320
2017/1/1 11:56:56	37	2017/1/1 12:46:34	101	993172764630
2017/1/1 21:10:28	14	2017/1/1 22:03:25	50	970672806438

began operation in Nanjing. In the past decade, Nanjing metro network expands rapidly. An overview of Nanjing metro network is extracted from Internet and shown in Figure 1. Line 4 of Nanjing metro system was opened on 18 January 2017, which is the latest line operated. After the opening of Line 4, there are 7 metro lines and 139 stations in operation in Nanjing. The operating mileage of Nanjing metro network is 258 km, and the average daily ridership in January 2017 is about 2.1 million.

To explore the impact of new metro line (i.e., Line 4) on metro passenger flow and travel time in densely populated urban areas, the data used in this study are records of smart cards used by metro passengers in Nanjing from January to February 2017, 61 million records in total. Data of the two months are used because the two months cover the periods before and after the opening of Line 4. Data of two consecutive months guarantee the stability of passengers' travel choice behaviour.

Smart card data contain complete entrance and exit records (including location and time) of each passenger, as shown in Table 1. Take the first record in Table 1 as an example; the record indicates that the passenger entered the metro system from station No. 14 (i.e., Nanjing Railway Station) at 10:10:21 on 1 January 2017 and exited from station No. 7 (i.e., Sanshanjie) at 10:30:21. Data cleaning is conducted to exclude incomplete and invalid data before analysis.

4. The Impact of New Metro Line on the Metro Network

The operation of the new metro line may influence metro passenger flow distribution. It can be seen from Figure 1 that Line 4 has intersections with three lines (i.e., Lines 1, 2,

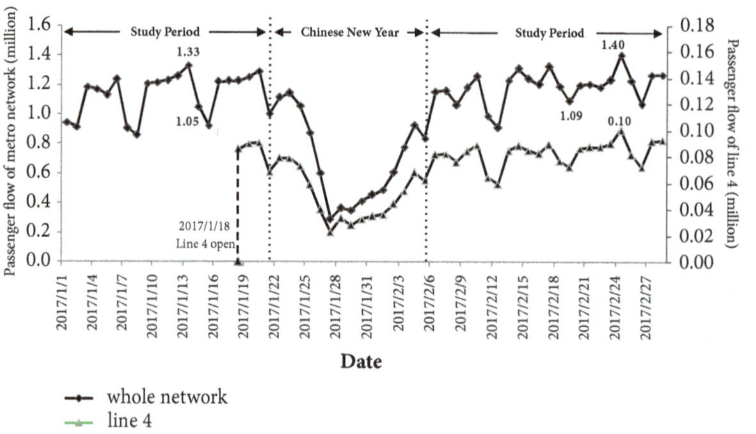

FIGURE 2: Daily metro passenger flow before and after Line 4 opening.

and 3) which pass through central areas of the city. Before the opening of Line 4, the three lines took the majority of passenger flows in the metro network. Line 4 is expected to share passenger flows with Lines 1-3 and alleviate the crowding effect at some metro stations in central business district (CBD). Thus, a need is evident to investigate the change of passenger flow in metro network after Line 4 opening.

In this section, spatial and temporal analysis of passenger flow are conducted to show the passenger flow distributions in the whole metro network before and after the opening of Line 4. Travel time and travel time reliability in the metro network are also investigated to show the impact of the new line.

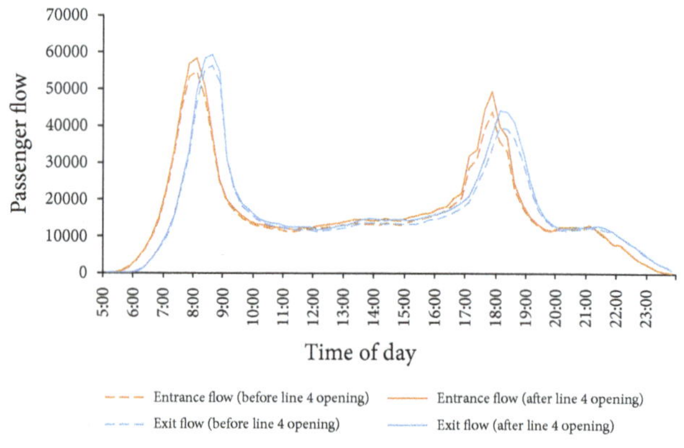

FIGURE 3: 15 min metro passenger flow distribution before and after Line 4 opening.

4.1. Metro Passenger Flow Distribution

4.1.1. Temporal Distribution of Metro Passenger Flow. On the basis of smart card data from 1 January to 28 February 2017, the daily flows of passengers using smart cards on Line 4 and on the whole metro network are shown in Figure 2. It is obvious from Figure 2 that the passenger flow varies from day to day. The passenger flow on workday is significantly higher than that at weekend, which implies that a large portion of metro passengers are commuters. Thus, in Section 5, explicit analyses of commuters' flow distributions and travel times are given.

It can also be seen from Figure 2 that the number of passengers increases after Line 4 opening. Before the opening of Line 4, the maximum daily passenger flow of the whole metro network is 1.33 million on workday and 1.05 million at weekend. After the opening of Line 4, the maximum passenger flows on workday and at weekend increase to 1.40 million and 1.09 million, respectively. The maximum passenger flow of Line 4 reaches 0.10 million on 24 February 2017.

Please note that passenger flow decreases dramatically during 21 January and 5 February as many residents leave Nanjing during Chinese New Year. Thus, data of the period affected by Chinese New Year are excluded from analysis in this study to facilitate essential ideas without loss of generality. The study period is then divided into two parts: (a) 1 Jan 2017 to 17 Jan 2017 (the period before Line 4 opening); (b) 18 Jan 2017 to 20 Jan 2017 and 6 Feb 2017 to 28 Feb 2017 (the period after Line 4 opening).

In addition to the daily fluctuation of passenger flow, the entrance/exit flow in metro network also varies greatly at different times of day. Figure 3 shows the average 15 min passenger flow on workdays before and after Line 4 opening. Increase of entrance/exit flow can be found after Line 4 opening due to the expansion of network capacity. It can also be seen from Figure 3 that there are two peaks of passenger flow on workdays, i.e., 7:00-9:00 in the morning and 17:00-19:00 in the afternoon. To capture the different conditions of metro network, it is necessary to analyse metro passenger flow and travel time in both peak hours and nonpeak hours.

4.1.2. Spatial Distribution of Metro Passenger Flow. Figure 4 shows the average daily exit flows of metro stations before and after the opening of Line 4 based on kernel density estimate.

The passenger flows at the stations highlighted in red circles vary significantly due to the opening of Line 4. It can be seen that the crowding at the stations in CBD (i.e., the red

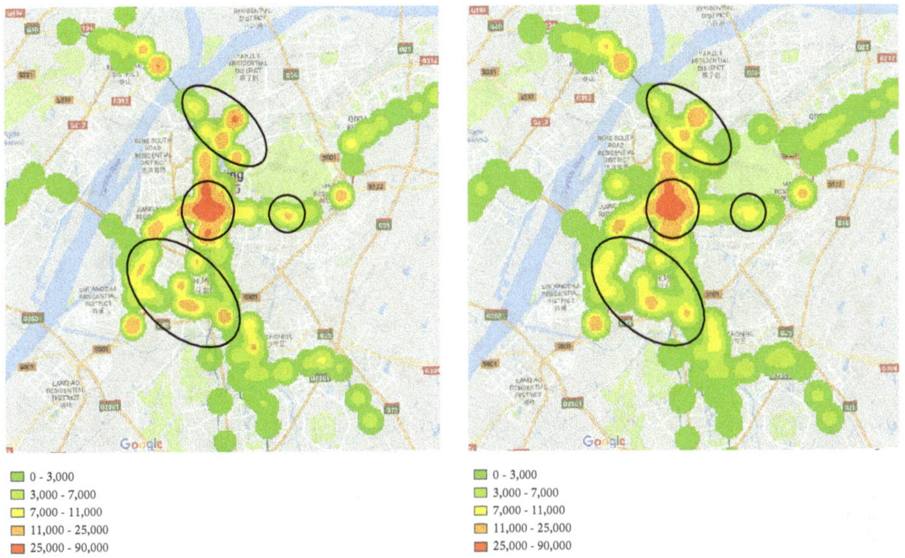

(a) Before Line 4 opening

(b) After Line 4 opening

FIGURE 4: Spatial distribution of metro passenger flow before and after Line 4 opening (map source: Google Maps).

circle in the centre of the figure) is alleviated after the opening of Line 4. Several stations on other lines (highlighted in other red circles) share fewer passenger flows after the opening of Line 4. Thus, Line 4 attracts more individuals to metro network while contributing to the whole network by sharing a large number of passengers and reducing the crowding at several existing metro stations.

To explicitly investigate the impact of Line 4 on spatial distribution of passenger flow, the metro stations significantly affected by Line 4 are identified as follows.

Denote the mean and the SD of daily passenger flow from station i to station j before Line 4 opening as f_{ij}^{b} and σ_{ij}^{b}, respectively:

$$f_{ij}^{\mathrm{b}} = \frac{\sum_{n=1}^{N^{\mathrm{b}}} f_{ij}^{n\mathrm{b}}}{N^{\mathrm{b}}}, \quad (1)$$

$$\sigma_{ij}^{\mathrm{b}} = \sqrt{\frac{\sum_{n=1}^{N^{\mathrm{b}}} \left(f_{ij}^{\mathrm{b}} - f_{ij}^{n\mathrm{b}}\right)^2}{\left(N^{\mathrm{b}} - 1\right)}}, \quad (2)$$

where N^{b} is the number of days of the study period before Line 4 opening, and $f_{ij}^{n\mathrm{b}}$ is the passenger flow on day n.

The mean of passenger flow from station i to station j after Line 4 opening (denoted as f_{ij}^{a}) can be obtained as

$$f_{ij}^{\mathrm{a}} = \frac{\sum_{n=1}^{N^{\mathrm{a}}} f_{ij}^{n\mathrm{a}}}{N^{\mathrm{a}}}, \quad (3)$$

where N^{a} is the number of days of the study period after Line 4 opening.

In this study, if f_{ij}^{a} exceeds the normal fluctuation ranges

$$f_{ij}^{\mathrm{a}} > f_{ij}^{\mathrm{b}} + \beta \cdot \sigma_{ij}^{\mathrm{b}}$$
$$\text{or } f_{ij}^{\mathrm{a}} < f_{ij}^{\mathrm{b}} - \beta \cdot \sigma_{ij}^{\mathrm{b}}, \quad (4)$$

the OD stations are viewed to be affected by Line 4. β is a parameter associated with the normal fluctuation ranges. In this study, the fluctuation range is set as 95%; thus β is 1.64. Setting the destination j as Gulou (a station located in CBD with a large number of passenger flows), Figure 5 shows the stations affected by Line 4 in peak and nonpeak hours.

As shown in Figure 5, the passenger flow travelling to CBD is significantly influenced by the operation of Line 4. From Figure 5(a), it can be seen that stations on Line 3 significantly attract more passengers to CBD, while the passenger flows of many stations on Line 1 and Line 2 decreased. The reason is that Line 2 is spatially parallel to the new metro Line 4, and passengers originally travelling by Line 2 might change to use Line 4. The shift of passenger flow from Line 1 to Line 3 can be explained by the decreased travel time from Line 3 to CBD, which will be demonstrated in Section 4.2. Comparing Figures 5(a) and 5(b), it can be found that the passenger flow distribution in nonpeak hours is relatively more stable than that in peak hours, indicating that commuting trips are significantly influenced by the new metro line.

4.2. Travel Time in the Metro Network. The introduction of a new metro line probably has impacts, as expected, on passengers' activity and travel choice behaviour as the accessibility/connectivity of activity locations and transit stations changes [44]. As a reliable transport mode which does not suffer from road traffic congestion, metro is one of the best choices for commuters who work in CBD. In a city with similar size of Nanjing, the 30-minute coverage area of metro stations in CBD is important to commuters. Figure 6 shows the travel times from different metro stations to CBD with a travel time less than 30 minutes. Apparently, the number of stations within the 30-minute coverage area greatly increases due to the opening of Line 4. The travel

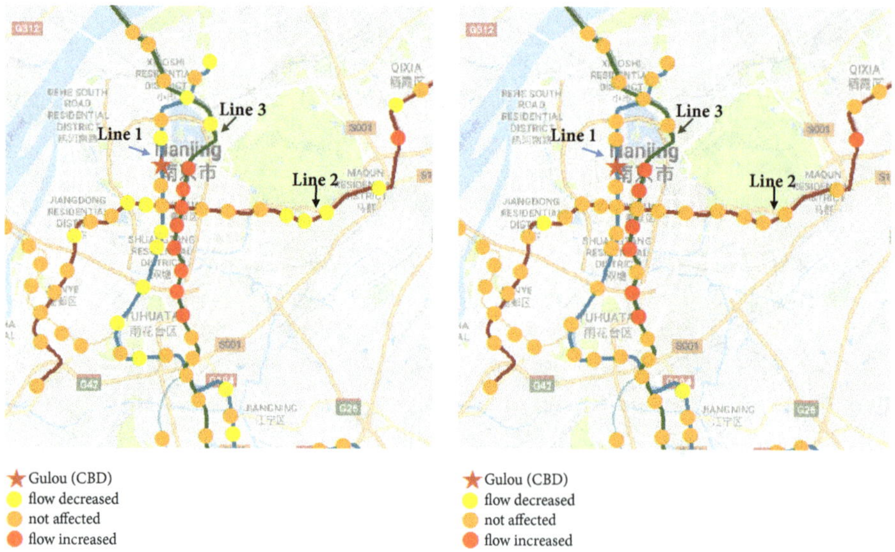

(a) Peak hours (b) Nonpeak hours

FIGURE 5: Change of passenger flow at metro stations after Line 4 opening (map source: Google Maps).

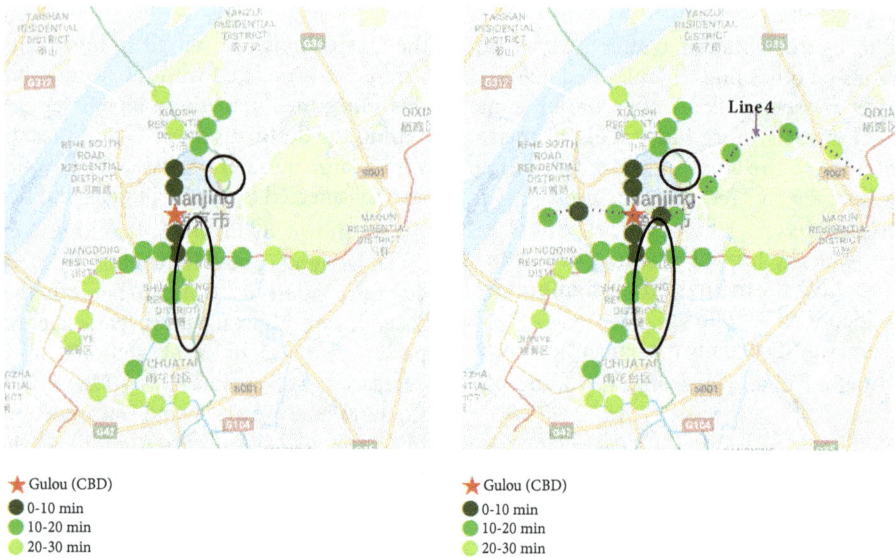

(a) Before Line 4 opening (b) After Line 4 opening

FIGURE 6: Travel times from metro stations to CBD (map source: Google Maps).

times from many stations on Line 3 (which are highlighted by red circles) to CBD also decrease after Line 4 opening, while the travel times from stations on Line 1 to CBD remain stable. Hence, many passengers choose Line 3 to travel to CBD instead of Line 1. The shift of passenger flow from Line 1 to Line 3 can be found in Figure 5.

In metropolitan areas, travel times in transit networks generally vary from day to day. Many empirical studies revealed that individuals, especially commuters, indeed consider the travel time uncertainty as a risk for their travels. To reduce the risk of late arrival, individuals may have concerns on the probability that a trip can be successfully fulfilled within a given travel time, referred to as travel time reliability in the literature. Therefore, the travel time reliability in metro network should be explicitly explored with real data.

In this study, the coefficient of variation (CV) of travel time is selected as the measure of travel time reliability. The travel time reliability within OD pair ij, cv_{ij}, can be expressed as

$$cv_{ij} = \frac{\sigma_{ij}}{t_{ij}}, \tag{5}$$

where σ_{ij} and t_{ij}, respectively, denote the SD and the mean of travel time between i and j. A higher cv_{ij} indicates lower travel time reliability.

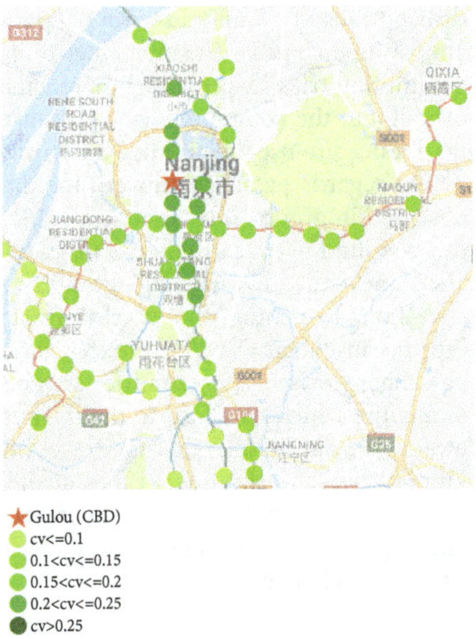

(a) Peak hours, before Line 4 opening

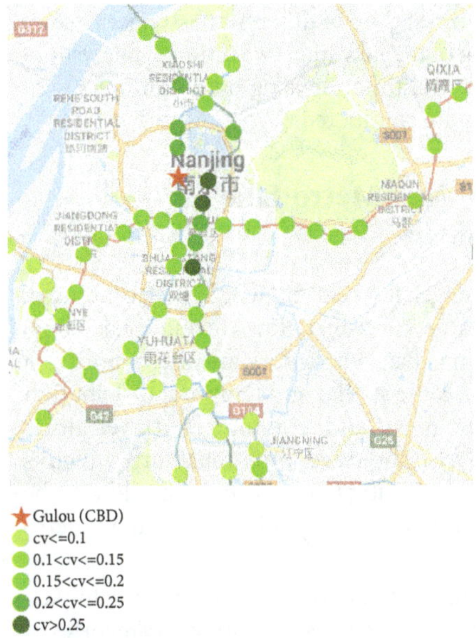

(b) Peak hours, after Line 4 opening

(c) Nonpeak hours, before Line 4 opening

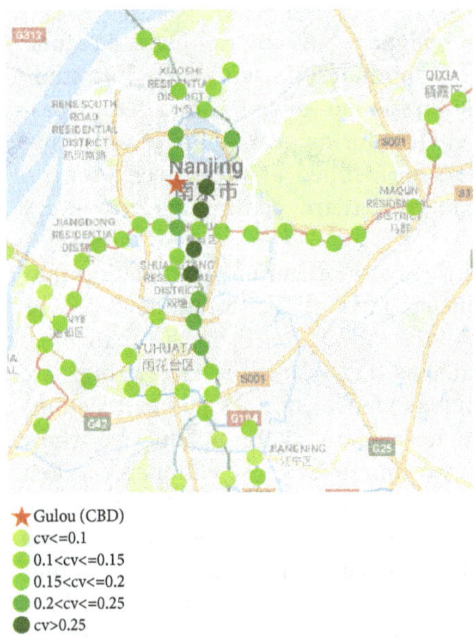

(d) Nonpeak hours, after Line 4 opening

FIGURE 7: Travel time reliability from metro stations to CBD (map source: Google Maps).

Based on the trip travel times extracted from smart card data before and after the opening of Line 4, the travel time reliability from all stations to CBD can be analysed. Figure 7 shows the CVs of travel times during peak and nonpeak hours before and after Line 4 opening. A light colour of a station is with a small value of CV, which indicates the travel time from this station to CBD is quite reliable. The crowding effect during peak hours may reduce travel time reliability because some individuals are not able to get on the first arriving vehicles; however, the frequency of metro service in peak hours is higher than that in nonpeak hours. Thus, comparing Figures 7(a) and 7(c) or comparing Figures 7(b) and 7(d), it can be found that there is no identical trend for all stations from peak to nonpeak hours in terms of travel time reliability.

Comparing Figures 7(a) and 7(b) or comparing Figures 7(c) and 7(d), it can be found that the opening of Line 4 surprisingly leads to decreased travel time reliability for many stations, especially the stations on Line 3. The finding can be attributed to the increased transfer choices introduced by Line 4. Many individuals prefer to transfer more times for shortening the travel time after Line 4 opening, while transfers bring random factors such as stochastic waiting and walking time.

The analysis in this section aggregates passengers in the whole metro network, which is unable to reveal the flow distribution and the travel time of different passengers before and after the opening of the new line. In Section 5, analyses of passenger flow and travel time focusing on commuters with different OD choices are conducted.

5. The Impact of New Metro Line on Metro Commuters

The analysis of passenger flow on the whole metro network level indicates the large passenger flow during peak hours in workdays, implying that commuters are the majority of metro users. Therefore, the spatial and temporal distribution of metro commuter flows and commuters' travel times deserve further analysis. The travel time reliability, which is a major concern for commuters' daily travel, should also be investigated before and after the opening of the new metro line.

In this section, analysis is conducted to investigate passenger flow and travel time changes for different categories of commuters.

5.1. Classification of Commuters. As commuting is a regular activity conducted by commuters in workdays, this study defines metro commuting as metro trips conducted by passengers between fixed ODs during peak hours of workdays. The metro commuters are defined as the passengers who conduct no less than one metro commuting trip every workday.

As the commuters' choices of entrance/exit station may change after introducing Line 4, the commuters can be further divided into three groups based on the changes of their entrance/exit stations (i.e., ODs). The three groups are as follows: (1) commuters whose entrance and exit stations are not changed and not on Line 4 throughout the whole study time period; (2) commuters whose entrance and exit stations are not changed but the entrance/exit stations are intersections of Line 4 and other lines; (3) commuters whose entrance/exit stations are changed from stations on other lines to stations on Line 4 after Line 4 opening. The three groups, respectively, represent the commuters whose ODs are not affected, indirectly affected, and directly affected by the opening of Line 4.

Based on above criteria, 8543, 588, and 639 passengers are identified as groups 1, 2, and 3 commuters, respectively. The distributions and travel times of the commuters in different groups are further investigated. The impacts of new metro line on different groups are discussed.

5.2. Passenger Flow Distribution of Different Groups of Commuters

5.2.1. Temporal Distribution of Different Groups of Commuters. The temporal distributions of the three groups of commuters over a day are presented in Figure 8. From the entrance/exit passenger flow profiles, it can be found that the operation of Line 4 exerts different degrees of impacts on the three groups of commuters. A significant growth in the flow of group 3 commuters can be found after Line 4 opening. However, the distributions of group 1 and group 2 passenger flows remain almost unchanged. The new metro line generates more commuting trips by providing more choices of O/D stations overall, but the commuters who have fixed OD stations and do not need to use the new line are hardly influenced.

As regards passenger travel time, hardly any change is observed in group 1 after Line 4 opening. A slight difference can be found between the travel times of group 2 before and after Line 4 opening. Although group 2 does not change ODs after Line 4 opening, the travel time of group 2 changes a little because their O/Ds are transfer stations with Line 4. The passenger flows in these transfer stations change after Line 4 opening which may lead to the travel time change. As for group 3, it can be seen that the travel time varies a lot after Line 4 opening. The travel times at different times of day are significantly different from the travel times before Line 4 opening. As group 3 changes O/Ds after Line 4 opening, the travel time changes accordingly.

5.2.2. Spatial Distribution of Different Groups of Commuters. Figure 9 depicts the cumulative number of trips with respect to the proportion of ordered OD pairs. It can be seen that a great number of trips are conducted between a small proportion of OD pairs. For instance, about 70% of the trips of group 1 are conducted between 10% OD pairs, implying that the majority of commuters' trips are between a few ODs. In addition, comparing to the trips in nonpeak hours, trips in peak hours are generated between fewer OD pairs, indicating that the spatial distribution of commuter flow concentrates in a limited number of ODs in peak hours. This phenomenon is more obvious in the morning peak than the afternoon peak, as commuters have more destination choices after work in the afternoon.

Comparing the distribution curves before and after Line 4 opening, we can find that the opening of Line 4 has different degrees of impact on the spatial distributions of different groups of commuters. The spatial distribution of group 1 is scarcely influenced. The spatial distribution of group 2 is less influenced in peak hours but more influenced in nonpeak hours, indicating that group 2 commuters have more OD choices in their spare time after Line 4 opening. Apparently, the impacts of Line 4 on the spatial distribution of group 3 are significant in both peak and nonpeak hours as group 3 changes O/D stations after Line 4 opening.

5.3. Travel Time Reliability of Different Groups of Commuters. It is widely recognized that the analysis of travel time reliability should focus on individual passengers [45, 46]. In this section, the travel time reliability of each commuter group is investigated. The impacts of new metro line on travel time reliability of different groups of commuters are discussed.

Being consistent with Section 4.2, the average CV of each group of commuter's travel time is obtained to indicate the metro travel time reliability. CV of individual k, cv_k, can be expressed as

$$cv_k = \frac{\sigma_k}{t_k}, \tag{6}$$

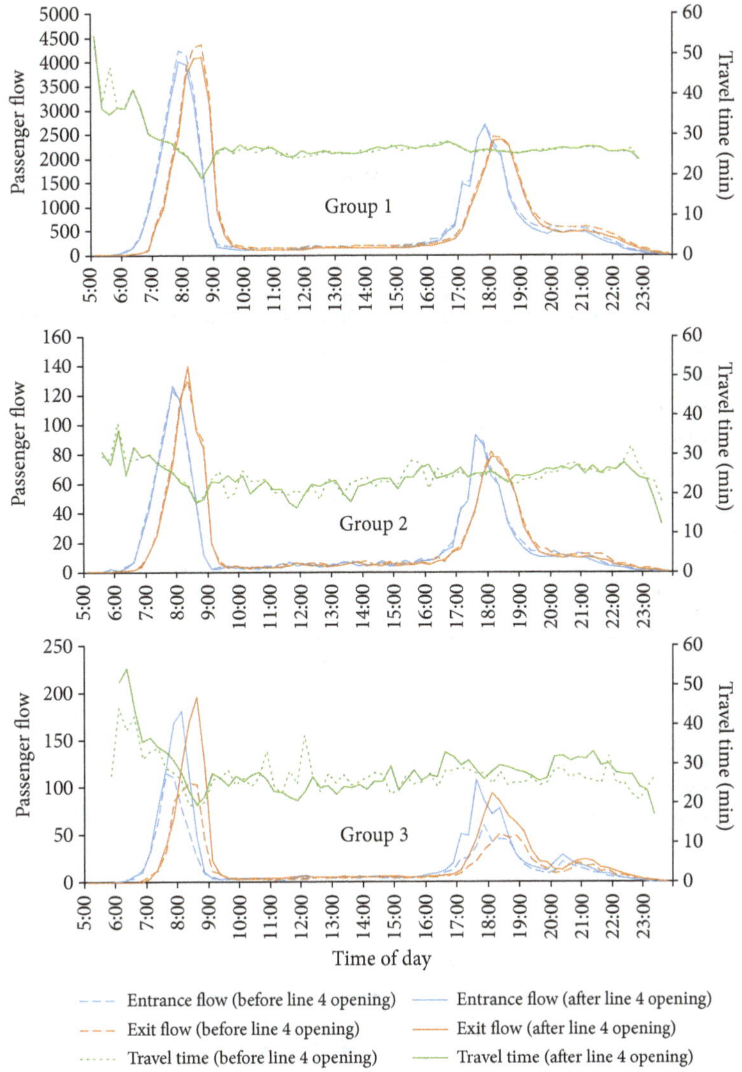

FIGURE 8: 15 min flow distributions of different groups of commuters and their travel times before and after Line 4 opening.

where σ_k and t_k, respectively, denote the SD and the mean of k's metro travel time between his commuting OD.

The mean and the SD of commuting time of each group before and after Line 4 opening are investigated and shown in Figure 10. As linear lines fit the trends well and the slope of line can intuitively depict the travel time reliability, linear trend lines are used in Figure 10. It can be found from Figures 10(a) and 10(b) that the mean and the SD of travel time of group 1 are hardly influenced by the opening of Line 4, as the OD choices of group 1 are not affected by Line 4.

Comparing Figures 10(c) and 10(d), it can be seen that the distribution of data points in Figure 10(d) is more concentrated around the trend line, indicating that the travel time variability of group 2 is more predictable. This is because Line 4 shares a considerate number of passengers travelling through the transfer stations (i.e., group 2's O/Ds). Thus, the variability of travel time is mainly associated with travel distance instead of other incidents such as overcrowdings at stations.

From Figures 10(e) and 10(f), it can be found that the slope of trend line after Line 4 opening is obviously lower than that before Line 4 opening, indicating that the introduction of Line 4 raises the travel time reliability of group 3 commuters. Group 3 shifts to use Line 4 instead of other lines. The passenger flow on Line 4 is relatively low compared to other lines as Line 4 is newly put into operation, which guarantees high travel time reliability.

The mean and the average CV of individual travel times in each commuter group before and after Line 4 opening are listed in Table 2. It can be found that after Line 4 opening the mean travel time of group 1 and group 2 hardly changes. It indicates that for passengers who do not change OD choices the effect of introducing new metro line on average travel time is little. Note that the mean travel time of group 2 commuters slightly decreases. This is because the operation of Line 4 results in reduced number of transfers for some group 2 commuters. However, the average travel time of group 3 significantly increases after Line 4 opening. It indicates that

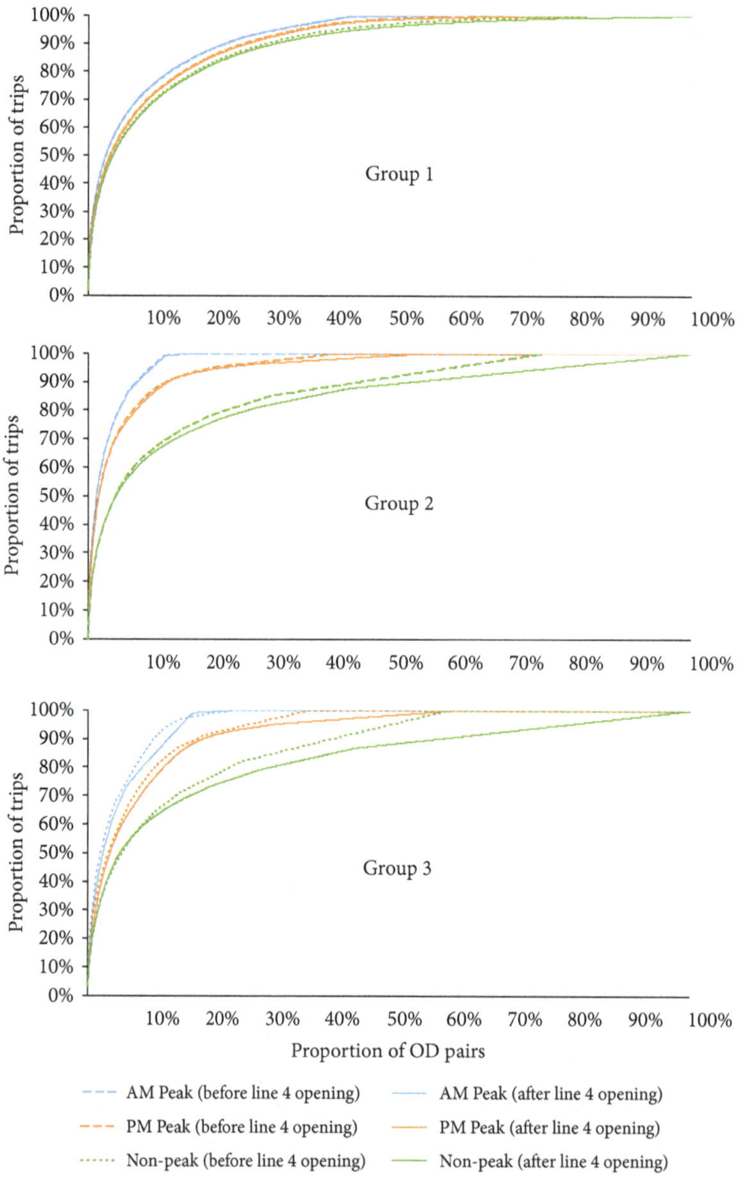

FIGURE 9: Cumulative number of trips with regard to proportion of ODs before and after Line 4 opening.

TABLE 2: Mean and average CV of commuting travel time.

	Before Line 4 opening		After Line 4 opening	
	Mean (min)	Average CV	Mean (min)	Average CV
Group 1	25.545	0.100	25.624	0.100
Group 2	25.630	0.084	25.063	0.082
Group 3	25.975	0.096	34.306	0.086

the new line makes group 3 commuters prefer travelling farther by metro, as the new stations provide commuters with more entrance/exit station choices.

It can be concluded from Table 2 that the new metro line has different levels of impacts on the travel time reliability of commuters in different groups. Introducing Line 4 hardly influences travel time reliability of commuters who do not change their OD choices. The average CV of group 1 travel time remains stable, while the average CV of group 2 travel time slightly decreases from 0.084 to 0.082. The impact of Line 4 on group 3 travel time reliability is significant. The average CV of group 3 travel time (i.e., the travel time of commuters who shift to use new stations on Line 4) drops from 0.096 to 0.086.

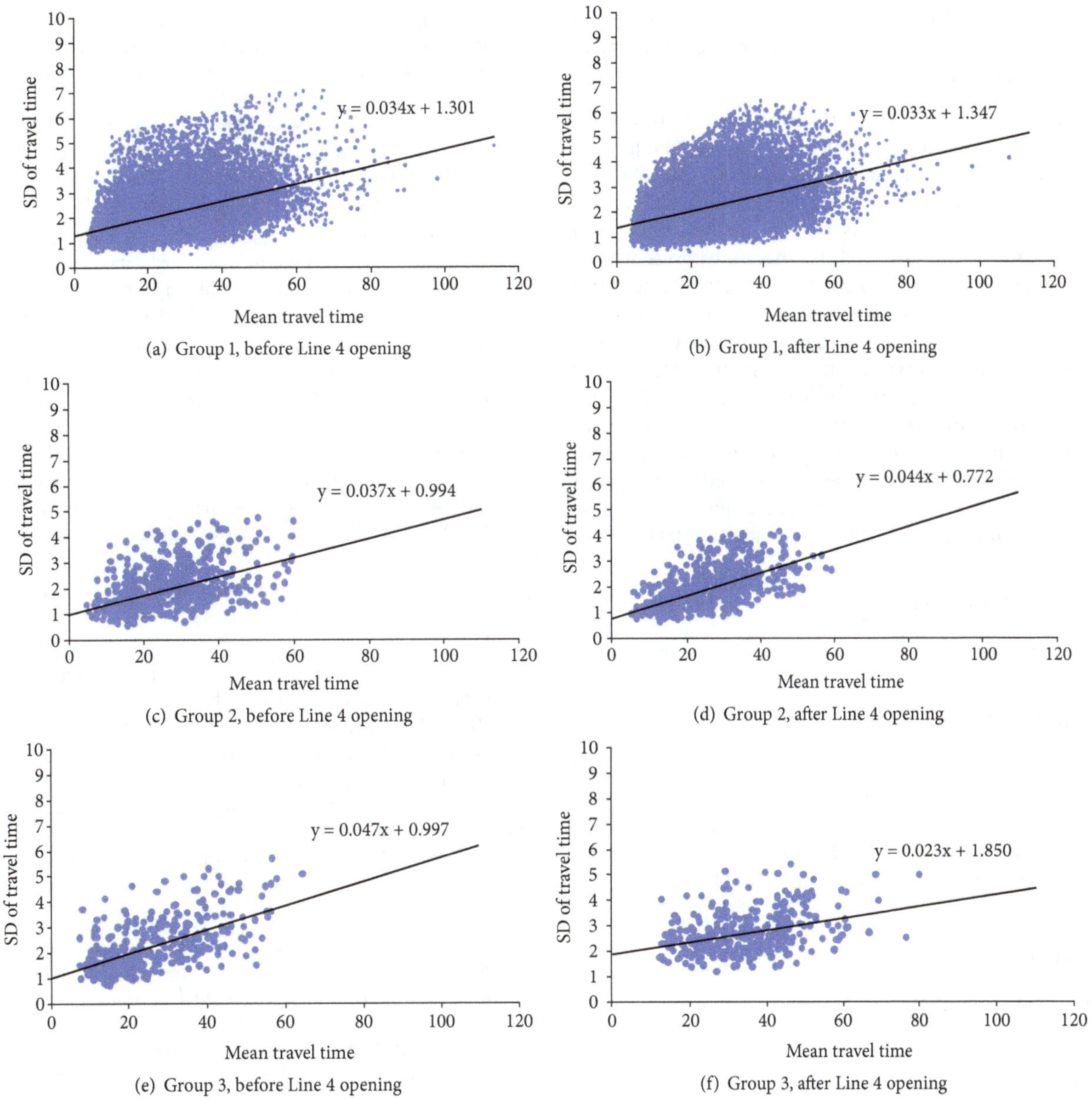

FIGURE 10: The mean and the SD of commuting travel time before and after Line 4 opening.

6. Conclusions

This paper presents a comprehensive analysis based on smart card data of metro system to demonstrate the effects of rail transit development in urban areas. The impacts of a new metro line (i.e., Line 4 of Nanjing metro) on both the whole metro network and different categories of commuters are investigated by analysing passenger flow distributions, travel times, and travel time reliability.

Analysis of passenger flow at all metro stations is conducted based on the smart card data before and after the opening of a new metro line. It is found that the passenger flow of the whole metro network increases after the new line opening, while the crowding at some metro stations reduced as the new line shares a large proportion of passengers. To intensively study the impact of the new metro line on the network level, this study investigates the stations with significant change of passenger flow after the opening of the new metro line. Additionally, the travel time and the travel time reliability in the metro network are intensively studied. It is found that the opening of Line 4 leads to changes of travel time and travel time reliability for many stations.

To demonstrate the impact of the new metro line on different categories of commuters, metro commuters are identified and classified according to their OD choices. Flow distribution, travel time, and travel time reliability of different groups of commuters are analysed. The results indicate that the new metro line influences commuters with different OD choices to different degrees. The results in this study provide reference for guiding and evaluating rail transit network planning in cities with large populations and have a potential to optimize and coordinate the operation between new transit lines and existing transit lines.

Several directions for further research are worth noting. First, although a big proportion of metro trips in Nanjing are made with smart card, it cannot be assumed that the flow pattern and travel times of passengers with one-way tickets are the same as smart card users. Due to the difficulty of collecting data of one-way tickets, the study presented in this paper can be extended by conducting activity and travel surveys. Second, in this study, smart card data from January to February 2017 are used to investigate the impact of the new metro line. In further research, more data after March 2017 are expected to be obtained to study passenger flow patterns and metro travel times to see how the new line contributes to the whole network after a long-term operation. Third, the analysis of travel time reliability presented in this paper can be extended to reliability-based accessibility by including detailed demographic information and land use data.

Conflicts of Interest

The authors declare that they have no conflicts of interest.

Acknowledgments

The work described in this paper was jointly supported by the Natural Science Foundation of Jiangsu Province (BK20160676), the National Natural Science Foundations of China (71601045, 71501038, and 41571375), and the Social Science Foundation of Jiangsu Province (16GLC008).

References

[1] M. Bagchi and P. R. White, "The potential of public transport smart card data," *Transport Policy*, vol. 12, no. 5, pp. 464–474, 2005.

[2] C. Chen, J. Ma, Y. Susilo, Y. Liu, and M. Wang, "The promises of big data and small data for travel behavior (aka human mobility) analysis," *Transportation Research Part C: Emerging Technologies*, vol. 68, pp. 285–299, 2016.

[3] C.-J. Cai, E.-J. Yao, S.-S. Liu, Y.-S. Zhang, and J. Liu, "Holiday destination choice behavior analysis based on AFC data of urban rail transit," *Discrete Dynamics in Nature and Society*, vol. 6, pp. 1–7, 2015.

[4] Y. Sun, J. Shi, and P. M. Schonfeld, "Identifying passenger flow characteristics and evaluating travel time reliability by visualizing AFC data: a case study of Shanghai Metro," *Public Transport*, vol. 8, no. 3, pp. 341–363, 2016.

[5] Y. Ma, W. Xu, X. Zhao, and Y. Li, "Modeling the hourly distribution of population at a high spatiotemporal resolution using subway smart card data: A case study in the central area of Beijing," *ISPRS International Journal of Geo-Information*, vol. 6, no. 5, 2017b.

[6] F. Zhou and R.-H. Xu, "Model of passenger flow assignmentfor Urban rail transit based on entryand exit time constraints," *Transportation Research Record*, no. 2284, pp. 57–61, 2012.

[7] J. Zhao, F. Zhang, L. Tu et al., "Estimation of Passenger Route Choice Pattern Using Smart Card Data for Complex Metro Systems," *IEEE Transactions on Intelligent Transportation Systems*, vol. 18, no. 4, pp. 790–801, 2017.

[8] S. Tao, J. Corcoran, F. Rowe, and M. Hickman, "To travel or not to travel: 'Weather' is the question. Modelling the effect of local weather conditions on bus ridership," *Transportation Research Part C: Emerging Technologies*, vol. 86, pp. 147–167, 2018.

[9] Y. Wang, G. H. D. A. Correia, E. D. Romph, and H. J. P. Timmermans, "Using metro smart card data to model location choice of after-work activities: An application to Shanghai," *Journal of Transport Geography*, vol. 63, pp. 40–47, 2017.

[10] W. Zhu, W. Wang, and Z. Huang, "Estimating train choices of rail transit passengers with real timetable and automatic fare collection data," *Journal of Advanced Transportation*, vol. 2017, Article ID 5824051, 12 pages, 2017.

[11] J. Kim, J. Corcoran, and M. Papamanolis, "Route choice stickiness of public transport passengers: Measuring habitual bus ridership behaviour using smart card data," *Transportation Research Part C: Emerging Technologies*, vol. 83, pp. 146–164, 2017.

[12] W. Jang, "Travel time and transfer analysis using transit smart card data," *Transportation Research Record*, no. 2144, pp. 142–149, 2010.

[13] X. Ma, Y. J. Wu, Y. Wang, F. Chen, and J. Liu, "Mining smart card data for transit riders travel patterns," *Transportation Research Part C: Emerging Technologies*, vol. 36, pp. 1–12, 2013.

[14] H. Faroqi, M. Mesbah, and J. Kim, "Spatial-temporal similarity correlation between public transit passengers using smart card data," *Journal of Advanced Transportation*, vol. 4, pp. 1–15, 2017.

[15] J. Zhao, Q. Qu, F. Zhang, C. Xu, and S. Liu, "Spatio-Temporal Analysis of Passenger Travel Patterns in Massive Smart Card Data," *IEEE Transactions on Intelligent Transportation Systems*, vol. 99, pp. 1–12, 2017a.

[16] K. Kim, K. Oh, Y. K. Lee, S. Kim, and J.-Y. Jung, "An analysis on movement patterns between zones using smart card data in subway networks," *International Journal of Geographical Information Science*, vol. 28, no. 9, pp. 1781–1801, 2014.

[17] L. Sun and K. W. Axhausen, "Understanding urban mobility patterns with a probabilistic tensor factorization framework," *Transportation Research Part B: Methodological*, vol. 91, pp. 511–524, 2016.

[18] Z. Zhao, H. N. Koutsopoulos, and J. Zhao, "Individual mobility prediction using transit smart card data," *Transportation Research Part C: Emerging Technologies*, vol. 89, pp. 19–34, 2018.

[19] C. Morency, M. Trépanier, and B. Agard, "Measuring transit use variability with smart-card data," *Transport Policy*, vol. 14, no. 3, pp. 193–203, 2007.

[20] S. Tao, D. Rohde, and J. Corcoran, "Examining the spatial-temporal dynamics of bus passenger travel behaviour using smart card data and the flow-comap," *Journal of Transport Geography*, vol. 41, pp. 21–36, 2014.

[21] L.-M. Kieu, A. Bhaskar, and E. Chung, "A modified density-based scanning algorithm with noise for spatial travel pattern analysis from smart card AFC data," *Transportation Research Part C: Emerging Technologies*, vol. 58, pp. 193–207, 2015.

[22] E. Manley, Z. Chen, and M. Batty, "Spatiotemporal variation in travel regularity through transit user profiling," *Transportation*, vol. 45, no. 3, pp. 703–732, 2016.

[23] L. M. Kieu, A. Bhaskar, and E. Chung, "Passenger segmentation using smart card data," *IEEE Transactions on Intelligent Transportation Systems*, vol. 16, no. 3, pp. 1537–1548, 2015a.

[24] X. Ma, C. Liu, H. Wen, Y. Wang, and Y. Wu, "Understanding commuting patterns using transit smart card data," *Journal of Transport Geography*, vol. 58, pp. 135–145, 2017.

[25] J. Y. Park, D.-J. Kim, and Y. Lim, "Use of smart card data to define public transit use in Seoul, South Korea," *Transportation Research Record*, no. 2063, pp. 3–9, 2008.

[26] Y.-S. Zhang and E.-J. Yao, "Splitting travel time based on AFC Data: Estimating walking, waiting, transfer, and in-vehicle travel times in metro system," *Discrete Dynamics in Nature and Society*, vol. 1, pp. 1–11, 2015.

[27] M. Bernal, E. W. Welch, and P. S. Sriraj, "The effect of slow zones on ridership: An analysis of the Chicago Transit Authority "El" Blue Line," *Transportation Research Part A: Policy and Practice*, vol. 87, pp. 11–21, 2016.

[28] J. Chan, *Rail transit OD matrix estimation and journey time reliability metrics using automated fare data. MS thesis*, Massachusetts Institute of Technology, Cambridge, 2007.

[29] D. Uniman, *Service reliability measurement framework using smart card data: application to the London underground. MS thesis*, Massachusetts Institute of Technology, Cambridge, 2009.

[30] Z. Liu, Y. Yan, X. Qu, and Y. Zhang, "Bus stop-skipping scheme with random travel time," *Transportation Research Part C: Emerging Technologies*, vol. 35, pp. 46–56, 2013.

[31] X. Fu, W. H. K. Lam, and B. Y. Chen, "A reliability-based traffic assignment model for multi-modal transport network under demand uncertainty," *Journal of Advanced Transportation*, vol. 48, no. 1, pp. 66–85, 2014.

[32] Z.-L. Ma, L. Ferreira, M. Mesbah, and A. T. Hojati, "Modeling bus travel time reliability with supply and demand data from automatic vehicle location and smart card systems," *Transportation Research Record*, vol. 2533, pp. 17–27, 2015.

[33] B. Y. Chen, H. Yuan, Q. Li et al., "Measuring place-based accessibility under travel time uncertainty," *International Journal of Geographical Information Science*, vol. 31, no. 4, pp. 783–804, 2017.

[34] M. L. Senior, "Impacts on travel behaviour of Greater Manchester's light rail investment (Metrolink Phase 1): evidence from household surveys and Census data," *Journal of Transport Geography*, vol. 17, no. 3, pp. 187–197, 2009.

[35] X. Cao and J. Schoner, "The influences of light rail transit on transit use: an exploration of station area residents along the Hiawatha Line in Minneapolis," in *Proceedings of the Transportation Research Board 92nd Annual Meeting*, 2013.

[36] J. Harvey, N. Thorpe, M. Caygill, and A. Namdeo, "Public attitudes to and perceptions of high speed rail in the UK," *Transport Policy*, vol. 36, pp. 70–78, 2014.

[37] J. Gadziński and A. Radzimski, "The first rapid tram line in Poland: How has it affected travel behaviours, housing choices and satisfaction, and apartment prices?" *Journal of Transport Geography*, vol. 54, pp. 451–463, 2016.

[38] S. Saidi, S. C. Wirasinghe, and L. Kattan, "Long-term planning for ring-radial urban rail transit networks," *Transportation Research Part B: Methodological*, vol. 86, pp. 128–146, 2016.

[39] A. Hong, M. G. Boarnet, and D. Houston, "New light rail transit and active travel: A longitudinal study," *Transportation Research Part A: Policy and Practice*, vol. 92, pp. 131–144, 2016.

[40] S. Li, "Influence of a New Subway Line's Opening on Passenger Flow Characteristics of an Urban Rail Transit Network," in *Proceedings of the 15th COTA International Conference of Transportation Professionals: Efficient, Safe, and Green Multimodal Transportation, CICTP 2015*, pp. 1756–1769, China, July 2015.

[41] S. Liu, E. Yao, and B. Li, "Exploring urban rail transit station-level ridership growth with network expansion," *Transportation Research Part D Transport & Environment*.

[42] M.-P. Kwan, "Beyond Space (As We Knew It): Toward Temporally Integrated Geographies of Segregation, Health, and Accessibility: Space-Time Integration in Geography and GIScience," *Annals of the Association of American Geographers*, vol. 103, no. 5, pp. 1078–1086, 2013.

[43] B. Y. Chen, Y. Wang, D. Wang, Q. Li, W. H. K. Lam, and S. L. Shaw, "Understanding the impacts of human mobility on accessibility using massive mobile phone tracking data," *Annals of the American Association of Geographers*, vol. 108, pp. 1115–1133, 2018.

[44] X. Fu and W. H. K. Lam, "A network equilibrium approach for modelling activity-travel pattern scheduling problems in multi-modal transit networks with uncertainty," *Transportation*, vol. 41, no. 1, pp. 37–55, 2014.

[45] K. M. N. Habib, L. Kattan, and T. Islam, "Model of personal attitudes towards transit service quality," *Journal of Advanced Transportation*, vol. 45, no. 4, pp. 271–285, 2011.

[46] H. Wakabayashi and Y. Matsumoto, "Comparative study on travel time reliability indexes for highway users and operators," *Journal of Advanced Transportation*, vol. 46, no. 4, pp. 318–339, 2012.

13

Metro Timetabling for Time-Varying Passenger Demand and Congestion at Stations

Keping Li,[1] Hangfei Huang,[1] and Paul Schonfeld[2]

[1] *State Key Laboratory of Rail Traffic Control and Safety, Beijing Jiaotong University, Beijing 100044, China*
[2] *Department of Civil & Environmental Engineering, University of Maryland, College Park, MD 20742, USA*

Correspondence should be addressed to Hangfei Huang; 14114200@bjtu.edu.cn

Academic Editor: Giulio E. Cantarella

For the train timetabling problem (TTP) in a metro system, the operator-oriented and passenger-oriented objectives are both important but partly conflicting. This paper aims to minimize both objectives by considering frequency (in the line planning stage) and train cost (in the vehicle scheduling stage). Time-varying passenger demand and train capacity are considered in a nonsmooth, nonconvex programming model, which is transformed into a mixed integer programming model with a discrete time-space graph (DTSG). A novel dwell time determining process considering congestion at stations is proposed, which turns the dwell times into dependent variables. In the solution approach, we decompose the TTP into a subproblem for optimizing segment travel times (OST) and a subproblem for optimizing departure headways from the shunting yard (OH). Branch-and-bound and frequency determining algorithms are designed to solve OST. A novel rolling optimization algorithm is designed to solve OH. The numerical experiments include case studies on a short metro line and Beijing Metro Line 4, as well as sensitivity analyses. The results demonstrate the predictive ability of the model, verify the effectiveness and efficiency of the proposed approach, and provide practical insights for different scenarios, which can be used for decision-making support in daily operations.

1. Introduction

With increasing concerns about urban congestion and climate change, urban metro rail transportation receives increasing attention due to its high capacity, punctuality and sustainability. Metro timetabling is the problem of assigning precise utilization times for infrastructure resources to every train in the metro system [1]. In large cities, a metro system is often the busiest public transportation system. To meet travel demand and passenger satisfaction, a timetable should be as compact and flexible as possible. Meanwhile, since many metro systems need government subsidies to cover operating expenses, the planners focus more on operating cost than on passenger factors when designing a timetable [2]. This motivates some recent studies to combine these two conflicting objectives in the train timetabling problem [3, 4].

Two types of timetables are used in metro systems: cyclic and noncyclic. In a cyclic timetable, the departure times of trains are scheduled in equally spaced cycles (e.g., a half hour cycle). In a railway system, the cyclic timetables are assumed to be preferred by both operators and passengers because such timetables are easy to operate and remember. However, in a metro system, passengers usually arrive randomly at stations and wait for the next available train, without checking precise timetables beforehand. Besides, as pointed out by Robenek et al. [5], a cyclic timetable provides an inefficient operation as there is a mismatch between the supply (determined by the timetables) and the demand (characterized by the time-varying demand). On the other hand, a noncyclic timetable imposes no special rules on the departure times of trains [6]. The noncyclic timetables are more flexible regarding time-varying passenger demand, especially when the demand is large [2]. Thus, the noncyclic timetabling problem is worth exploring [2, 3, 7].

The planning process in public transportation can be generally split into three stages: line planning, timetabling, and vehicle scheduling [8]. The line planning stage determines the route/station layouts and frequency. Timetabling is based on the output of line planning, after which the vehicle scheduling can be designed. It is pointed out by Schöbel [8] that going through all these stages sequentially (i.e., independently) leads to unsatisfactory solutions. Since they are interrelated, important factors and objectives at the line planning stage and the vehicle scheduling stage should be accounted for in the timetabling problem.

This paper proposes a rolling optimization algorithm to obtain noncyclic timetables considering time-varying passenger demand and the effects of congestion at stations. It integrates the objectives in the line planning (frequency), timetabling (conflicting objectives including passenger wait/in-vehicle time and energy), and vehicle scheduling (train cost). The main contributions of this paper are as follows.

(1) This paper specifically models the dynamic evolution of passenger loads on trains at each station, by considering passenger arrival rates, limited train capacity and actual passenger alighting/boarding rates associated with congestion. This is an extension of existing studies [9]. Particularly, Niu and Zhou [10] and Niu et al. [11] propose the concept of "effective loading time" that represents the actual time interval within which the arriving passengers can board a train. The detailed boarding and alighting processes are not discussed in their model. Wang et al. [3] set a lower bound (which accounts for details including the number of boarding/alighting passengers and their boarding/alighting rates) for dwell times but consider the dwell times as variables to be optimized, rather than parameters that depend on arrival/departure times and passenger demand. Impacts of gradually increasing passenger demand are analyzed in Robenek et al. [2]; however, in modelling they do not consider the effects of passenger congestion.

(2) This paper integrates objectives from different planning stages in the train timetabling problem, for which solution approaches are scarce [8]. Particularly, Schöbel [8] considers the line planning, timetabling and vehicle scheduling in an integrated way, with passenger- and cost-oriented objectives. However, since his eigenmodel is general and resulting approaches are generic, important objectives described in this paper are not considered in his work, that is, passenger wait time, dwell time, and train capacity. These objectives are included in [2, 5], whereas the frequency and dwell times are fixed in their studies. Besides, energy is not considered in [2, 5].

(3) This paper proposes a novel and effective solution approach, which decomposes the master timetabling model into two subproblems. The first subproblem is solved by branch-and-bound and frequency determining algorithms, and the second one is solved by a rolling optimization algorithm which optimizes the departure headways from the shunting yard (which are equivalent to arrival times of trains at their first station) and considers interdependent variables in each rolling step. The design of decision variables in this paper is similar to [2, 5], but the models and solution approaches are different. Compared to Wang et al. [3], which addresses the similar timetabling problem with traditional approaches such as sequential quadratic programming and a genetic algorithm, the approach proposed in this paper performs better computationally in terms of run time and solution quality.

The remainder of this paper is as follows. Section 2 reviews relevant studies on the train timetabling problem and summarizes the research gap in the literature. Section 3 first describes the problem, introduces formulation of different cost functions, and constructs the train timetabling model. Then the discrete time-space graph is proposed to transform the model into a mixed integer program (MIP). Section 4 discusses the solution approach for the optimization model. Section 5 presents different numerical experiments and sensitivity analyses on a short metro line and on Beijing Metro Line 4, to demonstrate the predictive ability of the model as well as the effectiveness and efficiency of the proposed approach. Section 6 summarizes the works done in this paper and presents the limitations and potential topics in future studies.

2. Literature Review

The train timetabling problem (TTP) aims to schedule trains to transport passengers (or goods) without conflicts, by specifying train arrival and departure times at stations. It is interrelated with other planning stages, that is, the line planning stage and the vehicle scheduling stage. We organize the related literature from the perspectives of operator-oriented objectives, passenger-oriented objectives, and their integration.

2.1. Operator-Oriented Objectives. In the literature, operator-oriented objectives can be found only for the noncyclic version of the TTP [2]. Most studies focus on the operation indicators, for example, energy [17, 18], train delay [19–21], and train travel time [22–24].

The total train delay and train travel time are usually considered in the TTP for railway networks, which aims to find a feasible timetable by minimizing the profit loss resulting from changes to the ideal/planned timetable. Energy is the focus and main cost for metro system operators. Most energy is consumed in train operations [25, 26], and thus obtaining an energy-efficient train timetable receives considerable attention [17, 18, 27]. Note that energy is affected by the travel time in the segment, and the segment travel time is determined by the arrival and departure times at adjacent stations, that is, the timetable. Hence, the energy-efficient train timetabling problem may be extended to include other objectives. For example, Yang et al. [23] consider both energy and train travel time as optimization objectives. Some recent studies also account for regenerative braking. Yang et al. [28] develop a scheduling approach to coordinate arrivals and departures of trains within the same electricity supply zones, thereby effectively recovering the regenerative energy. Generally, these studies focus solely on the timetabling stage, where frequency (in line planning) is fixed and train cost (in-vehicle scheduling) is not considered.

The efficient use of railway rolling stock (vehicle scheduling) is an important objective pursued by a railway agency or company because of intensive capital investment in rolling stock [29]. To this end, Lai et al. [29] develop an optimization model to improve the efficiency of rolling stock usage considering necessary regulations and practical constraints, where a hybrid heuristic process is designed to improve solution quality and efficiency. Haahr et al. [27] use CPLEX and a column and row generation approach to assign rolling stock units to timetable services in passenger railways, prepare daily schedules, and check their real-time applicability by testing different disruption scenarios. These vehicle scheduling studies require the trips as input data. Based on Schöbel [8], in a metro system, every trip describes the operation of a train between the start and end time of the line at its first and its last stations (given from the timetable). The objective function then aims at minimizing the number of trains and the costs for their movements.

Particularly, Schmid and Ehmke [30] and Schöbel [8] demonstrate that the integration of timetabling and vehicle scheduling is more beneficial than the sequential planning process. However, the studies on the integration of timetabling and vehicle scheduling usually adopt relatively general and generic models, and the TTP model is often simplified [31]. For example, Cadarso and Marin [32] focus on shunting operations and their timetable is calculated by readjusting frequencies, whereas specific details such as the dwell times and segment travel times are not considered. This motivates us to consider train cost in the timetabling stage, while focusing on the specified TTP formulation.

In addition, the conflict-free timetable is also pursued by the operators [33, 34]. It is worth mentioning that the discrete time-space graph proposed in Caprara et al. [6] is a directed multigraph in which nodes correspond to arrivals/departures at a certain station and at a given time stamp. This graph is widely used to formulate the TTP and derive different integer programming models that correspond to specific objectives. For example, Caprara et al. [6] use the graph to derive an integer linear programming (ILP) model with Lagrangian relaxation, which is embedded within a heuristic algorithm. Cacchiani et al. [12] use the graph to formulate an ILP with linear programming (LP) relaxation. Cacchiani et al. [35] use the time-space graph-based LP relaxation of an ILP to derive a dual bound in the TTP for a set of stations in an urban area interconnected by tracks, thus aiming to resolve the conflicts and evaluate the capacity saturation.

2.2. Passenger-Oriented Objectives. For a metro system, passenger-oriented timetables that consider reliability and reduction of passenger time are most desirable [11]. In some studies, the passenger demand is considered stable [36]. Assuming that passengers prefer easily memorable timetables, such timetables are usually cyclic. They are designed on the basis of a period event scheduling problem (PESP), aiming to minimize passenger travel and wait time [15, 37] and to maximize network stability [38]. However, as mentioned above, metro passengers arrive randomly and do not remember the timetable. Besides, cyclic timetables are less flexible than noncyclic ones in accounting for the passenger demand [2].

Recent studies consider time-varying demands due to the growing concerns for service level and congestion at stations. Such timetables are noncyclic, with objectives of improving passenger satisfaction [3, 7, 10]. Particularly, Barrena et al. [7] propose two nonlinear programming (NP) formulations to generalize noncyclic train timetables on a single line, which are solved by a fast adaptive large neighborhood search metaheuristic. The objective is to minimize passenger wait times at stations. Niu et al. [11] construct mathematically rigorous and algorithmically tractable nonlinear mixed integer programming (MIP) models for both real-time scheduling and medium-term planning applications, to jointly synchronize effective passenger loading time windows and train arrival and departure times at each station. Their work aims to minimize the total waiting times of passengers at stations. Robenek et al. [5] define a timetable as a set of departure times of every train from its origin station on every line and consider four attributes in passenger satisfaction: the in-vehicle-time, the waiting time at transfers, the number of transfers, and the schedule passenger delay. They describe the hybrid timetables through additional constraints that are imposed on the original passenger centric train timetabling problem [2], so that the passengers would obtain the same level of service as a cyclic timetable, with more flexibility. A specifically defined simulated annealing heuristic is proposed to solve the problem. Luan et al. [16] integrate the TTP and preventive maintenance time slots (PMTSs) planning problem on a general railway network. They propose an MILP model to jointly optimize train routes, orders and passing times at each station, and PMTSs. The objective is to minimize the sum of the absolute arrival time deviations of real trains at destinations between the ideal and actual timetables. A Lagrangian-based label correcting algorithm is designed for solving the time-dependent least cost path problem.

In addition, the delay management (DM) problem determines whether trains should wait for a delayed feeder train or should depart on time, while considering passenger-oriented objectives [13]. Some recent studies integrate the macroscopic DM with the microscopic TTP [39]. Particularly, Dollevoet et al. [14] formulate an integer programming (IP) model and propose an iterative optimization approach to solve such a bilevel problem that the macroscopic level is the delay management and the microscopic level is the train timetabling. The optimization approach repeats a process that uses DM to find a solution and uses TTP to validate it, until a feasible solution is found. It should be noted that the graph-based models are adopted in their work, which simplify the formulations: the TTP is formulated with the alternative graph [40], and the DM is formulated with an event-activity network [15].

2.3. Integrated Objectives. Since the operator- and passenger-oriented objectives are partly conflicting, focusing on only one side would yield undesirable solutions; for example, the best possible service for passengers may also be the most costly alternative for the operator [2]. Thus, many TTP studies integrate these objectives jointly [22, 41].

Wang et al. [3] aim to minimize passenger travel time and energy in their nonsmooth nonconvex programming (NSNCP) model. They propose an event-driven model that involves arrival and departure events and passenger arrival rates change events and use nonlinear programming approaches and evolutionary algorithms to solve the problem.

Qi et al. [42] focus on a line planning problem with budget constraints and evaluate the station layout with train timetable indicators. The objectives include constructing cost for additional tracks, total travel time, and network capacity. A mixed integer linear programming (MILP) model is formulated to address the TTP, and then a bilevel programming (BP) model is designed to address the integrated problem. Both CPLEX and a local search-based heuristic are developed to solve the models.

Robenek et al. [2] model the PCTTP as an MILP model with objectives of maximizing the operator's profit while maintaining passenger satisfaction. The model uses the output of the line planning problem as the input of the train timetabling problem, to address the gap found between them. CPLEX is used to construct a Pareto Frontier of the contradictory objectives.

Li et al. [43] combine dynamic train regulation and passenger flow control to minimize the timetable and the headway deviations for metro lines, thereby reducing the operator profit loss and the passenger delay. The model predictive control and a quadratic programming algorithm are proposed to solve the problem, and stability analysis is conducted to verify the system convergence.

However, the above studies do not consider the impacts of frequency. Schöbel [44] reviews some approaches to model and solve the line planning problem, which demonstrate the impacts of frequency to both operator- and passenger-oriented objectives. Particularly, Samanta and Jha [45] consider different objectives in the line planning stage which include minimizing user cost, operator cost, and location cost and maximizing the ridership or the service coverage of the line. The objectives are separately formulated with NP models, and a GA is developed to optimize the objective functions subject to constraints on the number and spacing of stations. Fu et al. [46] propose a BP model to determine line frequencies and the stopping patterns for a mix of fast and slow train lines, with the objectives of minimizing passenger travel time and the number of transfers and maximizing train capacity occupancy. The heuristics solve the problem by combining additional conflicting demand-supply factors. Cascetta and Coppola [47] specify both the frequency-based assignment models and the timetable-based models in the demand forecasting, to analyze their differences in modal split estimates and flows on individual trains. The results show that when passenger demand is time-varying or the timetable is noncyclic, the difference is significant. Thus, the integration of frequency and timetabling is worth investigating.

2.4. Research Gap. Table 1 highlights some publications related to the TTP. All the graph-based models use integer programs, including the model in this paper. Although different objectives are considered, very few studies integrate the objectives in all three stages. Particularly, Wang et al. [3] only consider the objectives in one stage, that is, timetabling stage; Robenek et al. [2] integrate passenger satisfaction and operating profit (including train cost) but do not consider frequency and energy; Schöbel [8] takes into account objectives in all stages with an iterating framework; however, crucial factors such as passenger wait time, dwell time, and train capacity are missing. Kroon et al. [15] integrate the objectives in the TTP and vehicle scheduling but omit energy.

Apart from the studies focusing on the delay management, most TTP-related studies optimize arrival and departure times of trains at each station; that is, the dwell times are optimized in the solution approach. However, the dwell times are heavily affected by the number of boarding and alighting passengers and their boarding and alighting rates, associated with congestion at stations. Thus, they should be considered as parameters that are dependent on other decision variables, that is, the departure times from the first station [2, 5] and the segment travel times. In this paper, we account for these characteristics and introduce specific equations to determine the dwell times in different scenarios. Such a process for determining dwell times has not been found in the literature.

Finally, heuristics and exact approaches are designed to find solutions based on the properties of the models. Particularly, rolling horizon [3] and iterative optimization [8, 14] are used to solve problems that have similar properties to our proposed model. However, the boundary constraints of the rolling horizon limit the solution flexibility of our model; the iterative optimization considers different stages with same importance, while this paper focuses only on the timetabling stage. Besides, the algorithms used in the rolling horizon and iterative optimization are relatively generic [3, 8]. Thus, a more specific algorithm is desirable for better solutions. In this regard, we develop a novel rolling optimization approach to solve the model, which combines the benefits of rolling horizon and iterative optimization.

3. Model Formulation

In this section, we present a mixed integer programming formulation for the TTP. It is a multiobjective optimization problem. The weighted-sum scalarization of different objective functions in terms of generalized monetary cost is used to resolve the tradeoffs among them [8].

3.1. Problem Formulation. Table 2 summarizes the parameters and variables used in this paper. Some are also explained further in the text. The model's simplifying assumptions are listed below and also explained in the problem description.

Assumption 1. Within one period, trains travel through each segment with the same speed profile; that is, the travel times for different trains in a segment are equal.

Assumption 2. We assume that the last train arrives at the first station at the end time of the period, to ensure that no late-arriving passengers will be left behind.

Assumption 3. Passenger arrival rate $\lambda_i(t)$ should not exceed total passenger boarding rate $\zeta_b \cdot n_d$. If $\lambda_i(t) > \zeta_b \cdot n_d$, the queue of waiting passengers increases even if a train is dwelling at

TABLE 1: Summary of publications related to the TTP.

Work	Model	Objectives	Decision variable	Solution approach
Caprara et al. [6]	ILP	Maximize profit	Arrival/departure times	Lagrangian-based heuristic
Cacchiani et al. [12]	ILP	Maximize profit	A full timetable of a train	Column generation, decomposition, and heuristic
Dollevoet et al. [13]	IP	Minimize total passenger delay	If a connection is maintained/used	Modified Dijkstra
Niu and Zhou [10]	IP	Minimize passenger wait time	Arrival/departure times	GA
Barrena et al. [7]	LP	Minimize passenger wait time	Arrival/departure times	Branch and bound
Dollevoet et al. [14]	IP	Minimize (1) passenger arrival times, (2) maximize consecutive delay	Connection and arrival/departure times	Iterative optimization approach
Kroon et al. [15]	MIP	Minimize the number of trains, passenger transfer time, and total travel time	Arrival/departure times	CPLEX
Niu et al. [11]	MIP	Minimize passenger wait time	Arrival/departure times	GAMS
Wang et al. [3]	NSNCP	Minimize passenger travel time and energy	Arrival/departure times	Rolling horizon with SQP and GA
Robenek et al. [2]	MILP	Maximize operating profit while maintaining passenger satisfaction	Departure times at first station	CPLEX
Robenek et al. [5]	MILP	Maximize passenger satisfaction	Departure times at first station	Simulated Annealing (SA) heuristic
Schöbel [8]	Eigenmodel	Passenger- and operator-oriented objectives in all three planning stages	Frequency, A/D times, trip assignment	Iterating algorithm
Luan et al. [16]	MILP	Minimize absolute arrival time deviations	A/D times, route selection	Lagrangian relaxation
Yin et al. [4]	MILP	Minimize passenger wait time and energy	A/D times, train control	Lagrangian-based heuristic
This paper	MIP	Minimize passenger wait and travel time, energy, and train cost	Segment travel times and arrival times at the first station	Branch and bound, Rolling optimization

the station. In practice, passenger entrance flow control* is taken to avoid such a case.

*Remark**. At most busy metro stations in China, passenger entrance flow control is adopted during peak hours to ensure operation safety, and thus the passenger arrival rate is a restricted number, which is expected not to exceed passenger boarding rate. Assumption 3 is introduced for consistency with practical experience.

Assumption 4. All passengers in train k who have reached their destination are assumed to exit the train at its arrival time of station i, that is, $t^0_{k,i}$. All waiting passengers at station i able to board train k are assumed to board the train at its departure time, that is, $t^b_{k,i}$. In other words, the number of in-vehicle passengers $P^v_{k,t}$ changes only at the arrival/departure times of train k at stations.

This paper focuses on a bidirectional metro line with a single track, as shown in Figure 1. Properties of such a metro line are described in [48]. The timetabling problem is studied within a given period. The time-varying passenger demand is given as input, and a series of trains (whose frequency is to be optimized) are scheduled to transport the passengers.

The arrival times at the first station are decision variables that must be optimized. The departure times at the first station are determined by the dwell times. As mentioned earlier, this paper obtains the dwell times based on trains' conditions when they arrive at stations. In this regard, optimizing arrival times at the first station is equivalent to optimizing departure times at the first station, as in [2, 5]. For the second station, the arrival times are determined by the segment travel time in the first segment. Practically, for the metro operators, uniform segment travel times are preferred within one period, because they not only make the system more stable and easier to manage but also simplify synchronized passenger transfers with connecting transit lines. Assumption 1 is introduced to consider this preference. Thus, within one period, the segment travel time in each segment needs to be optimized only once (for all trains within this period). The speed profile

TABLE 2: Parameters and variables.

Name	Description	Units	Purpose
$\mathcal{I}$	Set of stations	-	Set
$\mathcal{K}$	Set of trains	-	Set
$\mathcal{U}^i$	Set of time stamps in which trains can arrive at station i	-	Set
$\mathcal{W}^i$	Set of time stamps in which trains departure from station i	-	Set
θ	Shunting yard of the metro line	-	Symbol
i, j	Station index	-	Symbol
I	Total number of stations	-	Symbol
k	Train index	-	Symbol
K	Total number of trains within a given period/Frequency	-	Symbol
t	Time index	-	Symbol
u_k	Arrival node of train k	-	Symbol
w_k	Departure node of train k	-	Symbol
$\pi(u_k)$	Time stamp associated with node u_k (or w_k)	-	Symbol
$\Delta(u_k, w_{k'})$	Time duration between two nodes u_k and $w_{k'}$ ($k = k'$ is allowed)	-	Symbol
R^k	Path of train k	-	Symbol
T	Time duration	-	Symbol
$\lfloor a \rfloor$	The largest integer that is smaller than real number a	-	Symbol
$\lceil a \rceil$	The smallest integer that is larger than real number a	-	Symbol
$\overrightarrow{\mathbf{b}}$	A vector (represented by any bold alphabet)	-	Symbol
$\mathbb{M}$	A matrix (represented by any double-struck alphabet)	-	Symbol
$\mathbb{M}_{a:b,x:y}$	A sub-matrix whose elements correspond to those in matrix $\mathbb{M}$, from rows a to b, and from columns from x to y	-	Symbol
$\|\mathbb{M}\|$	Sum of all (numerical) elements in matrix $\mathbb{M}$	-	Symbol
c_e	Average energy cost	¥/Kwh	Parameter
c_o	Average train operating cost	¥ /(train·km)	Parameter
c_{pt}	Average passenger travel time cost	¥/(passenger·hour)	Parameter
c_{pw}	Average passenger wait time cost	¥/(passenger·hour)	Parameter
c_{tra}	Average capital cost of one train	¥/(train·hour)	Parameter
C_{total}	Total cost of the whole system	¥	Parameter
C^e_{total}	Total cost of energy consumption	¥	Parameter
C^p_{total}	Total cost of passenger time	¥	Parameter
$C^{\mathrm{tra}}_{\mathrm{total}}$	Total cost of train (rolling stock)	¥	Parameter
$\overline{d_i}$	Maximum allowable dwell time at station i	Second	Parameter
$\underline{d_i}$	Minimum allowable dwell time at station i	Second	Parameter
e_i	Energy consumption per unit mass for one train in segment i	Kwh/Kg	Parameter
f_t	Tractive force per unit mass	$\mathrm{m/s^2}$	Parameter
f_r	Running resistance per unit mass	$\mathrm{m/s^2}$	Parameter
F_s	Fleet size (the number of trains)	Trains	Parameter
$h_{\min}$	Minimum headway	Second	Parameter
M_k	Net weight of a train	Kg/train	Parameter
M_p	Average weight of a passenger	Kg/passenger	Parameter
n_d	Number of doors per train	-	Parameter
N_k	Capacity of train k	Passengers	Parameter

Table 2: Continued.

Name	Description	Units	Purpose
$P^{a}_{k,i}$	Number of passengers alighting train k at station i	Passengers	Parameter
$P^{b}_{k,i}$	Number of passengers boarding train k at station i	Passengers	Parameter
$P^{b}_{k,(i,j)}$	Number of passengers boarding train k at station i with destination j	Passengers	Parameter
$P_i(t_1, t_2)$	Number of passengers arriving at station i during $[t_1, t_2]$	Passengers	Parameter
$P_{(i,j)}(t_1, t_2)$	Number of passengers arriving at station i with destination j during $[t_1, t_2]$	Passengers	Parameter
$P^{w}_{i,t}$	Number of passengers waiting at station i at time t	Passengers	Parameter
$P^{v}_{k,t}$	Number of passengers in train k at time t	Passengers	Parameter
$s_l(i)$	Length of segment i (between station i and $i+1$)	Meter	Parameter
$s_t^{\max}(i)$	Maximum segment travel time for segment i	Second	Parameter
$s_t^{\min}(i)$	Minimum segment travel time for segment i	Second	Parameter
$t^{0}_{k,i}$	Arrival time of train k at station i	Second	Parameter
$t^{a}_{k,i}$	Alighting end time for train k at station i	Second	Parameter
$t^{b}_{k,i}$	Boarding end time for train k at station i/Departure time	Second	Parameter
t^{p}_{end}	End time of passenger demand for a period	Second	Parameter
t^{t}_{end}	End time of a complete timetable for a period	Second	Parameter
T^{k}_{round}	Round trip time (total travel time from the departure from the shunting yard to the arrival at the shunting yard) of train k	Second	Parameter
T_{pt}	Total passenger travel time	Second	Parameter
T_{pw}	Total passenger wait time	Second	Parameter
T_{turn}	Turnaround time	Second	Parameter
$v_k(t)$	Speed of train k at time t	m/s	Parameter
$x_k(t)$	Position of train k at time t	Meter	Parameter
$\tau_{i,j}(t)$	Passenger arrival rate at station i with destination j at time t	Passengers/second	Parameter
$\lambda_i(t)$	Passenger arrival rate at station i at time t	Passengers/second	Parameter
ζ_a	Passenger alighting rate	Passengers/second	Parameter
ζ_b	Passenger boarding rate	Passengers/second	Parameter
χ_x	Parameter for energy approximation	-	Parameter
χ_y	Parameter for energy approximation	-	Parameter
$s_t(i)$	Segment travel time of segment i	Second	Decision
$t^{0}_{k,1}$	Arrival time of train k at the first station	Second	Decision

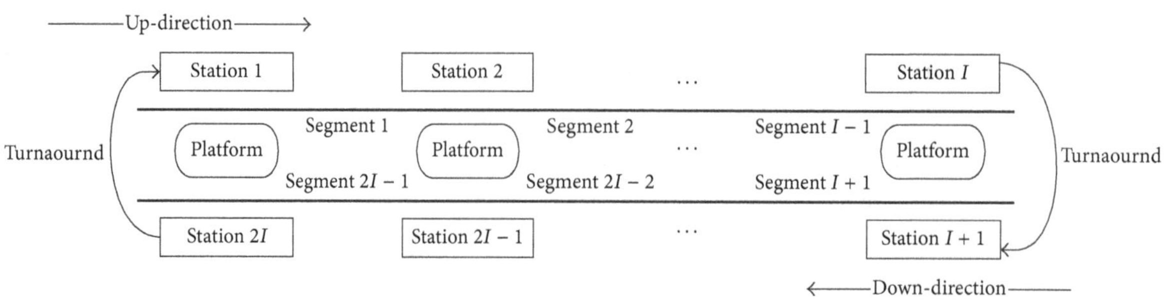

Figure 1: A bidirectional metro line with 2I stations.

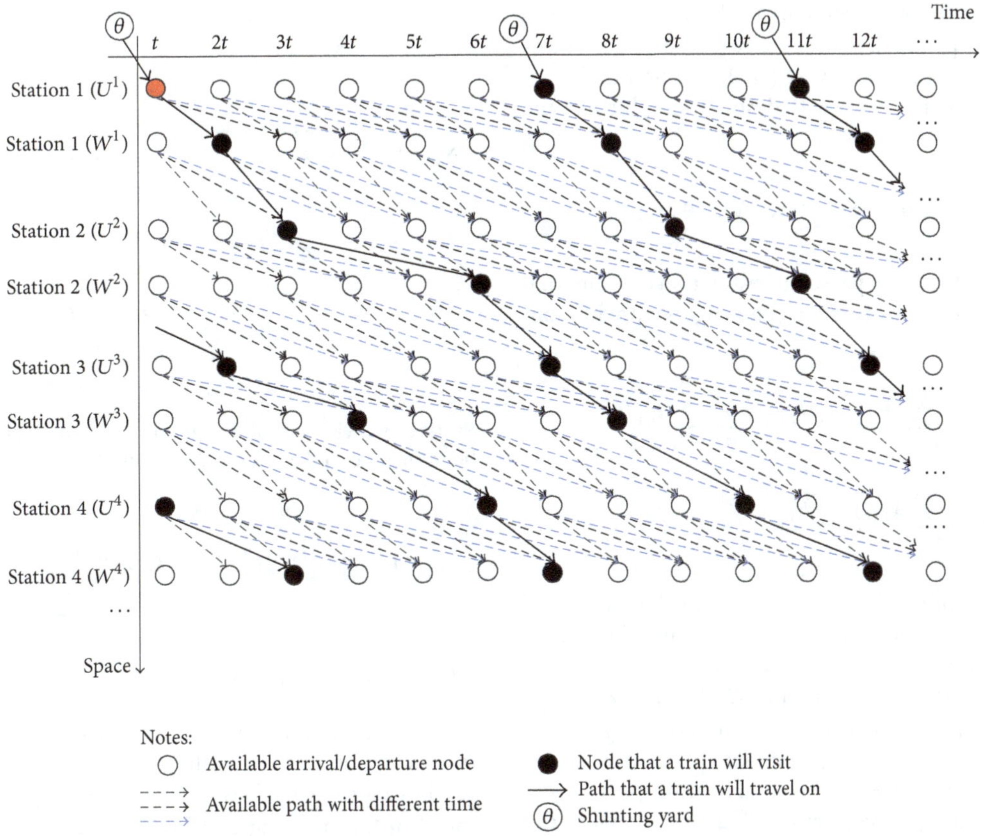

FIGURE 2: Representation of discrete time-space graph.

optimization is outside the scope of this paper. The departure times at the second station are also determined by the dwell time determining process.

Analogously, for subsequent train arrivals and departures, the arrival times are determined by the previous segment travel time, and the departure times are determined by the dwell times. Thus, within a period, the arrival times at the first station and the segment travel times determine the train timetable.

To represent the TTP, a directed multigraph called discrete time-space graph (DTSG) is proposed [6]. As shown in Figure 2, θ represents the shunting yard of the metro line and $\mathscr{U}^i$ ($\mathscr{W}^i$) corresponds to the set of time stamps in which trains can arrive at (depart from) station i. The nodes in $\mathscr{U}^i$ ($\mathscr{W}^i$) are called arrival (departure) nodes. Unlike the original graph that is designed for a railway network [6, 12], this paper considers a single-track metro line, and thus the constraints are different and implicitly imposed in the definition of the structure of the DTSG.

The continuous time is discretized into evenly spaced time intervals (1 second in this paper), and the period extends from 0 to t^t_{end}. Note that the values of t^t_{end} and t^p_{end} are different. t^p_{end} is the last time stamp of the passenger demand. The period mentioned above refers to the one extending from 0 to t^p_{end}. Assumption 2 imposes that the last train departs from the first station at time t^p_{end}. However, if the DTSG only considers the period from 0 to t^p_{end}, the arrival/departures of the last train at subsequent stations are not considered, and the passenger-oriented objectives are hard to formulate. Thus, the DTSG considers the period from 0 to t^t_{end}. The relation between t^t_{end} and t^p_{end} is set as:

$$t^t_{\text{end}} = t^p_{\text{end}} + \sum_{i=1}^{2I-1} s_t(i) + \sum_{i=1}^{2I} \overline{d_i}, \tag{1}$$

where $s_t(i)$ is the segment travel time of segment i and $\overline{d_i}$ is the maximum dwell time of station i. This setting ensures that even the slowest train can be completely included in the DTSG.

The construction of the DTSG is noncyclic. The objective functions of passenger time, energy, and train cost are associated with the chosen arcs and are described below.

3.2. Passenger Time

3.2.1. Passenger Arriving, Alighting, and Boarding

(1) Time-Varying Demand Information. The passenger demand at each station can be expressed by $\tau_{i,j}(t)$, which is defined as

$$\begin{aligned} &\tau_{i,j}(t) \geq 0, \quad \text{If } 1 \leq i < j \leq I \text{ or } I \leq i < j \leq 2I; \\ &\tau_{i,j}(t) = 0, \quad \text{Otherwise}; \\ &\forall i, j \in \mathscr{I}. \end{aligned} \tag{2}$$

Passenger arrival rates at stations are $\lambda_i(t) = \sum_{j=i+1}^{2I} \tau_{i,j}(t)$, $\forall i, j \in \mathcal{I}$. In this paper, demands are regarded as deterministic values, which are given by $\tau_{i,j}(t)$. Thus, the number of arriving passengers at station i with destination j during time interval $[t_1, t_2]$ is

$$P_{(i,j)}(t_1, t_2) = \sum_{t=t_1}^{t_2} \tau_{i,j}(t), \quad \forall i, j \in \mathcal{I}. \tag{3}$$

The number of arriving passengers at station i within time interval $[t_1, t_2]$ is the sum of arriving passengers with different destinations:

$$\begin{aligned} P_i(t_1, t_2) &= \sum_{j=i+1}^{2I} P_{(i,j)}(t_1, t_2) = \sum_{j=i+1}^{2I} \sum_{t=t_1}^{t_2} \tau_{i,j}(t) \\ &= \sum_{t=t_1}^{t_2} \lambda_i(t), \quad \forall i, j \in \mathcal{I}. \end{aligned} \tag{4}$$

(2) Determination of Dwell Arcs. Letting n_d be the number of doors per train, the average number of alighting (boarding) passengers per door when train k arrives at (departs from) station i is $P^a_{k,i}/n_d$ $(P^b_{k,i}/n_d)$. In addition, we denote the average number of boarding passengers per second (boarding rate) as ζ_b and alighting rate as ζ_a, which vary with passenger density in trains. For simplicity, we consider that boarding passengers wait until all alighting passengers get off the train; that is, the boarding process starts at time $t^a_{k,i}$.

The dwell time is the sum of alighting and boarding times. The calculation of alighting time is solely based on the number of alighting passengers:

$$t^a_{k,i} = t^0_{k,i} + \frac{P^a_{k,i}}{(n_d \cdot \zeta_a)}, \quad \forall k \in \mathcal{K},\ i \in \mathcal{I}. \tag{5}$$

The calculation of boarding time is slightly more complex. Figure 3 shows the relations of train remaining capacity, waiting passengers, boarding rate, and max/min dwell times. The horizontal axis represents time, and the vertical axis represents the number of passengers. These four factors determine the boarding time for train k at station i, and in Figure 3 they are represented by (1) remaining capacity of train k at station i: $N_k - P^v_{k,t^a_{k,i}}$, a horizontal line; (2) number of waiting passengers at station i at time t, a curve $P^w_{i,t}$; (3) maximum allowable number of boarding passengers, a diagonal function $f_k(t)$ with the slope of $n_d \cdot \zeta_b$; (4) minimum and maximum dwell times at station i, that is, $t^0_{k,i} + \underline{d_i}$ and $t^0_{k,i} + \overline{d_i}$. In Figure 3, the diagonal $P = f_k(t)$ intersects the horizontal line $P = N_k - P^v_{k,t^a_{k,i}}$ at time $t^{k,i}_1$ and intersects the curve $P = P^w_{i,t}$ at time $t^{k,i}_2$. The curve $P = P^w_{i,t}$ intersects the horizontal line $P = N_k - P^v_{k,t^a_{k,i}}$ at time $t^{k,i}_3$. Note that the time is discrete, and Assumption 3 ensures that the slope of $P = f_k(t)$ always exceeds the slope of $P = P^w_{i,t}$. Thus, we have

$$t^{k,i}_1 = \max\left\{t \mid \left(t - t^a_{k,i}\right) \cdot n_d \cdot \zeta_b \le N_k - P^v_{k,t^a_{k,i}}\right\}, \quad \forall k \in \mathcal{K},\ i \in \mathcal{I}, \tag{6}$$

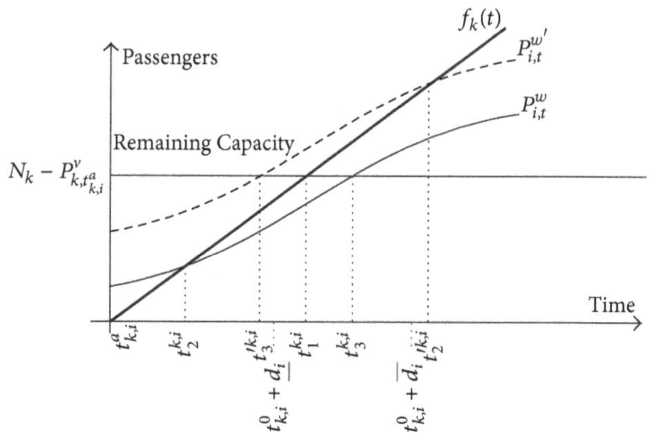

FIGURE 3: Relations of remaining capacity, waiting passengers, boarding rate, and max/min dwell times.

$$t^{k,i}_2 = \max\left\{t \mid \left(t - t^a_{k,i}\right) \cdot n_d \cdot \zeta_b \le P^w_{i,t}\right\}, \quad \forall k \in \mathcal{K},\ i \in \mathcal{I}, \tag{7}$$

$$t^{k,i}_3 = \max\left\{t \mid P^w_{i,t} \le N_k - P^v_{k,t^a_{k,i}}\right\}, \quad \forall k \in \mathcal{K},\ i \in \mathcal{I}. \tag{8}$$

Practically, in an oversaturated scenario, a train departs from a station as soon as the number of in-train passengers reaches its capacity. Based on this strategy, (6) shows that the boarding passengers who arrive after $t^{k,i}_1$ exceed train capacity, and they will be left behind; (7) shows that before $t^{k,i}_2$ there are passengers waiting at station, and after $t^{k,i}_2$ all passengers board the train; (8) points that the train is full at $t^{k,i}_3$, and passengers arriving after $t^{k,i}_3$ should wait for the next train.

The departure time of train k at station i ($t^b_{k,i}$) is determined based on the relations among $t^{k,i}_1$, $t^{k,i}_2$, $t^{k,i}_3$, $t^0_{k,i} + \underline{d_i}$, and $t^0_{k,i} + \overline{d_i}$. Based on Assumption 3, we consider two cases: (1) $t^{k,i}_1 > t^{k,i}_2$ and (2) $t^{k,i}_1 \le t^{k,i}_2$. Note that $\underline{d_i}$ and $\overline{d_i}$ should be considered in both cases. In case (1), we can further derive that $t^{k,i}_3 > t^{k,i}_1 > t^{k,i}_2$ (which is the solid curve in Figure 3), where the number of boarding passengers per second is determined by passenger arrival rate. Thus, the departure time is determined as $t^b_{k,i} = \min\{t^0_{k,i} + \overline{d_i}, \max\{t^{k,i}_2, t^0_{k,i} + \underline{d_i}\}, t^{k,i}_3\}$. In case (2), we can further derive that $t^{k,i}_3 \le t^{k,i}_1 \le t^{k,i}_2$ (which is the dashed curve in Figure 3), where the number of boarding passengers per second is determined by the actual passenger boarding rate. Thus, the departure time is determined as $t^b_{k,i} = \min\{t^0_{k,i} + \overline{d_i}, t^{k,i}_1\}$. The departure time of train k at station i is expressed as

$$t^b_{k,i} = \begin{cases} \min\left\{t^0_{k,i} + \overline{d_i}, \max\left\{t^{k,i}_2, t^0_{k,i} + \underline{d_i}\right\}, t^{k,i}_3\right\}, & t^{k,i}_1 > t^{k,i}_2 \\ \min\left\{t^0_{k,i} + \overline{d_i}, t^{k,i}_1\right\}, & t^{k,i}_1 \le t^{k,i}_2. \end{cases} \tag{9}$$

Finally, we consider the relations between alighting/boarding rates and congestion at stations. Here the congestion is described by the number of in-vehicle passengers.

Practically, the greater the passenger density in a train is, the slower the passengers can get off/on the train. Here, we design a monotonically decreasing function $\rho(P^{v}_{k,t}) \le 1$ with respect to the number of in-vehicle passengers, to approximate actual alighting/boarding rates according to the passenger density in trains. For simplicity, in this paper, the value of $\rho(P^{v}_{k,t})$ during the dwell process is fixed and is determined by the number of in-vehicle passengers at its arrival time. That is, ζ_a and ζ_b remain constant during the dwell process:

$$\begin{aligned}\zeta_a &= \rho\left(P^{v}_{k,t^0_{k,i}}\right)\cdot \zeta_a^{\text{ideal}}\\ \zeta_b &= \rho\left(P^{v}_{k,t^0_{k,i}}\right)\cdot \zeta_b^{\text{ideal}};\\ &\forall k \in \mathcal{K},\ i \in \mathcal{I},\end{aligned} \tag{10}$$

where ζ_a^{ideal} and ζ_b^{ideal} are the ideal alighting and boarding rates in the uncongested condition.

(3) Passenger Alighting and Boarding. Once the dwell arcs are determined, the alighting and boarding times for train k at station i are determined. Based on the results of dwell arcs, three cases should be considered: (1) $t_1^{k,i} < t_2^{k,i}$; (2) $t_1^{k,i} \ge t_2^{k,i}$ and $t_2^{k,i} \le t^0_{k,i} + \underline{d}_i$; (3) $t_1^{k,i} \ge t_2^{k,i}$ and $t_2^{k,i} > t^0_{k,i} + \underline{d}_i$. In case (1), waiting passengers board the train at rate ζ_b. In case (2), actual passenger boarding rate is equal to the passenger arrival rate. In case (3), the actual boarding rate is ζ_b. Thus, the number of boarding passengers for train k at station i is

$$P^b_{k,i} = \begin{cases} P^w_{i,t^b_{k,i}}; & \text{if } t_1^{k,i} \ge t_2^{k,i},\ t_2^{k,i} \le t^0_{k,i} + \underline{d}_i;\\ \left(t^b_{k,i} - t^a_{k,i}\right)\cdot \zeta_b \cdot n_d; & \text{otherwise}; \end{cases} \qquad \forall k \in \mathcal{K},\ i \in \mathcal{I}. \tag{11}$$

Wang et al. [3] point out that, in a busy metro line, passengers with different destinations are well mixed at each station. Thus, the ratio between the numbers of boarding passengers with destination j and of total boarding passengers is equivalent to the ratio between $\tau_{i,j}(t)$ and $\lambda_i(t)$:

$$P^b_{k,(i,j)} = \frac{\sum_{t=t^b_{k-1,i}}^{t^b_{k,i}} \tau_{i,j}(t)}{\sum_{t=t^b_{k-1,i}}^{t^b_{k,i}} \lambda_i(t)} \cdot P^b_{k,i}; \qquad \forall k \in \mathcal{K},\ j \in \mathcal{I}\backslash 1,\ i = 1, 2, \ldots, j-1. \tag{12}$$

The number of alighting passengers from train k at station j consists of all previously boarding passengers in train k with destination j:

$$P^a_{k,j} = \sum_{i=1}^{j-1} P^b_{k,(i,j)}; \quad \forall k \in \mathcal{K},\ j = 2, \ldots, I, I+2, \ldots, 2I. \tag{13}$$

There are no alighting passengers at the first station in each direction (station 1 and $I+1$).

3.2.2. Passenger Wait and Travel Time

(1) Passenger Wait Time. Total passenger wait time at a station can be calculated by summing up the products of the cumulative number of passengers and their time. Let k_{last} be the index of last train that has departed from station i before time t, that is, $t^b_{k_{\text{last}},i} \le t < t^b_{k_{\text{last}+1},i}$. The cumulative number of passengers at station i at time t is the difference between cumulative numbers of arrival passengers and boarding passengers within the period $[0, t]$:

$$P^w_{i,t} = P_i(0,t) - \sum_{k=1}^{k_{\text{last}}} P^b_{k,i}; \quad \forall i \in \mathcal{I}. \tag{14}$$

Since the continuous time is discretized into 1 second intervals, the passenger wait time between times t and $t+1$ is $P^w_{i,t}\cdot(t+1-t) = P^w_{i,t}$. Thus, the total passenger wait time is

$$T_{pw} = \sum_{i\in\mathcal{I}} \sum_{t=0}^{t'_{\text{end}}} P^w_{i,t}. \tag{15}$$

(2) Passenger Travel Time. The number of in-train passengers in train k at time t is the difference between cumulative numbers of boarding passengers and alighting passengers. Let i_{last} be the index of the last station that train k has visited before time t. We consider two cases: (1) train k is dwelling at station i_{last}, that is, $t^0_{k,i_{\text{last}}} \le t < t^b_{k,i_{\text{last}}}$; (2) train k is in the segment between station i_{last} and $i_{\text{last}}+1$, that is, $t^b_{k,i_{\text{last}}} \le t < t^0_{k,i_{\text{last}}+1}$. Based on Assumption 4, the boarding passengers at station i_{last} are not considered in case (1) but are included in case (2). Thus, the number of in-vehicle passengers in train k at time t is

$$P^v_{k,t} = \begin{cases} \sum_{i=1}^{i_{\text{last}}-1} P^b_{k,i} - \sum_{i=1}^{i_{\text{last}}} P^a_{k,i}; & t^0_{k,i_{\text{last}}} \le t < t^b_{k,i_{\text{last}}};\\ \sum_{i=1}^{i_{\text{last}}} \left(P^b_{k,i} - P^a_{k,i}\right); & t^b_{k,i_{\text{last}}} \le t < t^0_{k,i_{\text{last}}+1}; \end{cases} \qquad \forall k \in \mathcal{K}. \tag{16}$$

Analogously, the total passenger travel time is

$$T_{\text{pt}} = \sum_{k\in\mathcal{K}} \sum_{t=0}^{t'_{\text{end}}} P^v_{k,t}. \tag{17}$$

Finally, the total cost of passenger time is formulated as

$$C^P_{\text{total}} = \frac{\left(c_{\text{pw}}\cdot T_{\text{pw}}\right)}{3600} + \frac{\left(c_{\text{pt}}\cdot T_{\text{pt}}\right)}{3600}. \tag{18}$$

The cost unit is yuan (¥).

3.3. Energy. In general, for each segment i, with given segment travel time $s_t(i)$ and segment length $s_l(i)$, the speed profile determines energy use [49]. Thus, given a speed profile generating method, the energy per unit mass of different trains in segment i is the same. We denote the energy per unit mass for one train in segment i as e_i. Its calculation usually considers the following train dynamic equations [48]:

$$\begin{aligned} \frac{dv_k(t)}{dt} &= f_t - f_r, \\ \frac{dx_k(t)}{dt} &= v_k(t); \quad \forall t \in \left[t^b_{k,i}, t^0_{k,i+1}\right], \ i, i+1 \in \mathcal{I}, \\ f_r &= \alpha \cdot v_k^2(t) + \beta \cdot v_k(t) + \gamma, \\ v_k\left(t^0_{k,i}\right) &= v_k\left(t^b_{k,i}\right) = 0. \end{aligned} \tag{19}$$

The first two equations are train motion equations, where $v_k(t)$ and $x_k(t)$, respectively, represent traveling speed and distance. The third one is the Davis resistance function where Davis parameters are α, β, and γ. The last one requires that every train stops at each station. An optimal train control strategy can be obtained based on (19), in which energy per unit mass is

$$e_i = \int_{t^b_{k,i}}^{t^0_{k,i+1}} f_t \cdot v_k(t)\, dt; \quad \forall i, i+1 \in \mathcal{I}, \ k \in \mathcal{K}. \tag{20}$$

However, in (20), the relation between energy and decision variable $s_t(i)$ is unclear. It is also difficult to integrate the optimal train control with other objective functions [18]. Since the speed profile optimization is not considered, for simplicity, we use the method in [50] to generate practical speed profiles. Based on these speed profiles, we then adopt a linear piecewise curve to approximate the energy per unit mass of train k in segment i [48]:

$$e_i = \left(\chi_x \cdot s_t(i) + \chi_y\right) \cdot s_l(i); \quad \forall i \in \mathcal{I}. \tag{21}$$

We introduce here an estimation procedure for the values of χ_x and χ_y, which is based on mean squared error (MSE). First, we set several feasible values (e.g., $z = 1, 2, \ldots, 100$) of $s_l(z)$ and $s_t(z)$. Then, we obtain optimal speed profiles based on (19) and (20), thereby obtaining the optimal energy with respect to $s_l(z)$ and $s_t(z)$. Finally, we define the optimal energy per unit mass as e_z, and the estimated energy per unit mass obtained with (21) as $\widetilde{e_z}$. The MSE is formulated as

$$\text{MSE}\left(\chi_x, \chi_y\right) = \frac{1}{100}\sum_{z=1}^{100}\left(\widetilde{e_z} - e_z\right)^2. \tag{22}$$

The optimal values of χ_x and χ_y can be obtained when $\text{MSE}(\chi_x, \chi_y)$ reaches its minimum and are found as follows:

$$\begin{aligned} \frac{\partial \text{MSE}\left(\chi_x, \chi_y\right)}{\partial \chi_x} &= 0, \\ \frac{\partial \text{MSE}\left(\chi_x, \chi_y\right)}{\partial \chi_y} &= 0. \end{aligned} \tag{23}$$

With $\chi_x{}^*$ and $\chi_y{}^*$ obtained, the total energy can be calculated as

$$E_{\text{total}} = \sum_{i=1}^{2I-1}\sum_{k=1}^{K}\left[\left(\chi_x{}^* \cdot s_t(i) + \chi_y{}^*\right) \cdot s_l(i)\right] \cdot \left(M_k + M_p \cdot P^v_{k,t^b_{k,i}}\right). \tag{24}$$

Thus, the total cost of energy consumption is

$$C^e_{\text{total}} = c_e \cdot E_{\text{total}} = c_e \cdot \sum_{i=1}^{2I-1}\sum_{k=1}^{K}\left[\left(\chi_x{}^* \cdot s_t(i) + \chi_y{}^*\right) \cdot s_l(i)\right] \cdot \left(M_k + M_p \cdot P^v_{k,t^b_{k,i}}\right). \tag{25}$$

Note that when all tracks are straight and level, χ_x and χ_y suffice for estimating energy. However, if grades and curves are considered, we may have $e_i = (\chi^i_x \cdot s_t(i) + \chi^i_y) \cdot s_l(i)$, $\forall i \in \mathcal{I}$, where χ^i_x and χ^i_y are specified parameters to describe the characteristics in segment i. The parameter estimation procedure is the same.

3.4. Train Cost. In this paper, the train cost is estimated with empirical equations with respect to the interest rate, train price, and average round trip time, where long-term requirements (e.g., the consideration of the highest peak during a year and demand evolution over the planning horizon) are not considered.

Train cost consists of train capital cost and operating cost (excluding energy). Train capital cost is related to the price and economic life of a train, as well as interest rate. Let C_{annual} be the annual cost per train, C_{train} be the current price per train, r be the interest rate, and T_{ls} be the lifespan of a train. The relations among these four factors can be expressed as

$$C_{\text{annual}} = \frac{C_{\text{train}} \cdot r \cdot (1+r)^{T_{\text{ls}}}}{(1+r)^{T_{\text{ls}}} - 1}. \tag{26}$$

The capital cost per train hour c_{tra} can be then calculated by dividing the annual cost C_{annual} by the product of average working hours per day (D_{work}) and working days per year (Y_{work})

$$c_{\text{tra}} = \frac{C_{\text{annual}}}{D_{\text{work}} \cdot Y_{\text{work}}}. \tag{27}$$

In addition, a train can be reassigned after it finishes a round trip and returns to the shunting yard, and the round trip time T^k_{round} can be obtained as

$$T^k_{\text{round}} = t^b_{k,2I} - t^0_{k,1} + T_{\text{turn}}, \quad \forall k \in \mathcal{K}. \tag{28}$$

Thus, the actual period for train running is $\sum_{k=1}^{K} T^k_{\text{round}}$ ($\sum_{k=1}^{K} T^k_{\text{round}} \leq t^t_{\text{end}}$). The estimated fleet size F_s within period $[0, \sum_{k=1}^{K} T^k_{\text{round}}]$ depends on the round trip time T^k_{round} and the departure headway from the shunting yard (which is equivalent to the arrival time of train k at the first station $t^0_{k,1}$):

$$F_s = \frac{\left(\sum_{k=1}^{K} T^k_{\text{round}}\right)/K}{\left[\sum_{k=2}^{K}\left(t^0_{k,1} - t^0_{k-1,1}\right)\right]/(K-1)}, \quad (29)$$

where the term $(\sum_{k=1}^{K} T^k_{\text{round}})/K$ represents the average round trip time and the term $[\sum_{k=2}^{K}(t^0_{k,1} - t^0_{k-1,1})]/(K-1)$ represents the average headway. The total cost of trains is formulated as

$$C^{\text{tra}}_{\text{total}} = c_{\text{tra}} \cdot F_s \cdot \left(\frac{\sum_{k=1}^{K} T^k_{\text{round}}}{3600}\right) + c_o \cdot K \cdot \frac{\sum_{i=1}^{2I-1} s_l(i)}{1000}. \quad (30)$$

The first term $c_{\text{tra}} \cdot F_s \cdot (\sum_{k=1}^{K} T^k_{\text{round}}/3600)$ represents the capital cost of F_s trains for the actual running period $[0, \sum_{k=1}^{K} T^k_{\text{round}}]$. The second term $c_o \cdot K \cdot \sum_{i=1}^{2I-1} s_l(i)$ represents the operating cost with frequency K (trains) on a line with the length of the sum of $2I-1$ segments (Km). Both decision variables (segment travel times and arrival times at the first station) affect the train cost.

3.5. Mixed Integer Programming Formulation. The optimization model aims to minimize the weighted-sum of different cost functions, so that the conflicting objectives are generalized in terms of monetary cost. In this paper, the weights of different functions are reflected by their average costs, that is, c_e, c_o, c_{pt}, c_{pw}, and c_{tra}. For example, an operator-oriented timetable (neglecting passenger satisfaction) can be obtained by setting $c_{\text{pt}} = 0$ and $c_{\text{pw}} = 0$; a passenger-oriented timetable can be obtained by setting $c_e = c_o = c_{\text{tra}} = 0$. The weights for integrated objectives can be set similarly, with larger cost coefficients for larger weights. The model is formulated as follows:

Minimize

$$C_{\text{total}} = C^p_{\text{total}} + C^e_{\text{total}} + C^{\text{tra}}_{\text{total}}. \quad (31)$$

Subject to

$$t^0_{K,1} = t^p_{\text{end}}, \quad (32)$$

$$s_t(I) = T_{\text{turn}}, \quad (33)$$

$$s^{\min}_t(i) \le s_t(i) \le s^{\max}_t(i); \quad \forall i \in \mathcal{I}, \quad (34)$$

$$t^b_{k,i} - t^b_{k-1,i} \ge h_{\min}; \quad \forall k = 2, \dots, K, \ i \in \mathcal{I}, \quad (35)$$

$$P^v_{k,t} \le N_k; \quad \forall k \in \mathcal{K}, \ t \in \left[0, t^t_{\text{end}}\right]. \quad (36)$$

This model is a nonsmooth, nonconvex programming model, where the nonsmoothness is caused by the limited train capacity, and the nonconvexity is caused by the determination of dwell times [3]. Constraint (32) is equivalent to Assumption 2. Constraint (33) considers the turnaround from station I to $I+1$ as traveling in segment I, with fixed travel time T_{turn}. Constraint (34) ensures that the segment travel time should be optimized within a reasonable range $[s^{\min}_t(i), s^{\max}_t(i)]$. Constraint (35) imposes the minimum headway constraint for both departures and arrivals (based on Assumption 1). Constraint (36) requires that the number of in-train passengers must not exceed train capacity.

In Caprara et al. [6], the time-space graph is used to derive an integer linear programming model, thereby allowing a considerably faster solution. Analogously, this paper transforms the nonsmooth, nonconvex programming model into a mixed integer programming (MIP) model based on the DTSG. First, we construct two $(2I \times t^t_{\text{end}})$ matrixes $\mathbb{U}$ and $\mathbb{W}$, respectively, representing the selection of arrival and departure nodes in the DTSG:

$$\mathbb{U}_{i,t} = \begin{cases} 1; & \text{if } t = \pi(u_k),\ u_k \in \mathcal{U}^i,\ k \in \mathcal{K}; \\ 0; & \text{otherwise}; \end{cases}$$

$$\mathbb{W}_{i,t} = \begin{cases} 1; & \text{if } t = \pi(w_k),\ w_k \in \mathcal{W}^i,\ k \in \mathcal{K}; \\ 0; & \text{otherwise}; \end{cases} \quad (37)$$

$$\forall i \in \mathcal{I},\ t \in \left[0, t^t_{\text{end}}\right],$$

where $\mathbb{U}_{i,t} = 1$ ($\mathbb{W}_{i,t} = 1$) if a train arrives at (departs from) station i at time t. As mentioned earlier, the value of $\mathbb{U}_{1,t}$ is determined by $\pi(u_k) = t^0_{k,1}$, for $\forall u_k \in \mathcal{U}^1, k \in \mathcal{K}$. The value of $\mathbb{W}_{i,t}$ is determined by $\mathbb{U}_{i,t}$ and the dwell arcs, that is, $\Delta(u_k, w_k) = t^b_{k,i} - t^0_{k,i}$, for $\forall u_k \in \mathcal{U}^i$, $w_k \in \mathcal{W}^i$, $i \in \mathcal{I}$, $k \in \mathcal{K}$. The value of $\mathbb{U}_{i+1,t}$ is determined by $\mathbb{W}_{i,t}$ and the travel time in segment i, that is, $\Delta(w_k, u_k) = s_t(i)$, for $\forall w_k \in \mathcal{W}^i$, $u_k \in \mathcal{U}^{i+1}$, and $i \in \mathcal{I}\backslash\{2I\}$, $k \in \mathcal{K}$. Thus, the decision variables determine $\mathbb{U}$ and $\mathbb{W}$.

Based on the structures of $\mathbb{U}$ and $\mathbb{W}$, as well as Assumption 4, we introduce two $(2I \times t^t_{\text{end}})$ matrixes $\mathbb{P}_{\mathbb{A}}$ and $\mathbb{P}_{\mathbb{B}}$, respectively, representing passenger alighting and boarding information (number and time):

$$\mathbb{P}_{\mathbb{A}i,t} = \begin{cases} P^a_{k,i}; & t = \pi(u_k),\ u_k \in \mathcal{U}^i,\ k \in \mathcal{K}; \\ 0; & \text{otherwise}; \end{cases}$$

$$\mathbb{P}_{\mathbb{B}i,t} = \begin{cases} P^b_{k,i}; & t = \pi(w_k),\ w_k \in \mathcal{W}^i,\ k \in \mathcal{K}; \\ 0; & \text{otherwise}; \end{cases} \quad (38)$$

$$\forall i \in \mathcal{I},\ t \in \left[0, t^t_{\text{end}}\right].$$

The cumulative numbers of waiting passengers $\mathbb{Q}$ $(2I \times t^t_{\text{end}})$ and in-vehicle passengers $\mathbb{V}$ $(K \times t^t_{\text{end}})$ can be constructed based on $\mathbb{P}_{\mathbb{A}}$ and $\mathbb{P}_{\mathbb{B}}$:

$$\mathbb{Q}_{i,t} = Q_{i,t} - \left\|\mathbb{P}_{\mathbb{B}i,1:t}\right\|; \quad \forall i \in \mathcal{I},\ t \in \left[0, t^t_{\text{end}}\right], \quad (39)$$

$$\mathbb{V}_{k,t} = \begin{cases} \mathbb{V}_{k,t-1} + P^b_{k,i}; & \text{if } t = \pi(w_k),\ w_k \in \mathscr{W}^i,\ i \in \mathscr{I}; \\ \mathbb{V}_{k,t-1} - P^a_{k,i}; & \text{if } t = \pi(u_k),\ u_k \in \mathscr{U}^i,\ i \in \mathscr{I}; \\ \mathbb{V}_{k,t-1}; & \text{otherwise}; \end{cases} \quad (40)$$

$$\forall k \in \mathscr{K},\ t \in \left[1, t^t_{\text{end}}\right].$$

In (39), $Q_{i,t} = Q_{i,0} + \sum_{q=0}^{t} \lambda_i(t)$ represents the total waiting passengers at station i, where $Q_{i,0}$ is the number of waiting passengers at the start of this period. According to the definition of cumulative passengers, we have $\mathbb{Q}_{i,t} = P^w_{i,t}$. Thus, the total passenger wait time can be expressed by $\mathbb{Q}$: $T_{\text{pw}} = \sum_{i\in\mathscr{I}} \sum_{t=0}^{t^t_{\text{end}}} P^w_{i,t} = \sum_{i\in\mathscr{I}} \sum_{t=0}^{t^t_{\text{end}}} \mathbb{Q}_{i,t} = \|\mathbb{Q}\|$ (since $\mathbb{Q}_{i,t} \geq 0$). In (40), the initial number of in-vehicle passengers is zero, that is, $\mathbb{V}_{k,0} = 0$. The number of in-vehicle passengers changes only at arrival/departure times, which is consistent with Assumption 4. Analogously, the total passenger travel time can be expressed as $T_{\text{pt}} = \|\mathbb{V}\|$. Thus, the total cost for passenger time can be rewritten as

$$C^p_{\text{total}} = \frac{1}{3600}\left(c_{\text{pw}} \cdot \|\mathbb{Q}\| + c_{\text{pt}} \cdot \|\mathbb{V}\|\right). \quad (41)$$

For energy cost, the energy per unit mass in segment i can be rewritten as

$$e_i = \left(\chi_x \cdot \Delta(w_k, u_k) + \chi_y\right) \cdot s_l(i), \quad w_k \in \mathscr{W}^i,\ u_k \in \mathscr{U}^{i+1};\ \forall i \in \mathscr{I} \setminus \{2I\}. \quad (42)$$

The total energy can be rewritten as

$$E_{\text{total}} = \sum_{i=1}^{2I-1} \sum_{k=1}^{K} \sum_{t=0}^{t^t_{\text{end}}} \phi_t \cdot \left[\left(\chi_x \cdot \Delta(w_k, u_k) + \chi_y\right) \cdot s_l(i)\right] \cdot \left(M_k + M_p \cdot \mathbb{V}_{k,t}\right), \quad w_k \in \mathscr{W}^i,\ u_k \in \mathscr{U}^{i+1}, \quad (43)$$

where $\phi_t = 1$ if $t = \pi(w_k)$; otherwise $\phi_t = 0$. The total energy cost can be rewritten correspondingly.

For train cost, the round trip time T^k_{round} can be rewritten as

$$T^k_{\text{round}} = \Delta(u_k, w_k) + T_{\text{turn}}, \quad u_k \in \mathscr{U}^1,\ w_k \in \mathscr{W}^{2I};\ \forall k \in \mathscr{K}. \quad (44)$$

The departure headway from the shunting yard can be rewritten as

$$t^0_{k,1} - t^0_{k-1,1} = \Delta(u_{k-1}, u_k), \quad u_{k-1}, u_k \in \mathscr{U}^1;\ \forall k \in \mathscr{K} \setminus \{1\}. \quad (45)$$

The total cost of trains can be rewritten correspondingly.

Finally, the last departing train constraint (32) is equivalent to

$$\pi(u_K) = t^t_{\text{end}}, \quad u_K \in \mathscr{U}^1. \quad (46)$$

The turnaround constraint (33) is equivalent to

$$\Delta(w_k, u_k) = T_{\text{turn}}, \quad w_k \in \mathscr{W}^I,\ u_k \in \mathscr{U}^{I+1};\ \forall k \in \mathscr{K}. \quad (47)$$

The segment travel time window constraint (34) is equivalent to

$$s^{\min}_t(i) \leq \Delta(w_k, u_k) \leq s^{\max}_t(i), \quad w_k \in \mathscr{W}^{i-1},\ u_k \in \mathscr{U}^i;\ \forall k \in \mathscr{K},\ i \in \mathscr{I} \setminus \{1\}. \quad (48)$$

The minimum headway constraint (35) is equivalent to

$$\Delta(w_{k-1}, w_k) \geq h_{\min}, \quad w_{k-1}, w_k \in \mathscr{W}^i;\ k \in \mathscr{K} \setminus \{1\},\ i \in \mathscr{I}. \quad (49)$$

The train capacity constraint (36) is equivalent to

$$\mathbb{V}_{k,t} \leq N_k, \quad \forall k \in \mathscr{K},\ t \in \left[0, t^t_{\text{end}}\right]. \quad (50)$$

The selection of arcs determines the arrival nodes u_k and departure nodes w_k, thereby determining every element in the optimization model. Thus, the model is transformed into an MIP. The next section discusses how its solution can be found.

4. Solution Approach

In this section, we decompose the TTP into two subproblems. The first subproblem optimizes the segment travel times (OST), aiming at minimizing the costs for energy and passenger segment travel times. A branch-and-bound algorithm and a frequency determining algorithm are introduced to solve the OST. The second subproblem optimizes departure headways from the shunting yard (OH), aiming to minimize passenger wait time, dwell time and train cost. A novel rolling optimization algorithm is designed to solve the OH.

4.1. Decomposition. The rationale of the decomposition is explained as follows. The objectives in the optimization model include passenger wait time (at their original stations), passenger travel time (segment travel times and station dwell times), energy, and train cost. Based on (43), the total energy is related to the energy per unit mass and the number of passengers in each segment. The energy per unit mass is solely determined by segment travel times $s_t(i)$. Since passengers board at their origins and alight at their destinations regardless of which train they take, the sum of in-vehicle passengers in one segment i is determined by passenger demand. In this regard, the total energy can be expressed as

$$E_{\text{total}} = M_k \cdot K \cdot \sum_{i=1}^{2I-1} e_i + M_p \cdot \sum_{i=1}^{2I-1} \sum_{j=i+1}^{2I} \sum_{r=1}^{i} \sum_{t=0}^{t^p_{\text{end}}} \tau_{r,j}(t) \cdot e_i. \quad (51)$$

Therefore, the total energy cost depends only on segment travel times $s_t(i)$. The passenger segment travel times also depend only on segment travel times.

On the other hand, passenger wait time, dwell time, and train cost depend on the train arrival and departure times at stations. Based on Assumption 1, train arrival/departure times at stations are affected by the departure headways from the shunting yard. Thus, it is reasonable to use segment travel times as input and consider the costs for passenger wait time, dwell time, and trains as functions of the departure headways.

In general, the OST subproblem is formulated as follows:
Minimize

$$C_{\text{total}}^{\text{OST}} = \sum_{i=1}^{2I-1} \left\{ c_e \cdot M_k \cdot K \cdot e_i + \left(c_e \cdot M_p \cdot e_i + c_{\text{pt}} \cdot s_t(i)\right) \cdot \left[\sum_{j=i+1}^{2I} \sum_{r=1}^{i} \sum_{t=0}^{t_{\text{end}}^p} \tau_{r,j}(t) \right] \right\}. \tag{52}$$

Subject to constraints (47) and (48).

It should be noted that since segment travel times are given inputs, minimizing the cost of passenger dwell time is equivalent to minimizing the cost of total travel time. Since we already obtain the in-train passenger matrix $\mathbb{V}$, there is no need to further derive dwell times at stations. Thus, the objectives in the OH subproblem include the total passenger time cost (both $\mathbb{Q}$ and $\mathbb{V}$) instead of costs of passenger wait and dwell times. The OH subproblem is formulated as follows (with segment travel times as input):
Minimize

$$C_{\text{total}}^{\text{OH}} = C_{\text{total}}^{p} + C_{\text{total}}^{\text{tra}}. \tag{53}$$

Subject to constraints (46), (49) and (50).

4.2. Solution Approach for OST. The solution approach for OST consists of branch-and-bound (B&B) and frequency determining algorithms. On one hand, the B&B algorithm needs frequency as input. On the other hand, the frequency determining algorithm needs segment travel times obtained by the B&B to evaluate the performance. Thus, we iteratively evaluate the performance with different frequencies (each frequency is used as input for B&B to find a solution of segment travel times) and choose the one with the best performance. The segment travel times are then determined correspondingly.

4.2.1. Branch-and-Bound (B&B) Algorithm. The OST subproblem is an integer linear programming problem with $2I-1$ decision variables ($s_t(i)$, for $i = 1, 2, \ldots, 2I-1$). Here we use a generic B&B, which is widely used to solve linear programming problems [7]. Generally, a B&B needs an input model formulated as follows:

$$\min \quad \overrightarrow{\mathbf{c}} \cdot \overrightarrow{\mathbf{x}}^T, \tag{54}$$

$$\text{s.t.} \quad \mathbb{A} \cdot \overrightarrow{\mathbf{x}}^T \leq \overrightarrow{\mathbf{b}}^T$$
$$\mathbb{A}\text{eq} \cdot \overrightarrow{\mathbf{x}}^T = \overrightarrow{\mathbf{beq}}^T, \tag{55}$$

where $\overrightarrow{\mathbf{c}}$ is the coefficient vector and $\overrightarrow{\mathbf{x}}^T$ is the decision variable vector. Both $\overrightarrow{\mathbf{c}}$ and $\overrightarrow{\mathbf{x}}^T$ have $2I-1$ elements. $\mathbb{A}$ and $\mathbb{A}\text{eq}$ are coefficient matrixes, and $\overrightarrow{\mathbf{b}}^T$ and $\overrightarrow{\mathbf{beq}}^T$ are constant vectors representing the bounds. Here $\mathbb{A}$ is a $(4I-2)\times(2I-1)$ matrix, and $\mathbb{A}\text{eq}$ is a $1\times(2I-1)$ matrix. Correspondingly, $\overrightarrow{\mathbf{b}}^T$ has $2I-1$ elements, and $\overrightarrow{\mathbf{beq}}^T$ has one element. We define a vector with $2I-1$ elements as

$$\overrightarrow{\boldsymbol{\epsilon}}_{\mathbf{j}} = (0, 0, \ldots, 1, \ldots, 0, 0), \quad j = 1, \ldots, 2I-1. \tag{56}$$

The decision variable vector can be represented as $\overrightarrow{\mathbf{x}}_i = s_t(i)$ for $i = 1, 2, \ldots, 2I-1$, where $\overrightarrow{\mathbf{x}}_i$ is the i-th element in the vector $\overrightarrow{\mathbf{x}}$. We then define $\urcorner_i = \sum_{j=i+1}^{2I} \sum_{r=1}^{i} \sum_{t=0}^{t_{\text{end}}^p} \tau_{r,j}(t)$ for $i = 1, 2, \ldots, 2I-1$ as the total number of passengers that travel through segment i. Based on (52), $\overrightarrow{\mathbf{c}}_i$ can be represented as

$$\overrightarrow{\mathbf{c}}_i = c_e \cdot M_k \cdot K \cdot \chi_x \cdot s_l(i) + \urcorner_i \cdot \left(c_e \cdot M_p \cdot \chi_x \cdot s_l(i) + c_{\text{pt}}\right); \quad i = 1, 2, \ldots, 2I-1. \tag{57}$$

Constraints (55) are represented as follows. Based on constraint (47), we have $\mathbb{A}\text{eq} = \overrightarrow{\boldsymbol{\epsilon}}_{\mathbf{I}}$ and $\overrightarrow{\mathbf{beq}} = T_{\text{turn}}$. Based on constraint (48), we have $\mathbb{A}_{i,:} = -\overrightarrow{\boldsymbol{\epsilon}}_{\mathbf{i}}$, for $i = 1, 2, \ldots, 2I-1$; $\mathbb{A}_{i,:} = \overrightarrow{\boldsymbol{\epsilon}}_{\mathbf{i}}$, for $i = 2I, 2I+1, \ldots, 4I-2$, where $\mathbb{A}_{i,:}$ represents the i-th row of matrix $\mathbb{A}$. Analogously, $\overrightarrow{\mathbf{b}}_i = -s_t^{\min}(i)$, for $i = 1, 2, \ldots, 2I-1$; $\overrightarrow{\mathbf{b}}_i = s_t^{\max}(i)$, for $i = 2I, 2I+1, \ldots, 4I-2$. Since we use the standard B&B algorithm [40], the detailed algorithm description is not shown here.

4.2.2. Frequency Determining Algorithm. Here we consider cyclic timetables (train departure headways from the shunting yard are constant, e.g., 200 s) to discuss the frequency, because they have following properties: (1) The cyclic timetable with minimum cost (i.e., the best cyclic timetable) can be considered as an upper bound of the noncyclic timetable. (2) If passenger demand is stable, the best cyclic timetable is a good approximation to an optimal noncyclic one. (3) The cyclic timetables are easy to obtain.

First, the range of frequency should be estimated. We define the peak passenger volume in one segment as q_{peak}, which is the maximum value among the numbers of passengers traveling through the segment, that is, $q_{\text{peak}} = \max(\urcorner_i \mid i = 1, 2, \ldots, 2I-1)$. The minimum frequency $K_{\min}$ is determined by the train capacity N_k and q_{peak} : $K_{\min} = \lceil q_{\text{peak}}/N_k \rceil$. The maximum frequency $K_{\max}$ is determined by the minimum headway: $K_{\max} = \lfloor t_{\text{end}}^p / h_{\min} \rfloor + 1$.

Second, for each frequency $K \in [K_{\min}, K_{\max}]$, we use K as input of the B&B, to obtain the segment travel times. Since the frequency for a cyclic timetable determines the departure headways from the shunting yard ($h = t_{\text{end}}^p/(K-1)$, where h is the departure headway), the cyclic timetable that corresponds to frequency K can be obtained. Then, the total cost of the timetable can be calculated with (52).

Finally, we obtain different cyclic timetables with different costs, each corresponding to a frequency $K \in [K_{\min}, K_{\max}]$. We select the timetable with minimum cost as the best cyclic timetable, and its corresponding segment travel times as the OST solution within this period. The detailed algorithm is as shown in Algorithm 1.

4.3. Solution Approach for OH. Here we design a rolling optimization (RO) algorithm that optimizes the arrival time at the first station for one train at a time. The main difference between the RO and the rolling horizon (RH) approach is in the RH, the bound conditions (i.e., the start and end times of each horizon) require some of the original variables to have fixed, known values [3], while in the RO, each decision variable may still be optimized. Thus, the rolling optimization is more flexible.

4.3.1. The Rolling Optimization (RO) Algorithm. Based on (53), the objectives include the total costs of passenger time and trains. To obtain these costs, we need the segment travel times as input. Here, instead of considering the whole period, we consider a much shorter period $t_p \le h_{\min} + \sum_{i=1}^{2I-1} s_t(i) + \sum_{i=1}^{2I} \overline{d_i}$, e.g., 600 s, for solving the OST. For example, for each train $k \ge 2$, its previous train arrives at the first station at time $t^0_{k-1,1}$ (the first train arrives at time 0), and we consider the period $[t^0_{k-1,1}, t^0_{k-1,1} + t_p]$ to solve the OST (the constraint of t_p ensures that $t^0_{k-1,1} + t_p \le t^t_{\text{end}}$). In this way, the key ideas of "iterative optimization" (the frequencies, segment travel times, and headways are interdependent) and "rolling horizon" (the smaller period studied in each step) are adopted, and future passenger demand is considered.

Then, we discuss the costs for one train $k \ge 2$ if its arrival time at the first station is given. Since each train is scheduled in chronological order, the waiting passenger matrix $\mathbb{Q}$ and the in-train passenger matrix $\mathbb{V}$ can be constructed chronologically. For each train $k \ge 2$, the departure time of its preceding train at the first station is $t^b_{k-1,1}$, and train k departs from the first station at time $t^b_{k,1}$. We define the passenger wait time with respect to train k as $T_{\text{pw}}(k)$, which is formulated as

$$T_{\text{pw}}(k) = \sum_{i=1}^{2I} \sum_{t=t^b_{k-1,1}}^{t^b_{k,1}} \mathbb{Q}_{i,t}; \quad \forall k \in \mathcal{K}. \tag{58}$$

Analogously, we define the passenger travel time for train k as $T_{\text{pt}}(k)$, and we obtain

$$T_{\text{pt}}(k) = \sum_{t=t^0_{k,1}}^{t^b_{k,2I}} \mathbb{V}_{k,t}; \quad \forall k \in \mathcal{K}. \tag{59}$$

The estimated fleet size of train k (F^k_s) depends on its round trip time (T^k_{round}) and the departure headway from the shunting yard ($t^0_{k,1} - t^0_{k-1,1}$):

$$F^k_s = \frac{T^k_{\text{round}}}{t^0_{k,1} - t^0_{k-1,1}}; \quad \forall k \in \mathcal{K}. \tag{60}$$

Based on (18), (30), (53), (58), (59), and (60), we define the total costs in the OH for train k as $C^{\text{OH}}_{\text{total}}(k)$, which can be expressed as

$$C^{\text{OH}}_{\text{total}}(k) = \frac{c_{\text{pw}} \cdot T_{\text{pw}}(k) + c_{\text{pt}} \cdot T_{\text{pt}}(k) + c_{\text{tra}} \cdot F^k_s \cdot T^k_{\text{round}}}{3600} + c_o \cdot \frac{\sum_{i=1}^{2I-1} s_l(i)}{1000}; \quad \forall k \in \mathcal{K}. \tag{61}$$

The value of $C^{\text{OH}}_{\text{total}}(k)$ is a function of the arrival time for train k at the first station. Since the arrival time for the previous train is given, we consider the value of $C^{\text{OH}}_{\text{total}}(k)$ as a function of the headway h between train $k-1$ and train k ($h = t^0_{k,1} - t^0_{k-1,1}$), for consistency with other TTP studies.

Next we discuss how the optimal value for h can be obtained. Based on (58), (59), and (60), both the passenger time and train cost depend on passenger demand. Since the passenger demand is time-varying, the relation between $C^{\text{OH}}_{\text{total}}(k)$ and h cannot be described with a fixed function.

The expected objective function of OH is shown in Figure 4. Its value first decreases (the headway starts from $h_{\min}$) to a minimum value and then increases monotonically. We specify that this function reaches its minimum value at headway h^*. Since the DTSG divided the continuous time into seconds, we iteratively calculate the objective function of OH with respect to h (starting from $h = h_{\min}$, and $h = h+1$ in each following step), until the optimal headway h^* is found, or the total cost increases as h increases (in case that h^* cannot be reached). It should be noted that sometimes a headway is not feasible due to the model constraints (e.g., a short headway might violate the minimum headway constraint). In such cases, we define the value of $C^{\text{OH}}_{\text{total}}(k)$ to be $+\infty$ if it corresponds to an infeasible headway.

Finally, for each train $k \ge 2$, we can obtain its optimal headway h^* (between train $k-1$ and train k). By rolling optimization of departure headways for trains, the timetable can be obtained. Let $\overrightarrow{\mathbf{H}}$ be the vector of optimal headways for all trains $k \ge 2$. The detailed rolling optimization (RO) algorithm is described in Algorithm 2. The arrival and departure times of the first train should be obtained beforehand, and $t^0_{1,1} = 0$.

4.3.2. Other Approaches. As pointed out by Wang et al. [3], other approaches such as a pattern search [51] or a genetic algorithm [36] can be applied to solve the TTP. Here we briefly describe the generic structures of these two approaches.

A genetic algorithm (GA) is an iterative heuristic that seeks an optimal individual among a population of solutions in each generation, and uses selection, crossover, mutation and possibly other operations to obtain new generations. Such process is repeated until the termination condition is satisfied. Since the size of the chromosome should be preset, the frequency is considered here, a given and unchangeable input. Then, the segment travel times can be obtained with

Step 1: Estimate the frequency range, i.e., $[\mathbf{K_{min}}, \mathbf{K_{max}}]$. Set the initial frequency of the cyclic timetable as $\mathbf{K} = \mathbf{K_{min}}$, the best frequency found during the process as $\mathbf{K_{best}} = \mathbf{0}$, and the minimum total cost found during the process as $\mathbf{TC_{min}} = +\infty$. Go to Step 2.
Step 2: For frequency $\mathbf{K}$, obtain the segment travel times through the B&B algorithm, and obtain the uniform departure headways from the shunting yard $\mathbf{h} = \mathbf{t^{p}_{end}}/(\mathbf{K} - \mathbf{1})$. Go to Step 3.
Step 3: Calculate total cost $\mathbf{TC_K}$ corresponding to the frequency $\mathbf{K}$. If $\mathbf{TC_K} < \mathbf{TC_{min}}$, then set $\mathbf{TC_{min}} = \mathbf{TC_K}$, and $\mathbf{K_{best}} = \mathbf{K}$. Go to Step 4.
Step 4: If $\mathbf{K} = \mathbf{K_{max}}$, then go to Step 5. Otherwise, set $\mathbf{K} = \mathbf{K} + \mathbf{1}$, and return to Step 2.
Step 5: Output $\mathbf{K_{best}}$ and $\mathbf{TC_{min}}$. The segment travel times within this period are the solution of the B&B algorithm that corresponds to frequency $\mathbf{K_{best}}$.

ALGORITHM 1: Frequency determining algorithm.

Step 1: Set $\overrightarrow{\mathbf{H}} = (\mathbf{0}, \mathbf{0}, \ldots, \mathbf{0}, \mathbf{0})$. Set the period $\mathbf{t_p}$ for OST solution approach. Set $\mathbf{k} = \mathbf{2}$, $\mathbf{h} = \mathbf{h_{min}}$, and go to Step 2.
Step 2: Obtain the segment travel times for train $\mathbf{k}$ by OST solution approach. Obtain the arrival and departure times at stations for train $\mathbf{k}$. Go to Step 3.
Step 3: Calculate the value of $\mathbf{C^{OH}_{total}(k)}$ with respect to $\mathbf{h}$, and calculate the value of $\mathbf{C^{OH}_{total}}'(\mathbf{k})$ with respect to $\mathbf{h}' = \mathbf{h} + \mathbf{1}$. Go to Step 4.
Step 4: If $\mathbf{C^{OH}_{total}}'(\mathbf{k}) > \mathbf{C^{OH}_{total}(k)}$, then go to Step 5; Otherwise, set $\mathbf{h} = \mathbf{h}'$, and return to Step 2.
Step 5: Set the optimal headway for train $\mathbf{k}$ as $\mathbf{h^*} = \mathbf{h}$, and set $\overrightarrow{\mathbf{H}}(\mathbf{k} - \mathbf{1}) = \mathbf{h^*}$. Go to Step 6.
Step 6: If $\|\overrightarrow{\mathbf{H}}\| \geq \mathbf{t^{p}_{end}}$, then the rolling optimization is finished, and thus the timetable for the period $[\mathbf{0}, \mathbf{t^{t}_{end}}]$ can be obtained. Otherwise, set $\mathbf{k} = \mathbf{k} + \mathbf{1}$, $\mathbf{h} = \mathbf{h_{min}}$, and return to Step 2.

ALGORITHM 2: Rolling optimization algorithm.

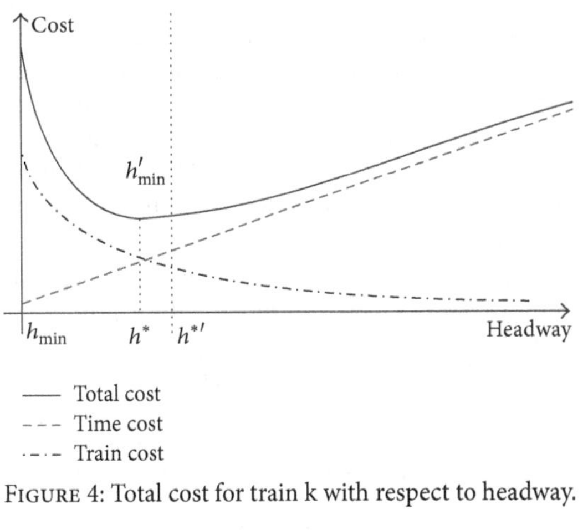

FIGURE 4: Total cost for train k with respect to headway.

the given frequency. An individual consists of headways for all trains ($K - 1$ headways), and its objective function is calculated based on (31). The initial individuals are generated as the cyclic headways. Random values are then added or subtracted to the headways, considering constraints (46) and (49). The details of standard GA [3, 36] are not shown here.

Compared with the proposed solution approach (RO) in this paper, generic GA (GGA) has following properties: (1) In terms of input, both GGA and RO require passenger demand as input. However, GGA requires fixed frequency as input, while RO determines optimal frequency during the process. (2) In terms of optimization process, GGA optimizes the solution (of all trains) in a random direction, while RO optimizes one train at a time. (3) In terms of computation complexity, GGA calculates the objective function of the whole system for each chromosome, while RO calculates the objective of one train at each step. These differences result in different computation performances, which will be further demonstrated in the numerical experiments.

For the numerical experiments, we set the parameter values in the generic GA as follows. The population size is 20, the maximum iteration number is 1000, the crossover rate is 0.7, and the mutation rate is 0.1. The algorithm terminates if the maximum number of iterations is reached or the best fitness value in this generation is the same as in the previous 100 generations ahead.

The pattern search (PS) algorithm is proposed to solve unconstrained optimization problems [52], which varies one variable (headway) at a time by steps of the same magnitude. When no such increase or decrease in any one variable further decreases the objective function, it decreases the step size and repeats the process until the steps are deemed sufficiently small. Here the unconstrained objective is constructed with the Augmented Lagrangian (AL) method, which adds the constraints as penalty terms to the objective function. The penalty term is the product of a Lagrangian multiplier and a constraint equation (e.g., from $g(x) \geq 0$ to $g(x) - y^2 = 0$, where y is a new variable). Let the number of constraints be m. Set the multiplier that corresponds to the j-th constraint as ω_j, and set a large enough parameter σ. By minimizing the objective function with respect to y (so that the intermediate

TABLE 3: Parameter values in the experiments.

Parameter	Value
$\mathbf{M_k}$ (Kg)	1.99×10^5
M_p (Kg)	75
$\mathbf{N_k}$ (Passenger)	1290
c_e (¥/Kwh)	0.79
$\mathbf{c_o}$ (¥ /(train·Km))	20
c_{pt} (¥/(passenger·hour))	10
$\mathbf{c_{pw}}$ (¥/(passenger·hour))	20
c_{tra} (¥/(train·hour))	800
$\mathbf{h_{min}}$ (s)	100
f_r (m/s)	$1.36\times10^{-4}\cdot v_k^2(t)+1.45\times10^{-2}\cdot v_k(t)+0.08$
$\boldsymbol{\chi_x}$	-1.944×10^{-4}
χ_y	0.4176

variable y does not influence the objective function), we can obtain the typical form of the penalty term as

$$\frac{1}{2\sigma}\sum_{j=1}^{m}\left\{\left[\max\left(0,\omega_j-\sigma\cdot g_j(x)\right)\right]^2-\omega_j^2\right\}. \tag{62}$$

The parameter value ω_j is optimized in each iterative step, until the termination condition is satisfied. In each step of AL, we use the PS to find a minimum value of the unconstrained objective function. In this way, we integrate AL and PS and define such approach as AL + PS. The same drawback exists that the frequency is preset and considered as input. The initial solution is the cyclic headways. For the numerical experiments, we set the values of parameters in the AL + PS as follows. The initial step size for PS is 32 s. The speeding factor for PS is 1. The shrinking factor for PS is 0.5. The allowance error for both PS and AL is 1 s (since the continuous time is discretized into seconds). The augmented factor for AL is 1.2. The large enough parameter σ in AL is 200.

5. Numerical Experiments

In this section, we present several numerical experiments to test the effectiveness and efficiency of the proposed solution approach. All experiments are performed with Matlab on a PC with 2 GHz Intel Core7 and 8 GB memory. Parameters in the examples are shown in Table 3. We adopt here the same energy calculation method as in [48]. To save space, we do not show the detailed calculations and directly provide the estimated values of parameters χ_x and χ_y in Table 3.

5.1. A Small Case Study. Here we consider a short metro line, as shown in Figure 5. The segment lengths are 1800 m, 1600 m, and 2000 m. For simplicity, the minimum and maximum dwell times for each station are all set as 30 s and 90 s, respectively.

The baseline passenger demand is shown in Table 4. The "Boarding (Alighting) Sum" represents the number of

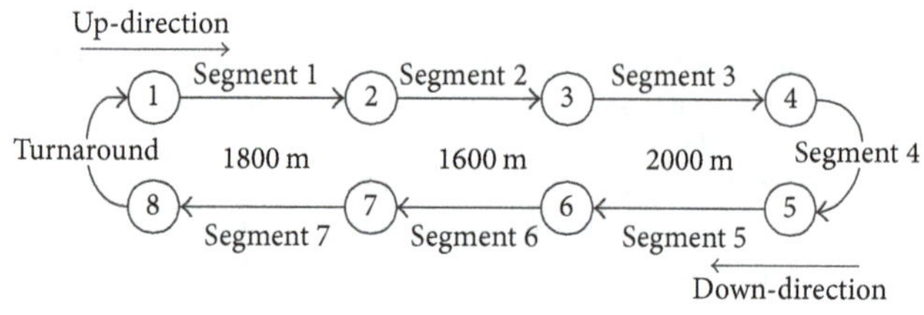

FIGURE 5: Representation of a short metro line.

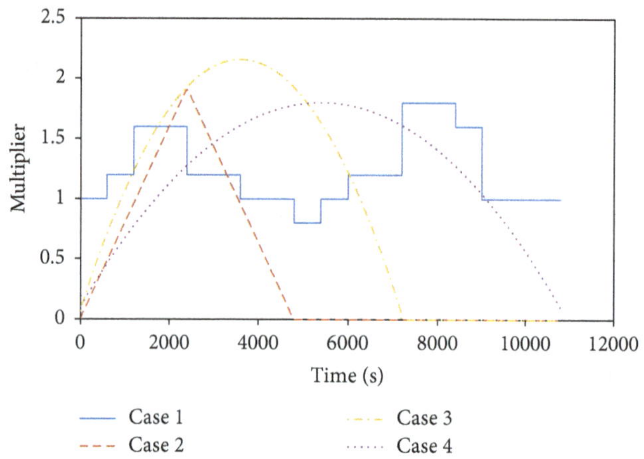

FIGURE 6: Passenger demand distributions in different cases.

total boarding (alighting) passengers per second at a station. Four cases are considered here. As shown in Figure 6, the curves represent demand variance over time, and "multiplier" represents the ratio between actual and baseline passenger demand. For example, in case 1, during period [1200, 2400] (s), the multiplier is 1.6, and passenger demand from station 2 to station 3 at $t \in [1200, 2400]$ is 0.96 (0.6×1.6) passengers/second. The four cases have different passenger demand distributions.

The computational results of different cases are shown in Table 5 and Figure 7. Specifically, in Table 5, the noncyclic result obtained by RO is denoted as $N_{K=K_N}$, which schedules K_N trains within a given time duration. The cyclic result obtained by Algorithm 1 is denoted as $U_{K=K_U}$, which schedules K_U trains within a given time duration. Particularly, the best cyclic result is denoted as $U^{\text{best}}_{K=K_U}$. The GGA result is denoted as $\text{GA}_{K=K_N}$, which uses frequency K_N from RO result as input. The objectives of different results are calculated, including passenger wait time (T_{pw}), passenger travel time (T_{pt}), total energy (E_{total}), fleet size (F_s), and total cost (C_{total}). Figure 7 specifically shows the numerical results in passenger satisfaction.

Table 5 and Figure 7 demonstrate that, under time-varying passenger demand, noncyclic timetables obtained by RO outperform the other timetables, while GGA results outperform cyclic ones obtained by Algorithm 1. This result verifies that the noncyclic timetable improves passenger satisfaction. Particularly, the passenger wait time (T_{pw}), passenger travel time (T_{pt}), and total cost (C_{total}) have their lowest values in the RO results, which demonstrates that the RO

TABLE 4: Baseline passenger demands for a short metro line (passengers per second).

Station index	1	2	3	4	5	6	7	8	Boarding sum
1	0	0.3	0.3	0.45	0	0	0	0	1.05
2	0	0	0.6	0.3	0	0	0	0	0.9
3	0	0	0	0.3	0	0	0	0	0.3
4	0	0	0	0	0	0	0	0	0
5	0	0	0	0	0	0.45	0.75	0.3	1.5
6	0	0	0	0	0	0	0.6	0.3	0.9
7	0	0	0	0	0	0	0	0.45	0.45
8	0	0	0	0	0	0	0	0	0
Alighting sum	0	0.3	0.9	1.05	0	0.45	1.35	1.05	-

TABLE 5: Comparison between cyclic and non-cyclic timetables.

Time duration (s)	Timetable type		T_{pw} (10^6)	T_{pt} (10^6)	E_{total} (10^3)	F_s	C_{total} (10^4)	Gap (%)
0-10800	Non-cyclic	$N_{K=60}$	7.2330	14.384	18.069	7	12.320	-
	Non-cyclic	$GA_{K=60}$	7.2786	14.402	18.069	7	12.329	0.07
Case 1	Cyclic	$U^{best}_{K=53}$	8.3360	14.523	16.148	6	12.496	1.41
	Cyclic	$U_{K=60}$	7.3561	14.406	18.069	7	$+\infty$	-
0-4800	Non-cyclic	$N_{K=24}$	2.3171	4.9470	7.1411	6	4.3586	-
	Non-cyclic	$GA_{K=24}$	2.5489	4.9758	7.1411	6	4.4954	3.04
Case 2	Cyclic	$U^{best}_{K=24}$	2.6102	4.9988	7.1411	6	4.5407	4.01
	Cyclic	$U_{K=23}$	2.7619	5.0347	6.8667	6	4.5681	4.59
	Cyclic	$U_{K=25}$	2.4989	4.9859	7.4156	7	4.5441	4.08
0-7200	Non-cyclic	$N_{K=45}$	4.3948	11.292	13.615	8	8.8022	-
	Non-cyclic	$GA_{K=45}$	4.5726	11.328	13.615	8	8.9245	1.37
Case 3	Cyclic	$U^{best}_{K=48}$	4.3975	11.319	14.438	8	9.0243	2.46
	Cyclic	$U_{K=45}$	4.7268	11.417	13.615	8	9.0345	2.57
0-10800	Non-cyclic	$N_{K=60}$	6.0280	14.301	18.069	7	11.621	-
	Non-cyclic	$GA_{K=60}$	6.3113	14.560	18.069	7	11.854	1.97
Case 4	Cyclic	$U^{best}_{K=60}$	6.4638	14.330	18.069	7	11.875	2.14
	Cyclic	$U_{K=59}$	6.5981	14.381	17.795	7	11.899	2.34
	Cyclic	$U_{K=61}$	6.3572	14.327	18.344	7	11.884	2.21

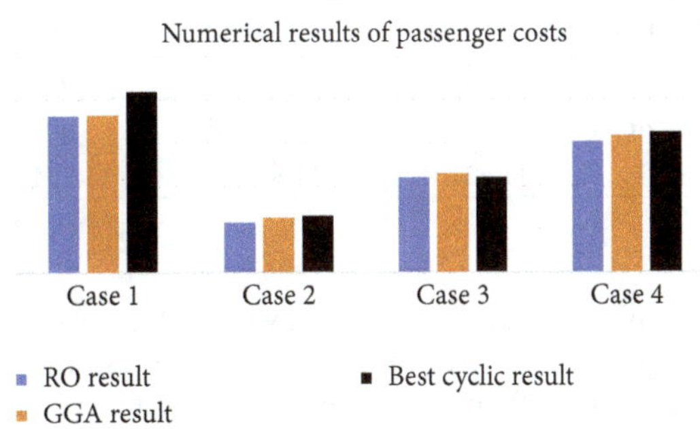

FIGURE 7: Passenger costs for different results in different cases.

method is effective in solving the proposed train timetabling problem. In case 1, the result of $U_{K=60}$ is infeasible because it violates the minimum headway constraint at intermediate stations (the dwell times are different from $U_{K=60}$ to $N_{K=60}$), and thus we have $C_{total} = +\infty$. It is interesting that $N_{K=60}$ and $U_{K=60}$ require the same energy and fleet size. This result verifies the reasonableness of model decomposition in this paper from two perspectives: (1) When all passengers reach their destinations, energy is solely determined by segment travel times (they are the same in $N_{K=60}$ and $U_{K=60}$). (2) With given segment travel times, the estimated fleet size is solely determined by headways. Similar results can be observed in other cases.

The gaps in Table 5 shows that case 2 has the largest gap, case 3 and 4 have similar gaps, and case 1 has the smallest gap. Note that, in case 2, passenger demand changes drastically; in cases 3 and 4, demand patterns are quite similar; in case 1, within certain subperiods, passenger demand is the same, which is the steadiest among all cases. Therefore, we can conclude that as the passenger demand varies more significantly over time, the gap between the cyclic and noncyclic timetables increases.

Figure 8 illustrates the obtained train timetables in different cases, where the departure headways vary in accordance with the fluctuating passenger demand. For example, in cases 2, 3, and 4, the headways are small in the peak period and

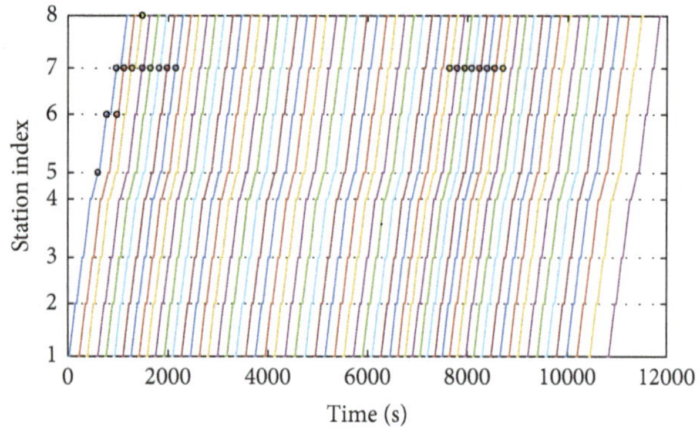

(a) Obtained timetable for case 1 ($N_{K=60}$)

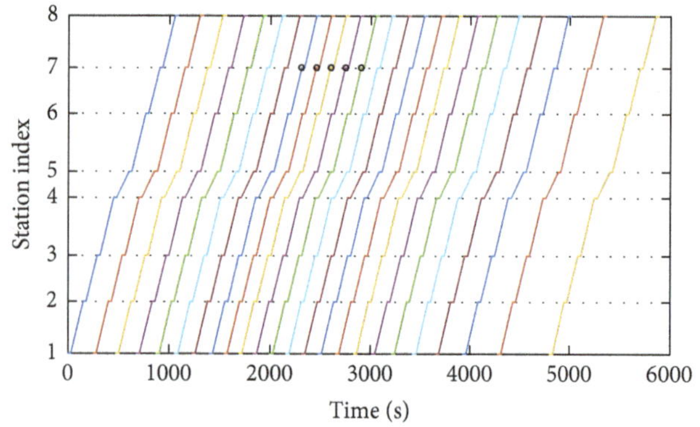

(b) Obtained timetable for case 2 ($N_{K=24}$)

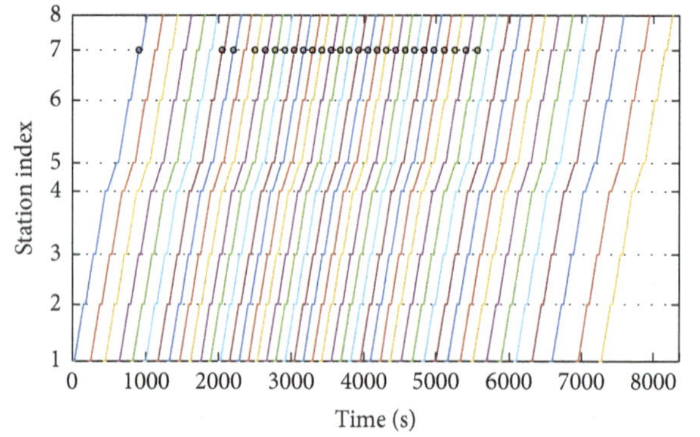

(c) Obtained timetable for case 3 ($N_{K=45}$)

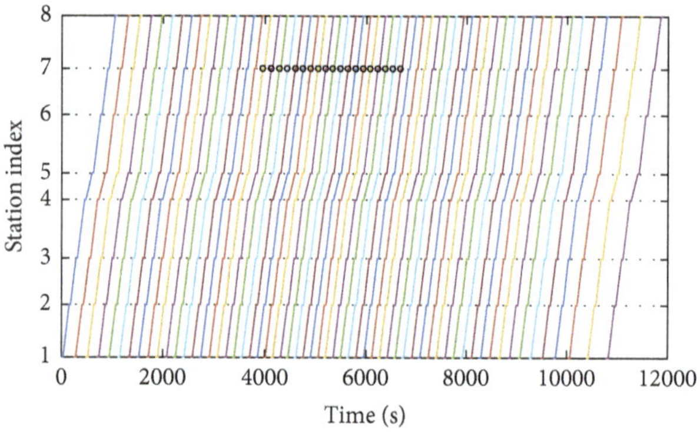

(d) Obtained timetable for case 4 ($N_{K=60}$)

FIGURE 8: Obtained timetables for different cases.

are larger before and after the peak period. In case 1, the small headways occur during subperiods [1200, 2400] (s) and [7200, 8400] (s), where the multipliers of passenger demand are largest (1.6 and 1.8). In Figure 8, the circles represent the dwell times that exceed the minimum dwell time (i.e., 30 s). We observe that the excess dwell times (circles) mostly occur at station 7. This is consistent with the OD pairs we present in Table 4. Specifically, the difference between the cumulative boarding sum and the cumulative alighting sum reaches its maximum value at station 7 (1.5 + 0.9 − 0.45 = 1.95). In this regard, the alighting and boarding rates are smallest at station 7. In addition, the total number of alighting and boarding passengers is largest (i.e., 1.8 passengers per second) at station 7, and thus the excess dwell times here are reasonable.

Moreover, despite the symmetry of passenger demand distributions, the headways at the ends of periods are larger than at their starts. This can be explained by that passenger demand increases at the start and decreases to zero at the end, which leads to different cost changes. Specifically, the passenger time cost increases more significantly when demand increases, while the train cost does not change with demand. Thus, the headway tends to be smaller at the start of periods than at the ends. In fact, it is often seen in actual operations that the headways between the last trains at the end of day exceed those between the first trains in the morning.

Finally, Table 6 shows the computational performances for different methods, including RO method, GGA, and AL + PS method. As described above, GGA and AL + PS require a preset fixed frequency, and thus we use the same frequency from RO result K_N as input. We terminate the algorithm when its computation time reaches 3 hours and output the best solution obtained by then.

TABLE 6: Computational performances for different methods.

	RO method	Generic GA	AL + PS
(Case 1) Time (s)	225.5	8299.4	10800
Total cost (¥)	12.32×10^4	12.33×10^4	12.42×10^4
(Case 2) Time (s)	86.1	4658.2	10800
Total cost (¥)	4.3586×10^4	4.4954×10^4	4.4872×10^4
(Case 3) Time (s)	120.6	5009.3	10800
Total cost (¥)	8.8022×10^4	8.9245×10^4	8.9567×10^4
(Case 4) Time (s)	240.717	8229.1	10800
Total cost (¥)	11.621×10^4	11.854×10^4	11.774×10^4

Table 6 demonstrates that RO is very efficient, obtaining a timetable very fast. In comparison, GGA takes more than 1.5 hours to obtain a timetable, and AL + PS takes more than 3 hours in all cases. In each case, the timetable obtained with the RO method has the lowest total cost, which further verifies its effectiveness. The GGA timetables perform better than AL + PS in cases 1 and 3 but are outperformed in cases 2 and 4.

TABLE 7: Real-world infrastructure data of Beijing Metro Line 4.

Station	Minimum dwell time (s)	Maximum dwell time (s)	Segment	Minimum travel time (s)	Maximum travel time (s)	Segment length (m)
1/48 AHQB	30	90	1	124	170	1363
2/47 BGM	30	90	2	114	156	1251
3/46 XY	30	90	3	152	209	1672
4/45 YMY	30	90	4	118	162	1295
5/44 BDDM	30	90	5	81	111	887
6/43 ZGC	30	90	6	82	113	900
7/42 HDHZ	30	90	7	97	133	1063
8/41 RMDX	30	90	8	96	131	1051
9/40 WGC	30	90	9	151	207	1658
10/39 NL	30	90	10	138	190	1517
11/38 BJZ	30	90	11	131	180	1441
12/37 XZM	30	90	12	93	128	1025
13/36 XJK	30	90	13	100	138	1100
14/35 PAL	30	90	14	100	138	1100
15/34 XS	30	90	15	79	109	869
16/33 LJHT	30	90	16	92	126	1011
17/32 XD	30	90	17	74	102	815
18/31 XWM	30	90	18	105	144	1152
19/30 CSK	30	90	19	109	150	1200
20/29 TRT	30	90	20	149	205	1643
21/28 BJSRS	30	90	21	135	185	1480
22/27 MJP	30	90	22	75	103	827
23/26 JMX	30	90	23	90	124	989
24/25 GYXQ	30	90	-	-	-	-

Combining all the computational results in this case study, we can conclude some disadvantages of GGA and AL + PS compared to RO. (1) Although the timetables obtained by them are noncyclic, train departure headways in their results differ greatly from one another without clear correspondence to the passenger demand. In comparison, RO timetables are more regular (headways change gradually) and capture well the passenger demand patterns. (2) Due to the computation complexity, the computation time for running GGA and AL + PS once is far greater than RO. (3) They both require a preset frequency as input. If we do not know the best frequency beforehand, which is true in most real-world scenarios, we must run these methods with different frequencies and select the best solution, thereby further increasing substantially computation time.

In general, the RO method is superior to GGA and AL + PS in obtaining a noncyclic timetable with time-varying passenger demand.

5.2. Case Study of Beijing Metro Line 4. This section aims to test the applicability and effectiveness of RO in a real-world case study based on Beijing's Metro Line 4 (BML4). The infrastructure information and some operational parameters of BML4 are shown in Table 7. The turnaround time is 120 s. To satisfy the large passenger demand, we set train capacity

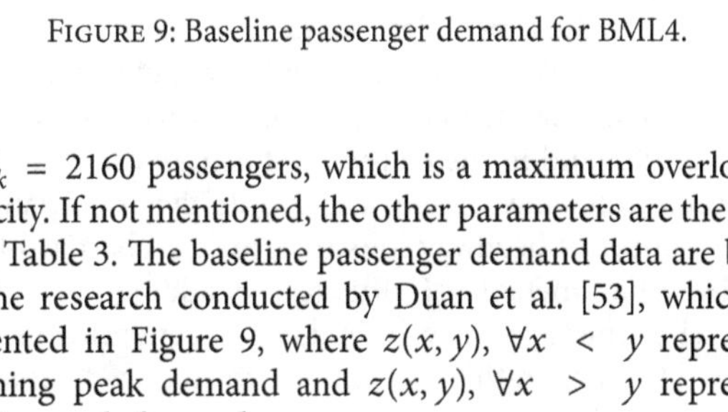

FIGURE 9: Baseline passenger demand for BML4.

as $N_k = 2160$ passengers, which is a maximum overloaded capacity. If not mentioned, the other parameters are the same as in Table 3. The baseline passenger demand data are based on the research conducted by Duan et al. [53], which are presented in Figure 9, where $z(x, y)$, $\forall x < y$ represents morning peak demand and $z(x, y)$, $\forall x > y$ represents evening peak demand.

Here we consider three types of scenarios whose passenger demands are large and unsteady, which are difficult to manage: (a) period of 3 hours (i.e., 10800 s) including morning peak; (b) period of 3 hours including evening peak; and (c) a shopping festival hosted in the shopping centers next to the XD station (whose index is 8/41). Representations of

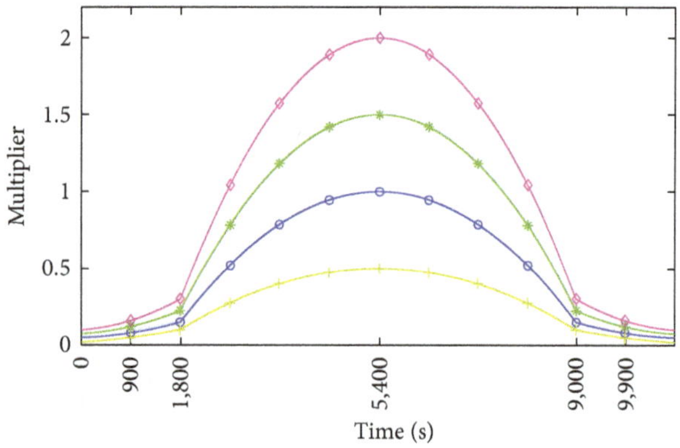

FIGURE 10: Passenger demand patterns in different cases.

passenger demands are the same as in the small case study (as shown in Figure 10): the actual passenger demand is the product of baseline passenger demand (Figure 9) and the multipliers (Figure 10). To test RO with more instances, we design three subcases of scenario (a) with different peak multipliers, and two subcases of scenario (c): without the special event (c-1) or with it (c-2). In case (c-2), the passenger demand at station XD is tripled.

It should be noted that the real operation data of BML4 is dynamic and confidential. Besides, the operator today prefers to schedule trains uniformly. Therefore, in this experiment, we use cyclic timetables with best performance to approximate the real-world timetables. The notations are the same as in the small case study.

Table 8 and Figure 11 show the computational results in different cases. Note that, for a major metro line such as BML4, the computation times of GGA and AL + PS are extremely large, and in many cases they do not yield a satisfactory solution within a reasonable time. Thus, we do not present their results in this experiment. The results of $U_{K=46}$ in case (a-1), $U_{K=54}$ in case (a-2), and $U_{K=54}$ in case (a-3) are infeasible, and thus their total costs are $+\infty$. Figure 11 shows that considerable costs are reduced by the noncyclic timetables obtained by RO. Interestingly, in the cases where cyclic timetables are feasible, that is, cases (a-1), (b), (c-1), and (c-2), the frequency of the noncyclic timetable is always larger than the best cyclic ones. This is because the passenger demand in the real-world fluctuates so drastically that cyclic timetables cannot capture its dynamic patterns, and the minimum headway constraint makes it hard to obtain feasible cyclic timetables. In this regard, adding more trains in a cyclic timetable may be either uneconomic (i.e., the increased train operating cost exceeds the reduced passenger time value) or infeasible.

In comparison, the noncyclic timetables obtained by RO respond well to demand changes. For example, in case (a-1), the result of $N_{K=46}$ substantially reduces passenger wait time compared to $U_{K=44}^{\text{best}}$ and $U_{K=46}$. The adaptation for time-varying demand is further demonstrated in cases (c-1) and (c-2). In case (c-1), the demand distribution pattern is

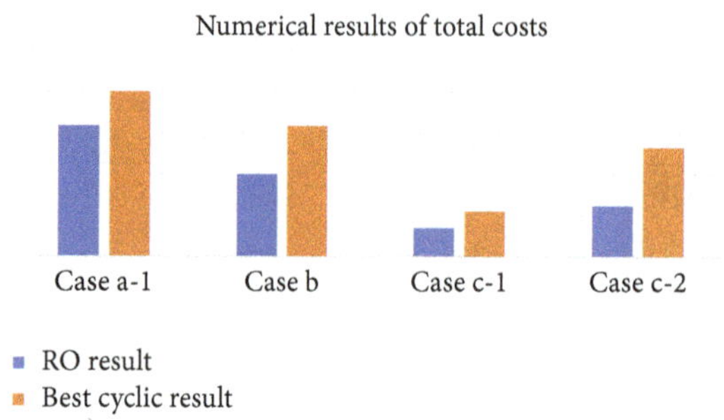

FIGURE 11: Total costs for different results in different cases.

similar to cases (a-1) and (b), and the cost reduction of the noncyclic timetable is closer to the other cases (36.08%). In case (c-2), the special event leads to a more heterogeneous demand distribution, and the cost reduction reaches 53.35%. Besides, the computation times (shown below each case) are acceptable for such a major and busy metro line.

However, in cases (a-2) and (a-3), the passenger wait times for noncyclic timetables are larger than for cyclic ones. This occurs because the RO method considers the objective function that corresponds to an infeasible headway as $+\infty$. With a large demand, the first departure headway that is feasible may greatly exceed h^* (as can be seen in Figure 4), whereas the cyclic timetable does not consider the feasibility and divides headways uniformly. In other words, the increase of departure headways (leading to larger T_{pw}) is due to safety issues, which is consistent with the actual operational requirements that put safety ahead of costs. In addition, with parameters given in this experiment, the largest frequency for BML4 is $K_N = 54$, as can be concluded by comparing cases (a-2) and (a-3). Therefore, real-time dispatching and passenger flow control methods should be applied with passenger demand at or above the levels in case (a-2).

Figures 12, 13, and 14 illustrate the noncyclic timetables obtained in case (a-1), case (b), and case (c-2), respectively. The circles here represent the occurrences of maximum dwell times. These figures illustrate that the headways vary in accordance with demand patterns, which is consistent with the results in the small case study. Particularly in Figure 14, the dense headways in the middle are set to transport many passengers due to the special event. This shows the flexibility of noncyclic timetables in special scenarios.

In addition, from the distribution of the circles, we can see that case (a-1) is the busiest scenario among these figures. The many circles in case (a-1) help explain the results in cases (a-2) and (a-3) (which have larger demands than case (a-1)). With so many trains taking maximum dwell times to load passengers, real-time dispatching and passenger flow control methods are necessary. In case (b), the maximum dwell times mostly occur at stations 7, 11, and 12. Specifically, the neighborhood of station 7 (HDHZ) includes famous schools, fashion shopping streets, and dense office buildings. The neighborhood of station 11 (BJZ) includes a major bus hub, the Beijing Zoo, and one of the largest wholesale markets. The neighborhood of station 12 (XZM), which has transfers with two other busy metro lines, includes a railway station, large

TABLE 8: Computational results for BML4.

Scenario			T_{pw} (10^8)	T_{pt} (10^8)	E_{total} (10^4)	F_s	C_{total} (10^6)	Gap (%)
Case (a-1)	Non-cyclic	$N_{K=46}$	3.8665	3.0555	7.4898	31	3.1788	-
3869 s	Cyclic	$U^{best}_{K=44}$	5.3237	3.0943	7.2385	30	3.9925	20.38
	Cyclic	$U_{K=46}$	4.9805	3.0985	7.4898	31	$+\infty$	-
Case (a-2)	Non-cyclic	$N_{K=54}$	11.426	4.3898	9.2418	37	7.7869	-
1544 s	Cyclic	$U_{K=54}$	10.653	4.4060	9.2418	37	$+\infty$	-
Case (a-3)	Non-cyclic	$N_{K=54}$	19.568	4.7953	9.4566	38	12.426	-
1133 s	Cyclic	$U_{K=54}$	18.891	4.816	9.4958	38	$+\infty$	-
Case (b)	Non-cyclic	$N_{K=38}$	2.2416	2.1501	5.9624	25	1.9902	-
3932 s	Cyclic	$U^{best}_{K=30}$	4.4170	2.1632	4.9379	21	3.1748	37.31
Case (c-1)	Non-cyclic	$N_{K=30}$	0.5482	1.0256	4.3421	19	0.7004	-
6848 s	Cyclic	$U^{best}_{K=22}$	1.2744	1.0946	3.3695	14	1.0957	36.08
Case (c-2)	Non-cyclic	$N_{K=41}$	1.1415	1.5061	6.0564	26	1.2342	-
4068 s	Cyclic	$U^{best}_{K=25}$	3.8100	1.5533	3.9911	17	2.6459	53.53

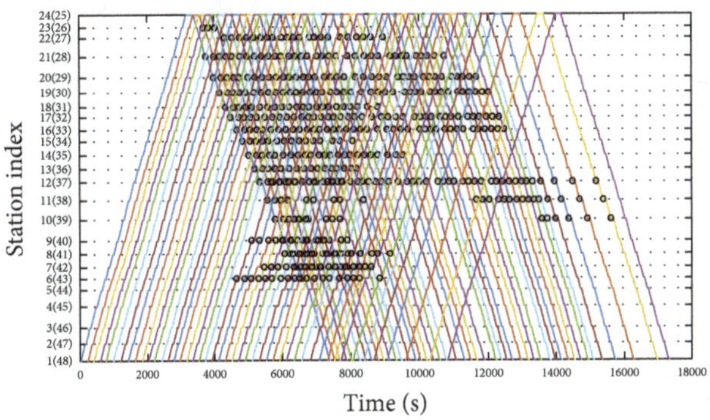

FIGURE 12: Timetable for morning peak (case a-1).

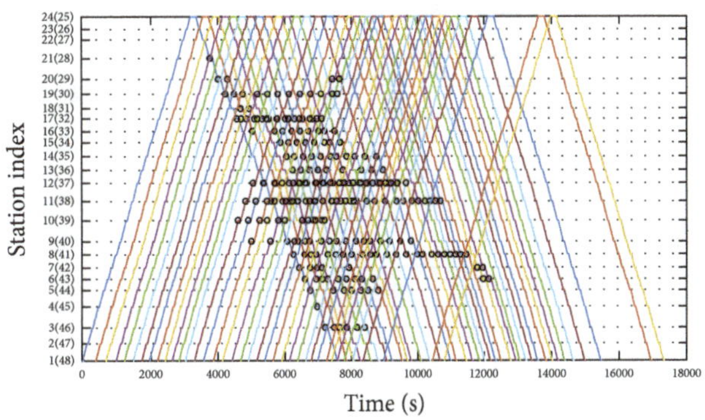

FIGURE 13: Timetable for evening peak (case b).

shopping malls, and office buildings. They are well known for the large numbers of passengers during peak hours. Thus, we find that the model is reasonable and consistent with real-world scenarios. Analogously, in case (c-2), the circles are mostly seen at stations near the station XD, where the special event is held.

It should be noted that the last two trains have an unusually small headway in case (b). The reason is that we assume that the last train must arrive at the first station at the end of the period. It is expected that if there is still demand after 10800 s, the headways will be optimized by our approaches.

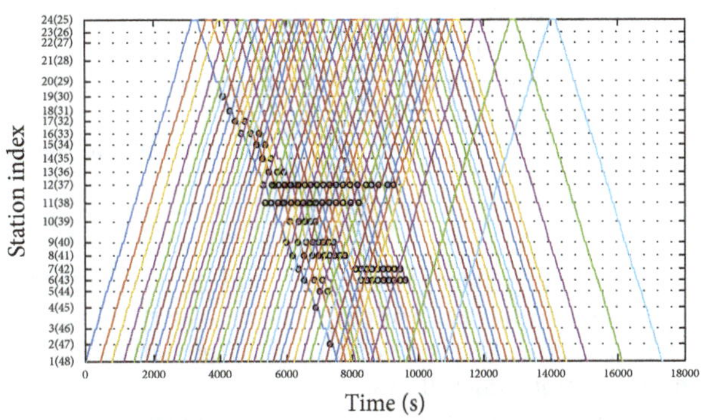

FIGURE 14: Timetable for a special event (case c-2).

5.3. Additional Experiments. In this section, we implement three sets of additional experiments to investigate the relations between the values of costs and the objectives. The first experiment tests the sensitivity of passenger time and peak frequency to unit passenger time cost, based on the demand in case (b) of BML4 study. In practice, the unit time value of passenger travel time is often estimated at half of passenger wait time. Thus, we set $c_{pt} = 0.5 \cdot c_{pw}$ and obtain the results that correspond to different values of c_{pw}, as shown in Figure 15. Results show that as time value increases, total passenger wait time tends to decrease, while the frequency within the peak hour tends to increase. They demonstrate the tradeoffs between passenger-oriented objectives (service level) and operator-oriented objectives (train cost). It may be expected that if the average value of time continues growing, the system manager should add more trains to improve passenger satisfaction.

In addition, neither passenger wait time nor peak hour frequency changes are strictly monotonic. This occurs because the minimum headway constraints would somehow limit the flexibility of the RO method, especially based on a major metro line (as described above), where very different

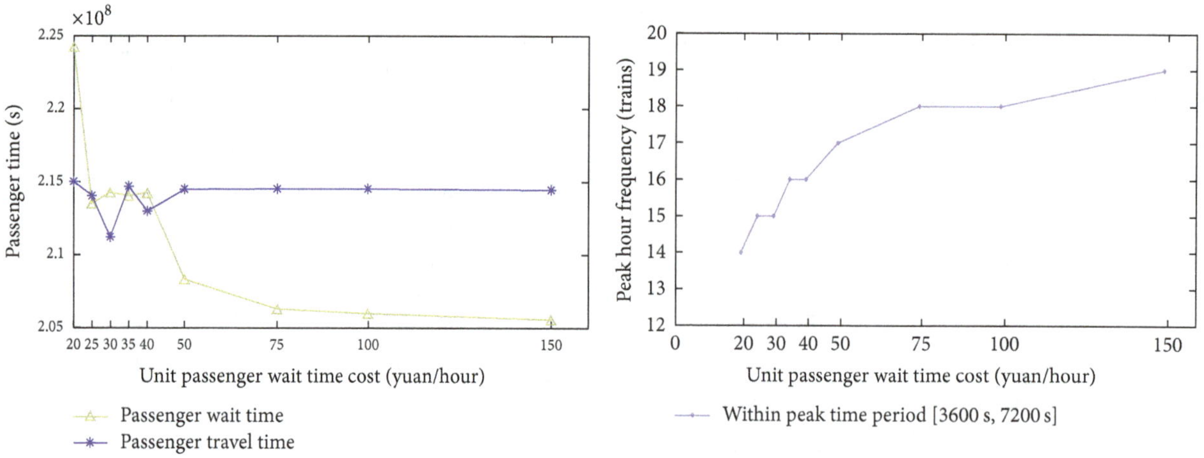

(a) Passenger time with different unit time costs

(b) Peak hour frequency with different unit time costs

FIGURE 15: Sensitivity to unit passenger time cost.

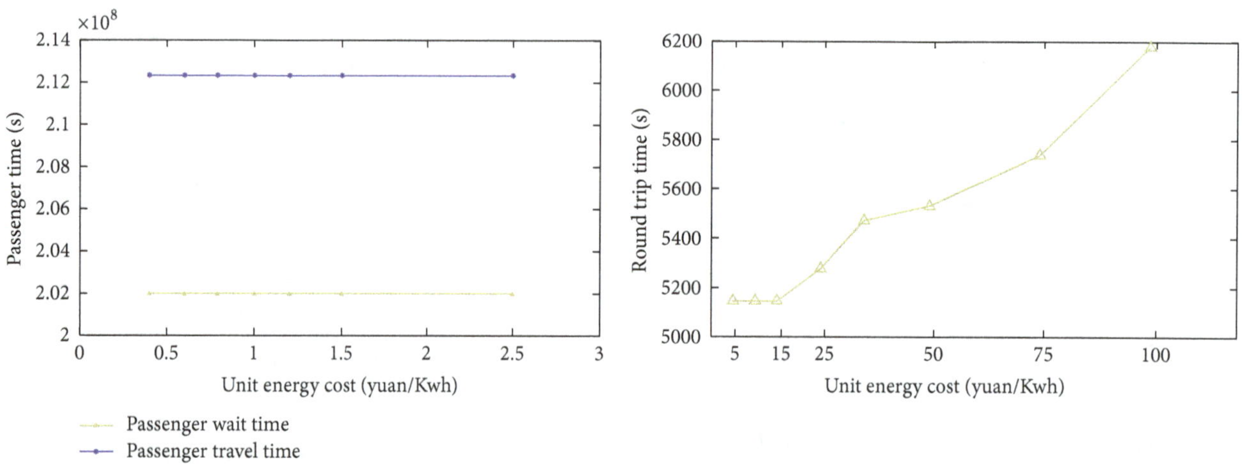

(a) Passenger time with different unit energy costs

(b) Round trip time with different unit time costs

FIGURE 16: Sensitivity to unit energy cost.

distributions of passenger demand lead to varying dwell times. Thus, in the real-world operations, safety issues may limit improvements in service level. The timetables respond well to different values of time.

Besides, we find that total passenger travel time barely changes. In fact, all the segment travel times reach minimum values. The reason will be specifically discussed in the second experiment.

The second experiment tests how unit energy cost affects passenger time and train round trip time, based on the passenger demand in case (c-1) of BML4. Figure 16(a) demonstrates that when unit energy cost changes within a practical range, neither passenger wait time nor travel time is affected. The minimum segment travel times are obtained in all cases. This can be seen in daily operations that, with large enough passenger demand, the metro operators would run trains at the fastest speed, that is, with minimum segment travel times.

To extend the discussion, we run more experiments to show the relations between larger unit energy costs and round trip time and obtain the results shown in Figure 16(b). Although such large unit energy costs are unrealistic, we can see that the round trip time increases; that is, the average speed decreases, as energy cost increases. In other words, as the ratio between unit energy cost and passenger demand increases, round trip time increases. This result is consistent with the practical operation strategy that trains should run faster with more passengers and slower when they carry fewer passengers (where the unit energy cost is fixed, and thus the ratio depends on the passenger demand).

The third experiment aims to test the sensitivity of the passenger wait time and the total cost to train capital cost, based on the passenger demand in case (c-1). Figure 17 illustrates that, with the increased train capital cost, the passenger wait time and total cost tend to increase. The tradeoffs between passenger wait time and train cost, that is,

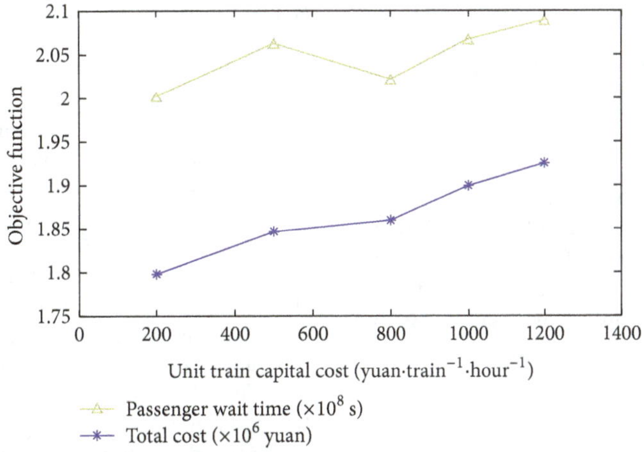

FIGURE 17: Sensitivity to train capital cost.

the passenger-oriented and operator-oriented objectives, is again demonstrated. The fluctuation of passenger wait time is caused by the minimum headway constraints, as explained for the first experiment. In addition, the monotonically increasing total cost demonstrates that technologies which increase the train's economic life help reduce total cost in a metro system.

Generally, we obtain reasonable and practical results with the proposed solution approach. These results are consistent with practical experiences and can be used as decision-making support in daily operations. Since similar results are obtained with different passenger demand patterns and on different metro lines, the predictive ability of the model is demonstrated. The effectiveness of the proposed approach is further verified.

6. Conclusions

This paper investigates the train timetabling problem (TTP) with time-varying passenger demand and accomplishes the following:

(1) It accounts for both the passenger-oriented and operator-oriented objectives. Factors in the line planning stage, the timetabling stage, and the vehicle scheduling stage are integrated in our model and solution approach. The noncyclic timetable obtained in this paper outperforms the cyclic timetable, the noncyclic timetables obtained by the generic GA, and a classic direct optimization approach.

(2) The dwell time determining process is specifically designed to consider both congestion at stations and passenger demand. Besides, limited train capacity and varying passenger alighting/boarding rates are accounted for. Thus, the noncyclic timetable obtained in this paper is more practical in depicting passenger loading evolution. For this reason, some timetables with otherwise good properties would violate the minimum headway constraint at intermediate stations and thus be infeasible. Passenger flow control methods and real-time dispatching can be employed to make these timetables applicable.

(3) Based on a general scheduling scheme, the TTP is decomposed into the OST and OH subproblems. The solution approach includes a branch-and-bound (B&B) algorithm, a frequency determining algorithm, and a novel rolling optimization (RO) method. Then, computational results based on a short metro line and the Beijing Metro Line 4 are obtained. They verify the effectiveness and efficiency of the proposed solution approach and demonstrate that the noncyclic timetables obtained in this paper respond well to passenger demand characteristics. Since the results are based on different cases with different passenger distributions, the predictive ability of the proposed solution approach can also be demonstrated.

(4) Additional experiments are presented whose results demonstrate the following points: (a) As passenger time values increase, passenger wait time may be further reduced by scheduling more trains or by shifting departure times for riders. (b) Average train speed decreases as passenger demand increases, due to longer dwell times at stations. This dwell time effect outweighs the speed increase justified by more riders per train. (c) Longer train economic life not only reduces total system cost, but also indirectly improves the service level.

This paper has some limitations. Specifically, the linear piecewise approximation simplifies the energy calculation. Assumption 1 is consistent with a common strategy in practice but does not account for the relevant effects of driver behaviors. Assumption 4 may be extended to a more accurate model which distributes the passenger boarding and alighting over the station dwell times. A more systematic comparison with real data of BML4 in the case study should be conducted to fully validate RO method.

These drawbacks can be addressed by a more complex model which considers driver behaviors (by estimating the magnitude of their relevance), line layouts, and physics of real train run process, as well as simultaneously optimizing departure headways and segment travel times under continuous time. Future studies can investigate the improvement of our solution approach, to reduce the computation time needed for such a complex model. Besides, it would also be interesting to adjust the model's parameters with respect to different real-world operation data, to obtain more realistic timetables.

Conflicts of Interest

The authors declare that they have no conflicts of interest.

Acknowledgments

This work was supported by National Natural Science Foundation of China (Grant no. U1434209), National Key Research and Development Program of China (2017YFB1201105), and Research Foundation of State Key Laboratory of Railway Traffic Control and Safety, Beijing Jiaotong University (Grant no. RCS2018ZZ003).

References

[1] S. Burggraeve, S. H. Bull, P. Vansteenwegen, and R. M. Lusby, "Integrating robust timetabling in line plan optimization for railway systems," *Transportation Research Part C: Emerging Technologies*, vol. 77, pp. 134–160, 2017.

[2] T. Robenek, Y. Maknoon, S. S. Azadeh, J. Chen, and M. Bierlaire, "Passenger centric train timetabling problem," *Transportation Research Part B: Methodological*, vol. 89, pp. 107–126, 2016.

[3] Y. Wang, T. Tang, B. Ning, T. J. J. van den Boom, and B. De Schutter, "Passenger-demands-oriented train scheduling for an urban rail transit network," *Transportation Research Part C: Emerging Technologies*, vol. 60, pp. 1–23, 2015.

[4] J. Yin, L. Yang, T. Tang, Z. Gao, and B. Ran, "Dynamic passenger demand oriented metro train scheduling with energy-efficiency and waiting time minimization: Mixed-integer linear programming approaches," *Transportation Research Part B: Methodological*, vol. 97, pp. 182–213, 2017.

[5] T. Robenek, S. Sharif Azadeh, Y. Maknoon, and M. Bierlaire, "Hybrid cyclicity: Combining the benefits of cyclic and non-cyclic timetables," *Transportation Research Part C: Emerging Technologies*, vol. 75, pp. 228–253, 2017.

[6] A. Caprara, M. Fischetti, and P. Toth, "Modeling and solving the train timetabling problem," *Operations Research*, vol. 50, no. 5, pp. 851–861, 2002.

[7] E. Barrena, D. Canca, L. C. Coelho, and G. Laporte, "Single-line rail rapid transit timetabling under dynamic passenger demand," *Transportation Research Part B: Methodological*, vol. 70, pp. 134–150, 2014.

[8] A. Schöbel, "An eigenmodel for iterative line planning, timetabling and vehicle scheduling in public transportation," *Transportation Research Part C: Emerging Technologies*, vol. 74, pp. 348–365, 2017.

[9] A. Berbey, R. Galan, J. D. Sanz Bobi, and R. Caballero, "A fuzzy logic approach to modelling the passengers' flow and dwelling time," in *Proceedings of the 18th International Conference on Urban Transport and the Environment, UT12*, pp. 359–369, Spain, May 2012.

[10] H. Niu and X. Zhou, "Optimizing urban rail timetable under time-dependent demand and oversaturated conditions," *Transportation Research Part C: Emerging Technologies*, vol. 36, pp. 212–230, 2013.

[11] H. Niu, X. Zhou, and R. Gao, "Train scheduling for minimizing passenger waiting time with time-dependent demand and skip-stop patterns: nonlinear integer programming models with linear constraints," *Transportation Research Part B: Methodological*, vol. 76, pp. 117–135, 2015.

[12] V. Cacchiani, A. Caprara, and P. Toth, "A column generation approach to train timetabling on a corridor," *4OR*, vol. 6, no. 2, pp. 125–142, 2008.

[13] T. Dollevoet, D. Huisman, M. Schmidt, and A. Schöbel, "Delay management with rerouting of passengers," *Transportation Science*, vol. 46, no. 1, pp. 74–89, 2012.

[14] T. Dollevoet, F. Corman, A. D'Ariano, and D. Huisman, "An iterative optimization framework for delay management and train scheduling," *Flexible Services and Manufacturing Journal*, vol. 26, no. 4, pp. 490–515, 2014.

[15] L. G. Kroon, L. W. P. Peeters, J. C. Wagenaar, and R. A. Zuidwijk, "Flexible connections in PESP models for cyclic passenger railway timetabling," *Transportation Science*, vol. 48, no. 1, pp. 136–154, 2014.

[16] X. Luan, J. Miao, L. Meng, F. Corman, and G. Lodewijks, "Integrated optimization on train scheduling and preventive maintenance time slots planning," *Transportation Research Part C: Emerging Technologies*, vol. 80, pp. 329–359, 2017.

[17] S. Su, X. Li, T. Tang, and Z. Gao, "A subway train timetable optimization approach based on energy-efficient operation strategy," *IEEE Transactions on Intelligent Transportation Systems*, vol. 14, no. 2, pp. 883–893, 2013.

[18] X. Li and K. Lo Hong, "An energy-efficient scheduling and speed control approach for metro rail operations," *Transportation Research Part B: Methodological*, vol. 64, pp. 73–89, 2014.

[19] A. Higgins, E. Kozan, and L. Ferreira, "Optimal scheduling of trains on a single line track," *Transportation Research Part B: Methodological*, vol. 30, no. 2, pp. 147–161, 1996.

[20] J. Wang, Y. Yu, R. Kang, and J. Wang, "A novel space-time-speed method for increasing the passing capacity with safety guaranteed of railway station," *Journal of Advanced Transportation*, vol. 2017, Article ID 6381718, 2017.

[21] X. Xu, K. Li, and L. Yang, "Scheduling heterogeneous train traffic on double tracks with efficient dispatching rules," *Transportation Research Part B: Methodological*, vol. 78, pp. 364–384, 2015.

[22] X. Zhou and M. Zhong, "Single-track train timetabling with guaranteed optimality: branch-and-bound algorithms with enhanced lower bounds," *Transportation Research Part B: Methodological*, vol. 41, no. 3, pp. 320–341, 2007.

[23] L. Yang, K. Li, Z. Gao, and X. Li, "Optimizing trains movement on a railway network," *Omega* , vol. 40, no. 5, pp. 619–633, 2012.

[24] S. Yang, K. Yang, Z. Gao, L. Yang, and J. Shi, "Last-Train Timetabling under Transfer Demand Uncertainty: Mean-Variance Model and Heuristic Solution," *Journal of Advanced Transportation*, vol. 2017, pp. 1–13, 2017.

[25] L. Yang, Y. Zhang, S. Li, and Y. Gao, "A two-stage stochastic optimization model for the transfer activity choice in metro networks," *Transportation Research Part B: Methodological*, vol. 83, pp. 271–297, 2016.

[26] G. Malavasi, P. Palleschi, and S. Ricci, "Driving and operation strategies for traction-energy saving in mass rapid transit systems," *Proceedings of the Institution of Mechanical Engineers, Part F: Journal of Rail and Rapid Transit*, vol. 225, no. 5, pp. 475–482, 2011.

[27] J. T. Haahr, J. C. Wagenaar, L. P. Veelenturf, and L. G. Kroon, "A comparison of two exact methods for passenger railway rolling stock (re)scheduling," *Transportation Research Part E: Logistics and Transportation Review*, vol. 91, pp. 15–32, 2016.

[28] X. Yang, A. Chen, X. Li, B. Ning, and T. Tang, "An energy-efficient scheduling approach to improve the utilization of

regenerative energy for metro systems," *Transportation Research Part C: Emerging Technologies*, vol. 57, pp. 13–29, 2015.

[29] Y.-C. Lai, D.-C. Fan, and K.-L. Huang, "Optimizing rolling stock assignment and maintenance plan for passenger railway operations," *Computers & Industrial Engineering*, vol. 85, pp. 284–295, 2015.

[30] V. Schmid and J. F. Ehmke, "Integrated timetabling and vehicle scheduling with balanced departure times," *OR Spectrum*, vol. 37, no. 4, pp. 903–928, 2015.

[31] V. Guihaire and J. K. Hao, "Transit network design and scheduling: a global review," *Transportation Research Part A: Policy and Practice*, vol. 42, no. 10, pp. 1251–1273, 2008.

[32] L. Cadarso and A. Marin, "Integration of timetable planning and rolling stock in rapid transit networks," *Annals of Operations Research*, vol. 199, pp. 113–135, 2012.

[33] J. M. Mera, E. Carabaño, M. Soler, and E. Castellote, "Increasing metro line capacity by optimisation of track circuit in a speed code Automatic Train Protection system," *Proceedings of the Institution of Mechanical Engineers, Part F: Journal of Rail and Rapid Transit*, vol. 230, no. 1, pp. 165–180, 2016.

[34] S. Chiusolo, A. Dicembre, S. Ricci, and F. Sorace, "Automation of high density metro lines: Rome line a case study," in *Proceedings of the 13th International Conference on Automated People Movers and Transit Systems 2011: From People Movers to Fully Automated Urban Mass Transit*, pp. 491–501, France, May 2011.

[35] V. Cacchiani, F. Furini, and M. P. Kidd, "Approaches to a real-world Train Timetabling Problem in a railway node," *OMEGA - The International Journal of Management Science*, vol. 58, pp. 97–110, 2016.

[36] X. Xu, K. Li, and X. Li, "A multi-objective subway timetable optimization approach with minimum passenger time and energy consumption," *Journal of Advanced Transportation*, vol. 50, no. 1, pp. 69–95, 2016.

[37] L. G. Kroon and L. W. P. Peeters, "A variable trip time model for cyclic railway timetabling," *Transportation Science*, vol. 37, no. 2, pp. 198–212, 2003.

[38] D. Sparing and R. M. P. Goverde, "A cycle time optimization model for generating stable periodic railway timetables," *Transportation Research Part B: Methodological*, vol. 98, pp. 198–223, 2017.

[39] F. Corman, A. D'Ariano, A. D. Marra, D. Pacciarelli, and M. Samà, "Integrating train scheduling and delay management in real-time railway traffic control," *Transportation Research Part E: Logistics & Transportation Review*, 2016.

[40] A. D'Ariano, D. Pacciarelli, and M. Pranzo, "A branch and bound algorithm for scheduling trains in a railway network," *European Journal of Operational Research*, vol. 183, no. 2, pp. 643–657, 2007.

[41] K. Ghoseiri, F. Szidarovszky, and M. J. Asgharpour, "A multi-objective train scheduling model and solution," *Transportation Research Part B: Methodological*, vol. 38, no. 10, pp. 927–952, 2004.

[42] J. Qi, L. Yang, Y. Gao, S. Li, and Z. Gao, "Integrated multi-track station layout design and train scheduling models on railway corridors," *Transportation Research Part C: Emerging Technologies*, vol. 69, pp. 91–119, 2016.

[43] S. Li, M. M. Dessouky, L. Yang, and Z. Gao, "Joint optimal train regulation and passenger flow control strategy for high-frequency metro lines," *Transportation Research Part B: Methodological*, vol. 99, pp. 113–137, 2017.

[44] A. Schobel, "Line planning in public transportation: models and methods," *OR Spectrum*, vol. 34, no. 3, pp. 491–510, 2012.

[45] S. Samanta and M. K. Jha, "Modeling a rail transit alignment considering different objectives," *Transportation Research Part A: Policy and Practice*, vol. 45, no. 1, pp. 31–45, 2011.

[46] H. L. Fu, L. Nie, L. Y. Meng, B. R. Sperry, and Z. H. He, "A hierarchical line planning approach for a large-scale high speed rail network: the China case," *Transportation Research Part A: Policy and Practice*, vol. 75, pp. 61–83, 2015.

[47] E. Cascetta and P. Coppola, "Assessment of schedule-based and frequency-based assignment models for strategic and operational planning of high-speed rail services," *Transportation Research Part A: Policy and Practice*, vol. 84, pp. 93–108, 2016.

[48] J. Yin, T. Tang, L. Yang, Z. Gao, and B. Ran, "Energy-efficient metro train rescheduling with uncertain time-variant passenger demands: an approximate dynamic programming approach," *Transportation Research Part B: Methodological*, vol. 91, pp. 178–210, 2016.

[49] A. P. Cucala, A. Fernández, C. Sicre, and M. Domínguez, "Fuzzy optimal schedule of high speed train operation to minimize energy consumption with uncertain delays and drivers behavioral response," *Engineering Applications of Artificial Intelligence*, vol. 25, no. 8, pp. 1548–1557, 2012.

[50] H. Huang, K. Li, and Y. Wang, "A simulation method for analyzing and evaluating rail system performance based on speed profile," *Journal of Systems Science and Systems Engineering*.

[51] R. Hooke and R. Jeeves, "Direct search solution of numerical and statistical problems," *Journal of the ACM*, vol. 8, pp. 212–229, 1961.

[52] W. C. Davidon, "Variable metric method for minimization," *SIAM Journal on Optimization*, vol. 1, no. 1, pp. 1–17, 1991.

[53] W. Duan, Y. Chen, and J. Lai, *Passenger flow characteristics in Beijing Subway Line 4. Urban Rapid Rail Transit*, vol. 26, 43-46, 26, 2013.

14

Trend Analysis of Pakistan Railways Based on Industry Life Cycle Theory

Xuemei Li,[1] Khalid Mehmood Alam ,[1] and Shitong Wang[1,2]

[1]*School of Economics and Management, Beijing Jiaotong University, Beijing 100044, China*
[2]*School of Economics and Management, Tianjin Chengjian University, Tianjin 300384, China*

Correspondence should be addressed to Khalid Mehmood Alam; 15119006@bjtu.edu.cn

Academic Editor: Zhi-Chun Li

The core purpose of this paper was to analyze the trend analysis of Pakistan railways from the year 1950 to 2015, using the principal component analysis method and industrial life cycle theory. Industrial life cycle theory, the development trend analysis of Pakistan railway industry, entails four stages: introduction, growth, maturity, and decline. The results indicated that railway industry in Pakistan was at its pinnacle in the middle of the seventies and thereafter the decline of railway industry was observed. The main reasons behind the decline were underinvestment, political interference, and the rise of the same-sector competitor, the National Logistics Cell (NLC). From the year 2011, it experienced an upward trend of combined utility curve and showed a new round of industry life cycle. To revive its previous glory, the current government has proposed a development document for Pakistan railways, Vision 2025. It is envisaged that railways share in transportation will be increased from current 4% to 20% by the year 2025, which seems to be an onerous challenge for the organization.

1. Introduction

It can be undoubtedly asserted that the economic development of a country is mainly subjected to its efficient transport system. Logistics is one of the key inputs in the production process and adequate provision of logistics and transport infrastructure helps in increasing productivity and lowering the production overheads. In comparison to other modes of transport, rail transport is relatively reliable, safe, and fast. Railways consist of many parts and operations [1] and appropriate employment of this mode to its full potential can help to reduce the congestion and anomalies on the roads, safeguarding a further safer and flat traveling. Rail is a comfortable, environment-friendly, and cheapest mode of transportation in Pakistan [2], whereas the large volume of goods can be transported over long distances quickly with insignificant impact upon the environment. Rail is the most emissions-efficient major mode of transport, and electric trains powered by renewable energy can offer practically carbon-free journeys and transport [3].

In 1947, when Pakistan achieved independence, 8,122 kilometers of railway tracks were adopted. Pakistan Railways is a public-sector organization, earlier known as the Pakistan Western Railway (1947 to 1974). From 1948, usage of railways increased and the network became profitable. From the year 1950 to 1955, it improved its operations by connecting West and East Pakistan (now Bangladesh). Railways used to be the predominant mode of transport in Pakistan, which, at its peak in the 1950s and 60 s, handled 73% of the freight traffic, compared to less than 4% by the year 2011. During the 1970s, Pakistan Railways also had the largest passenger carrier share in transportation. Over the years, it has been criticized for its deteriorating quality of services and fading number of passengers, freight, and trains. Unfortunately, its role as a catalyst for economic development has received a setback due to significant underinvestment by successive governments who preferred investment in road infrastructure at the cost of railways. In the late 1980s and early 1990s, Pakistan Railways shut its numerous branch lines and minor stations due to declining passenger numbers and financial sufferings. To enhance the number of passengers and freight volume, Pakistan Railways had to make technological innovations, improve the quality of services, and control mismanagements.

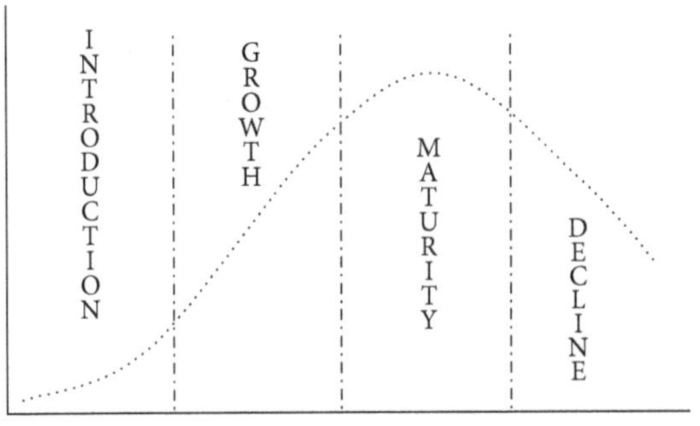

FIGURE 1: Industry life cycle curve.

2. Concept of Industry Life Cycle

An industry life cycle analysis allows for firms within the industry to frame policies [4] and nature of an industry changes as the industry moves through its life cycle [5]. The main views of how firms and industries progress are as follows: as some businesses mature and the source of competition shifts from product innovation to process innovation [6–9]. The idea of industry life cycle was presented by Michael in 1980 [10] and, according to this concept, the industry is the most important part of a firm. The concept of Porter (1980) is continued to explain certain behaviors of stages of industry life cycle with Abernathy/Utterback model [11]. It distinguishes life cycle of an industry with respect to the degree of dynamics of development of process and product innovations. The criteria for differentiating changes of individual stages are an increase or decrease in product and innovation process. The dynamics of development of the industry are defined by two types of threats: obsolescence threat to main events and threat to main assets [12]. The threat to main events refers to a major threat to industrial activities that have realized profits for the industry in the past. The threat to main assets is a threat to resources, knowledge, and brand which made the organization distinctive in the market. More generally, a firm's expertise and its ability to feat opportunities determine the organizational form [13–15].

2.1. Different Phases of Industry Life Cycle. The basic idea of life cycle theory comes from biology parallel with the life cycle of an organism [16]. According to industry life cycle, all industries go through four different phases, where each particular phase requires a policy that will effectively absorb the specific situations of the phase [17]. Industrial life cycle curve can usually be a sales revenue on the vertical axis and time on the horizontal axis represents S-shaped curve shown in Figure 1. It is generally divided into four stages, introduction, growth, maturity, and decline [16], and the industry life cycle analysis allows the management to adjust the business according to the needs of industry [18–20].

2.1.1. Introduction Stage. Evolving industries are new and product appears on the market in the introduction stage. There is small market demand and customers are not aware of its potential benefits, so advertising is usually necessary and investments are needed to build upmarket. The basic distinctive of this phase of the life cycle is that there are no rules. The main reward is the biggest market share for those who are entering the market first [21]. Numerous advantages can be achieved in several ways, by designing a new product, using advanced business processes, and entering new markets. Over time, as new firms enter the market, the initial reward is reduced, compelling pioneers to take actions to raise or defend their market share. Pioneers reward depends on benefit from the difference between the cost of innovation and artificial costs, cost saving, mistakes of pioneers, and benefit from economies of scale [22]. There are two options to be a pioneer or to be a follower. Followers also have to decide whether to be early or late entrants and all options have benefits and drawbacks [23]. The followers may slightly be better than pioneers but, logically, the conclusion should depend on the specific features of each new market [24].

2.1.2. Growth Stage. In growth stage, the product has been successful and there is rapid spread of awareness and use of the product. With the development of new technologies, the unit cost is reduced, quality is improved, and demand is increasing, in addition to growth in profits. Most of the firms are planning their business not only by taking into account profit and sales but also by keeping the market share in mind, which is regarded as key to long-run profitability. However, the firm's goal should be not only to maximize the market share but also to achieve optimum market share [25]. Once the firm determines its optimal market share to achieve the market leader position, the firm should increase the overall market demand, defend its current market share, and try to increase market share.

2.1.3. Maturity Stage. In the maturity stage, the market share reaches a certain level but, at the same time, market demand is relatively slow and stable and industry is in recession. This transition from a phase of growth to maturity is an almost critical period for firms in the industry. This is the period during which fundamental changes are taking place in the company and are induced and also followed by a number of trends.

2.1.4. Decline Stage. Decline phase is characterized by falling profit margins, reduction in production, lower investment, and fewer competitors. In this phase of the industry life cycle, the strategic choice is the harvesting strategy as a strategic management decision to reduce investment in the business with the hope of reducing costs or improving cash flow. Customers have been attracted to other products, due to the fact that products and services cannot meet the market demand or lower prices of alternative products, making profit margins decline and outputs start to decrease and industry will enter the decline stage of the industry life cycle.

However, if the industry after recession comes through certain industrial innovations and upgradations, the output

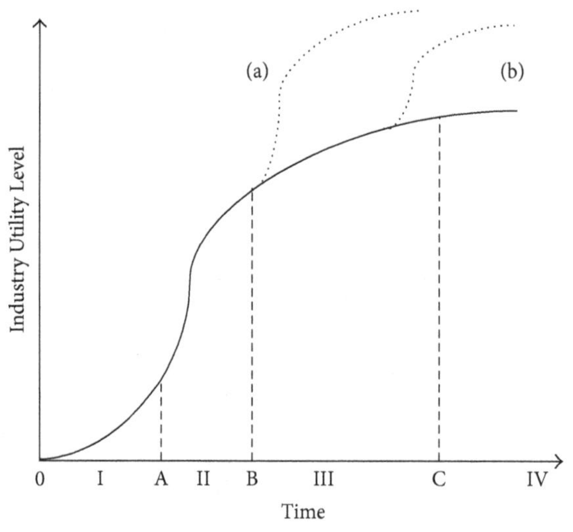

FIGURE 2: Industry life cycle with innovation.

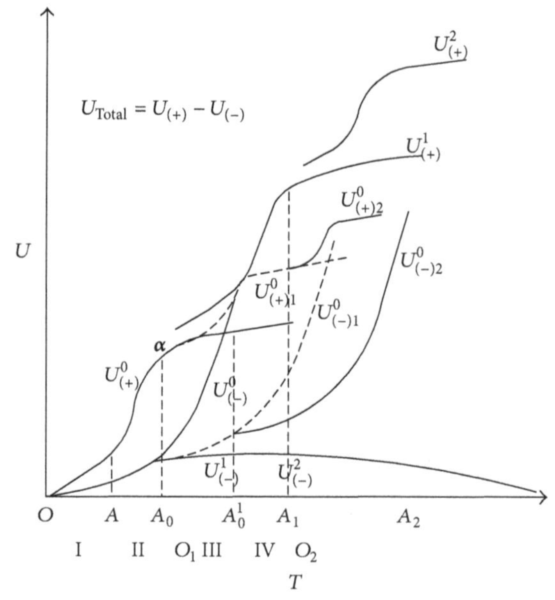

FIGURE 3: Long-term development trend of railway industry.

of the industry will stop to decline and may revive the industry and may bring a new round of industry life cycle [26]. Few industries will survive at the initial stage of maturity, while some industries survive at the end of maturity stage with the adaptation of advanced technologies, modernization of equipment, and management and service innovation as shown by dotted lines (a) and (b) in Figure 2.

2.2. Long-Term Development Trend of Railway Industry Based on Industrial Life Cycle Theory. In the long run, the sustainable development of railway industry is achieved by technological advancement and industrial modernization and advancement through management innovation, improvement in the quality of service, and adjustment according to market needs. The different approaches to decision-making may have different effects on firm performance. A response to opportunities is an appropriate way for firms in growth stage industries, where conditions are rapidly changing and opportunities for development are abundant. However, such environments may not favour the kind of typical competitive aggressiveness [27].

The marginal value of output is positive in the introduction, growth, and maturity periods of the industry and has a positive impact on the society, and marginal output of industry is significantly reduced or is even zero when the industry is in recession at the end of maturity period, indicating that the industrial development potential has been obviously insufficient and needs to take urgent countermeasures to achieve industrial improvement. When the industry enters the recession period, the marginal output of the industry is reduced sharply or is even negative, indicating that the contribution of the industry to the society cannot offset its negative effect on the society and the positive effects during the early stages of industry life cycle are replaced by the negative effect in the latter stages of industry life cycle [28]. If there are no effective industrial upgradation measures, the industry should shut down immediately.

The ideal long-term development trend of the railway industry is shown in Figure 3.

Following the theory of industrial life cycle, the function of measuring the output value of the railway can be expressed as the marginal utility of industrial development.

$$U_{\text{Total}} = U_{(+)} - U_{(-)}, \tag{1}$$

where U_{Total} is marginal revenue for output, $U_{(+)}$ is marginal positive utility for the level of industrial development, and $U_{(-)}$ is marginal disutility for the level of industrial development. T is for time, A, A_0, and A_0^1 are demarcation points for introduction, growth, and maturity stages, respectively, O_i is for a new round of industrial innovation and upgradation point ($i = 1, 2, \ldots$), A_j is beginning point of industrial innovation and upgradation ($j = 0, 1, 2, \ldots$), $U_{(+)}^n$ is change in marginal utility of railway industry for the gradual introduction of new technologies and management innovations ($n = 1, 2, \ldots$), $U_{(-)}^m$ is change in marginal disutility for the gradual adoption of new technologies and management innovations ($m = 1, 2, \ldots$), and $\boldsymbol{\alpha}$ identifies the time to adjust the industry.

In order to avoid the railway industry into the recession, the railway should determine $\boldsymbol{\alpha}$ even earlier in a maturity period, start to take scientific and technological progress, and improve the speed, service improvement, industrial upgradation, corporate restructuring, market expansion, and other revolutionary measures to improve industrial output and social impact. So, the utility curve $U_{(+)}$ along the new curve $U_{(+)}^1$ entered a new round of industrial life cycle. The growth trend of $U_{(+)}^1$ is similar to $U_{(+)}$, the first (I) stage will experience slow growth, the second (II) stage will grow rapidly, the third (III) stage has steady growth, and the fourth (IV) stage has gradual decay. It still needs to discover transition point that can make the industry growth. Similarly, it is also important to reduce the negative effects $U_{(-)}$ on output to achieve growth. The negative impact of the railway industry on the level of social development can be reduced by reducing internal costs, reduction in consumption of energy, improving internal organization and management. The positive impact of railway industry can be assumed from the positive growth, while the negative social impact continued downward deflection of the long-term level of development trend curve.

3. Trend Analysis

Pakistan Railways started its operations with 8,122 kilometers of railway tracks in the year 1947 [29]. At the initial stage, the overall capacity of railways was limited to provide a better transport facility to the customers. From 1948, railway industry improved its service quality, people used to travel with railways, and usage of railways increased. Data related to railway industry was available from the year 1950 to 2015; therefore the year 1950 was used as starting or base point to examine and study the development trends. For this paper, data were collected from Pakistan Railway Year Book 2014-15 and SPSS software was used for statistical analysis. At the initial stage, seven indicators, passenger volume, average passenger mileage, passenger turnover, freight volume, average freight mileage, freight turnover, and number of employed persons, were included for the analysis.

3.1. Principal Component Analysis. The central idea of principal component analysis (PCA) is to reduce the dimensionality of a data set consisting of a large number of interrelated variables, while retaining as much as possible the variation present in the data set [30]. PCA aims to detect the correlation between variables. If a strong correlation between variables exists, the attempt to reduce the dimensionality only makes sense [31]. In this paper, the variables are interrelated. The correlation coefficients are presented in Table 1.

Commonly used rotation methods are orthogonal rotation and oblique rotation. Orthogonal rotation method assumes that the factors in the analysis are uncorrelated, while oblique rotation method assumes that the factors are correlated. According to [32], varimax, quartimax, and equimax are orthogonal rotation methods and direct oblimin and promax are oblique rotation methods. To explore the rotation of components, varimax and quartimax of orthogonal rotation method and oblimin of oblique rotation method are employed. Table 2 represents the rotation of components and results appear to make very little difference.

3.2. Determination of Comprehensive Utility Factors. The comprehensive utility factor can reflect the overall utility of the railway industry development level. The contribution rate of the common factor indicates that the common factor reflects the amount of information of the original index, and the cumulative contribution rate represents the amount of information in which the corresponding common factors reflect the original index. The principal component analysis is used to identify the patterns in data [33] and to calculate the eigenvalues and contribution rates of the common factors of the seven variables are shown in Table 3.

TABLE 1: Correlation coefficients.

	X_1	X_2	X_3	X_4	X_5	X_6	X_7
X_1	1						
X_2	−.873**	1					
X_3	.411*	−.018	1				
X_4	.931**	−.918**	.256	1			
X_5	−.174	.415*	.596**	−.185	1		
X_6	.944**	−.855**	.312	.970**	−.196	1	
X_7	.832**	−.855**	.113	.885**	−.357	.918**	1

*Correlation is significant at the 0.05 level (2-tailed). **Correlation is significant at the 0.01 level (2-tailed). X_1 is number of passengers carried; X_2 is average number of kilometers travelled by passengers; X_3 is passenger turnover; X_4 is total freight carried; X_5 is average kilometers travelled by tonne; X_6 is freight turnover; X_7 is number of employed persons.

Many methods have been proposed to determine the number of components to retain. The complication of the methods is diverse; some use simple graphs, while others are computationally demanding. These methods include scree plot and cumulative percentage of total variance. The traditional approach is to use the first few principal components in data analysis, since they capture most of the variation in the original data set. It has been suggested that [34, 35] one can use enough components to include most of the variation in the data.

The number of components to keep was based on the Kaiser criterion, for which only the principal components with the total variance greater than 85% were considered important [36]. On the basis of cumulative variance >85%, factor analysis developed first two factors explaining about 92.36% of the total variance (Table 3). Therefore, the seven indicators shown in Table 3 can be integrated into the first two main factors and can be factor load matrix. The first factor quantifies the maximum possible variation among different indicators and explains the largest proportion (67.620%) of the total variance and the second factor explains a significant proportion (24.120%) of the total variance.

A scree plot in Figure 4 displays the eigenvalues associated with a component or factor in descending order versus the number of the components or factors. This scree plot shows that two of those factors explain most of the variability because the line starts to straighten after the second factor. The remaining factors explain a very small proportion of the variability and are likely insignificant.

Commonality in principal component analysis is equal to the sum of all the squared factor loadings for all the factors related to the observed variable and this value is the same as R^2 in multiple regression [37]. The value ranges from zero to one, where one indicates that the variable can be fully defined by the factors and has no uniqueness. In contrast, a value of zero indicates that the variable cannot be predicted at all from any of the factors. In order to analyze the importance of included indicators in the development of

TABLE 2: Rotation of components.

Variables/components	Varimax rotation components		Oblimin rotation components		Quartimax rotation components	
	1	2	1	2	1	2
X_1	.963	.151	.964	.163	.961	.160
X_2	−.933	.220	−.931	.208	−.935	.212
X_3	.260	.904	.266	.907	.251	.906
X_4	.979	.050	.979	.063	.978	.059
X_5	−.280	.880	−.274	.877	−.288	.878
X_6	.979	.082	.980	.095	.979	.091
X_7	.934	-.133	.933	−.121	.935	−.125

TABLE 3: Principal component analysis explains total variance.

Component	Initial eigenvalues	Total % of variance	Cumulative %
1	4.733	67.620	67.620
2	1.688	24.120	91.740
3	.297	4.244	95.984
4	.172	2.460	98.444
5	.068	.978	99.422
6	.037	.523	99.945
7	.004	.055	100.000

TABLE 4: Communality of indicators.

Variable	Extraction
Passenger volume	0.950
Average passenger mileage	0.919
Passenger turnover	0.884
Freight volume	0.960
Average freight mileage	0.853
Freight turnover	0.966
Number of employed persons	0.890

railway industry, principal component analysis was used to get the communality values (Table 4). The communality value of all the seven indicators, passenger volume, average passenger mileage, passenger turnover, freight volume, average freight mileage, freight turnover, and number of employed persons, was above 80%, and it indicates that all the seven indicators have greater impact in the development of railway industry.

Rotated component matrix was obtained through principal component analysis to achieve the purpose of simplifying the structure so that each variable has a high load on a single factor and only a small to medium load on the rest of the factor. The most commonly used rotation method was the variance of the maximum orthogonal rotation method. Table 5 shows that freight turnover, freight volume, passenger volume, average passenger mileage, and number of employed persons load heavily on the first factor and the other two variables, passenger turnover and average freight mileage, load on the second factor.

TABLE 5: Rotated component matrix.

Variables	Components	
	1	2
Freight turnover	0.979	0.082
Freight volume	0.979	0.050
Passenger volume	0.963	0.151
Average passenger mileage	0.934	−0.133
Number of employed persons	−0.933	0.220
Passenger turnover	0.260	0.904
Average freight mileage	−0.280	0.880

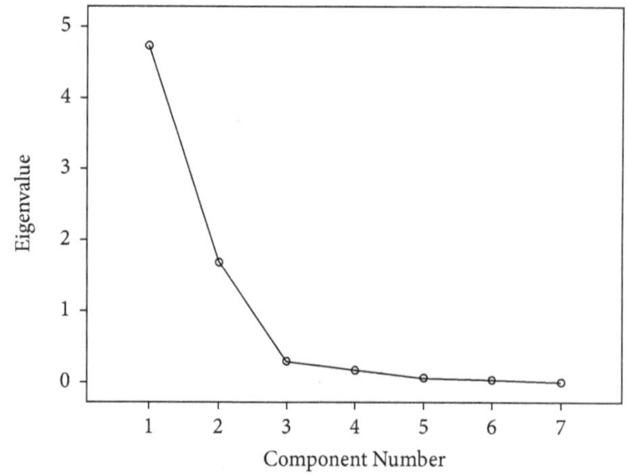

FIGURE 4: Scree plot of industry life cycle of railways.

The expression of the annual utility factor can be obtained:

$$\begin{aligned} Y_1 &= 0.979X_1 + 0.979X_2 + 0.963X_3 + 0.934X_4 \\ &\quad - 0.933X_5 + 0.26X_6 - 0.28X_7, \\ Y_2 &= 0.082X_1 + 0.05X_2 + 0.151X_3 - 0.133X_4 \\ &\quad + 0.22X_5 + 0.904X_6 + 0.88X_7, \\ Y &= 0.6762Y_1 + 0.2412Y_2, \end{aligned} \tag{2}$$

where $X_1, X_2, \ldots, X_7$ are freight turnover, freight volume, passenger volume, average passenger mileage, number of employed persons, passenger turnover, and average freight mileage, respectively, and Y value represents the annual level of development of railway industry.

3.3. Evaluation of Factor Scores. In order to evaluate the indexes, the factor scores are calculated for the two principal factors, and the factor scores are obtained, and the linearly weighted average sum is obtained with the respective contribution rate as the weight. The comprehensive scores of Y, Y_1, and Y_2 are obtained as shown in Table 6.

In order to better analyze the development trend of the railway industry, the time series regression analysis method was used for forecast. The number of time periods is represented by t, where $t = 1, 2, \ldots, 29$. The first time period represents the years 1950–55 and the last time period represents the years 2014-15.

In order to better predict the development trend of the railway, the paper uses the time series regression to fit the railway comprehensive utility level curve from the year 1950 to 1999 and from 1999 to 2015 to the corresponding utility development level curves Y_1 and Y_2. Y_1 represents the time period from year 1950 to 1999 and Y_2 represents the period from 1999 to 2015 and the result of time series regression analysis is represented in Tables 7 and 8. The results are as follows:

$$\begin{aligned} Y_1 &= -0.2514328 + 0.1282887t - 0.0027294t^2, \\ Y_2 &= 6546 - 464.1t - 12.29t^2 + -0.1441t^3 - 0.0006306t^4, \\ \widehat{Y} &= \begin{cases} -0.2514328 + 0.1282887t - 0.0027294t^2 & t \in (1950, 1999) \\ 6546 - 464.1t + 12.29t^2 - 0.1441t^3 + 0.0006306t^4 & t \in (2000, 2015). \end{cases} \end{aligned} \tag{3}$$

Seven variables, passenger volume, average passenger mileage, passenger turnover, freight volume, average freight mileage, freight turnover, and number of employed persons, were used for factor analysis to calculate the comprehensive utility curve of Pakistan railway industry over the years from the year 1950 to 2015, as shown in Figure 5. Y_1 represents the time period from year 1950 to 1999, Y_2 represents the period from 1999 to 2015, and Y represents the comprehensive utility curve of the railway using the factor analysis. It can be seen from Figure 5 that the comprehensive utility level of the Pakistan railways reached its peak in 1975, and the development of the railway after 1975 showed a downward trend until the lowest level in 1999.

3.4. Industrial Life Cycle Analysis of Pakistan Railway Industry. The railway development stage is divided into the following stages:

1st Stage. From the year 1950 to 1975, the level of railway development is on the rise.

2nd Stage. From the year 1975 to 1999, the level of railway development showed a downward trend.

3rd Stage. From the year 1999 to 2005, the level of railway development is on the rise and started a new round of railway industry life cycle, but the pace of development is slow.

4th Stage. From the year 2005 to 2011, the level of railway development was experiencing a downward trend.

Final Stage. From the year 2011 to 2015, the level of railway development has increasing trend and started a new second round of railway industry life cycle.

4. Detailed Trend Analysis

Pakistan railways experienced introduction, growth, maturity, and decline stage and a new round of industry life cycle over the years. The detailed trend analysis is as follows.

4.1. First Round of Industrial Life Cycle (1950–1999). The figure represents that Pakistan railways experienced the first round of industrial life cycle from the year 1950 to 1999 and completed four stages of introduction, growth, maturity, and decline. Detailed analysis of first round is as follows.

4.1.1. Start of Railway Industry (1947–1955). Railway services started with an adaptation of 8122 kilometers of railway tracks when Pakistan achieved its independence in the year 1947. This was the beginning of the Pakistan railway industry; its prominent feature was the lack of railway infrastructure, most needed to be reconstructed and repaired, but the rail transport has greatly improved over the years. After independence, Government of Pakistan invested huge amount for the development of railways. At the initial stage, railways were the only transport mode to connect the main cities and were used for passenger and freight transportation. In 1950s, railways industry increased its connectivity by extension of railways lines to other cities and started to provide better facilities by conversion of railway tracks from meter gauge to broad gauge. In 1954, the railway line was extended to Mardan and Charsadda (cities towards the northwest of the country), and in 1956 the Jacobabad-Kashmore (in the south) line was converted to broad gauge from meter gauge [29].

4.1.2. Growth and Maturity of Pakistan Railways (1955–1975). The rail industry has made considerable progress in expanding railway network and improved the quality of service to passengers and freight by new equipment and technology, management innovation, and increase in transport capacity. The overall level of effectiveness of railways industry has entered a rapid growth stage by an increase in passengers and freight carried. Because of the geographical shape of the country with a length of more than 2000 km from one

TABLE 6: Factor scores.

Years	Factor 1 (Y_1)	Factor 2 (Y_2)	Factors (Y)
1950–55, average	0.408	−1.970	−0.199
1955–60, average	1.024	−1.546	0.320
1960–65, average	1.756	−0.868	0.978
1965–70, average	1.987	−0.633	1.191
1970–75, average	1.799	−0.234	1.159
1975–80, average	2.098	0.779	1.606
1980–85, average	1.308	0.468	0.997
1985–90, average	0.911	0.123	0.645
1990–95, average	0.158	−0.147	0.071
1995-96	−0.263	−0.123	−0.208
1996-97	−0.381	−0.175	−0.300
1997-98	−0.448	−0.220	−0.355
1998-99	−0.490	−0.122	−0.361
1999-00	−0.382	0.072	−0.241
2000-01	−0.473	0.248	−0.260
2001-02	−0.384	0.490	−0.142
2002-03	−0.295	0.622	−0.049
2003-04	−0.276	0.834	0.015
2004-05	−0.304	1.719	0.209
2005-06	−0.204	1.461	0.214
2006-07	−0.150	1.207	0.189
2007-08	−0.171	1.359	0.212
2008-09	−0.425	0.831	−0.087
2009-10	−0.850	−0.259	−0.637
2010-01	−1.236	−2.234	−1.374
2011-02	−1.318	−1.780	−1.320
2012-13	−1.226	−0.707	−0.999
2013-14	−1.120	0.287	−0.688
2014-15	−1.051	0.518	−0.586

TABLE 7: Time series regression analysis for 1950–1999.

Coefficients	Estimate	Std. error	T value
Intercept	−0.2514328	0.1717710	−1.464
Year	0.1282887***	0.0154574	8.299
Year^2	−0.0027294***	0.0002836	−9.625

∗∗∗, ∗∗, and ∗ indicate 1%, 5%, and 10% significance level, respectively; $Y_1 = -0.2514328 + 0.1282887t - 0.0027294t^2$; $R^2 = 0.9254$.

corner to another, Pakistan railways was considered to be an excellent, efficient, and cost-effective means for transporting goods and services from Karachi (south) to Peshawar (north-west). The Kot Addu-Kashmore line was completed in 1973, providing an alternative route from Karachi to upcountry, along with a right bank of Indus River [29]. During the 1970s, Pakistan railways also had the largest passenger carrier share in transportation. This was the golden period of Pakistan railways as its efficiency was at highest level. Railways used to be the predominant mode of transportation in Pakistan, which, at its peak in the 1950s and 60s, handled 73% of the freight traffic, compared to less than 4% by 2011 [38].

TABLE 8: Time series regression analysis for 1999–2015.

Coefficients	Estimate	Std. error	T value
Intercept	6546**	1579	4.147
Year	−464.1**	109.7	−4.229
Year^2	12.29**	2.853	4.307
Year^3	−0.1441***	0.03288	−4.381
Year^4	0.0006306***	0.0001417	4.450

∗∗∗, ∗∗, and ∗ indicate 1%, 5%, and 10% significance level, respectively; $Y_2 = 6546 - 464.1t - 12.29t^2 + -0.1441t^3 - 0.0006306t^4$; $R^2 = 0.8631$.

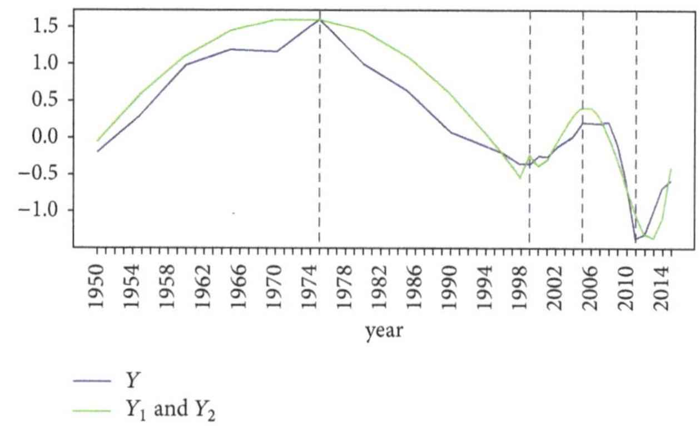

FIGURE 5: Trend analysis of Pakistan's railway industry from 1950 to 2014.

4.1.3. Decline of Railway Industry (1975–1999). Once considered to be the largest and most efficient transit system of freight (Magazines & Review, 2011), railway industry after the year 1975 had a decline in the level of railway development because its share in passenger and freight carried started to decay and was at bottom level by the year 1999.

(1) Under Investment and Political Interference. Railways role as a catalyst for economic development, in Pakistan, has received a setback due to significant underinvestment by successive governments that preferred investment in road infrastructure at the cost of railways. Apart from low investment in the railway industry, it also suffered from various issues that may have contributed to its decline. Some of the more projecting factors are political interference, poor human resource, decrepit infrastructure, and inability to develop a strategic business plan. Political intrusion in the recruitment process, nepotism, and poor human resource management have also backed a great deal in its decline. Posting and transfers on personal pleas or on political basis have caused a sense of job insecurity among the officers and thereby seriously affected the efficiency of the organization. Additionally, frequent changes in the appointment of senior officials have deteriorated the issue of long-term planning and its implementation.

(2) Establishment of National Logistic Cell (NLC). The main reason behind the decline of Pakistan Railways is the rise of National Logistics Cell (NLC), another competitor in the market. In 1978, National Logistics Cell was established as state organization to provide logistics management services

and logistics protection and to reduce congestion at Karachi seaport. NLC successfully accomplished the task but Pakistan Railways was the biggest sufferer. It lost not only the advantage enjoyed in moving public sector freight but also the logistic business contracts of the Pakistan army. According to statistics released by Pakistan Railways, the number of freight wagons loaded was 680,696 by the years 1975–80, which was decreased to 325,308 by the years 2005–10. In the same period, the number of freight wagons owned by Pakistan Railways crashed from 36,515 to 18,569. Road transport now carries 96% of the inland freight. Having the country's largest fleet of dry and liquid cargo freight service, the share of the NLC alone was three times that of the PR (Pervez Tahir, 2017).

(3) Preference to Road Infrastructure. On the other hand, the government spent three times more on road sector. This combination of neglect to railways and preference to roads was the root cause of downfall railways. Ministry of Railways, which was responsible for providing funds for the maintenance and development of the railway network, did not have any long-term framework for capital support or further development.

(4) Other Factors. After 1973, Pakistan Railways' budget was combined with the national budget with the result that the profit Pakistan Railways earned was diverted to other heads, leaving less space for its maintenance, expansion, and improvement. The other reasons for the decline of were decrease in assets like locomotives, wagons, and track kilometers. Maximum of assets including the physical infrastructure, rolling stock, passenger and freight wagons have lived their productive lives. The poor physical infrastructure not only interrupted the efficiency of services provided by railways but also increased the probability of unexpected events such as rail calamities.

At the initial stage, complete deficit of transport infrastructure in the country gave full opportunities to railway industry to fill the gap and it transported millions of people and goods; however, with the passage of time, road network improved and private transport sector started providing better alternatives for travel in even remote areas of the country. Railways gradually lost their business after shift of passenger and goods transportation to other modes like buses, wagons, and truck services.

4.2. The Second Round of Industrial Life Cycle of Pakistan Railways (1999–2011). Pakistan Railways experienced a second round of industrial life cycle from 1999 to 2011. The level of railway development was on the rise from year 1999 to 2005, but the pace of development was slow and the level of railway development was experiencing a downward trend from 2005 to 2011.

4.3. The Third Round of Life Cycle of the Railway Industry (2011–Present). From the year 2011, the upward trend of combined utility curve shows a new round of industry life cycle of railway industry by increase in number of passengers and freights carried. Pakistan Railways has acquired new locomotives, wagons, and number of passengers and tonnes of freight have increased. To revive Pakistan Railways, the current government has proposed Pakistan Railways Vision 2025. It is expected that railways share in transportation will be increased from 4% to 20% by 2025 [39].

4.3.1. China-Pakistan Economic Corridor (CPEC). In line with the government's Vision 2025 for infrastructure development, Pakistan Railways is to undertake necessary steps to increase its share in the overall transport sector of Pakistan during the next 10 years. A priority project in this regard is China-Pakistan Economic Corridor (CPEC) and upgradation of ML-1 route that connects Karachi, Lahore, and Peshawar (main cities of Pakistan). China and Pakistan have signed memorandum of understanding (MoU) to increase the speed of the railway connecting Karachi and Peshawar as well as to upgrade its signal system and railway stations [40, 41]. This also includes efforts of doubling the train tracks to control traffic. Doubling of track from Khanewal to Raiwind is near to complete. Provision of signaling system on Lodhran-Khanpur-Kotri section and centralized traffic control are also underway.

4.3.2. Modernization of Infrastructure. Pakistan will receive a total of 55 modern locomotives from USA. A consignment of seven locomotives reached Pakistan from the United States to reinforce Pakistan Railways' existing fleet and the remaining 48 are expected to be handed over to Pakistan Railways by the end of 2017. The modern locomotives are designed for a maximum speed of 120 km per hour and are expected to provide better fuel efficiency and longer maintenance intervals [40, 41]. Currently, the fastest speed trains can achieve is 105 km/h only, where tracks are in good condition. The target for current management is to take this up to 160 km/h. Pakistan Railways has also undertaken renovation of several railways stations and also plans to strengthen and rehabilitate 159 weak bridges by June 2017 [42].

4.3.3. Improvement in Services and Performance. Over the period of the past two years, Pakistan Railways has managed to achieve a punctuality rate of 70% in 2015, in comparison to 53% in 2013 [42]. This is a marked improvement when compared to 70% punctuality rate in India and 87% punctuality rate in UK. Seemingly improved services and comforts offered to passengers are one of the reasons contributing towards the increase in number of passengers and freights carried. Pakistan Railways claims that the improved services include improved food quality, water dispensers, and Wi-Fi facilities.

5. Conclusion

This paper examined the rise and fall of Pakistan railways industry using industry life cycle and principal component analysis from year 1950 to 2014. Data were collected from Pakistan Railway Year Book 2014-15 and SPSS software was used for statistical analysis. Principal component analysis was used to calculate the eigenvalues and contribution rates of the common factors and seven indicators, passenger volume, average passenger mileage, passenger turnover, freight volume, average freight mileage, freight turnover, and number of employed persons, were included for the analysis.

The results represented that Pakistan railways experienced the first round of industrial life cycle from 1950 to 1999 and completed four stages of introduction, growth, maturity, and decline. Pakistan railways story is a dramatic history of rise and decline. At the beginning of railway industry, its prominent feature was the lack of railway infrastructure, most needed to be reconstructed and repaired, but with the passage of time rail transport had greatly improved. After 1955, the rail industry had made considerable progress in expanding railway network and improved the quality of service to passengers and freight by new equipment and technology, management innovation, and increase in transport capacity. Railway business was at the peak between mid-sixties and early seventies.

The level of railways started to decline after 1975 and it was at its minimum level in the year 1999. Role of Pakistan Railways as a catalyst for economic development has received a setback due to significant underinvestment by successive governments that preferred investment in road infrastructure at the cost of railways and also developed other organizations in the same sector, that is, NLC.

Pakistan railways experienced a new round of industry life cycle after the year 2011 by increase in number of passengers and freights carried. To revive the fame of Pakistan Railways, the current government has developed a vision paper named "Vision 2025." The vision document aims to improve (including improved quality services) the share in transportation sector from 4% to 20% by 2025, which seems to be an onerous challenge for the organization.

Conflicts of Interest

The authors declare that there are no conflicts of interest regarding the publication of this paper.

Acknowledgments

This research was supported by Fundamental Funds for Humanities and Social Sciences of Beijing Jiaotong University (B16JB00200).

References

[1] H. Stripple and S. Uppenberg, "Life Cycle Assessment of Railways and Rail Transports," 2010.

[2] S. M. Irfan, D. M. H. Kee, and S. Shahbaz, "Service quality and rail transport in Pakistan: a passenger perspective," *World Applied Sciences Journal*, vol. 18, no. 3, pp. 361–369, 2012.

[3] CER and UIC, "Rail Transportation and Environment: Facts & Figures," 2015.

[4] W. L. Hill Charles and R. Gareth, *Industry Life-Cycle Analysis. Strategic Management: An Integrated Approach*, CengageBrain.com, 2012.

[5] E. V. Karniouchina, S. J. Carson, J. C. Short, and D. J. Ketchen Jr., "Extending the firm vs. industry debate: Does industry life cycle stage matter?" *Strategic Management Journal*, vol. 34, no. 8, pp. 1010–1018, 2013.

[6] W. J. Abernathy, *Mastering the Dynamics of Innovation*, Harvard Business School Press, Boston, MA, USA, 1994.

[7] J. M. Utterback and F. F. Suárez, "Innovation, competition, and industry structure," *Research Policy*, vol. 22, no. 1, pp. 1–21, 1993.

[8] S. Klepper, "Industry life cycles," *Industrial and Corporate Change*, vol. 6, no. 1, pp. 145–181, 1997.

[9] R. Adner and D. Levinthal, "Demand heterogeneity and technology evolution: implications for product and process innovation," *Management Science*, vol. 47, no. 5, pp. 611–628, 2001.

[10] P. Michael, *Competitive Strategy: Techniques for Analyzing Industries and Competitors*, The Free Press, 1980.

[11] W. J. Abernathy and J. M. Utterback, "How Industries Change," *MIT Technology Review*, vol. 80, no. 7, pp. 40–47, 1978.

[12] A. M. McGahan, "How industries change," *Harvard Business Review*, vol. 82, no. 10, pp. 86–94, 2004.

[13] J. M. Campa and K. Simi, "Explaining the diversification discount," *Journal of Finance*, vol. 57, no. 4, pp. 1731–1762, 2002.

[14] V. Maksimovic and G. Phillips, "Do conglomerate firms allocate resources inefficiently across industries?" *Journal of Finance*, vol. 57, no. 2, pp. 1731–1762, 2003.

[15] B. Villalonga, "Diversification discount or premium? new evidence from the business information tracking series," *Journal of Finance*, vol. 59, no. 2, pp. 479–506, 2004.

[16] Z. Tomes, *Long-Term Structural Decline of Railways*, Masaryk University, 2006.

[17] A. Sabol, M. Šander, and Đ. Fučkan, "The concept of industry life cycle and development of business strategies," in *Proceedings of the Management, Knowledge and Learning International Conference*, pp. 635-42, 2013.

[18] M. Cusumano, S. Kahl, and F. Suarez, "Product, process, and service: a new industry lifecycle model," A research and education initiative at the MIT Sloan School of Management, Working Paper, 2006, pp. 1–39.

[19] N. Argyres and L. Bigelow, "Does transaction misalignment matter for firm survival at all stages of the industry life cycle?" *Management Science*, vol. 53, no. 8, pp. 1332–1344, 2007.

[20] A. Macdonald and C. M. Hartt, "A strategic industry life cycle analysis for nova scotia's new world wines," *American International Journal of Social Science*, vol. 3, no. 7, pp. 117–132, 2014.

[21] G. L. Urban, T. Carter, S. Gaskin, and Z. Mucha, "Market share rewards to pioneering brands: an empirical analysis and strategic implications," *Management Science*, vol. 32, no. 6, pp. 645–659, 1986.

[22] R. A. Kerin, P. R. Varadarajan, and R. A. Peterson, "First-mover advantage: a synthesis, conceptual framework, and research propositions," *Journal of Marketing*, vol. 56, no. 4, pp. 33–52, 1992.

[23] Đ. Kalicanin, "Extending the firm vs. industry debate: does industry life cycle stage matter?" *Strategic Management Journal*, vol. 15, no. 177, pp. 89–102, 2008.

[24] M. B. Lieberman and D. B. Montgomery, "First-mover (dis)advantages: Retrospective and link with the resource-based view," *Strategic Management Journal*, vol. 19, no. 12, pp. 1111–1125, 1998.

[25] P. N. Bloom and K. Philip, "Strategies for High Market-Share Companies," *Harvard Business Review*, vol. 53, no. 6, pp. 63–72, 1975.

[26] X. Li and W. Yan, "Analysis on the development trend of China railway based on industry life cycle theory," *China Railway Science*, vol. 32, no. 1, pp. 127–132, 2011.

[27] G. T. Lumpkin and G. G. Dess, "Linking two dimensions of entrepreneurial orientation to firm performance: The moderating role of environment and industry life cycle," *Journal of Business Venturing*, vol. 16, no. 5, pp. 429–451, 2001.

[28] D. B. Audretsch and M. P. Feldman, "Innovative clusters and the industry life cycle," *Review of Industrial Organization*, vol. 11, no. 2, pp. 253–273, 1996.

[29] Railway Year Book, "Ministry of Railways Pakistan," 2014, http://iqbalkalmati.blogspot.hk/2016/03/pakistan-railways-year-book-2014-15-in.html.

[30] Springer, *Principal Component Analysis*, Springer Series in Statistics, Springer, New York, NY, USA, 2nd edition, 2002.

[31] S. Raschka, "Principal Component Analysis in 3 Simple Steps," 2016.

[32] R. L. Gorsuch, *Factor Analysis*, LEA, Hillsdale, NJ, USA, 2nd edition, 1983.

[33] I. Smith Lindsay, "A tutorial on principal components analysis introduction," *Cornell University, USA*, vol. 51, no. 52, p. 65, 2002.

[34] J. Khan, J. S. Wei, M. Ringnér et al., "Classification and diagnostic prediction of cancers using gene expression profiling and artificial neural networks," *Nature Medicine*, vol. 7, no. 6, pp. 673–679, 2001.

[35] J. Landgrebe, W. Wurst, and G. Welzl, "Permutation-validated principal components analysis of microarray data," *Genome Biology*, vol. 3, no. 4, pp. 1–19, 2002.

[36] Q. Wang and X. Wang, "Assessment of agricultural modernization level in Zhejiang using SPSS software," in *Proceedings of the The 5th International Conference on Computer Science Education (ICCSE)*, pp. 1069–1073, Hefei, China, August 2010.

[37] R. Beaumont, "An Introduction to Principal Component Analysis Factor Analysis Using SPSS 19 and R (Psych Package)," *Journal of Geophysical Research: Atmospheres*, 2012.

[38] K. Khaleeq, "Poor Infrastructure a Major Hurdle to Growth," 2013, Dawn News, https://www.dawn.com/news/1044734.

[39] Pakistan 2025 One Nation—One Vision, "Ministry of Planning, Development & Reform, Government of Pakistan," 2015.

[40] The Express Tribune, "China, Pakistan to Upgrade 1,600kms of Railway Lines," 2017.

[41] The Express Tribune, "Consignment of Modern US Locomotives Reaches Karachi," The Express Tribune, 2017.

[42] Pakistan Institue of Legislative Development And Transparency, *Pakistan Railways: A Performance Analysis*, 2015.

15

PULSim: User-Based Adaptable Simulation Tool for Railway Planning and Operations

Yong Cui ,[1] Ullrich Martin ,[1] and Jiajian Liang [2]

[1] *Institut fuer Eisenbahn- und Verkehrswesen der Universitaet Stuttgart, Pfaffenwaldring 7, 70569 Stuttgart, Germany*
[2] *China Academy of Railway Sciences, Daliushu Road 2, Haidian District, Beijing 100081, China*

Correspondence should be addressed to Yong Cui; yong.cui@ievvwi.uni-stuttgart.de

Academic Editor: Lingyun Meng

Simulation methods are widely used in the field of railway planning and operations. Currently, several commercial software tools are available that not only provide functionality for railway simulation but also enable further evaluation and optimisation of the network for scheduling, dispatching, and capacity research. However, the various tools are all lacking with respect to the standards they utilise as well as their published interfaces. For an end-user, the basic mechanism and the assumptions built into a simulation tool are unknown, which means that the true potential of these software tools is limited. One of the most critical issues is the lack of the ability of users to define a sophisticated workflow, integrated in several rounds of simulation with adjustable parameters and settings. This paper develops and describes a user-based, customisable platform. As the preconditions of the platform, the design aspects for modelling the components of a railway system and building the workflow of railway simulation are elaborated in detail. Based on the model and the workflow, an integrated simulation platform with open interfaces is developed. Users and researchers gain the ability to rapidly develop their own algorithms, supported by the tailored simulation process in a flexible manner. The productivity of using simulation tools for further evaluation and optimisation will be significantly improved through the user-adaptable open interfaces.

1. Introduction

In order to improve the efficiency and effectiveness of railway planning and operations, the relationships between the critical components of railway systems and the behaviour of the studied system must be investigated. A specific investigated problem usually presents many possible alternatives for design. In order to enable evaluation and comparison among the various possible alternatives, their respective effects and outputs will be predicted through experiments. However, it is not practical to examine a design alternative directly based upon an actual system, because either such a system is not existing or experimentation upon it is prohibitively expensive.

For a simple railway network with a limited number of train runs, an analytical approach is practical in order to evaluate the output of a design alternative and even to find an optimal solution in a closed-form expression. For a large-scale network with a high density of train movements, the computational complexity of such an analytical model becomes considerably high. In this case, the simulation approach is often applied.

Several simulation tools are available for railway planning and operations purposes. The application of personal computer based simulation approaches for railway systems began in the 1990s, at which time most tools were originally developed as laboratory versions. Along with continuous development and applications, these tools have been widely accepted for research and commercial purposes. Meanwhile, a large number of data and models of railway systems have become available. In this paper, the existing simulation tools and applications will be presented in Section 2.

However, the potential of simulation approaches has not yet been sufficiently utilised. The limits of the current simulation tools, especially their lack of capability for extension and customisation, are discussed in Section 2. To overcome these limits, a user-based, adaptable simulation tool called PULSim has been developed in recent years. In Section 3, the model, the workflow, and the main features of PULSim are introduced. The software's capability to provide users and third-party applications with an open interface for dynamic interaction and flexible extension is presented in Section 4, along with several use cases and scenarios. Finally, the perspective of further development is discussed in Section 6.

2. Applications of Railway Simulation and the Limitations of Existing Simulation Tools

For railway planning and operations, simulation approaches have been widely used by researchers, railway infrastructure companies, and railway operating companies. In German-speaking countries and Europe in general, the simulation software RailSys (developed by Rail Management Consultants GmbH, [1]), OpenTrack (from OpenTrack Railway Technology Ltd., [2]), and LUKS (by VIA Consulting & Development GmbH, [3]) are very practical for railway planning and operations.

The potential areas of application of simulation tools include technical feasibility studies, determination of conflicts, evaluation of the quality of a timetable, capacity research, and dispatching. Simulation tools can be used, for example, to build a test environment for the evaluation of the technical feasibility and the benefits of using the European Train Control System (ETCS) [4]. To investigate potential conflicts, a timetable simulation can be carried out, and the resulting waiting time can be determined as well [5, 6]. In certain situations, various optimisation functions can be added to a timetable simulation in order to reduce hindrances and waiting time of train runs. As a deterministic process in principle, the possible influence and variances during railway operations are not considered in a timetable simulation. If the robustness and stability of a timetable are matters of concern, a so-called operational simulation can be carried out, in which stochastic influences will be introduced. The robustness and stability of the investigated timetable can be derived according to statistical indicators, which are calculated based on the outputs of several rounds of simulation with different randomly generated influences [7, 8]. In the area of capacity research, system performance can be evaluated through simulating randomly generated timetables for a certain operating program, a process which can, in addition, identify bottlenecks and determine the quality of an operating program [9]. The approach of capacity research is implemented in the software tool PULEIV, developed by the IEV (Institut für Eisenbahn- und Verkehrswesen der Universitaet Stuttgart) for flexible analysis, evaluations and reports [10]. Simulation approaches can not only be used to identify conflicts, but also to generate a feasible dispatching timetable to resolve conflicts [11, 12], as well as to study the relationship between systems performance and dispatching measurements [13, 14].

At the beginning of development of a certain simulation tool, its further possible applications are often unknown by its developers. Therefore, it is difficult to foresee and to provide specifically desired functions and outputs, even if the simulation tool would theoretically able to fulfil said requirements. Hence, an open interface that enables customisation for user-based, adaptable simulation processes is a vital factor for the success of a simulation tool. Unfortunately, currently available simulation tools for railway planning and operations lack, in large part, transparent standards and open interfaces. Some internal workflows, such as the applied dispatching algorithm and the implementation of signalling systems, are not sufficiently documented for users and third party individuals and, as such, it is difficult for the end-user to understand the internal assumptions and simplifications made within these tools. The outputs of various simulation tools are insufficient and not standardised with insufficient documentation, which prevents users from carrying out further evaluation and optimisation. In addition, it is almost impossible to organise user-defined workflows and to integrate new functions into the existing simulation tools, which leads to constraints in their usability and applicability. The high efforts required for further evaluation decrease the efficiency and effectiveness of these simulation approaches. In addition, some well-known issues, such as deadlock problems in synchronous simulations, have not been addressed sufficiently. Once deadlocks take place, the simulation process must be cancelled or be solved manually, which greatly limits the usability of the simulation tool as well.

A user-based, adaptable simulation tool, PULSim, has been developed in recent years. In [15, 16], the tool was known as DoSim. The name has been changed to PULSim in order to conform to the naming convention of software tools in the IEV. PULSim provides a platform for railway simulation with a unified model, a transparent workflow, and open interfaces. Critically, PULSim provides the possibility for third-party applications to be flexibly integrated into the software. The functionality of PULSim can be extended thanks to its open interface, and the software has been tested with several railway networks in Germany. At the moment, the software is available for download from GitHub (https://github.com/herrcui/RailView/wiki), with the instructions for installing and using it provided with demo data. The user manual and the open interfaces are planned to be comprehensively documented and published in further development. The Continuous Integration (CI) for open access and updates will be implemented as well. In this paper, the model, the workflow, and the Graphical User Interface (GUI) are presented. Three special features of PULSim are outlined in particular:

(i) Dispatching mechanism

(ii) Deadlock avoidance

(iii) Open interfaces for user-based adaptable simulation.

An introduction of the model, the workflow, and the GUI of PULSim is provided in Section 3.1. In Sections 3.2 and 3.3, the design of the dispatching mechanism and deadlock avoidance are explained. The open interface for user-based adaptable simulation and its applications are introduced in Sections 4 and 5.

3. Introduction of Simulation Tool PULSim

3.1. The Model, the Workflow, and the GUI of PULSim. In PULSim, the infrastructure, rolling stocks, and railway

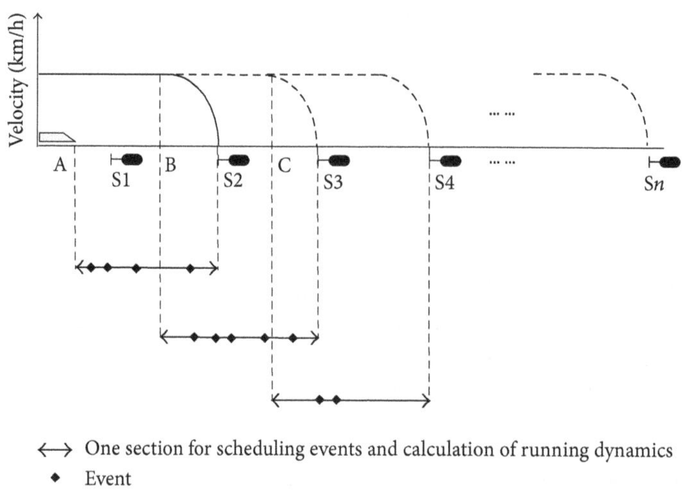

FIGURE 1: An example of event scheduling and calculation of running dynamics.

operations are modelled as components of railway systems. Attributes are used to describe the static and the dynamic information. The static information refers to the invariant values of the railway systems, for example, the length of a track, the configuration of a train, and the scheduled departure time of a train run. The dynamic information will vary along the duration of the simulation. For example, the aspect of a signal, the running dynamics of a train, and the occupancy situation of a track are updated continuously during the simulation. The model applied in PULSim provides a basis for multiscale simulation on the microscopic, mesoscopic, and macroscopic levels [17].

An internal class model of railway systems comes built-in with PULSim. This model is the further development and extension from the IEV core model [18]. The publication of the detailed design of this model, and its workflow, is planned for 2018. It will also consist of the details of the open interfaces for interacting with the model along with the software. Particularly, the basic structures and the path components based on the infrastructure element model are developed in the core model, so that the macroscopic, mesoscopic, and the microscopic models are fully integrated (the infrastructure element model is a microscopic model to model the basic infrastructure elements including tracks, turnouts, crossings, and single and double slips. A basic structure is defined as a basic occupancy element in which all parts should be equally occupied, regardless of the design of the operating program. A path component is a directed edge inside a block section used for train runs, which can be released separately as a directed occupancy element. Basic structures and path components are modelled for the mesoscopic level). This model can be used for various purposes, including capacity research, bottleneck analysis, and train dispatching [19]. It is independent of a concrete data format, which allows users to concentrate on the workflow and the business logic of railway planning and operations without requiring knowledge relating to a certain data format. Additional parsers are required to convert different data formats into the class model. At the moment, the data format used by RailSys [1] is supported by PULSim. The development of the parser to support railML [20] is in progress.

The workflow for an event-driven simulation is applied in PULSim. An event is defined as an occurrence that may change the attributes of the system at a certain point in time. During the simulation process, a series of time points for the occurrence of an event are identified and continuously updated. The system evolves over the series of time points. During a simulation process, the attributes (dynamic information) of the system (e.g., the position of a train) may be changed continuously. In PULSim, only that change of attribute, which causes interactions between two or more components or triggers another event, is considered a discrete event. Otherwise, the changes of the running dynamics, for example, the continuous variation of train velocity, will be handled at each discrete time point by the train itself, without requiring an explicit event. Hence, the event-driven process in PULSim can be organised efficiently with a limited number of events.

During the simulation process, a running time calculation is carried out for each individual train run. The change of position and velocity of each train will not be treated as an event. Instead, the events will be scheduled with respect to the signalling system. An example of organising an event-driven simulation is shown in Figure 1. After a train has received a Movement Authority (MA) at point A, the complete speed profile will be calculated until the End of Authority (EOA) at main signal S2. The corresponding events from point A to EOA S2 are scheduled. At point B, a new MA ending at main signal S3 is granted. The train movement from B to S3 will be calculated. Meanwhile, the events already scheduled between B and S2 will be discarded, since the running dynamics from B to S2 will be updated with the newly calculated speed profile. The events from B to S3 will be rescheduled and updated again.

Once a request of infrastructure resources is sent, the current operational situation will be observed by the dispatching module. The requested infrastructure resources can only be granted at a conflict-free and deadlock-free situation.

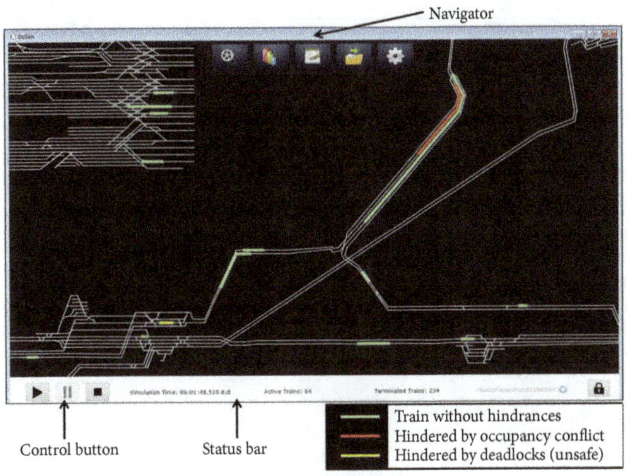

FIGURE 2: Screenshot of "Network/Train Run" view in PULSim.

A simple First Come First Served (FCFS) algorithm is provided by PULSim as a default module to solve occupancy conflicts. It can be replaced with a customised dispatching module provided by the user (see Section 3.2). A standard solution for deadlock avoidance, combined with the approach of searching for feasible resources to reduce false-positive situations, is provided in PULSim as well (see Section 3.3).

The design of the GUI in PULSim aims to provide the user with an insightful view during the process of railway simulation and to support efficient user interaction. Therefore, the concepts of visualisation and interaction are considered critical areas of focus. Various user interfaces are organised in different views. The movement of train runs in the railway network (the view of "Network/Train Run"), the information behind the scenes ("Analyse and Evaluation"), and the interaction between the simulation tool and users ("Open Interface") are provided.

In Figure 2, the screenshot of the "Network/Train Run" view is presented. This is a standard function provided by almost all railway simulation tools. A control panel can start, pause, or stop a simulation. Trains are marked in green, red, or yellow, which represent a train without hindrances, with hindrances due to occupancy conflicts, or with hindrances due to deadlocks, respectively. In order to enable a quick switch among different views, a floating navigator is set on top of the screen.

The function of tracing the occurrence of events is provided by PULSim. Within this view, users can observe the entire workflow of the simulation within the diagram of the blocking time stairway and running dynamics for each train run (Figure 3). Other evaluation views for the illustration of occupancy and hindrance of the running simulation are also provided. In addition, an "Open Interface" view to achieve a user-based, adaptable simulation is built into PULSim (see Section 4).

3.2. Dispatching Mechanism of PULSIM. With conventional dispatching approaches [21], the dispatching algorithm is activated once one or more potential conflicts are identified. However, it might be too late to start train dispatching at the time of the occurrence of conflicts. Furthermore, it is necessary to check for potential conflicts within the conventional dispatching systems periodically. The timing for identification of conflicts is critical. On one hand, some conflicts may be overlooked using a long time interval between two rounds of checks; on the other hand, overly frequent checks might lead to performance issues.

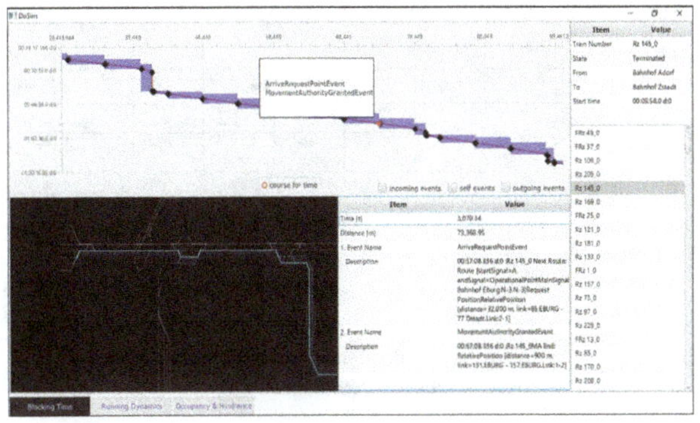

FIGURE 3: Screenshot of tracing events and train run.

In PULSim, a new conflict identification/resolution mechanism is implemented. In contrast with other conventional dispatching approaches, the activation of a designed algorithm for train dispatching is not initiated at the time of facing conflicts, but at an earlier time, at which a train has the chance to occupy the requested infrastructure resources. For example, if the action to occupy the requested resources would produce significant hindrances on other trains, the train will give up the chance of occupancy. Therefore, potential conflicts from potential hindrances can be prevented in advance. In addition, this mechanism suits the workflow of event-driven simulation. A periodical examination of conflicts in a fixed time interval is not necessary; any potential conflicts can be identified and resolved at the time a Movement Authority is able to be granted.

In order to better illustrate this phenomenon, Figure 4 provides an example of giving up the opportunity of occupancy [15]. Within it, Train T1 is requesting occupancy of the route from signals S1 to S3. This route is not currently occupied by other trains (conflict-free), and its occupancy will not cause deadlocks (deadlock-free). Deadlocks may, indeed, be regarded as a type of conflict in train dispatching. In order to differentiate between occupancy conflicts and deadlocks, the term conflict-free and deadlock-free are used in this paper. There is a train T0 occupying the route from S3 to S5. Its occupancy time along this route is assumed to be very high due to a technical failure of train T0. If train T1 occupies the route from S1 to S3, trains T2 and T3, which will take the route from S2 to S4, must wait until the technical failure of T0 has been resolved. It would be more efficient if train T1 were to give up the chance of occupancy and wait before signal S2. Therefore, potential hindrances and conflicts can be reduced or avoided in advance.

The workflow of the dispatching mechanism is shown in Figure 5. As usual, a request will be pended in case of conflicts or deadlocks. A special feature for train dispatching

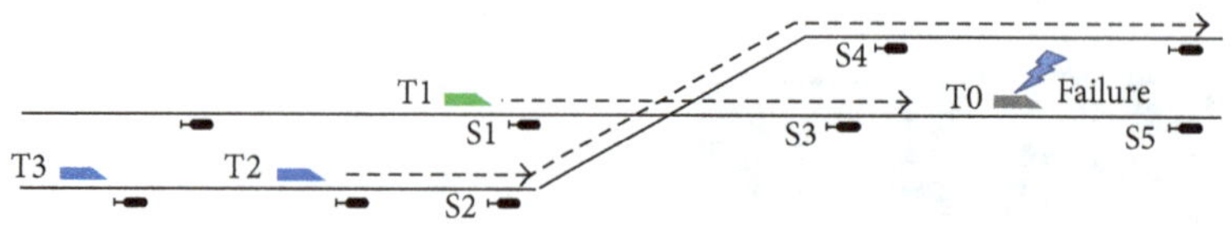

FIGURE 4: Example of giving up of occupancy [15].

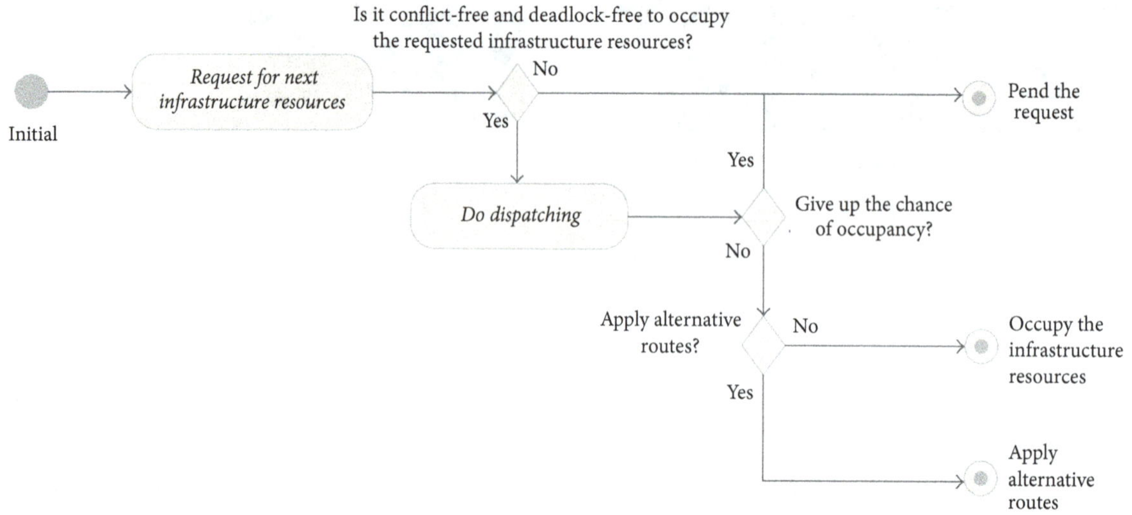

FIGURE 5: The dispatching mechanism in PULSim.

is designed when a request for infrastructure resources is conflict-free and deadlock-free. In this case, the train still has to evaluate the operational situation before it occupies the requested infrastructure resources. The decision of whether or not to give up the chance of occupancy depends on the applied dispatching algorithm, which is illustrated as the activity "do dispatching" (see Figure 5). On the basis of a FCFS strategy, the train will always take the chance to occupy the requested infrastructure resources. The FCFS strategy is implemented in PULSim by default, and if a dispatching strategy other than FCFS is applied, the train may give up the chance of occupancy according to certain dispatching objectives (e.g., to reduce total hindrances). The request will be then pended for the next round of dispatching. In addition, the dispatching mechanism also supports the implementation of applying alternative routes.

Although FCFS is currently the only strategy implemented in PULSim, users can customise and integrate their own dispatching algorithms into the dispatching process. This is achieved through the open interface and a standard workflow (see Section 4). Inside the workflow, users can define their own implementation to calculate the predicted effects for all possible dispatching actions (to occupy, to give up, or to choose an alternative route). The action with the best effects (e.g., with the minimal waiting time) will be chosen as the dispatching decision.

An example of using user-customised dispatching algorithms is published on GitHub (https://github.com/herrcui/RailView/blob/master/PULSimReleases/scripts/railapp.dispatching.services.ExternalDispatchingService.py). Within this example, the delays experienced by conflicting trains are retrieved from the simulation model by comparing the current departure/arrival times with the scheduled departure/arrival times. The example utilises a simple form of logic in order to illustrate the usage of the open interface. That is, the train experiencing the longest delay will receive the highest priority.

A Java interface is defined for users to provide customised train dispatching. Through the interface, a list of resulting states according to the generated possible actions will be given from PULSim as input. A state is a multidimensional list, which represents the resulting occupancy situation of an action for each train. The predicted occupancy situation includes the lists of the occupied infrastructure resources, as well as the start and end of the blocking time for each infrastructure resource. It should be noted that a state is not conflict-free. Users can implement their own algorithm to identify and to resolve conflicts. As a basis, all trains predict their further occupancy situation as if other trains do not exist. The predicted occupancy situation will then be shifted or changed according to a certain action. For the action to occupy the requested infrastructure resources by a train T, the blocking time of the infrastructure resources for other trains should be shifted until the infrastructure resources are released the train T. Similarly, if a train T gives up the chance of occupancy to wait for another train T′, the blocking time of other related trains should also be shifted, until the infrastructure resources are released by T′. Hence, the current operational situation and the action to be taken are represented in the form of the resulting states.

Users can implement the method "determineAction" to determine a dispatching action. The method returns an

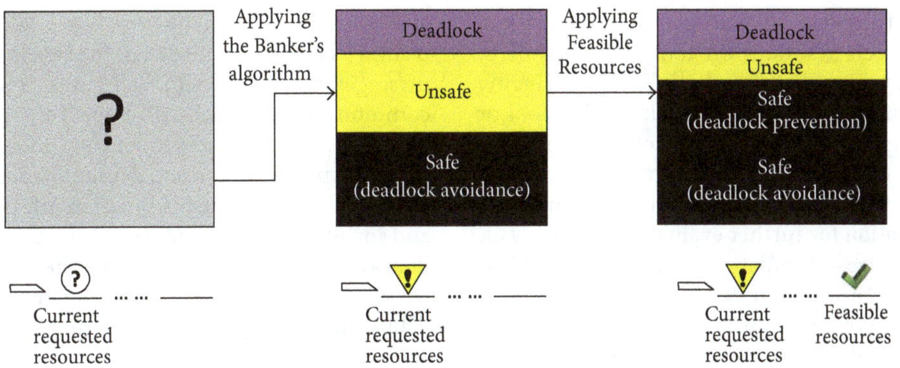

FIGURE 6: Searching feasible resources to reduce false-positive situations with the Banker's algorithm [15].

index of the given actions, which indicates the chosen action according to a specific dispatching objective (e.g., with minimal waiting time). At the moment, an algorithm based on reinforcement learning for train dispatching is being developed. A preliminary investigation to evaluate the effects for a certain state and a given action is described in Section 5.

3.3. Deadlock Avoidance Based on Banker's Algorithm. Deadlock problems [22] are well-known in the area of synchronous simulation. Some approaches of deadlock avoidance are proposed in [23–26]. PULSim applies a method based on the Banker's algorithm [27], which was initially used for the purposes of railway simulation in [11]. Once a train requests to occupy one or several infrastructure resources, a deadlock-free test will be carried out based on the Banker's algorithm. The workflow of this deadlock-free test is straightforward; if all the required infrastructure resources along the path of a train are available (in other words, not occupied by other trains), the train will be identified as a passed train. All the resources occupied by the passed train will be then returned to the system. The deadlock-free test will be executed iteratively for every train, until all trains have been identified as passed trains. In this case, a deadlock-free situation can be always guaranteed, since the feasibility of resource allocation has been proven.

If there are some trains remaining, whose required infrastructure resources are not available, the situation will be regarded as being in an unsafe state. It should be noted that an unsafe state does not necessarily result in a deadlock situation. A false-positive situation is defined in [11] as a deadlock-free test result that is read as positive but is actually negative. False-positive situations should be reduced to avoid unnecessary waiting time due to the overly strict rules in deadlock-free tests.

A method to search for feasible resources for reducing false-positive situations with the Banker's algorithm is implemented in PULSim (Figure 6). Feasible resources are defined as one or more connected and not yet granted infrastructure resources in the further path of a train. If the train occupies the feasible resources, a circular wait situation should not take place. The circular wait situation is one of the necessary conditions for deadlocks [25]. Once the circular wait situation is eliminated, the feasible resources can be granted to the train without deadlocks, even if the train has not passed the deadlock-free test through the Banker's algorithm. In [15], the approaches to identify feasible tracks as well as the effects of reducing false-positive situations are presented in detail.

In this method, a request for infrastructure resources will be at first analysed in accordance with the Banker's algorithm. Upon passing this test, the requested infrastructure resources can be safely granted to the requesting train. In case a request fails to pass the test with the Banker's algorithm, the method of searching for feasible resources will commence. If feasible resources are indeed available, the train's request may still be approved.

The aforementioned combination of the Banker's algorithm and the method of searching for feasible resources can ensure a high applicability and efficiency for deadlock avoidance. This method has been proven in PULSIM for several practical applications and case studies. For example, a real railway network in Germany is used for the case studies, consisting of 129 stations, including a large terminus station, and 2,388 train runs. Even with randomly shuffled departure times of train runs, deadlocks can be successfully avoided. Another extreme example is also available on GitHub. For a highly congested network with a high potential of deadlocks, the software PULSim is still able to simulate all the train runs without deadlocks.

4. User-Based Adaptable Simulation

4.1. Open Interface of Railway Simulation Tools. A simulation tool serves as the basis for further investigations into railway planning and operations, in which the output of a simulation can be used as the input of other applications. Evaluation of the simulation output is the most popular use case of these applications. Therefore, the interaction between a simulation tool and other applications for evaluation of simulation output should be enabled.

The interactions for evaluation can be categorised as either offline or online evaluations. In the case of offline evaluation, the data generated by the simulation tool will be initially stored and will be analysed afterwards. This process is suitable for a middle-term or long-term evaluation with a very large amount of data. The saved log files can be reused in the future for other purposes of evaluation without requiring an additional run of simulation. With an online evaluation, the evaluation between a simulation tool and third-party applications can be achieved efficiently through exchanging real-time data. A built-in package in the simulation tool or an open interface supporting direct data access is required for online evaluation. Sometimes, the output of the third-party applications will be fed back to the simulation tool. In order to exchange information for further evaluation and analysis, open interfaces between a simulation tool and other third-party applications are provided in PULSim.

When conducting offline evaluation, it is very common to define the format and the schematic structure of the saved log files as an open interface. Popular formats of the log files include Extensible Markup Language (XML) files, Comma Separated Values (CSV), files or plain text files. As an open interface, the format and the schematic definition of the output files should be provided and published in advance. At the time that a simulation tool is developed, its possible usage cases may still be unknown. Therefore, it is a challenge to ensure that the tool provides open interfaces for further possible applications at an early stage. Currently, an open and standardised interface for the output of railway simulation is not available, with each simulation tool possessing its own definition for the output of simulation. Attempts to unify various outputs generated from different simulation tools require considerable effort. Certain simulation tools output log files without a published interface, which can be used in internal testing for various special purposes. It is not recommended to use these data due to the lack of documentation and support, although the information may be inferred from the contents of the files.

For online evaluation, the output of simulation can be retrieved and analysed directly through an open interface. The process of railway planning and operations can be evaluated and optimised by tuning the parameters of the simulation tools in real time. The open interface can be provided either from an open Application Programming Interface (API) or in the form of a scripting language.

An open API enables a dynamic data exchange via directly accessing the simulation tool from other third-party applications. In the simulation tool OpenTrack, the open API is provided in the form of web services [28], from which a third-party application can send messages as commands to OpenTrack and retrieve the output as status messages back to the application. The commands and the status messages are transferred in a machine readable format, according to the Simple Object Access Protocol (SOAP) over Hypertext Transfer Protocol (HTTP). An example use case of using the OpenTrack API for train dispatching is presented in [28]. New dispatching algorithms and prototypes are tested in a simulation environment, in which the realistic situation is replaced by the simulation tool OpenTrack. The commands and status messages in the same format used in reality are exchanged between the dispatching software and OpenTrack. Hence, the newly developed dispatching algorithms can be evaluated in an inexpensive experimental environment.

The system performance and efforts required for the development and maintenance of an open API solution should be considered. The amount of data in railway planning and operations is usually very large. Taking the example of using a web service with SOAP, the enveloped and transferred messages via XML data will cause a high overhead for communication, which will impact the performance of the system. Large software vendors and tools are often required for the implementation and development of web service solutions, which makes such a situation infeasible for researchers and institutes who are interested in rapid development with lightweight solutions. In addition, the complexity of learning the web service-based system and the induced efforts for debugging, testing, and maintaining the entire system are considerable.

Another option for an open interface is to use scripting language, which can combine the advantages of both offline evaluation and online evaluation. A scripting language is a programming language in the form of a series of imperative commands, which are executed in a certain run-time environment. These commands are interpreted during the run-time without needing to be compiled. Hence, sophisticated processes can be customised by users in an adaptable way. With scripting languages, the output of simulation can be either exported as log files for offline evaluation, or be accessed online via the provided interfaces (commands). Upon the demands and the required system performance, a user can flexibly decide the mode of interaction. A scripting language hides the internal structure and the implementation details of a complex system, which enables the user to concentrate on the core functions and the integration of the functions. The complexity of learning the system and the efforts for development, deployment, and maintenance are therefore significantly reduced.

Using open interfaces with scripting language can provide additional advantages in balancing the requirements of transparency and the complexity in development. It is important for users to understand the mechanism and the assumptions of a simulation tool. The open-source method would be an option to promote transparency and open collaboration. However, an excessive amount of implementation details of the complete source code will increase the complexity in comprehension. It should be noted that most users of railway simulation tools are interested principally in the domain logics related to railway planning and operations. Therefore, open interfaces will provide a moderate level of details for researchers to enable rapid development and integration of their own algorithm. As a precondition, the mechanism of the simulation tool should be published, so that users can customise their own logic based on the utilised mechanism. Hence, the internal assumptions and mechanism can be naturally understood by users according to the well-specified open interfaces. For example, the internal mechanism of dispatching systems should be specified in advance (see Section 3.2). As long as a customised dispatching algorithm is implemented, the mechanism of train dispatching is understood by users as well. In the future, the mechanism of running time calculations and signalling systems will be published in the form of open interfaces to users as well.

Other simulation tools also provide some open access to their internals. For example, the parameters used in RailSys and OpenTrack can be viewed and set through system

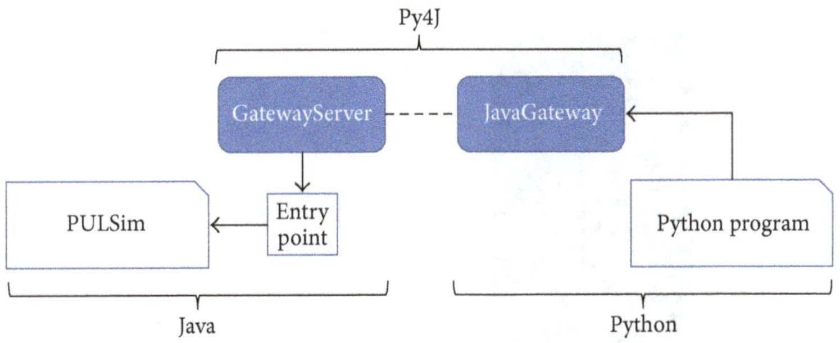

FIGURE 7: Py2J: accessing Java objects and services from Python.

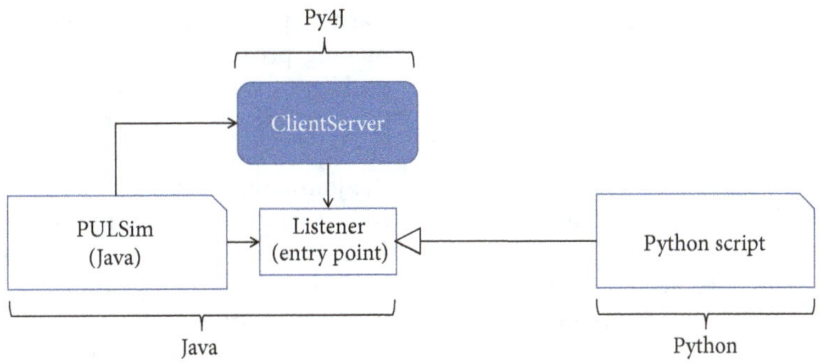

FIGURE 8: J2Py: accessing Python script from Java.

configuration. Users can also interact with OpenTrack through web services. PULSim provides additional support for customised simulation workflow and integration of their own algorithm. In Sections 4.2 and 5, the implementation of an open interface with scripting language and the applications of using the open interface provided by PULSim are presented.

4.2. The PULSim Implementation. In PULSim, a user-based adaptable interface for the simulation, evaluation, and optimisation for different applications is built using scripting language. Scripting languages can be categorised as either domain-specific languages or general-purpose languages. A domain-specific language is specifically designed for a certain application and platform, and an external parser is needed to interpret its commands. Additional efforts are required for users to learn the language. With general-purpose languages, and plenty of support, libraries and documentation are available, which usually allows users to obtain sufficient knowledge of general-purpose languages with lower learning efforts. In addition, most general-purpose languages are easily extended for certain special purposes.

Popular general-purpose scripting languages include Python, Perl, and Ruby. Among others, Python has been widely used for rapid and efficient development with plenty of extensions. Particularly, most implementations of machine learning in recent years have been written in Python. Therefore, Python is utilised as the scripting language in PULSim, and the open-source framework Py4J [29] is integrated within the software. Py4J enables Java programs to access Python code and also provides interfaces for Python programs to access Java objects. It works in a client-server mode for both directions.

In order to access Java objects from a Python program, the class GatewayServer should be at first initiated and started in Java code, taking an entry class as an entry point. The Python program will call the methods provided by the entry point through JavaGateway. Thereby, the Python program can access Java objects and services provided by PULSim. The mechanism of accessing Java objects and services from Python (Py2J) is shown in Figure 7.

To enable the Java code running in PULSim to access a Python script, a ClientServer class is initialised to start the Python script, which implements a listener interface defined in PULSim. Hence, the listener will be used as an entry point for Java code to access the functions provided by Python. The mechanism of accessing Python script from Java (J2Py) is shown in Figure 8.

The examples and the applications of using this bidirectional communication for user-based, adaptable simulation are described in Section 5.

5. Applications of User-Based Adaptable Simulation

A very common requirement for user-based, adaptable simulation is the ability to organise and customise several rounds of simulation as well as the desired output of simulation. For this purpose, PULSim provides an open interface for timetable simulation for Python. A simulator for timetable simulation can be obtained through the interface, in which users can designate the investigated data of infrastructure,

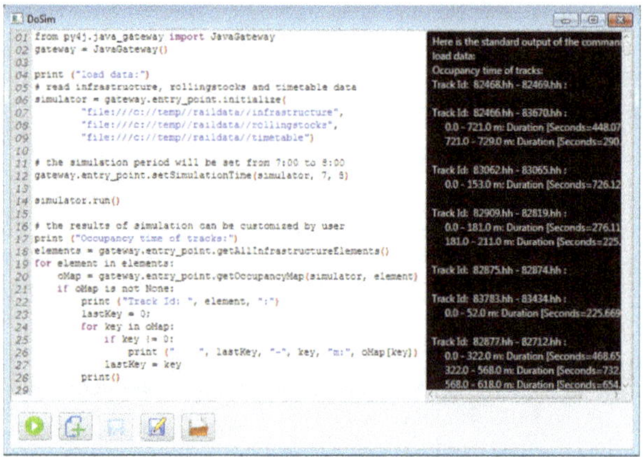

FIGURE 9: User interface for accessing Java services from Python in PULSim (Py2J).

rolling stocks, and the timetable in the Python environment. Other configurations, for example, the time period of the simulation, can also be optionally specified for the simulator. With a few lines of code, a timetable simulation can be carried out in the Py2J mode (Figure 7).

The results of the simulation can be further analysed. For example, the occupancy and hindrance values during the simulation are matters of interest for purposes of capacity research. In Figure 9, the user interface for a timetable simulation with the outputs of occupancy and hindrance is shown. The editor for the Python program, the Py4J libraries, and the output window of the running Python program are integrated into PULSim as well.

In addition to being saved in log files for an offline evaluation, the output can also be directly retrieved by the Python program for further calculation and online evaluation. This is especially useful in order to customise a complicated workflow with many rounds of simulation. For example, to derive the recommended area of traffic flow within the scope of capacity research (source [10]) automatically, it is necessary to seamlessly integrate the software for capacity research with the simulation tools. Otherwise users would have to shift between different software tools manually to simulate several densified timetable variants. The advantage of the user-based, adaptable simulation can also be gained in the process of calibration of the parameters for operational simulation. In [16], an automatic process for calibration is implemented by the authors using PULSim. It is possible in this case, since the developers for the calibration system are also the developers of the simulation tool. However, if developers from a third party want to implement the calibration process, it is difficult if they do not have access and understanding of the source code of the simulation tools. A Py2J interface can provide the developer with the opportunity to adapt the disturbance parameters through many rounds of operational simulation iteratively, without the additional effort of interacting with the simulation tool.

Another advantage of user-based, adaptable simulation is the enhancement of the capabilities of the simulation tool through third-party software and algorithms. Today, machine

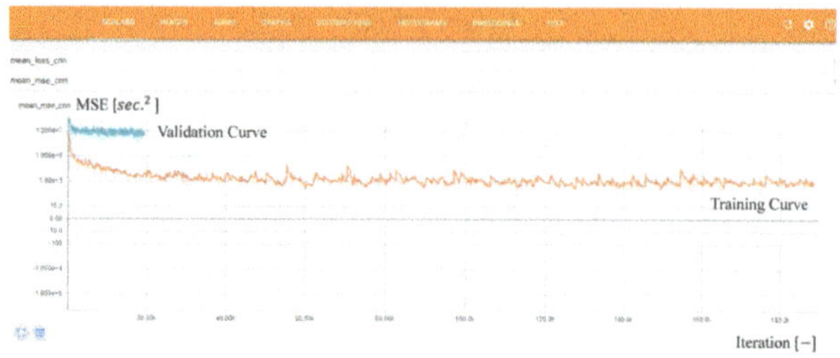

FIGURE 10: Prediction of total waiting time with TensorFlow (J2Py).

learning plays an increasingly important role in learning and making predictions about data. Nowadays, many software packages for machine learning are implemented in Python, for example, Scikit-Learn [30], Theano [31], and Google TensorFlow [32]. PULSim is able to access these available Python-based software tools and to perform a learning process through J2Py communication (Figure 8). The desired functions for machine learning can be initially defined in a listener in PULSim. A Python script will implement the listener interface in order to carry out the learning process. PULSim will provide the training data for the Python-based software tools, which enables the learning process to be carried out either offline or online.

The authors are carrying out a case study, which integrates Google TensorFlow with PULSim to predict the total waiting time of a dispatching action. The case study is taken from a real railway network in Germany with 71 stations. There are in total 1350 trains running in the entire investigated day. The simulated timetable is not conflict-free; a dispatching action has to be taken for a certain state with occupancy conflicts. For a given state, there are several possible dispatching actions to be taken. The task of the dispatching system is to decide on a certain action as the dispatching decision, which results in the minimum possible waiting time. Therefore, the resulting total waiting time for a given state and a certain action has to be predicted exactly. This can be achieved through supervised learning. The object of learning is to minimise the Mean Squared Error (MSE) between the real waiting time and the predicted waiting time. During the simulation process, the states and the actions to be taken are recorded in the form of resulting states by PULSim as input (see Section 3.2), and the resulting total waiting time are learned as output. The total waiting time is learned and predicted by the Python script within the framework of TensorFlow.

With the case study, 12,000 resulting states and the total waiting time are fed into a Convolutional Neural Network (CNN) as training data, and 1,000 resulting states and the total waiting time are used as validation data. The CNN is at first trained with the training data and then validated with the validation data. The average value of the real waiting time for the training and validation data is 44.63 minutes. In Figure 10, the results are shown in Google TensorBoard, which is a tool provided for the visualisation of the learning process. The training curve represents the change of MSE for the training data along iterations, and the validation curve represents the MSE for the validation data along iterations. After ten iterations of training on batch data, an epoch of validation is carried out. Hence, the iterations for training are ten times the iterations of validation. At the moment, the mean error of the predicted total waiting time from the training data is around 1 to 2 minutes, and the mean error of the predicted total waiting time from the validation data is around 8 minutes. The accuracy should be further improved through tuning the hyperparameter of the model.

6. Conclusion and Perspectives

In this paper, a user-based, adaptable simulation tool PULSim for railway planning and operations is presented. Supported by its open interface, users and researchers can rapidly develop their own algorithms to be integrated with PULSim. The productivity of using simulation tools for evaluation and optimisation of railway planning and operations will be improved significantly.

More interfaces will be provided in the further developments of PULSim. Special focus will be placed on the integration of varied signalling systems, dispatching algorithms, and the flexible configuration of operational simulation. The further developed interfaces will be built according to the demands and the feedback from users of the software. For example, in order to find an optimised setting of block sections, an interface to allow flexible placement of signals can be provided. The position of signals can be adjusted dynamically, so that the effects for different variants can be evaluated and optimised by the user-defined optimisation algorithm. In further development, additional dispatching actions, including the extension or shortening of train paths, along with the reordering or cancellation of train runs, should be integrated into PULSim for both railway simulation and real operations control.

Having the large amount of data generated by PULSim, various machine learning algorithms can be further applied to reveal and predict the system behaviour of complex railway systems. PULSim enables a simple integration of the simulation platform and popular machine learning software. Furthermore, it is a growing trend to carry out machine learning combined with the technology of big data. Since PULSim is developed in Java, it can be naturally integrated with Apache Hadoop [33], which is a Java-based software framework for managing and processing large amounts of data. The efficiency and the effectiveness of railway planning and operations will be continuously improved through integrating cutting-edge technologies with PULSim.

Conflicts of Interest

The authors declare that there are no conflicts of interest regarding the publication of this paper.

Acknowledgments

During the work, Mr. Kai Chen made his contribution to GUI programming. The text and the English have been proofread and improved by Mr. Syed Murtaza Hasan.

References

[1] RMCON, "RailSys 7 User Manual. Preparations for Multiple Simulation," 2010.

[2] D. Hürlimann, *Open Track Betriebssimulation von Eisenbahnnetzen*, Zürich, Switzerland, 1.6 edition, 2010.

[3] VIA Con, LUKS Handbuch Version 2.1, 2011.

[4] B. Kogel, N. Nießen, and T. Büker, *Influence of the European Train Control Sys-tem (ETCS) on the Capacity of Nodes*, International Union of Railways, Paris, France, 2010.

[5] M. Luethi, D. Huerlimann, and A. Nash, "Understanding the timetable planning process as a closed control loop," in *Proceedings of the 1st International Seminar on Railway Operations Modelling and Analysis*, I. A. Hansen, F. M. Dekking, R. M. P. Goverde, B. Heidergott, and L. E. Meester, Eds., Delft, Netherlands, 2005.

[6] A. Radtke, *EDV-Verfahren zur Modellierung des Eisenbahnbetriebs*, vol. 64, Eurailpress Tetzlaff-Hestra (Wissenschaftliche Arbeiten für den Schienenver-kehr, 64), Hamburg, Germany, 2005.

[7] A. Radtke and J. Bendfeldt, "Handling of railway operation problems with RailSys," in *Proceedings of the 5th World Congress on Rail Research*, Cologne, Germany, 2001.

[8] F. Corman, A. D'Ariano, and I. A. Hansen, "Evaluating disturbance robustness of railway schedules," *Journal of Intelligent Transportation Systems: Technology, Planning, and Operations*, vol. 18, no. 1, pp. 106–120, 2014.

[9] U. Martin, Y. Cui, F. Hantsch, Z. Chu, and X. Li, "Knotenkapazität - Bewertungs-verfahren für das Mikroskopische Leistungsverhalten und die Engpasserkennung im Spurgeführten Verkehr (RePlan)," in *VWI Neues verkehrswissenschaftliches Journal - Band 8*, Books on Demand GmbH Norderstedt, Norderstedt, Deutschland, 2014.

[10] U. Martin, C. Schmidt, and Z. Chu, "PULEIV Anwendungsleitfaden zur PULEIV-Version 2.1," *Anleitung zur Ermittlung des Leistungsverhaltens von Eisenbahninfra-struktur mithilfe von Simulationsprogrammen*, 2011.

[11] Y. Cui, "Simulation-Based Hybrid Model for a Partially-Automatic Dispatching of Railway Operation," *Norderstedt: Books on Demand GmbH (Neues verkehrswissenschaftliches Journal, 4)*, 2010.

[12] A. D'Ariano, M. Pranzo, and I. A. Hansen, "Conflict resolution and train speed coordination for solving real-time timetable perturbations," *IEEE Transactions on Intelligent Transportation Systems*, vol. 8, no. 2, pp. 208–222, 2007.

[13] I. A. Hansen and J. Pachl, *Railway Timetabling & Operations: Analysis - Modelling - Optimisation - Simulation - Performance Evaluation*, DW Media Group, Hamburg, Germany, 2014.

[14] J. Liang, *Metaheuristic-based Dispatching Optimization Integrated in Multi-scale Simulation Model of Railway Operation*, Institut für Eisenbahn- und Verkehrswesen der Universität Stuttgart, Stuttgart, Germany, 2017.

[15] Y. Cui, U. Martin, and J. Liang, "Searching feasible resources to reduce false-positive situations for resolving deadlocks with the Banker's algorithm in railway simulation," *Journal of Rail Transport Planning and Management*, vol. 7, no. 1-2, pp. 50–61, 2017.

[16] Y. Cui, U. Martin, and W. Zhao, "Calibration of disturbance parameters in railway operational simulation based on reinforcement learning," *Journal of Rail Transport Planning & Management*, vol. 6, no. 1, pp. 1–12, 2016.

[17] Y. Cui and U. Martin, "Multi-scale Simulation in Railway Planning and Operation," *PROMET - Traffic & Transportation*, vol. 23, no. 6, 2011.

[18] M. Wörner and Y. Cui, *Begriffslexikon - Datenmodell*, Institut für Eisenbahn- und Verkehrswesen der Universität Stuttgart, Stuttgart, Germany, 2008.

[19] U. Martin and J. Liang, "The Influence of Dispatching on the Relationship between Capacity and Operation Quality of Railway Systems (DFG Research Project MA 2326/15-1)," in *VWI Neues verkehrswissenschaftliches Journal - Band 20*, Books on Demand GmbH, Norderstedt, Deutschland, 2017.

[20] A. Nash, D. Huerlimann, J. Schütte, and V. P. Krauss, "RailML A Standard Data Interface for Railroad Applications," *WIT Transactions on The Built Environment*, vol. 74, 2004.

[21] W. Fang, S. Yang, and X. Yao, "A Survey on Problem Models and Solution Approaches to Rescheduling in Railway Networks," *IEEE Transactions on Intelligent Transportation Systems*, vol. 16, no. 6, pp. 2997–3016, 2015.

[22] E. G. Coffman, M. Elphick, and A. Shoshani, "System Deadlocks," *ACM Computing Surveys*, vol. 3, no. 2, pp. 67–78, 1971.

[23] G. Mills and P. Pudney, "The Effects of Deadlock Avoidance on Rail Network Capacity and Performance," in *Proceedings of the 2003 Mathematics-in-Industry Study Group*, Brisbane, Australia, 2003.

[24] E. R. Petersen and A. J. Taylor, "Line block prevention in rail line dispatch and simulation models," *Information Systems and Operations Research*, vol. 21, no. 1, pp. 46–51, 1983.

[25] J. Pachl, "Deadlock Avoidance in Railroad Operations Simulations," in *Proceedings of the 90th Annual Meeting of Transportation Research Board*, Washington, DC, USA, 2011.

[26] R. Mittermayr, J. Blieberger, and A. Schöbel, "Kronecker algebra-based deadlock analysis for railway systems," *Promet - Traffic - Traffico*, vol. 24, no. 5, pp. 359–369, 2012.

[27] E. W. Dijkstra, "The mathematics behind the Banker's algorithm," in *Selected Writings on Computing: A Personal Perspective*, pp. 308–312, Springer-Verlag, New York, NY, USA, 1982.

[28] B. Seybold and D. Huerlimann, "OpenTrack - Simulation of railway systems. Presentation of the OpenTrack API. OpenTrack," http://www.opentrack.ch/opentrack/downloads/OpenTrack.API.pdf 2016.

[29] Py4J, "Py4J – A Bridge between Python and Java," https://www.py4j.org/, 2007.

[30] F. Pedregosa, G. Varoquaux, and A. Gramfort, "Scikit-learn: machine learning in Python," *Journal of Machine Learning Research*, vol. 12, pp. 2825–2830, 2011.

[31] J. Bergstra, O. Breuleux, P. Lamblin et al., "Theano: Deep learning on gpus with python," in *NIPS 2011, Big Learning Workshop*, vol. 3, Granada, Spain, 2011.

[32] M. Abadi, A. Agarwal, P. Barham et al., "Tensorflow: Large-scale machine learning on heterogeneous distributed systems," https://arxiv.org/abs/1603.04467, 2016.

[33] J. Nandimath, E. Banerjee, A. Patil, P. Kakade, and S. Vaidya, "Big data analysis using Apache Hadoop," in *Proceedings of the 2013 IEEE 14th International Conference on Information Reuse and Integration, IEEE IRI 2013*, pp. 700–703, San Francisco, CA, USA, August 2013.

Permissions

All chapters in this book were first published in JAT, by Hindawi Publishing Corporation; hereby published with permission under the Creative Commons Attribution License or equivalent. Every chapter published in this book has been scrutinized by our experts. Their significance has been extensively debated. The topics covered herein carry significant findings which will fuel the growth of the discipline. They may even be implemented as practical applications or may be referred to as a beginning point for another development.

The contributors of this book come from diverse backgrounds, making this book a truly international effort. This book will bring forth new frontiers with its revolutionizing research information and detailed analysis of the nascent developments around the world.

We would like to thank all the contributing authors for lending their expertise to make the book truly unique. They have played a crucial role in the development of this book. Without their invaluable contributions this book wouldn't have been possible. They have made vital efforts to compile up to date information on the varied aspects of this subject to make this book a valuable addition to the collection of many professionals and students.

This book was conceptualized with the vision of imparting up-to-date information and advanced data in this field. To ensure the same, a matchless editorial board was set up. Every individual on the board went through rigorous rounds of assessment to prove their worth. After which they invested a large part of their time researching and compiling the most relevant data for our readers.

The editorial board has been involved in producing this book since its inception. They have spent rigorous hours researching and exploring the diverse topics which have resulted in the successful publishing of this book. They have passed on their knowledge of decades through this book. To expedite this challenging task, the publisher supported the team at every step. A small team of assistant editors was also appointed to further simplify the editing procedure and attain best results for the readers.

Apart from the editorial board, the designing team has also invested a significant amount of their time in understanding the subject and creating the most relevant covers. They scrutinized every image to scout for the most suitable representation of the subject and create an appropriate cover for the book.

The publishing team has been an ardent support to the editorial, designing and production team. Their endless efforts to recruit the best for this project, has resulted in the accomplishment of this book. They are a veteran in the field of academics and their pool of knowledge is as vast as their experience in printing. Their expertise and guidance has proved useful at every step. Their uncompromising quality standards have made this book an exceptional effort. Their encouragement from time to time has been an inspiration for everyone.

The publisher and the editorial board hope that this book will prove to be a valuable piece of knowledge for researchers, students, practitioners and scholars across the globe.

List of Contributors

Ziyan Luo
State Key Laboratory of Rail Traffic Control and Safety, Beijing Jiaotong University, Beijing 100044, China

Xiaoyu Li and Naihua Xiu
Department of Mathematics, Beijing Jiaotong University, Beijing 100044, China

Yingze Yang, Zheng Xu, Weirong Liu, Heng Li, Rui Zhang and Zhiwu Huang
School of Information Science and Engineering, Central South University, Changsha 410083, China

Qin Luo, Wei Li and Xiongfei Zhang
College of Urban Traffic and Logistics, Shenzhen Technology University, Lantian Road 3002, Shenzhen, China

Qin Luo and Yufei Hou
Shenzhen Key Laboratory of Urban Rail Transit, Shenzhen University, Nanshan Ave 3688, Shenzhen 518000, China

Wei Li
Key Laboratory of Optoelectronic Devices and Systems of Ministry of Education and Guangdong Province, College of Optoelectronic Engineering, Shenzhen University, Nanshan Ave 3688, Shenzhen, China

Jin Wang, Leishan Zhou, Yixiang Yue, Jinjin Tang and Zixi Bai
School of Traffic and Transportation, Beijing Jiaotong University, Beijing 100044, China

Fabrizio Cerreto, Otto Anker Nielsen and Steven S. Harrod
Department of Management Engineering, Technical University of Denmark, 2800 Kongens Lyngby, Denmark

Bo Friis Nielsen
Department of Applied Mathematics and Computer Science, Technical University of Denmark, 2800 Kongens Lyngby, Denmark

Xinyuan Chen and Inhi Kim
Monash Institute of Transport Studies, Australia

Yuhe Zhou, Yun Bai, Jiajie Li and Baohua Mao
MOE Key Laboratory for Urban Transportation Complex Systems Theory and Technology, Beijing Jiaotong University, Beijing 100044, China

Tang Li
Department of Civil and Environmental Engineering, Imperial College London, London SW7 2AZ, UK

Chuntian Zhang, Yuan Gao, Lixing Yang and Ziyou Gao
State Key Laboratory of Rail Traffic Control and Safety, Beijing Jiaotong University, Beijing 100044, China

Wenjie Li
Sid and Reva Department of Civil, Environmental and Infrastructure Engineering, George Mason University, Fairfax, VA 22030, USA

Dewei Li and Yonghao Yin
State Key Lab of Rail Traffic Control and Safety, Beijing Jiaotong University, Beijing 100044, China

Hong He
Guangzhou Railway Polytechnic, Guangzhou, Guangdong 510430, China

Xuemei Wang
Department of Transportation Engineering, Tongji Zhejiang College, Jiaxing 314000, China

Ning Zhang and Zhuangbin Shi
Intelligent Transportation System Institute of Ministry of Education, Southeast University, Nanjing 211189, China

Yunlong Zhang
Zachry Department of Civil Engineering, Texas A&M University, 3136 TAMU, College Station, USA

Shasha Liu and Enjian Yao
MOE Key Laboratory for Urban Transportation Complex Systems Theory and Technology, Beijing Jiaotong University, Beijing, China

Toshiyuki Yamamoto
Institute of Materials and Systems for Sustainability, Nagoya University, Nagoya, Japan

Xiao Fu
School of Transportation, Southeast University, Nanjing 210096, China

Yu Gu
Jiangsu Key Laboratory of Urban ITS, Jiangsu Province Collaborative Innovation Centre of Modern Urban Traffic Technologies, School of Transportation, Southeast University, Nanjing 210096, China

Keping Li and Hangfei Huang
State Key Laboratory of Rail Traffic Control and Safety, Beijing Jiaotong University, Beijing 100044, China

Paul Schonfeld
Department of Civil & Environmental Engineering, University of Maryland, College Park, MD 20742, USA

Xuemei Li, Khalid Mehmood Alam and Shitong Wang
School of Economics and Management, Beijing Jiaotong University, Beijing 100044, China

Shitong Wang
School of Economics and Management, Tianjin Chengjian University, Tianjin 300384, China

Yong Cui and Ullrich Martin
Institut fuer Eisenbahn- und Verkehrswesen der Universitaet Stuttgart, Pfaffenwaldring 7, 70569 Stuttgart, Germany

Jiajian Liang
China Academy of Railway Sciences, Daliushu Road 2, Haidian District, Beijing 100081, China

Index

U

V

www.ingramcontent.com/pod-product-compliance
Lightning Source LLC
LaVergne TN
LVHW080912210926
848485LV00004B/639
* 9 7 8 1 6 3 9 8 7 4 7 1 2 *